Soft Nanotechnology

The Royal Society, London, UK
15–17 June 2009

FARADAY DISCUSSIONS
Volume 143, 2009

SC Publishing

The Faraday Division of the Royal Society of Chemistry, previously the Faraday Society, founded in 1903 to promote the study of sciences lying between Chemistry, Physics and Biology.

EDITORIAL STAFF

Editor
Philip Earis

Assistant editor
Anna Roffey

Publishing assistant
Kate Bandoo

Team leader, Informatics
Elinor Richards

Technical editor
Christina Hodkinson

Publisher
Janet Dean

Faraday Discussions (Print ISSN 1359-6640, Electronic ISSN 1364-5498) is published 4 times a year by the Royal Society of Chemistry, Thomas Graham House, Science Park, Milton Road, Cambridge, UK CB4 0WF. Volume 143 ISBN-13: 978 1 84755 8381

2009 annual subscription price: print+electronic £592, US $1,160; electronic only £533, US $1045. Customers in Canada will be subject to a surcharge to cover GST. Customers in the EU subscribing to the electronic version only will be charged VAT. All orders, with cheques made payable to the Royal Society of Chemistry, should be sent to RSC Distribution Services, c/o Portland Customer Services, Commerce Way, Colchester, Essex, UK CO2 8HP.
Tel +44 (0) 1206 226050;
E-mail sales@rscdistribution.org

If you take an institutional subscription to any RSC journal you are entitled to free, site-wide web access to that journal. You can arrange access via Internet Protocol (IP) address at www.rsc.org/ip. Customers should make payments by cheque in sterling payable on a UK clearing bank or in US dollars payable on a US clearing bank. Periodicals postage is paid at Rahway, NJ and at additional mailing offices. Airfreight and mailing in the USA by Mercury Airfreight International Ltd., 365 Blair Road, Avenel, NJ 07001, USA.

US Postmaster: send address changes to *Faraday Discussions*, c/o Mercury Airfreight International Ltd., 365 Blair Road, Avenel, NJ 07001. All despatches outside the UK by Consolidated Airfreight.

PRINTED IN THE UK

Faraday Discussions documents a long-established series of *Faraday Discussion* meetings which provide a unique international forum for the exchange of views and newly acquired results in developing areas of physical chemistry, biophysical chemistry and chemical physics.

ORGANISING COMMITTEE, Volume 143

Chair
Professor Wilhelm Huck (University of Cambridge, UK)
(Co-chair)
Dr Joachim Steinke (Imperial College London, UK)
(Co-chair)

Professor Ullrich Steiner (University of Cambridge, UK)
Professor Tony Cass (Imperial College London, UK)
Professor Ulrich Wiesner (Cornell University, USA)
Professor Dr David Reinhoudt (University of Twente, The Netherlands)

FARADAY STANDING COMMITTEE ON CONFERENCES

Chair
D E Heard (Leeds, UK)

W A Brown (UCL, UK)
H M Colquhoun (Bath, UK)
G Jackson (Imperial, UK)
A Rodger (Warwick, UK)

© The Royal Society of Chemistry 2009. Apart from fair dealing for the purposes of research or private study, or criticism or review, as permitted under the Copyright, Designs and Patents Act 1988 and Related Rights Regulations 2003, this publication may only be reproduced, stored or transmitted, in any form or by any means, with the prior permission in writing of the Publishers or in the case of reprographic reproduction in accordance with the terms of licences issued by the Copyright Licensing Agency in the UK. US copyright law applicable to users in the USA. The Royal Society of Chemistry takes reasonable care in the preparation of this publication but does not accept liability for the consequences of any errors or omissions.

Royal Society of Chemistry: Registered Charity No. 207890.

∞The paper used in this publication meets the requirements of ANSI/NISO Z39.48-1992 (Permanence of Paper).

Soft Nanotechnology

Faraday Discussions
www.rsc.org/faraday_d

A General Discussion on Soft Nanotechnology was held at The Royal Society, London, UK on 15th, 16th and 17th June 2009.

RSC Publishing is a not-for-profit publisher and a division of the Royal Society of Chemistry. Any surplus made is used to support charitable activities aimed at advancing the chemical sciences. Full details are available from www.rsc.org

CONTENTS

ISSN 1359-6640; ISBN 978-1-84755-838-1

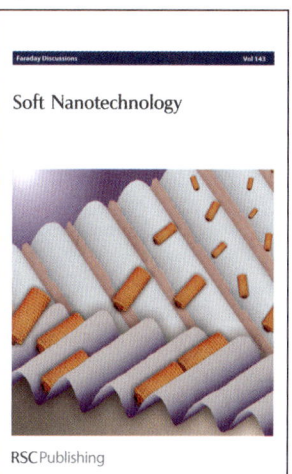

Cover
See Fery, Böker et al., Faraday Discuss., 2009, **143**, 143–150.

The cover image shows a schematic of the deposition of TMV nanoparticles on a wrinkled PDMS/silicon oxide surface (foreground, not to scale). The background displays a 3D scanning force microscopy image of the virus particles forming an array of TMV-lines in the grooves of the wrinkles.

Image reproduced by permission of Professor Dr Alexander Böker, from Faraday Discuss., 2009, **143**, 143.

INTRODUCTORY LECTURE

9   **Challenges in soft nanotechnology**
      Richard A. L. Jones

PAPERS AND DISCUSSIONS

15   **Chemo and phototactic nano/microbots**
      Ayusman Sen, Michael Ibele, Yiying Hong and Darrell Velegol

29   **The efficiency of encapsulation within surface rehydrated polymersomes**
      A. J. Parnell, N. Tzokova, P. D. Topham, D. J. Adams, S. Adams, C. M. Fernyhough, A. J. Ryan and R. A. L. Jones

47   **Molecular control of ionic conduction in polymer nanopores**
      Eduardo R. Cruz-Chu, Thorsten Ritz, Zuzanna S. Siwy and Klaus Schulten

63   **Nanomechanics of organic/inorganic interfaces: a theoretical insight**
      Maria L. Sushko

Enjoyed this discussion?
How about some others...

Faraday Discussions provide a unique discussion forum for original research in physical chemistry, chemical physics and biophysical chemistry.

Previous volumes of interest include:

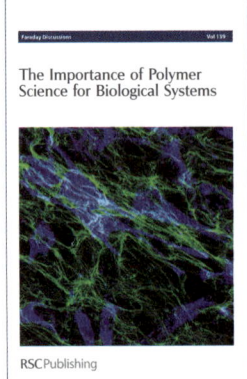

The Importance of Polymer Science for Biological Systems
Faraday Discussion 139

This volume focuses on polymer science and its role in understanding how biological systems behave. So many of the key molecules in biological systems are polymers; proteins, polysaccharides and nucleic acids. Their long chain behaviour is a crucial contributing factor to their function in living systems.

Nanoalloys: From Theory to Application
Faraday Discussions 138

This volume focuses on nanoalloys, which have lately been of increasing interest for a number of reasons. One of the major reasons is the fact that their chemical and physical properties can be tuned by varying the composition and atomic ordering, as well as the size of the clusters.

Molecular Wires and Nanoscale Conductors
Faraday Discussions 131

This volume concentrates on the measurement, synthesis, assembly and manipulation of molecular wires and nanostructures. It incorporates the assembly and use of molecules into nanostructures (nanoparticles, quantum dots, catenanes, conducting oligomers, DNA) for electronic switching.

To view the full contents and, where available, purchase an article or a volume visit the website.

RSCPublishing   www.rsc.org/faraday

Registered Charity Number 207890

81   General discussion

95   **Solid state nanofibers based on self-assemblies: from cleaving from self-assemblies to multilevel hierarchical constructs**
Olli Ikkala, Robin H. A. Ras, Nikolay Houbenov, Janne Ruokolainen, Marjo Pääkkö, Janne Laine, Markku Leskelä, Lars A. Berglund, Tom Lindström, Gerrit ten Brinke, Hermis Iatrou, Nikos Hadjichristidis and Charl F. J. Faul

109 **Nanoparticles at electrified liquid−liquid interfaces: new options for electro-optics**
M. E. Flatté, A. A. Kornyshev and M. Urbakh

117 **Free-standing porous supramolecular assemblies of nanoparticles made using a double-templating strategy**
Xing Yi Ling, In Yee Phang, David N. Reinhoudt, G. Julius Vancso and Jurriaan Huskens

129 **Polymer crystallization under nano-confinement of droplets studied by molecular simulations**
Wenbing Hu, Tao Cai, Yu Ma, Jamie K. Hobbs, O. Farrance and Günter Reiter

143 **Nanostructured wrinkled surfaces for templating bionanoparticles—controlling and quantifying the degree of order**
Anne Horn, Heiko G. Schoberth, Stephanie Hiltl, Arnaud Chiche, Qian Wang, Alexandra Schweikart, Andreas Fery and Alexander Böker

151 **Silica nano-particle super-hydrophobic surfaces: the effects of surface morphology and trapped air pockets on hydrodynamic drainage forces**
Derek Y. C. Chan, Md. Hemayet Uddin, Kwun L. Cho, Irving I. Liaw, Robert N. Lamb, Geoffrey W. Stevens, Franz Grieser and Raymond R. Dagastine

169 General discussion

187 **Counterion-activated polyions as soft sensing systems in lipid bilayer membranes: from cell-penetrating peptides to DNA**
Toshihide Takeuchi, Naomi Sakai and Stefan Matile

205 **Recognition of sequence-information in synthetic copolymer chains by a conformationally-constrained tweezer molecule**
Howard M. Colquhoun, Zhixue Zhu, Christine J. Cardin, Michael G. B. Drew and Yu Gan

221 **DNA self-assembly: from 2D to 3D**
Chuan Zhang, Yu He, Min Su, Seung Hyeon Ko, Tao Ye, Yujun Leng, Xuping Sun, Alexander E. Ribbe, Wen Jiang and Chengde Mao

235 **Self-assembly of liquid crystal block copolymer PEG-*b*-smectic polymer in pure state and in dilute aqueous solution**
Bing Xu, Rafael Piñol, Merveille Nono-Djamen, Sandrine Pensec, Patrick Keller, Pierre-Antoine Albouy, Daniel Lévy and Min-Hui Li

251 **A novel self-healing supramolecular polymer system**
Stefano Burattini, Howard M. Colquhoun, Barnaby W. Greenland and Wayne Hayes

265 General discussion

277 **Quantitative approaches to defining normal and aberrant protein homeostasis**
Michele Vendruscolo and Christopher M. Dobson

293 **Evolving nanomaterials using enzyme-driven dynamic peptide libraries (eDPL)**
Apurba K. Das, Andrew R. Hirst and Rein V. Ulijn

305 **Rational design of peptide-based building blocks for nanoscience and synthetic biology**
Craig T. Armstrong, Aimee L. Boyle, Elizabeth H. C. Bromley, Zahra N. Mahmoud, Lisa Smith, Andrew R. Thomson and Derek N. Woolfson

319   **The influence of viscosity on the functioning of molecular motors**
        Martin Klok, Leon P. B. M. Janssen, Wesley R. Browne and Ben L. Feringa

335   **Template sol–gel synthesis of mesostructured silica composites using metal complexes bearing amphiphilic side chains: immobilization of a polymeric Pt complex formed by a metallophilic interaction**
        Wataru Otani, Kazushi Kinbara and Takuzo Aida

345   **Self-assembled interpenetrating networks by orthogonal self assembly of surfactants and hydrogelators**
        Aurelie M. Brizard, Marc C. A. Stuart and Jan H. van Esch

359   **General discussion**

CONCLUDING REMARKS

373   **Soft nanotechnology: "structure" vs. "function"**
        George M. Whitesides and Darren J. Lipomi

ADDITIONAL INFORMATION

385   **Poster titles**
389   **List of participants**
393   **Index of contributors**

PAPER

Challenges in soft nanotechnology

Richard A. L. Jones*

Received 6th August 2009, Accepted 12th August 2009
First published as an Advance Article on the web 26th August 2009
DOI: 10.1039/b916271m

Introduction

Why "soft nanotechnology"? The word betrays its origins in the confluence of two ideas. The first is the emergence of soft condensed matter as an interdisciplinary area of physics, chemistry and materials science. The term "soft matter" is associated with the late Pierre Gilles de Gennes, as an umbrella term for all those states of matter—polymers, colloids, liquid crystals *etc.*—in which typical energies of interaction are comparable to thermal energies. The second idea is the notion of "nanotechnology" itself; this word is associated with the endeavour of making potentially useful structures and devices from components on the scale of atoms and molecules. Thus, in soft nanotechnology, we seek to use our knowledge of the behaviour of soft matter to make from such components useful nanostructures and nano-scale devices.

The relationship between soft nanotechnology and cell and molecular biology is important and should be stressed at the outset. The structures and mechanisms of cell biology present compelling proof from their existence that sophisticated, highly functional nano-scale devices are possible:[1,2] cell biology is indeed nanotechnology that works. However, to understand the mechanisms of cell biology we need ideas and concepts from soft matter physics, together with an appreciation that biological systems possess a complexity not found in synthetic systems.

This suggests that there are two complementary ways of thinking about soft nanotechnology. On the one hand, we can ask what useful nano-scale constructs and devices can we make from the repertoire of components familiar from soft matter science—for example, polymers, amphiphiles, block copolymers, polymer brushes, colloidal particles, *etc*. On the other hand, we might also look at the functional features of living cells, and ask which of those features we might hope to emulate in synthetic systems.[3]

Whichever way one frames the challenges of soft nanotechnology, one has to appreciate the nature of the physical environment in which one is trying to operate. Assuming that we are operating in water or another liquid solvent, at around 300 K, the dominant feature will be Brownian motion. Thus transport will be essentially diffusive in character, and we expect extended objects such as polymer chains to show a high degree of conformational flexibility. At the nano-scale, strong surface forces will be the rule. The physics of these situations is characterised by Langevin equations; hydrodynamics is at very low Reynolds numbers and any charge interactions are likely to be strongly screened. A variety of forces of entropic origin will be in play, such as the entropic elasticity of polymer chains, Helfrich forces between fluctuating membranes, and the osmotic effects that underlie phenomena such as depletion forces. Together, these effects add up to an operating environment very different from anything encountered in macroscopic engineering, and this dictates the need to embrace entirely different design principles.

Department of Physics and Astronomy, University of Sheffield, Sheffield, UK S3 7RH. E-mail: r.a.l.jones@sheffield.ac.uk; Fax: +44 (0)114 222 3555; Tel: +44 (0)114 222 9820

Some design principles of soft nanotechnology

One ubiquitous theme of soft nanotechnology is the importance of self-assembly as a powerful and scalable method of making nano-scale structures. Equilibrium self-assembly, exemplified by the complex phase diagrams of amphiphiles and block copolymers, is now well understood theoretically (at least in principle, though considerable practical difficulties may still stand in the way of calculating phase diagrams of complex systems). Some of the most elegant and powerful implementations of this principle are now to be found in the field of DNA nanotechnology, where the simplicity and tractability of the base-pair interaction allows complex structures in two and three dimensions to be designed and executed.[4] Biological inspiration also lies beneath the increasing use of proteins and designed synthetic peptides to exploit the motifs of protein folding.[5]

Self-assembly can very usefully be thought of in terms of information. Equilibrium self-assembly is defined by the condition that all the information required to make the structure must be encoded in the molecules themselves. Many powerful variants of self-assembly relax this principle in various ways. Templating methods, precursor routes, layer-by-layer assembly, and combinations of self-assembly with top-down patterning, all use external interventions in different ways to impose extra information on the system, yielding considerable extra flexibility in terms of the kinds of structures that can be formed. As always, biology offers powerful models: one example is the way tough and insoluble collagen fibrils are formed from the hierarchical self-assembly and subsequent chemical modification of soluble pre-collagen precursors.

To be distinguished both from self-assembly at equilibrium, and in various conditions of restricted equilibrium, are a number of methods of forming nano-scale structures by various types of non-equilibrium pattern formation. These include the intricate structures formed in bio-mineralisation and its synthetic analogues by the interaction of growing crystals with adsorbing macromolecules, and structures that arise as a result of reaction–diffusion systems.[6]

Soft matter is characterised by weak interactions—interactions whose energy scale is comparable to that of thermal energy—and it is the shifting balance between different weak interactions in the face of subtle changes in external conditions that gives soft matter is characteristic mutability, leading to organisational and conformational changes in response to changes in the environment. In aqueous systems, hydrogen bonding plays a central role, both in its direct importance for molecular recognition, and more indirectly through the hydrophobic interaction. It is the subtle interplay of these interactions, together with screened charge interactions, which underlie the phenomenon of protein folding. For an example of a much simpler macromolecular conformational transition, which still illustrates the complexity of these kinds of problem, consider a responsive polyelectrolyte brush—a layer of weak poly-acid or poly-base molecules tethered by their ends to a planar surface, and immersed in an aqueous solution of controlled pH and ionic strength.[7] In outline, the behaviour of such a system is simple to understand—a poly-acid brush in conditions of low pH will be un-ionised; the chains will be relatively hydrophobic and will tend to form a dense layer, collapsed close to the substrate. As the pH is increased, the ionisation equilibrium will shift, the chains will become charged and will stretch away from the surface to form a diffuse and extended layer. However, to account for this behaviour in detail is surprisingly complex; one has to take into account the screening of the charge interactions by the counter-ions, the osmotic pressure of those counter-ions, the entropic elasticity of the chains, all in the light of the fact that the degree of ionisation of the chains can vary spatially along the chain, as well as in a global way in response to the applied bulk solution conditions.[8]

Mimicking the features of cell biology

There are, of course, a huge variety of living cells in biology which between them display great diversity of structures and capabilities. A very incomplete list of the sorts of features of cell biology one might, in nanotechnology, wish to emulate might begin with *containment*. Cells are defined by a membrane, which encloses an interior space in which chemical components and systems can be maintained out of equilibrium with the external environment. Containment cannot be complete, of course; the operations of the cell require that both energy and molecules can enter and exit the cell. This traffic must be selective and controlled. The simplest way of achieving some selectivity is by relying on the difference in diffusion coefficient through the membrane between molecules of different sizes and chemical types.

Much more sophisticated control of traffic is obtained by selective pores and mechanisms for active transport, as well as mechanisms such as endocytosis by which nanoscale objects are engulfed by invaginations of the membrane, often triggered by very specific molecular recognition events.

Within the cell, the contained chemical species are not merely inert cargo, but undertake a series of complex and linked chemical reactions, which together define the cell's metabolism. An important part of the metabolism is devoted to creating more of the molecules that form the components of the cell. This network of reactions needs a continuous source of free energy.

A living cell, then, is defined by constant flows of energy and matter. Flows of information are important, too; all but the most rudimentary organisms are able detect aspects of their environment and respond to this. This response may take the form of modifications of their own metabolism (the classic example is the lac repressor), of modifications of the environment itself (for example, by the formation of a biofilm) or by the cell physically taking itself off to find a new and more suitable environment, if it is capable of autonomous motility. Typically such a response begins with a sensor molecule, using molecular recognition to detect a certain chemical species.

The response to an environmental cue is mediated by chemical signals, and the information carried by these signals is itself processed by other molecules. Bray pointed out some years ago[9] that many proteins in cells seem to have as their purpose the processing of information rather than the catalysis of chemical reactions or to be structural components; the property of allostery means that an individual protein molecule can behave as a logic gate, with its catalytic activity being turned on or off by the binding of a regulatory molecule. Such logic gates can be linked together in networks—chemical circuits that can carry out computational tasks of some complexity in response to the original detection of one or more environmental signals.

What progress has been made in mimicking some of these features of cell biology? The prototype of a biomimetic containment system is the phospholipid vesicle, or liposome, which is now very well studied and used in practice. Analogues of liposomes made from amphiphilic block copolymers (polymersomes) have been attracting increasing interest recently.[10] The variety of different chemistries available and the possibility of controlling the degree of polymerisation of the blocks make possible the rational design of polymersomes. For example, the wall thickness, and thus the permeability to molecular species of various sizes, is directly related to the degree of polymerisation of the hydrophobic block.[11,12] This is another example of the way that the specification of objects built by self-assembly is encoded in the architecture of the component of molecules. An interesting feature of all types of vesicles is that, while self-assembly controls the thickness of the walls, there is not a strong selection mechanism for the overall size of the object itself. Thus self-assembly is not by itself sufficient for making a population of vesicles with controlled size distribution; this distribution depends on the details of the preparation technique and is often very broad. One way of achieving a narrow and controlled size distribution of polymersomes is to combine self-assembly with top-down patterning.[13] A block copolymer film is cast onto a substrate with hydrophobic and oleophobic patches; this pattern is reproduced in the polymer film by dewetting. When the film is rehydrated, the surface area of each vesicle is set by the size of the patches on the patterned substrate.

As already mentioned, some degree of control of transport in and out of vesicles can be achieved by varying the thickness and the chemical properties of the wall. To go beyond this, one can envisage incorporating pores in the walls which it might be possible to open and close in response to chemical signals. An exemplar of this approach encapsulated a cell-free protein expression system derived from *Escherichia coli* within phospholipid vesicles incorporating pore-forming proteins.[14] A wider variety of nano- and micro-scale enclosed reaction systems is reviewed in ref. 15.

Moving from metabolism to molecular information processing, some examples of synthetic molecular logic devices have been reported.[16] These are likely to lead to new sensors of increasing sophistication. However, many of these systems are characterised by the fact that their output takes the form of a fluorescent signal, rather than a chemical signal. This limits the extent to which such logic elements could be built up into large-scale networks like the cell signalling networks of biology. Synthetic DNA-based systems currently seem to offer the best hope for building molecular logic systems from first principles.[17]

The particular problems of motility at the micro- and nano-scale stem from the special features of hydrodynamics at low Reynolds number, as emphasised in Purcell's classic paper.[18] This emphasises the need to break time symmetry in order to achieve motion; recently a number of elegant theoretical papers have explored various ways of achieving this. The most interesting types of synthetic micro- and nano-scale motors will use chemical energy to drive directional motion, as biological motor proteins use the energy of ATP.[19] One potential class of synthetic systems is built on the fascinating chemistry of catenanes and rotaxanes,[20] while exciting progress is being made demonstrating motors based on DNA,[21] which provide a different implementation of basic idea underlying the operation of protein molecular motors—a coupling of the conformational change of a macromolecule with the catalysis of a chemical reaction by that macromolecule. The coupling of macromolecular conformational change with a cyclic chemical reaction also underlies experiments in which responsive polymers, such as weak polyelectrolytes, change shape in response to a cyclic chemical reaction.[22,23] In these systems, however, the coupling between the chemical reaction and the conformational change is only indirect in contrast to both the DNA-based systems and biological motors.

One other class of systems that can convert chemical energy into the mechanical motion of micro-scale objects relies on a phoretic response to a self-generated chemical gradient that arises from an asymmetrically localised chemical reaction.[24-26] The mechanisms of this motion may be electrophoretic or diffusiophoretic in character; in the case of motion driven by self-diffusiophoresis there is some theoretical understanding which is at least consistent with experimental data.[27] These autophoretic motions result in the propulsion of a particle at a velocity which depends on the rate at which reaction products are generated, but it is important to recall that this process takes place in a Brownian environment, in which the orientation of the particle randomly changes over a rotational diffusion time which has a strong dependence on particle radius. This means that if one characterises the motion of such particles, one sees a cross-over in the type of motion.[26] At short times, transport is ballistic, but at the rotational diffusion time there is a crossover to diffusive transport, resulting from a random walk with a step size proportional to the product of the propulsion velocity and a rotational diffusion time, with an effective diffusion coefficient that is substantially enhanced over the classical Stokes–Einstein value. The degree of this enhancement, and the length of the window of time in which ballistic behaviour is observed, depends strongly on the size of the particle. Thus to use these mechanisms for particles whose size starts to fall below the micro- to the nano-scale will require chemical reactions that generate products at a considerably higher rate than the reactions which have been studied so far. The other great challenge is to achieve some degree of directionality and purpose to the motion. Mimicking the ability of some bacteria to undergo chemotaxis, for example, poses an attractive target for which some progress has already been reported.[28]

What soft nanotechnology can and cannot now do

One way in which the comparison between soft nanotechnology and the structures and mechanisms of cell biology is helpful, though sobering, is that it emphasises the gulf between what must be possible in principle, as demonstrated by the example of biology, and what we can actually do.

The use of self-assembly in its various forms to generate useful and interesting nanostructures is now well developed, backed by considerable theoretical understanding and a growing set of design rules. There is progress towards designing nano- and micro-scale encapsulating systems, while the principles of using conformational change and osmotic effects such as phoresis to generate motility are beginning to be explored. However, the development of analogues of biological systems for chemical sensing and information processing has only just begun. A number of fundamental theoretical and practical issues in soft nanotechnology remain to be addressed; the theoretical basis for understanding small systems driven far from equilibrium remains underdeveloped. It seems likely that the design of complex bio-mimetic nano-systems will require the use of evolutionary design methods, given the size of the configuration spaces that need to be explored. Finally, the intricate mechanisms that underlie the ability of biological systems to self-replicate seem, currently, to be quite out of reach to any synthetic system.

To conclude, it is worth reflecting on the "technology" aspects of soft nanotechnology—those areas of potential application that will drive the development of some of these ideas to become the basis of useful products. The science of soft matter originally found its applications in the chemical industry, and in applications such as home and personal care. In these areas, the scalability of self-assembly is what makes it possible to contemplate what is really quite sophisticated control of nanostructure, and in some cases a degree of environmental responsiveness, in products that are sold at very low cost. Higher margins are possible for materials that are used in information technology, and we are seeing the use of self-assembling structures in materials like high-performance dielectrics and materials for information storage. The drive to decarbonise our energy economies will put a premium on being able to make scalable and cheap nanostructures for applications such as batteries, fuel cells and new photovoltaic materials. That said, one of the most compelling answers to the question "why nano?" must take us back again to biology. The most basic operations of cell biology take place at the nano-scale, so in one sense the nano-scale is the most appropriate length scale for intervening in biology—this is the fundamental motivation for the idea of nanomedicine. Thus we can expect to see the most compelling applications of soft nanotechnology in medicine. We are already seeing applications in areas such as drug delivery, regenerative medicine and sensors and diagnostics. However, one should not underestimate the difficulties, and the timescales over which the applications of some of these ideas will come to full fruition may be long.

Acknowledgements

It is a pleasure to acknowledge the many contributions of my colleagues at Sheffield, in particular Tony Ryan, Ramin Golestanian, Mark Geoghegan, Jon Howse, Andy Parnell, Steve Armes, and Beppe Battaglia, to the development of the ideas explored in this article.

References

1 R. A. L. Jones, *Soft Machines: nanotechnology and life*, Oxford University Press, Oxford, 2004.
2 D. S. Goodsell, *Bionanotechnology: lessons from nature*, Wiley-Liss, Hoboken, NJ, 2004.
3 S. Mann, *Angew. Chem., Int. Ed.*, 2008, **47**, 5306–5320.

4 N. C. Seeman and A. M. Belcher, *Proc. Natl. Acad. Sci. U. S. A.*, 2002, **99**, 6451–6455.
5 E. H. Bromley, K. Channon, E. Moutevelis and D. N. Woolfson, *ACS Chem. Biol.*, 2008, **3**, 38–50.
6 B. Grzybowski, K. Bishop, C. Campbell, M. Fialkowski and S. Smoukov, *Soft Matter*, 2005, **1**, 114.
7 A. J. Parnell, S. J. Martin, R. A. L. Jones, C. Vasilev, C. J. Crook and A. J. Ryan, *Soft Matter*, 2009, **5**, 296.
8 E. B. Zhulina, T. M. Birshtein and O. V. Borisov, *Macromolecules*, 1995, **28**, 1491.
9 D. Bray, *Nature*, 1995, **376**, 307–312.
10 D. E. Discher and A. Eisenberg, *Science*, 2002, **297**, 967–973.
11 G. Battaglia and A. J. Ryan, *J. Am. Chem. Soc.*, 2005, **127**, 8757–8764.
12 G. Battaglia, A. J. Ryan and S. Tomas, *Langmuir*, 2006, **22**, 4910–4913.
13 J. R. Howse, R. A. L. Jones, G. Battaglia, R. E. Ducker, G. J. Leggett and A. J. Ryan, *Nat. Mater.*, 2009, **8**, 507–511.
14 V. Noireaux and A. Libchaber, *Proc. Natl. Acad. Sci. U. S. A.*, 2004, **101**, 17669–17674.
15 D. Vriezema, M. Aragones, J. Elemans, J. Cornelissen, A. Rowan and R. Nolte, *Chem. Rev.*, 2005, **105**, 1445–1489.
16 A. P. de Silva and S. Uchiyama, *Nat. Nanotechnol.*, 2007, **2**, 399–410.
17 G. Seelig, D. Soloveichik, D. Y. Zhang and E. Winfree, *Science*, 2006, **314**, 1585–1588.
18 E. M. Purcell, *Am. J. Phys.*, 1977, **45**, 3.
19 R. D. Vale and R. A. Milligan, *Science*, 2000, **288**, 88–95.
20 E. R. Kay, D. A. Leigh and F. Zerbetto, *Angew. Chem., Int. Ed.*, 2007, **46**, 72–191.
21 J. Bath and A. J. Turberfield, *Nat. Nanotechnol.*, 2007, **2**, 275–284.
22 R. Yoshida, T. Takahashi, T. Yamaguchi and H. Ichijo, *J. Am. Chem. Soc.*, 1996, **118**, 5134–5135.
23 J. R. Howse, P. Topham, C. J. Crook, A. J. Gleeson, W. Bras, R. A. L. Jones and A. J. Ryan, *Nano Lett.*, 2006, **6**, 73–77.
24 W. F. Paxton, S. Sundararajan, T. E. Mallouk and A. Sen, *Angew. Chem., Int. Ed.*, 2006, **45**, 5420–5429.
25 W. F. Paxton, A. Sen and T. E. Mallouk, *Chem.–Eur. J.*, 2005, **11**, 6462–6470.
26 J. R. Howse, R. A. L. Jones, A. J. Ryan, T. Gough, R. Vafabakhsh and R. Golestanian, *Phys. Rev. Lett.*, 2007, **99**, 048102.
27 R. Golestanian, T. B. Liverpool and A. Ajdari, *Phys. Rev. Lett.*, 2005, **94**, 220801.
28 Y. Hong, N. M. K. Blackman, N. D. Kopp, A. Sen and D. Velegol, *Phys. Rev. Lett.*, 2007, **99**, 178103.

Chemo and phototactic nano/microbots

Ayusman Sen,[*a] Michael Ibele,[a] Yiying Hong[a] and Darrell Velegol[*b]

Received 16th January 2009, Accepted 23rd March 2009
First published as an Advance Article on the web 21st July 2009
DOI: 10.1039/b900971j

One of the more interesting recent discoveries has been the ability to design nano/microparticles which catalytically harness the chemical energy in their environments to move autonomously. These "nanomotors" can be directed by externally applied magnetic fields, or optical and chemical gradients. Our group has now developed two systems in which chemical secretions from the translating micro/nanomotors initiate long-range, collective interactions among the particles *via* self-diffusiophoresis. Herein, we discuss two different approaches to model the complex emergent behavior of these particles, the first being a qualitative probability-based model with wide applicability, and the second being a more quantitative Brownian dynamics simulation specific to the self-diffusiophoretic phenomenon.

1 Introduction

One of the grand challenges in science is: *How can we master energy and information on the nano/microscale to create new technologies with capabilities rivaling those of living things?* We have initiated a study of emergent behavior and the consequent formation of organized material assemblies by using catalytic nano/micromotors.[1–7] Catalytic motors are a novel class of nano- and microscale materials that convert chemical to mechanical energy. Their collective behavior offers a means of using the free energy of chemical fuels to fabricate organized systems of particles. They may thus offer a route to functional assemblies that cannot be accessed by other means. Individually the motors move in a *random* direction according to physics that has been well-established by our groups.[3,8–10] But *collectively*, the motors give complex behaviors similar to the chemotaxis, phototaxis, or predator–prey phenomena normally seen only in biological systems.[9]

The construction of functional nano/microbots capable of emergent collective behavior requires the following design elements: (1) autonomous movement through catalysis; (2) control of directionality through chemical gradients; and (3) inter-motor communication *via* chemical signals. As discussed below, we have "proof of principle" examples of each of the above design elements, both separately and in combination.

2 Background

Several breakthroughs from our laboratory make the design of micro/nanobots based on catalytic motors possible. The first of these advances is our ability to fabricate catalytically active multi-metallic rods or aggregates. We have demonstrated that platinum/gold rods (2 µm length, 300–500 nm diameter) can "swim" by self-electrophoresis up to 20 µm s^{-1}.[1,3] Outside of biological systems, this was the first

---

[a]Departments of Chemistry, The Pennsylvania State University, University Park, PA, 16802, USA. E-mail: asen@psu.edu
[b]Chemical Engineering, The Pennsylvania State University, University Park, PA, 16802, USA. E-mail: velegol@psu.edu

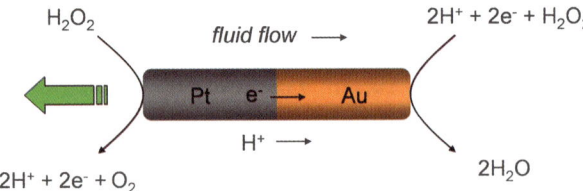

Fig. 1 Schematic illustrating self-electrophoresis. Hydrogen peroxide is oxidized to generate protons in solution and electrons in the wire on the platinum end. The protons and electrons are then consumed with the reduction of H_2O_2 on the gold end. The resulting ion flux induces an electric field and motion of the particle relative to the fluid, propelling the particle towards the platinum end with respect to the fluid.[3]

example of microscale objects moving by catalysis. They do this by catalyzing the decomposition of hydrogen peroxide (H_2O_2) (Fig. 1), and the rods move along their axis with platinum end forward, as our electrokinetic model predicts.[3,8,10,11] Several important capabilities have been added to the basic auto-mobile device. Among these are (1) steering; (2) the ability to carry cargo; (3) chemotaxis; and (4) collective behavior.

Steering of catalytic nanomotors

"Steering" capability is added by incorporating nickel (magnetic) stripes. The nanomotors can thus be "remote-controlled" by weak magnetic fields.[12]

We further demonstrated control over catalytic motors by designing platinum/ gold structures with different geometries, such as "microgears" that rotate in hydrogen peroxide solutions (Fig. 2).[13] We have also expanded our work to include immobilized catalyst systems that can be used as micropumps. These silver/gold pumps (Fig. 3) are capable of "pumping" tracer particles due to the electrochemical decomposition of hydrogen peroxide (or other redox fuels[14]) on the bimetallic surface.[3,8,10]

Cargo-carrying catalytic nanomotors

We were able to attach a prototypical cargo: polystyrene microspheres to platinum/ gold nanomotors, which can then be transported (Fig. 4).[15] Assuming

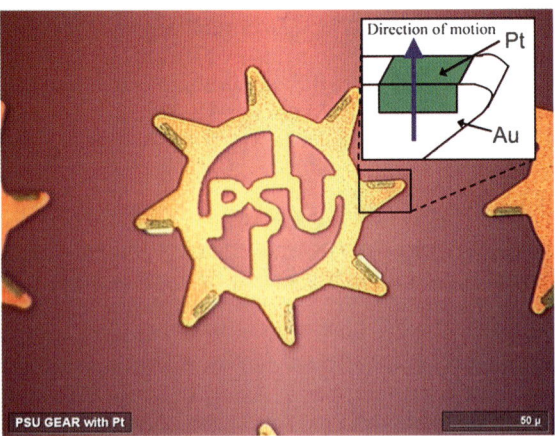

Fig. 2 100 μm diameter gold "microgears" with platinum "teeth" can rotate ~360° s^{-1} in aqueous hydrogen peroxide systems.[13]

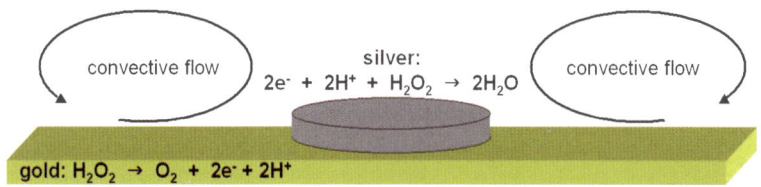

Fig. 3 A catalytic micropump consisting of a silver disk on a gold substrate. The electrochemical decomposition of hydrogen peroxide establishes a weak electric field. This field causes tracer particles to migrate *towards* or *away* from the silver depending on their *surface charge*. Particles migrating towards the silver follow *electroosmotic convection* (arrows) near the catalyst surface.[8]

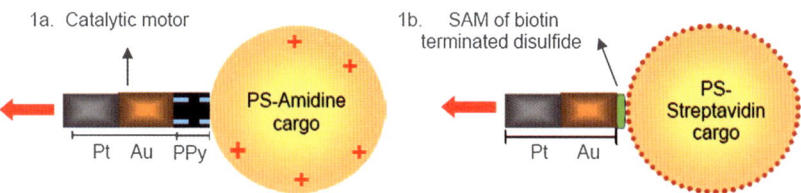

Fig. 4 Cargo attachment by (a) electrostatic interaction between the negative polypyrrole (PPy) end of a platinum/gold/polypyrrole motor and a positively charged polystyrene (PS)–amidine micro-sphere and (b) biotin–streptavidin binding between the gold tips of platinum/gold rods functionalized with a biotin-terminated disulfide and streptavidin-coated cargo.[15]

a cargo-independent motive force, the speeds are inversely proportional to the Stokes resistance, which we compute using a double-layer boundary integral equation. Magnetic Ni segments incorporated into the rods quench the rotational diffusion (and biased rotation) in the presence of an external magnetic field, making such motors ideal for cargo pick-up and delivery. Chemotaxis of the rods was also used to transport cargo to a region of high hydrogen peroxide concentration (see below). Many interesting applications can be envisioned for such cargo bearing motors in the mesoscale. For instance, this provides a simple method for concentrating colloids against their entropic tendency to disperse.

Chemotaxis of catalytic nanomotors

We have discovered that the platinum/gold nanorods exhibit chemotaxis (Fig. 5),[9] traditionally defined as the movement of "organisms" toward or away from a chemical attractant or toxin by a biased random walk process.[16] Again, this is the first example outside living systems. Our work also reveals that chemotaxis does not require a sophisticated "temporal sensing" mechanism commonly attributed to small organisms. Rather, the nanoparticles move up a fuel gradient as a result of faster "powered diffusion" in higher fuel concentrations; a straightforward extension is movement towards or away from a signaling molecule.[9] The signaling molecule could be a promoter or an inhibitor of the catalytic reaction. This behavior provides a novel way to direct particle movement towards specific targets, even while allowing them to sample a large region of fluid by powered diffusive motion. This discovery is potentially important in the design of "smart" autonomous nano-robots, which could move independently in the direction they are needed, perhaps by harvesting energy from glucose or other abundant fuels in biological or organic systems. Chemotaxis also offers a novel method of sorting and separating particles of similar mass and size. Only those particles that are catalytically active move in response to the chemical gradient. From a fundamental standpoint, our work could be the starting point for the design of new motors for collective functions, such as catalytically

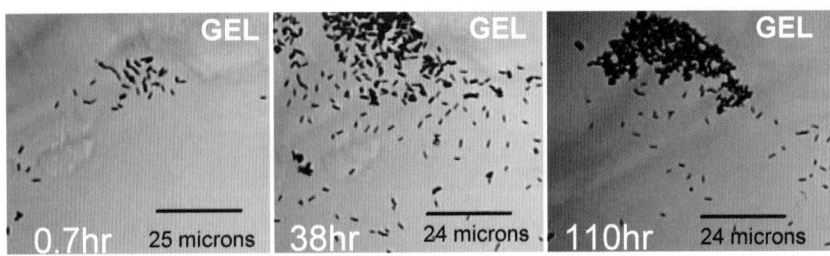

Fig. 5 The changing distribution of platinum/gold rods in a hydrogen peroxide concentration gradient. The gel (soaked in 30% hydrogen peroxide) appears in the upper left part. The images were taken at 0.7 h, 38 h, and 110 h.[9]

driven swarming and pattern formation. This aspect of our work is elaborated in the next section.

Results and discussion

Collective behavior of particles undergoing self-diffusiophoresis

The biological world is rife with examples of organisms using chemotaxis to communicate between spatially isolated cells and accomplish collective tasks. For instance, the unicellular slime mold amoebae *Dictyostelium discoideum* secretes a signaling chemical, 3′-5′-cyclic adenosine monophosphate (cAMP), into the environment when stressed. Nearby slime molds detect this chemical, respond by secreting additional cAMP, thus amplifying the signal, and move up the global cAMP chemical gradient to assemble as a multi-cellular fruiting body.[16]

The key aspect of this collective behavior among slime molds is that the organisms both produce and are translated by the cAMP chemical gradient. We have recently developed a synthetic system operating on that very same principle. Namely, we have observed that micrometer-sized silver chloride (AgCl) particles move under UV illumination in deionized water *via* self-diffusiophoresis (Fig. 6).[17] The AgCl particles move in response to self-imposed salt gradients, a phenomenon which although obviously mechanistically different to the slime mold system, exhibits many qualitative similarities. Like the *Dictyostelium discoideum*, each AgCl particle secretes chemicals (in this case, ions) as it moves, to which the other particles respond. Over the course of a few minutes, this causes the AgCl particles to organize into discrete regions with higher particle concentrations, or "schools," a pattern that fits the known 2D model of the slime mold's behavior (Fig. 7).[17,18]

Although oxygen is produced during the dissolution of the AgCl, a bubble propulsion mechanism for these particles is unlikely because bubble evolution is not observed over the course of the experiment due to the relatively slow reaction rate. Bubble propulsion also cannot account for the long range interaction between particles. Although particle organization from optical trapping is well known, it is unlikely to be the case in this system since the phenomenon was not observed for photo-insensitive polystyrene colloids of similar refractive index. Furthermore, the map of the UV light intensity produced by our microscope was roughly Gaussian in nature and contained no "hot spots." If such a beam resulted in optical trapping, particles would be shuttled to the beam center and not form dozens of isolated "schools".

Also, because diffusiophoresis[19,20] is a physical phenomenon, even photo-inactive silica particles respond to the chemical secretion of the AgCl by swimming towards and surrounding individual AgCl particles, a predator–prey behavior not unlike that of neutrophils (Fig. 8).[21] This ability to fabricate systems of synthetic particles which can interact over relatively long distances demonstrates new design principles for

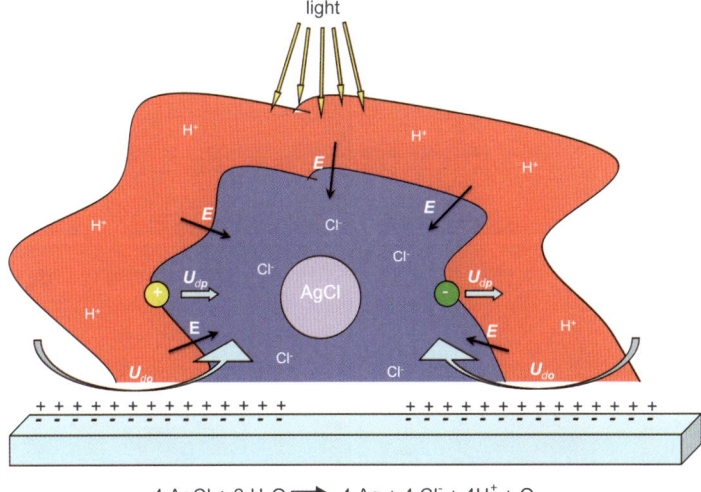

$$4\ AgCl + 2\ H_2O \longrightarrow 4\ Ag + 4\ Cl^- + 4H^+ + O_2$$

Fig. 6 Photochemistry of silver chloride. Briefly, the photolysis of a AgCl particle results in the formation of H^+ and Cl^- ions. The H^+ diffuses out significantly faster than the Cl^-, resulting in an electric field. This electric field acts (a) phoretically on any other nearby charged particles (shown as yellow and green spheres) giving them a velocity U_{dp}; (b) osmotically on ions in the double layer of any nearby wall surfaces to create a fluid flow near the surface U_{do}; and (c) phoretically back on the particle itself in the case of an asymmetric field.[19,20] When there is a high enough concentration of particles, the particles move cooperatively in response to each others' ion gradients to form macroscopic schools (Fig. 7).

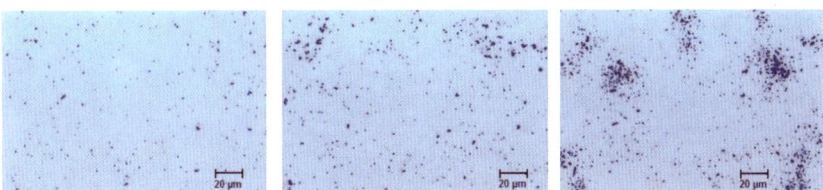

Fig. 7 AgCl particle "schooling". AgCl particles in deionized water (left) before UV illumination, (centre) after 30 s of UV exposure, and (right) after 90 s.[17]

"intelligent" synthetic nano/micromachines that function collectively. There are many possible ways of designing ion-producing nano/microparticles, including particles with attached catalysts or enzymes that form ions as products. These particles should then interact with each other or with inert nano/microparticles through their ionic products. In addition, the design principle allows the coordinated movement of dissimilar particles that are not attached to each other, making it easier to transport and deliver cargo at designated areas.

A simple model for collective behavior

In order to explore the dynamics of such collective behavior among individual particles, a simple simulation was developed from first principles. Let us first consider that the particles (white circles in Fig. 9a) are infinitesimally small and each of the (N) particles occupies a discrete site (x, y) on a finite two-dimensional square lattice with periodic boundaries. The particles are then allowed to "hop" to one of their four nearest neighbor sites with some probability every time-step. There is no limit to the number of particles which can occupy any one site. Purely Brownian motion,

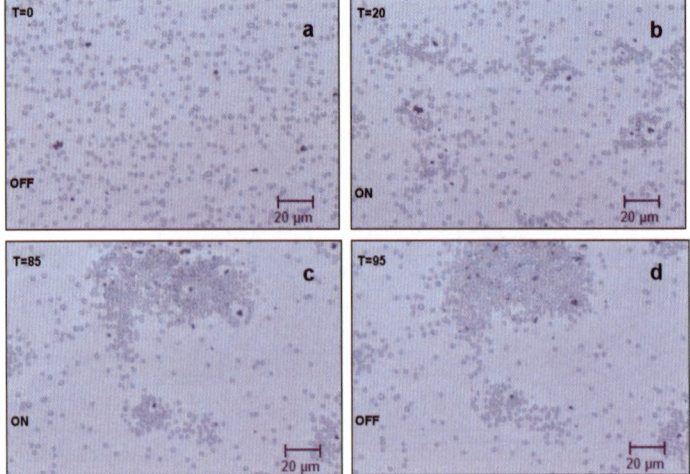

Fig. 8 AgCl particles ~7 μm in diameter (darker objects) have been mixed with 2.34 μm silica spheres and placed in deionized water (a). When illuminated with UV light (b and c) the silica spheres actively seek out the AgCl particles and surround them. While the UV light is on, an exclusion zone is seen around the AgCl particles; this exclusion zone disappears when the UV light is turned off (d). Times in seconds are listed in the upper left hand corner. Status of the UV light is listed in the bottom left.[17]

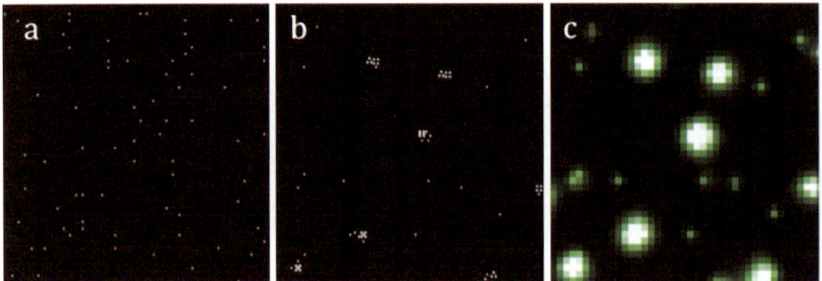

Fig. 9 A typical simulation of particle collective behavior. Image represents one-fourth of the total simulation space and has been zoomed in for clarity. (a) Initially randomly distributed particles. (b) Particles after 2000 times-steps. (c) A plot of the chemical values of every lattice site at 2000 time-steps. Brighter shades of green approaching white indicate higher chemical concentrations.

for instance, can then be considered as a uniform hopping probability λ_B for every particle in each of the four directions.

In the AgCl system described above, particles interact by secreting chemical signals into their environments. These chemical signals diffuse out into the solution over time, allowing for long range interactions. To incorporate such interactions into our model, we propose that each lattice site also has a numerical "concentration" value ($C_{x,y}$) associated with it (Fig. 9c). This value is initially zero for all sites. During the simulation, for each time-step, the concentration value of the lattice site is increased in proportion (by a factor of a in eqn (1)) to the number of particles currently occupying that site ($n_{x,y}$).

The concentration value associated with each lattice site is then allowed to "diffuse" outward to neighboring lattice sites—i.e., the concentration value from one site gets redistributed to its four nearest neighbors in a manner which is dictated by Fick's Second Law. Specifically, each time-step every cell loses some small portion of its current chemical concentration [$D \times C_{x,y}(t)$] to each of its four nearest neighbors, and gains a small portion of their chemical in return. In addition to this strict diffusion, each lattice site loses some small proportion (b in eqn (1)) of its total chemical value every time-step. This small loss is included to allow the two-dimensional simulation to more closely approximate conditions found in real systems which are three-dimensional. More specifically, it helps account for the fact that any chemical signal produced by the particles will not only diffuse in the two-dimensional field of view, but also some portion of the chemical will diffuse into the bulk solution and be lost. Therefore, in the simulation, the chemical concentration of any site ($C_{x,y}$) at any time-step ($t + \Delta t$) is given by eqn (1), where a and b are constants and are members of $[0, \infty]$ and $[0,1]$, respectively.

$$\underbrace{C_{x,y}(t+\Delta t)}_{\text{New Conc.}} = \underbrace{(1-b)}_{\text{Bulk Loss}} \Big\{ \underbrace{C_{x,y}(t)}_{\text{Original Conc.}} + \underbrace{an_{x,y}}_{\text{Chemical Production}} + \underbrace{D\big[C_{x+1,y}(t) + C_{x-1,y}(t) + C_{x,y+1}(t) + C_{x,y-1}(t) - 4C_{x,y}(t)\big]}_{\text{Diffusion}} \Big\} \quad (1)$$

The final component of the simulation is then just the addition of a bias which causes the particles to have a preference to hop towards regions with higher chemical values. This bias is introduced as extra hops (in addition to the Brownian hops) in the x and the y direction which depend on the gradient of chemical value across the particle's host site. To do this, the gradient in the chemical value across the host site of a given particle is evaluated in both the x and y directions (ψ_x and ψ_y). Individual hops are then taken in the x and y directions with probabilities (λ_x and λ_y) proportional to the gradients in those respective directions, as depicted by eqn (2) and (3). Here α is a constant representing the sensitivity to the chemical gradient. The directions of these two hops are determined by the signs of ψ_x and ψ_y, and both λ_x and λ_y are of course constrained to be less than one by an appropriate choice of α.

$$\psi_x = C_{x+1,y}(t) - C_{x-1,y}(t) \quad (2a)$$

$$\psi_y = C_{x,y+1}(t) - C_{x,y-1}(t) \quad (2b)$$

$$\lambda_x(t) = \alpha |\psi_x| \quad (3a)$$

$$\lambda_y(t) = \alpha |\psi_y| \quad (3b)$$

In the physical world, this gradient-related hopping bias may take a variety of forms. It may correspond to a diffusiophoretic force on a particle (*e.g.*, the AgCl system), a chemotactic tendency to travel up fuel gradients (*e.g.*, catalytic nanorods in H_2O_2 gradient), or any one of a number of complex signaling pathways found in biological systems.

As we can see from Fig. 10(a)–(e), our model shows that a number of factors can influence the ability of these modeled particles to form collective schools. In each of these graphs, one parameter of the system was changed, holding all other parameters constant. The simulation was run for 1000 time-steps, and the propensity of the particles to be in schools at $t = 1000$ was plotted. Here the schooling propensity is defined as the proportion of particles occupying lattice sites which are occupied by four or more particles (an arbitrary cut-off) at the end of the simulation. Each data point represents ten simulations with error bars of one standard deviation.

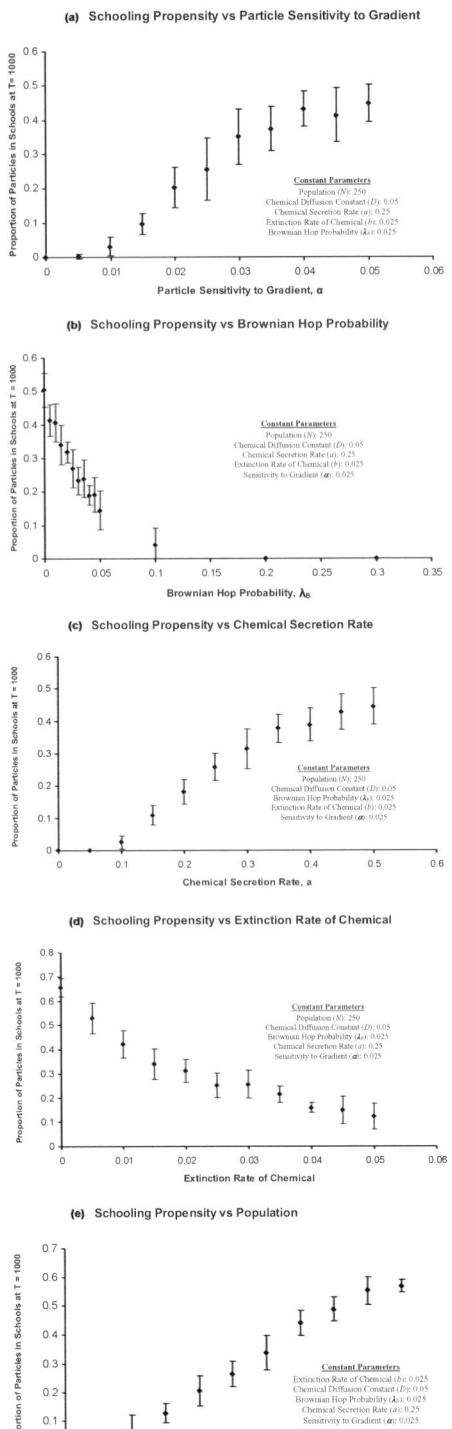

Fig. 10 Plots summarizing the particle location statistics for the qualitative probability-model of collective behavior. In each graph, the proportion of particles residing in lattice sites with

Certainly, the most obvious variable of note is the sensitivity α to the chemical gradient. For a given gradient, the particle's propensity to hop up the gradient (Fig. 10(a)) must be large enough to overcome the randomizing effects of Brownian motion (Fig. 10(b)). Similarly, the steepness of the chemical gradient also plays a role. This can be achieved by increasing the "secretion rate" a of the chemical from the colloids, or by lowering the "extinction rate" b of the chemical from the environment (Fig. 10(c) and 10(d)).

The number density of particles in the system (Fig. 10(e)) is also an important parameter. For small populations of particles, the inter-particle spacing is relatively large. As a result, the chemical signal from one particle becomes significantly diluted before reaching its nearest neighbor. The resultant gradient is therefore gentler, and thus the hopping probability (λ) of the second particle towards the first is relatively small.

Photo-patterning of particles

Thus far, our discussion of self-diffusiophoresis has assumed a uniform illumination of the ion-producing photoactive particles. Yet this need not be the case. Recently we have observed that an asymmetric gradient of light impinging on photocatalytic particles can result in pattern formation. An example is shown in Fig. 11 where silica–silver (SiO_2–Ag) Janus particles move away from the UV illuminated area due to a chemical gradient created by a photolysis reaction.

In this example, the SiO_2–Ag Janus particles, with a 60~80 nm thickness of Ag evaporated onto one side of the SiO_2 (d, 2.34 µm), are placed in 0.5% H_2O_2 solution in a sealed imaging chamber. A photolysis reaction between Ag and H_2O_2 is initiated upon UV irradiation through the microscope objective, producing Ag^+ and OOH^- ions. The differential diffusion coefficients of these ions induce an electric field localized at the particle length scale, which in turn causes the individual particles to move by the mechanism of self-diffusiophoresis.[19,20] A macro-scale ion gradient is also produced corresponding to the different UV illumination levels within the system. This results in a circular ion gradient, centered on the microscope objective's focal point, causing the particles to move away from the highest ion concentration area, *i.e.*, the center of the field. The action is again emergent, analogous to that of quorum sensing,[22,23] where a threshold concentration of the signaling chemical is required to trigger the migration of the organisms. We found that a minimum particle density of 10^6 mL^{-1} is required in our SiO_2–Ag phototaxis experiment to overcome the randomizing effect of Brownian motion for an observation time of 1–2 h.

Control experiments performed without H_2O_2 showed a slight accumulation of particles at the light spot after 24 h, indicating the existence of a thermal effect, which is in the opposite direction to the phototactic effect and takes longer. The negative phototaxis (*i.e.*, a hole at the UV spot) of SiO_2–Ag Janus particles in the presence of H_2O_2 is significant because the particles have to *overcome* the thermal effect to be able to escape from the high UV intensity region.

Earlier in this paper we described a general model for the examination of collective particle interactions. Although the general nature of the model allows for high throughput and wide application, its quantitative predictive power is limited. Therefore, we have developed a second, independent model which specifically addresses particle collective interactions from a standpoint of self-diffusiophoresis.

This system-specific Brownian Dynamic Simulation (BDS)[24] addresses the diffusiophoretic motion of settled particles close to a surface, a diffusioosmotic motion due to the underlying surface, and a Brownian motion of the particles (eqn (4)).

---

four or more total particles at $T = 1000$ is plotted *vs.* one independently modified parameter. Each data point represents ten simulations with error bars of one standard deviation.

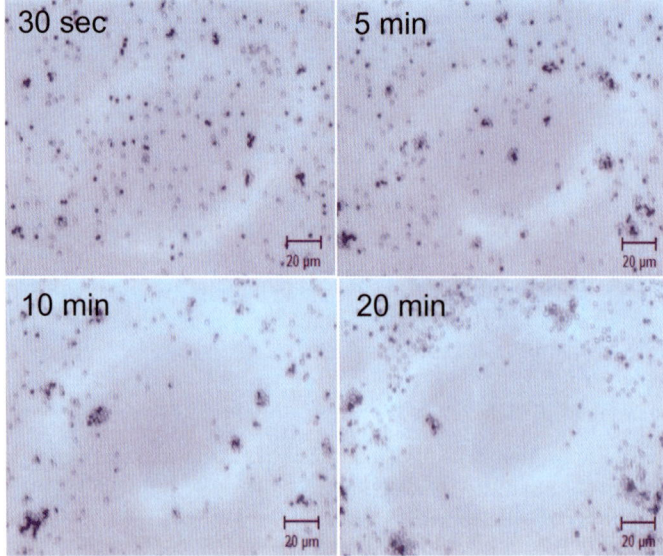

Fig. 11 Time-lapse images show the phototactic process of SiO$_2$–Ag Janus particles in 0.5% H$_2$O$_2$ under the irradiation of a circular spot by ultraviolet light.

Just as before, we modeled N particles in 2 dimensions, in a fluid over an area ($A = L \times L$). However, in this model, particle positions were treated as floating-point locations on top of a 100 × 100 integer grid which contained the chemical concentration data. Each grid site corresponds to a 30 μm^2 area of fluid. Also, in this model the particle motions were treated not as hops of unit distance with finite probability but as hops of variable distance which occur every time-step.

To simplify the applicable hydrodynamic equations, spherical particles were assumed, and any change in the hydrodynamics near the surface due to lubrication or multi-body effects was neglected. The overall change in position, Δx, for a particle in a given time-step can then be described as a superposition of the displacements caused by the diffusiophoretic velocity (U_{dp}), the diffusioosmotic velocity (U_{do}), and the Brownian diffusion (Δx_B), as seen in eqn (4).

$$\Delta x = U_{dp}\Delta t + U_{do}\Delta t + \Delta x_B \quad (4)$$

The diffusiophoretic velocity U_{dp} of a particle with a thin double layer is given by eqn (5),

$$U_{dp} = \frac{2Ze}{\kappa^2 \eta}\left[\left(\frac{D_+ - D_-}{D_+ + D_-}\right)\zeta_p - \frac{2kT}{Ze}\ln(1 - \gamma_p^2)\right]\nabla C \quad (5)$$

where Z is the valence of the ions; e is the proton charge; κ^{-1} is the solution Debye length (178 nm for water in equilibrium with air); ζ_p is the particle ζ-potential (ζ_p, approx. −30 mV for most of our particles); η is the solution viscosity (0.89 cP for water at 25 °C); $\gamma_p = \tanh(Ze\zeta_p/4kT)$; and ∇C is the concentration gradient of the ions, which by electroneutrality must have the same bulk concentration at all locations. The diffusion coefficients are known at 25 °C for the positive (Ag$^+$, $D_+ = 1.65 \times 10^{-9}$ m^2 s^{-1})[25] and negative (OOH$^-$, $D_- = 0.3 \times 10^{-9}$ m^2 s^{-1}) ions.[26] The diffusioosmotic velocity (U_{do}) of the fluid carrying the particles, where this movement is due to the underlying surface, is given by eqn (6), where the wall ζ-potential (ζ_w, approx. −70 mV) is considered.

$$U_{do} = -\frac{2Ze}{\kappa^2 \eta} \left[\left(\frac{D_+ - D_-}{D_+ + D_-} \right) \zeta_w - \frac{2kT}{Ze} \ln\left(1 - \gamma_w^2\right) \right] \nabla c \qquad (6)$$

The illumination spot was defined as having a radius of 10 grid points. All particles that were within the central laser spot produced Ag^+ and OOH^- ions at a turnover rate of 9.4 per catalytic site per second (catalytic site size ~ 0.25 nm^2), which was determined from separate experiments. No account was taken to assess changes in reaction rate with H_2O_2 concentration or illumination power. And the change in H_2O_2 concentration over the time span of our experiment was negligible due to its relatively high concentration. The particles were assumed to be "point sources" producing these ions, which were then allowed to diffuse in two dimensions by Fick's second law. From the final concentration field at each time step, we calculated ∇c.

The results show that a hole develops in the center of the UV spot over time. We use a relatively low particle density ($N = 1000$) to show the time-wise effect. As can be seen in Fig. 12, the hole grows over time, but stops growing after about 5 h ($\tau = 200$). The size of the hole is dependent on the UV spot size. In the real experiments, thermal effects take over after 4–6 h and cause aggregation of particles. Thus, we focus on the phototactic effect shown within 1–2 h.

In order to simulate the emergent behavior, we set the reaction time to $\tau = 50$ ($= 75$ min), which is about the time required to observe the phototactic effect experimentally (~ 1 h). We vary the particle number and find that below $N = 1000$, the phototactic effect is no longer observed (Fig. 13). The threshold is determined by the balance between the reaction rate (how fast the ion gradient builds up) and the dissipation rate, which is mainly contributed by Brownian diffusion. Knowing that the actual UV spot is 600 μm in diameter, $N = 1000$ in the model translates to a particle density of 2×10^5 mL^{-1}. This is in agreement with our experimental observation that a minimum particle density of 10^6 mL^{-1} is needed for the patterning to occur.

Conclusion

We have engineered the first examples of nano/micro-objects outside living systems that move autonomously by converting chemical energy into mechanical force. With

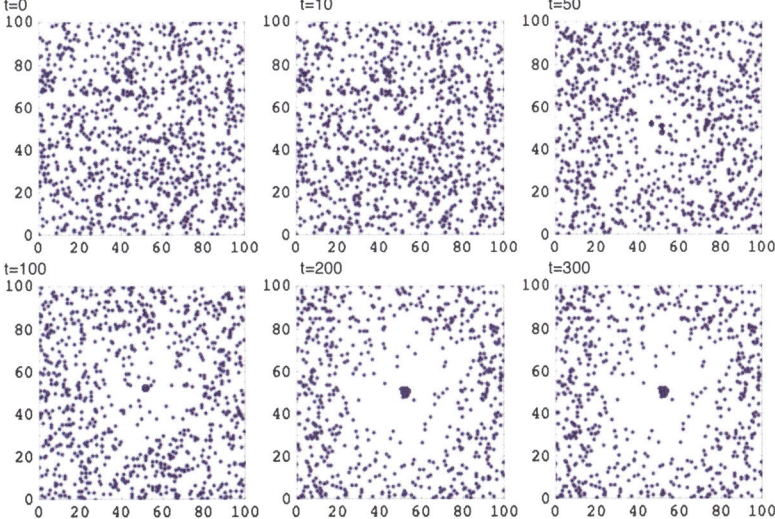

Fig. 12 Brownian dynamics simulation shows the development of the central hole over time at a particle number of $N = 1000$.

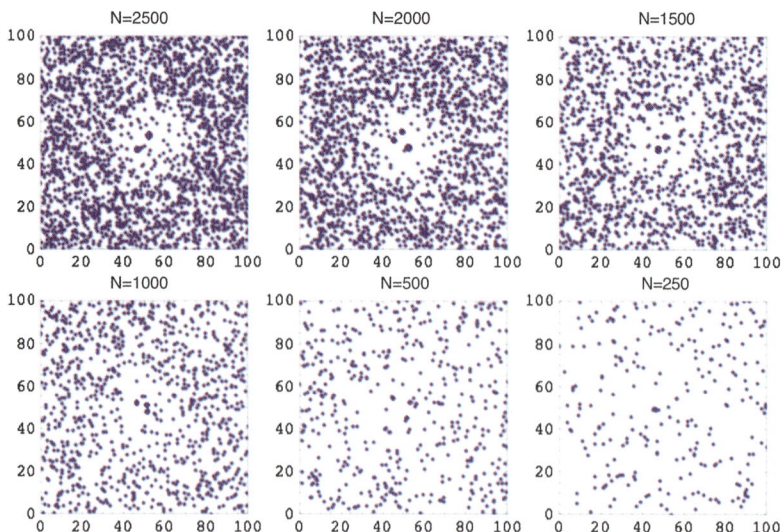

Fig. 13 Brownian dynamics simulation shows how particle density affects the phototactic behavior.

very little "information" input (in the form of chemical gradients), these objects begin to display emergent collective behaviors that were thought to lie solely in the realms of biology. Liberated of the usual biological constraints, we now have an unprecedented opportunity to probe the ultimate limits of self-organization in these dynamic systems. If successful, our work should lead to novel design paradigms for assembly and patterning of particles.

Acknowledgements

We thank Dr Thomas Mallouk, Dr Vincent Crespi and Dr Paul Lammert for helpful discussions. The work is funded by the Penn State Center for Nanoscale Science (NSF-MRSEC, DMR-0820404), NSF NIRT CTS-0506967, and NSF CBET-0651611. Support was also given by the NSF-supported Penn State University's National Nanotechnology Infrastructure Network (NNIN).

References

1. W. F. Paxton, K. C. Kistler, C. C. Olmeda, A. Sen, S. K. St. Angelo, Y. Cao, T. E. Mallouk, P. E. Lammert and V. H. Crespi, *J. Am. Chem. Soc.*, 2004, **126**, 13424.
2. (a) S. Fournier-Bidoz, A. C. Arsenault, I. Manners and G. A. Ozin, *Chem. Commun.*, 2005, 441; (b) G. A. Ozin, I. Manners, S. Fournier-Bidoz and A. Arsenault, *Adv. Mater.*, 2005, **17**, 3011.
3. W. F. Paxton, P. T. Baker, T. R. Kline, Y. Wang, T. E. Mallouk and A. Sen, *J. Am. Chem. Soc.*, 2006, **128**, 14881.
4. J. Vicario, R. Eelkema, W. R. Browne, A. Meetsma, R. M. La Crois and B. L. Feringa, *Chem. Commun.*, 2005, 3936.
5. N. Mano and A. Heller, *J. Am. Chem. Soc.*, 2005, **127**, 11574.
6. J. Howse, R. A. L. Jones, A. Ryan, T. Gough, R. Vafabakhsh and R. Golestanian, *Phys. Rev. Lett.*, 2007, **99**, 048102.
7. W. F. Paxton, S. Sundararajan, T. E. Mallouk and A. Sen, *Angew. Chem., Int. Ed.*, 2006, **45**, 5420.
8. T. R. Kline, W. F. Paxton, Y. Wang, D. Velegol, T. E. Mallouk and A. Sen, *J. Am. Chem. Soc.*, 2005, **127**, 17150.
9. Y. Hong, N. M. K. Blackman, N. D. Kopp, D. Velegol and A. Sen, *Phys. Rev. Lett.*, 2007, **99**, 178103.

10 T. R. Kline, J. Iwata, P. E. Lammert, T. E. Mallouk, A. Sen and D. Velegol, *J. Phys. Chem. B*, 2006, **110**, 24513.
11 W. F. Paxton, A. Sen and T. E. Mallouk, *Chem.–Eur. J.*, 2005, **11**, 6462.
12 T. R. Kline, W. F. Paxton, T. E. Mallouk and A. Sen, *Angew. Chem., Int. Ed.*, 2005, **44**, 744.
13 J. Catchmark, S. Subramanian and A. Sen, *Small*, 2005, **1**, 202.
14 M. Ibele, Y. Wang, T. R. Kline, T. E. Mallouk and A. Sen, *J. Am. Chem. Soc.*, 2007, **129**, 7762.
15 S. Sundararajan, P. E. Lammert, A. W. Zudans, V. H. Crespi and A. Sen, *Nano Lett.*, 2008, **8**, 1271.
16 (*a*) P. Devreotes and C. Janetopoulos, *J. Biol. Chem.*, 2003, **278**, 20445; (*b*) H. C. Berg, *Phys. Today*, 2000, 53; (*c*) J. Adler and W. W. Tso, *Science*, 1974, **184**, 1292; (*d*) R. M. Macnab and D. E. Koshland, *Proc. Natl. Acad. Sci. U. S. A.*, 1972, **69**, 2509.
17 M. Ibele, T. Mallouk and A. Sen, *Angew. Chem., Int. Ed.*, 2009, **48**, 3308.
18 U. Wilensky, 1997. *NetLogo Slime Model from the Center for Connected Learning and Computer-Based Modeling*, Northwestern University, Evanston, IL http://www.ccl.northwestern.edu/netlogo/models/Slime.
19 J. L. Anderson, *Annu. Rev. Fluid Mech.*, 1989, **21**, 61.
20 S. Dukhin, and B. Derjaguin, *Electrokinetic Phenomena*, John Wiley, New York, 1974.
21 D. Billadeau, *Nat. Immunol.*, 2008, **9**, 716.
22 S. R. Chhabra, B. Philipp, L. Eberl, M. Givskov, P. Williams and M. Cámara, *Top. Curr. Chem.*, 2005, **240**, 279.
23 P. Devreotes, *Science*, 1989, **245**, 1054.
24 W. B. Russell, D. A. Saville and W. R Schowalter, *Colloidal Dispersions*, Cambridge University Press, 1989. Sections 3.5 and 3.6 outline how the BDS is obtained from the Langevin equation.
25 *CRC Handbook of Chemistry and Physics*, 87th edn., ed. D. R. Lide, Taylor & Francis, Boca Raton, New York, 2006–2007.
26 T. R. Kline and A. Sen, *Langmuir*, 2006, **22**, 7124.

PAPER

The efficiency of encapsulation within surface rehydrated polymersomes

A. J. Parnell,[*a] N. Tzokova,[b] P. D. Topham,[c] D. J. Adams,[d] S. Adams,[e] C. M. Fernyhough,[b] A. J. Ryan[b] and R. A. L. Jones[a]

Received 6th February 2009, Accepted 1st April 2009
First published as an Advance Article on the web 28th July 2009
DOI: 10.1039/b902574j

The key to the use of polymersomes as effective molecular delivery systems is in the ability to design processing routes that can efficiently encapsulate the molecular payload. We have evaluated various surface rehydration mechanisms for encapsulation, in each case characterizing the morphologies formed using DLS and confocal microscopy as well as determining the encapsulation efficiency for the hydrophilic dye Rhodamine B. In contrast to bulk methods, where the encapsulation efficiencies are low, we find that higher efficiencies can be obtained by the rehydration of thin films. We relate these results to the non-equilibrium mechanisms that underlie vesicle formation and discuss how an understanding of these mechanisms can help optimize encapsulation efficiencies. Our conclusion is that, even considering the good encapsulation efficiency, surface methods are still unsuitable for the massive scale-up needed when applied to commercial "mass market" molecular delivery scenarios. However, targeting more specialized applications for high value ingredients (like pharmaceuticals) might be more feasible.

1. Introduction

Polymersomes – vesicles based on self-assembled bilayers composed of amphiphilic copolymers – are good candidates for molecular delivery systems;[1,2] hydrophilic molecules can be enclosed within the aqueous core, to be released by a trigger which disrupts the vesicle's wall.[3,4] Polymersomes are the macromolecular analogue of liposomes, in which the vesicle wall is composed of a phospholipid bilayer. Liposomes are already used extensively for molecular delivery, but polymersomes have a number of promising advantages. Potentially, they are more robust than liposomes,[5] and their wall thicknesses and compositions can be tuned precisely like liposomes to control the diffusion of molecular species in and out of the vesicle.[6] There is also considerable flexibility in designing chemical systems whose state of aggregation depends strongly on the environment, allowing the triggered release of their contents. The practical utility of polymersomes, or any other molecular delivery system, is crucially dependent on the ease and efficiency with which it is possible to encapsulate their payload. The aim of this paper is to measure quantitatively

[a]*Department of Physics and Astronomy, University of Sheffield, Sheffield, UK S3 7RH. E-mail: a.j.parnell@sheffield.ac.uk*
[b]*Department of Chemistry, University of Sheffield, UK S3 7HF*
[c]*Chemical Engineering and Applied Chemistry (CEAC), Aston University, Aston Triangle, Birmingham, UK B4 7ET*
[d]*Department of Chemistry, University of Liverpool, UK L69 7ZD*
[e]*Unilever Research Colworth, Sharnbrook, Bedfordshire, UK MK44 1LQ*

the efficiency with which a model payload is encapsulated in polymersomes when prepared by a number of variants of the surface rehydration method and to explore what can be done to optimise that encapsulation efficiency.

Amphiphilic block copolymers – which consist of two or more chemically different blocks, one of which has a significantly more favourable interaction with water than the other, form a diverse array of nanostructures spontaneously in aqueous solution.[7] The most common structures formed in solution consist of cylindrical or spherical micelles, or lamellae (sheets); these basic units may then be arranged with higher levels of order. The equilibrium nanostructure is determined by the thermodynamic interactions of the blocks with each other and with water and the architecture of the copolymer, all of which determine the natural curvature of the interfaces in the nanostructure, and the copolymer concentration.[8] Vesicles are generally formed from systems that form lamellae (*i.e.* in which the interfaces have small natural curvature); to make a vesicle these sheets need to curve round and close up to separate an inner space from the environment. This process generally takes place under non-equilibrium conditions, and it is the details of this non-equilibrium process that determine how much of any payload molecule – which may be present in solution or in a more concentrated copolymer phase – is encapsulated within the vesicle.

There have been a variety of techniques and methods used to load vesicles with an active compound, either inside the aqueous core or within the vesicle membrane.

In the *solvent switch route*[9,10] the polymer amphiphile is first dissolved in a water-miscible solvent that both blocks are soluble in, typically ethanol or THF, resulting in molecular dissolution of the polymer, together with the molecule to be encapsulated. This solution is then diluted with a selective solvent (generally water) causing the amphiphile to self-assemble into vesicles, encapsulating some of the active compound. The organic solvent is then removed from the aqueous vesicle solution by subsequent dialysis. This route tends to produce vesicles consistent in size and lamellarity that depend on the initial concentration, stirring speed and rate of addition.

The problem with these bulk processing routes is that they do not proceed *via* lamellar sheet wrap up. Instead they form by rather complex routes where no quick equilibration of the encapsulated volume with the continuous phase can occur.[11,12] Therefore the solution inside the vesicles does not correspond to a sampling of the aqueous environment in which it was formed, as is expected for the lamellar sheet wrap up mechanism.

The *reverse phase evaporation (RPV) technique*[9,13,14] also begins with a molecularly dissolved solution of the amphiphile in a solvent which solvates both blocks, but in contrast to the solvent switch route a water-immiscible solvent such as chloroform is used. The aqueous solution containing the hydrophilic molecules that are to be encapsulated is then added and the mixture is mixed or sonicated to make a stable emulsion. The amphiphiles will segregate to the interfaces between the water droplets and the solvent. The solvent is removed by evaporation, leaving an aqueous vesicle dispersion. RPV enables the encapsulation of hydrophilic and hydrophobic molecules. However this method may be unsuitable for the loading of molecules that could be damaged by the organic solvent and also the solvent could remain after formation making dialysis necessary.

Other promising routes rely on the use of microfluidic devices, which can be scaled out so that there are many in parallel to produce viable industrial quantities. The *double emulsion route*[13,15] involves making stable water-in-oil-in-water (W/O/W) droplets, which has mainly been carried out using small scale capillaries or microfluidic devices.[16] This formation technique enables direct loading of a hydrophilic species into the vesicle as the emulsion is formed. Another route to forming relatively monodisperse and unilamellar vesicles is the *inkjet-printing route*, demonstrated by Hauschild *et al.*[17] This technique uses a conventional inkjet printing cartridge to eject block copolymer (and hydrophilic dye) in a molecularly dissolved state into water whereupon the block copolymer will spontaneously form small unilamellar vesicles with a narrow size distribution.

The ease and uniformity of vesicle formation using these techniques make them attractive technologies but again they would need filtering or dialysing to remove any residual organic solvents.

The technique we discuss is this publication – *the thin film rehydration method*[18–20] – is illustrated schematically in Fig. 1 along with a few other vesicle forming techniques.

This method starts by dissolving the amphiphilic polymer in organic solvent. A thin film or layer is then formed on a glass surface or roughened PTFE surface and the solvent in the layer is removed by rotary evaporation or high vacuum. This leaves a dry polymer film that can be rehydrated in an aqueous solution; this can be pure water or a buffered salt solution. As water is added the lamellae will become increasingly more ordered and swollen.[21] The rehydration can be assisted by heat, shaking or sonicating the polymer film to aid vesicle formation. The vesicles formed using this technique tend to be multilamellar vesicles with a large spread of vesicle sizes in the micron range. Vesicle processing methods like extrusion and sonication can be used to normalize the vesicles to a narrow size distribution and also reduce the vesicle lamellarity. The molecule to be encapsulated can either be incorporated in the film to be rehydrated by co-dissolving it in the original amphiphile solution, or it can be added to the rehydrating solution. The main problem with this formation route is the non-uniformity of the vesicles that are produced as well as the polymer systems that can be used need to be readily rehydrated to form vesicles. Any lengthy rehydration stage or processing step would make it non-viable industrially.

Despite a number of papers describing the use of polymersomes for encapsulation,[4,22] surprisingly few actually quantify the encapsulation efficiency[12,23] – that is to say, what is the proportion of the molecules to be encapsulated that end up inside polymersomes, as a fraction of the molecules used in the process? If this fraction is low, this may limit the practicality and economic viability of the encapsulation process, particularly if the molecules to be encapsulated are expensive or difficult to make, by requiring difficult additional recycling steps.

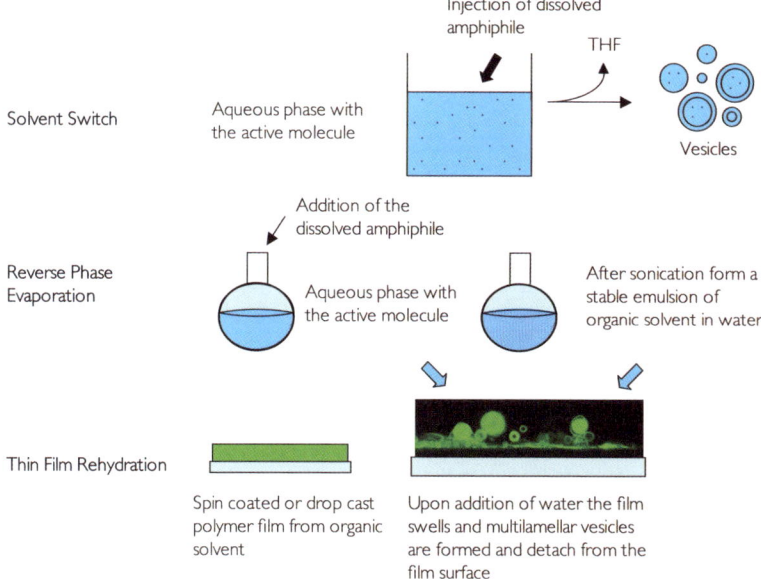

Fig. 1 Schematic summary of some of the methods used to encapsulate molecules within polymersomes and lipids.

In this paper we quantify encapsulation efficiencies for polymersomes prepared by the surface rehydration route. We use a model system consisting of a vesicle forming diblock copolymer, poly(ethylene oxide)-b-poly(butylene oxide),[18,24,25] with the hydrophilic fluorescent dye Rhodamine B as the molecule to be encapsulated.

2. Experimental

2.1 Materials

The copolymer used in this study, with nominal composition $PEO_{28}PBO_{36}$, was prepared by anionic polymerization (sequential monomer addition) and characterized by gel permeation chromatography and ^1H NMR spectroscopy as described elsewhere.[26] The block copolymer had a number-average molecular weight, M_n, of 3800 g mol^{-1} with a polydispersity index of 1.10 and a hydrophilic volume fraction of 0.29.

2.2 Confocal microscopy

Confocal laser scanning microscopy was performed using a Zeiss LSM 510M.

Excitation of Rhodamine B was performed by 543 nm laser radiation and the fluorescence was collected at wavelengths longer than 620 nm using a long-pass filter. Several hundred images were collected over a 72-hour period with a pinhole collecting data over a narrow vertical slice (~50 μm) through the sample.

The columned or dye separated vesicles were imaged in imaging chambers (Coverwell) with the most populous vesicle fractions (from the counts per second) being imaged by DLS measurements. This was normally fraction number 7 or 8 of the 1 mL fractions collected after columning.

To visualize the vesicles forming from a surface, the same block copolymer poly(ethylene oxide)-b-poly(butylene oxide) was used. The polymer's hydrophobic domains were labelled using Rhodamine B octadecyl ester perchlorate (Aldrich) amphiphilic fluorescent dye. A solution of the block polymer was made in chloroform (5 wt %) to which was added a few drops of a dilute solution (0.05 wt %) of Rhodamine B octadecyl ester perchlorate in chloroform.

The chamber was sealed with vacuum grease (Dow) to prevent evaporation over the long experimental run. The polymer-coated silicon was placed into a sample cell at a slight incline to the vertical and filled with filtered 18.2 MΩ^{-1} water.

2.3 Dynamic light scattering

Dynamic light scattering (DLS) measurements were performed on a Brookhaven Instruments 200SM laser light scattering goniometer using a HeNe (125 mW, 633 nm) laser. Single five minute exposure scans were performed and particle sizes were estimated using the CONTIN multiple pass method of data analysis at an angle of 90°.

2.4 Static light scattering

Static light scattering was carried out using a Brookhaven Instruments 200SM laser light scattering goniometer with a HeNe (125 mW, 633 nm) laser over an angle range from 10°–135° in 5° steps. Solutions at concentrations in the range from 30 μg mL^{-1}– 10 μg mL^{-1} were measured. The vesicle solution was extruded 15 times using the LiposoFast-Basic extruder with a polycarbonate membrane of well-defined pore size (100 nm) prior to the measurements. All solutions were maintained at 25 °C throughout the measurements. Care was taken to ensure that the solutions were free from dust or other contaminants by filtering the solvent through 0.45 μm PTFE filters (Goodfellow) prior to analysis.

2.5 Encapsulation efficiency assay

The assay used to measure the total absolute amount of dye encapsulated within the vesicles is shown schematically in Fig. 2, for the case of the thin film rehydration route.

For column chromatography, two Sephadex PD10 (GE healthcare) desalting columns were used in series. The two columns were purged with deionised water prior to introduction of the vesicle containing solution. A total of 1 mL of the vesicle solution was introduced to the top of the first column and the displaced solution was collected in individual small sealable Eppendorf containers. After the first 1 mL solution fraction was collected a further 1 mL of deionised water was added to the top column and collected in a second Eppendorf container. This process continued until 20 individual 1 mL fractions had been collected from the columns.

To measure the true dye loading and efficiency for the various techniques the vesicle membranes needed to be disrupted and the effect of quenching removed.

The fluorescence for each fraction was correlated with the derived count rate from the DLS measurements to determine which fractions contained vesicles and their level of fluorescence. The fluorescent dye Rhodamine B was excited at 545 nm and the fluorescence was measured from 550 nm–700 nm using a Varian fluorimeter with the solutions being measured using quartz cuvettes (Starna scientific). Prior to, and after, measuring the fluorescence for each of the collected fractions the cuvette was rinsed with clean deionised water and the fluorescence checked to be that of water, in order to ensure that there was no contamination between the measurements.

The presence of vesicles in the fractions was detected by DLS measurements on the Brookhaven instrument by plotting the derived count rate against the column fraction. For vesicles with a narrow size distribution the derived count rate gave an estimate of the relative concentrations. The dilution factor in passing the solution through two columns was measured to be a factor of 2.4 times more dilute than the original concentration using DLS count measurements.

From the 1 mL fractions containing vesicles (fractions 5–13), 0.5 mL from each of these fractions was combined. This solution was then rotary evaporated to remove

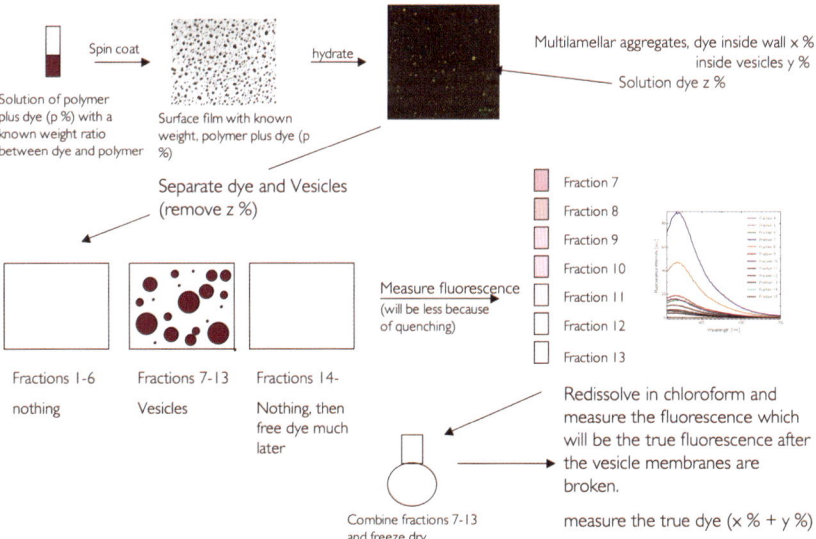

Fig. 2 The procedure for evaluating the encapsulated dye within the vesicles by separation and disruption in chloroform.

most of the water and then freeze dried to remove the remaining water to leave a dry vesicle powder containing dye. This was then re-dissolved in chloroform and the fluorescence measured. A known Rhodamine B dye concentration series was measured in chloroform to give a linear region for assessing the redissolved dye and vesicle powder true concentrations.

2.6 Surface rehydration

A polymer solution of 160 mg $PEO_{28}PBO_{36}$ was dissolved in 0.5 mL chloroform and 0.5 mL ethanol. The solution was mixed thoroughly to ensure complete dissolution of the polymer and the dye. The dye-loaded polymer solution was then spin coated onto a square of PTFE (Goodfellow UK) at 3000 rpm for 30 seconds to produce a thin polymer layer that spontaneously began to dewet into droplets.

The layers were dried in a vacuum oven at room temperature and 100 mbar pressure for over 12 hours to remove any traces of residual organic solvent.

Rehydration of the droplets produced a distribution of giant multilamellar vesicles, as seen by confocal and DLS measurements.

This mixture was spin cast onto a hydrophobic silicon substrate at 3000 rpm for 10 seconds to generate a film of several hundred nanometers in thickness (measured atomic force microscopy). The polymer film spontaneously dewetted and broke up to form droplets on the surface of the PTFE film.

(a) **In 125 μM dye solution.** This method was used to evaluate the efficiency of thin film rehydration from a surface with no dye loaded into the dry polymer film. The solution used was 160 mg of the diblock in a solvent mixture of 0.5 mL chloroform and 0.5 mL ethanol. A small amount of this solution was spin coated onto a PTFE substrate and after spin coating the layer dewetted to form isolated droplets. The amount of material on the surface was measured by weighing by difference of the PTFE layer before and after spin coating. The layer was rehydrated in 3 mL of filtered (0.45 μm pore size) 125 μM aqueous Rhodamine B solution.

(b) **Surface rehydration in a pre-loaded film.** The solution used was a mixture of 160 mg of the diblock in 0.5 mL chloroform to which was added 0.5 mL of an ethanol dye stock solution. The three stock dye solutions were made up in 2 mL of ethanol with masses of 51 mg, 105 mg and 200 mg of Rhodamine B respectively. The three different dye loadings allowed an investigation of the dye to diblock mass ratio. In the same order the mass ratios were 7%, 16% and 23% by mass for 51 mg, 105 mg and 205 mg. The same weighing by difference method was used as before and the ratio of dye to diblock in the dry film was assumed to be that of the original spin coated solution. The mass of diblock copolymer for the 16% mass loading dewetted film varied from 1.7 mg–2.8 mg. The layer was rehydrated in 2 mL of filtered (0.45 μm pore size) deionised water.

(c) **In a pre-loaded film (left over 48 hours at ambient temperature).** The same procedure was used as in (b) for making the dye loaded dewetted droplets and using the 105 mg dye (16% by mass). The film was rehydrated in 2 mL of deionised water at room temp. over 48 hours.

(d) **In a pre-loaded film (left over 48 hours at 60 °C).** This film used the same procedure as in (c) although the solution was drop cast onto the surface. The film was rehydrated in 2 mL of deionised water at 60 °C for 48 hours.

(e) **In a pre-loaded film annealed in a water saturated environment (extruded).** The same procedure for forming the dewetted layer was used as in (c). The dry film was placed in a Petri dish and surrounded by water droplets (less than 0.5 mL), although not immersed in water. The Petri dish was left in an oven at 35 °C for 2 hours. This produced a humid atmosphere inside the Petri dish. The film was rehydrated in a total of 2 mL deionised water and then extruded (see section 2.5).

(f) **In a pre-loaded film (sonicated upon rehydration).** The same procedure for forming the dewetted layer was used as in (c). The dry film was placed in a Petri dish. This was placed in a sonicator and rehydrated in 2 mL of deionised water. Water was added at the same time as sonication commenced.

(g) **In a pre-loaded film (sonicated upon rehydration and extruded).** In addition to the method used in (f) the solution was extruded through a 100 nm polycarbonate membrane and extruded (see section 2.5).

2.7 Dehydration-rehydration route

This is a variant of the surface rehydration route, in which the surface formed vesicles are subjected to an additional cycle of dehydration and rehydration. The solution used was a mixture of 160 mg of the diblock in 0.5 mL chloroform to which was added 0.5 mL of an ethanol dye stock solution (52 mg mL^{-1} Rhodamine B). The same surface formed vesicle route was used as before, although this time the solution was left uncolumned. The dye and vesicles were concentrated down to approximately 0.5 mL using a Buchi rotary evaporator attached to a Buchi vacuum pump V-700 at a pressure of 10^{-3} Torr. The solution was then freeze-dried using a high vacuum pump (10^{-5} Torr) to remove the remaining water from the mixture. The remaining consisted of the encapsulated dye within the vesicle-forming polymer as well as the unencapsulated dye. This material was again rehydrated and analyzed using the fluorescence assay described above.

3. Results

3.1 Vesicle characterisation

The aggregation number of the PEO$_{28}$PBO$_{36}$ vesicles was determined using static light scattering (SLS) and, independently, from the diameter obtained from DLS measurements together with geometrical arguments.

The vesicles were extruded through a 100 nm size polycarbonate membrane to give a near-monodisperse vesicle diameter as it is important to have a well defined

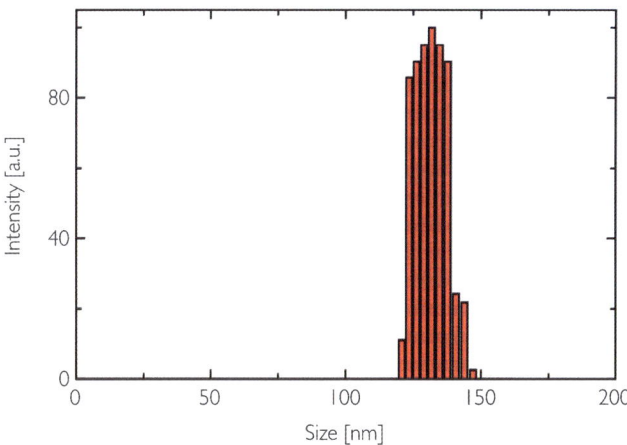

Fig. 3 Dynamic light scattering data for the extruded vesicles, where the correlation function has been fitted using the CONTIN routine.[28]

size distribution to measure an accurate aggregation number. The DLS data for the extruded vesicles in Fig. 3 shows a narrow distribution around 135 nm in diameter. It is assumed that the vesicles are unilamellar as previous work using extrusion through similar size membranes has shown them to be unilamellar after a series of extrusions through small membranes.[27]

To calculate the aggregation number using literature values we used eqn 1, 2 and 3, where r is the radius of the vesicle, t is the membrane thickness and A_0 is the surface area per head group of 0.48 nm calculated using eqn 2 and 3.

$$M_W = \frac{4\pi\left(r^2 + (r-t)^2\right)}{A_0} \quad (1)$$

$$A_0 = \frac{N\nu_{PBO}}{t} \quad (2)$$

N is the number of PBO units in the block copolymer, ν_{PBO} is the volume of the hydrophobe unit determined from eqn 3

$$\nu_{PBO} = \frac{N M_{WBO}}{\rho_{PBO} N_{Avagadro}} \quad (3)$$

ρ_{PBO} is the density of PBO (0.97 g cm^{-3}), M_{WBO} is the molecular weight of an individual butylene oxide monomer unit and $N_{Avagadro}$ is Avogadro's number.

The membrane layer thickness t, of 3.7 nm was extrapolated from the data published by Battaglia and Ryan[24] on the change in membrane thickness for a series of PEO–PBO polymers. Together this gives rise to an aggregation number for this system of 226 000 corresponding to the above mentioned vesicular mass of 8.6×10^8 Daltons.

The Debye plot in Fig. 4 shows a total mass for the vesicles of 1×10^9 Daltons for the intercept. This is close to the result of 8.6×10^8 Daltons obtained from molecular and physical data for the block-copolymers together with the diameter of 135 nm measured by DLS[29] as discussed above.

With this result we can now estimate the aggregation number for the much larger giant vesicles. The surface-formed vesicles that were formed from thin film rehydration had an average diameter of 600 nm as measured using DLS, giving rise to a maximum encapsulation volume of 2.3 µL where the total number of polymer chains making up the aggregate is 2.1×10^{10}. Measuring the fluorescence after breaking the vesicles gives a total value of 8 µM, which gives an encapsulated mass of 3.75 µg (0.78 mmol). Therefore, the internal concentration of Rhodamine B before breaking the vesicles was 3.4 mM, compared to 107 µM in the surrounding solution. Therefore the dye inside the vesicles is 32 times more concentrated than the dye in the free solution, due to the non-equilibrium distribution of dye as rehydration proceeds. It should also be noted that this is an upper bound for the encapsulated volume as the calculation involves the assumption that the surface-formed vesicles are unilamellar. However, the confocal microscopy image in Fig. 8 clearly shows that the total vesicle population is made up of both single and multi-lamellar vesicles.

3.2 Encapsulation efficiency

The overall aim of this work was to optimize the surface rehydration technique to encapsulate the fluorescent dye Rhodamine B within polymersomes, and to measure the efficiency of this loading, expressed as the amount of dye that ends up encapsulated in the polymersomes as a fraction of the total amount of dye in the system. However; at high local concentration of dye the fluorescence quenches, resulting in a fluorescence intensity that may be only 5–10% of the value that would be

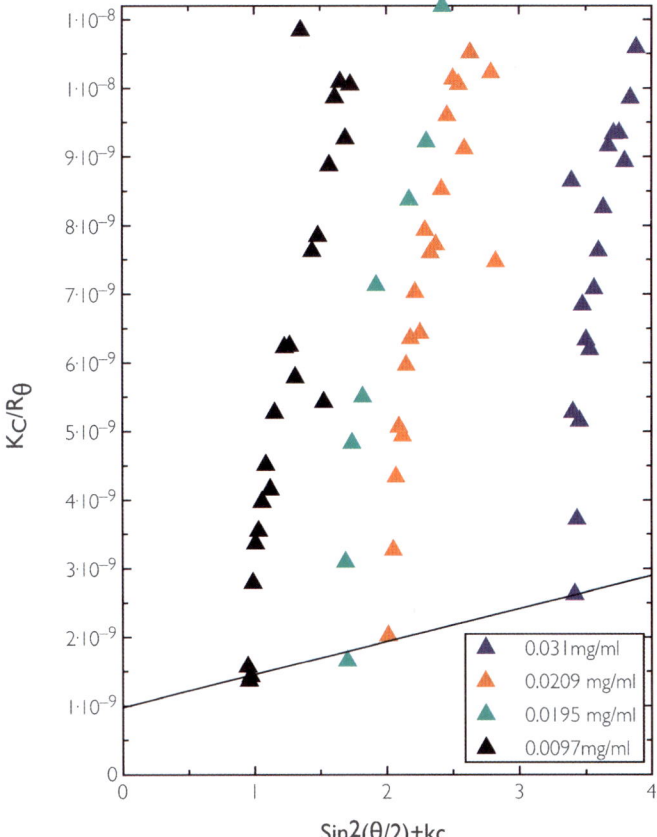

Fig. 4 Debye plot for a series of solution concentrations for $PEO_{28}PBO_{36}$ vesicles.

measured if the same amount of dye was dispersed in free solution. This effect is seen in Fig. 5 for our loaded vesicles; as the dye concentration is increased the fluorescence spectrum shifts to the right. Hence, to measure the true dye loading and efficiency for the various techniques, the assay method described previously was used (see section 2.5), and the effect of quenching removed.

One potential drawback of using column chromatography is that some dye could be lost on the column during the time taken to separate the vesicles and dye. This is only the case if the vesicles are unilamellar and leaky, whilst the confocal microscopy shows that they are mostly multilamellar. In our summary of results, Table 1, any loss during separation is not included in the final encapsulated amounts as there is no estimate for how much is lost during this process yet.

3.3 Encapsulation efficiency of surface rehydration method

The initial experiments looking at surface formed vesicles revealed that rehydrated dye-loaded films are more efficient at encapsulating the dye than some of the bulk processing routes. For the preliminary surface rehydration measurements (see section 3.3(b)), this was around 4% of the total dye in the system.

Encouraged by this preliminary success, we attempted to modify the process to improve the encapsulation efficiency. This optimisation was guided by the hypotheses that the encapsulation efficiency was likely to be strongly affected by the state of ordering of the block copolymer phase, and the degree of aggregation of the

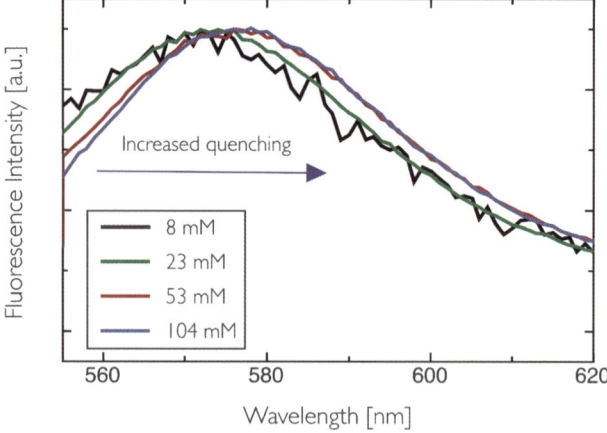

Fig. 5 Normalised fluorescence for vesicle containing fractions for a series of different loading concentrations showing an increase in the fluorescence quenching as the loading concentration is increased.

dye molecule. In addition, we built on earlier observations that laterally patterning the surface polymer film produces smaller vesicles, with a tighter size distribution, and speeds up the process of vesicle formation.

We induced lateral structure in the thin film of the amphiphilic polymer by depositing it on a surface that is both hydrophobic and oleophobic – PTFE. During the deposition of the film from a co-solution of the polymer with the dye, the spin-coated layer dewets[30] to form droplets in which dye and copolymer are uniformly mixed, as shown by confocal microscopy (Fig. 6). The average droplet size in this image is 16 μm with a large size distribution as there are two size populations for the droplets around 4 μm and 80 μm. In practice the droplet size will template the final vesicle diameter as this surface area will limit the size of vesicle that is able to form.[31]

It is likely that some dye remains excluded from the polymer film, and this will reduce the final encapsulation efficiency. Polymersome formation from laterally patterned films has also been studied by Howse *et al.*,[31] who used a diblock copolymer with a similar volume fraction but with a smaller molecular weight

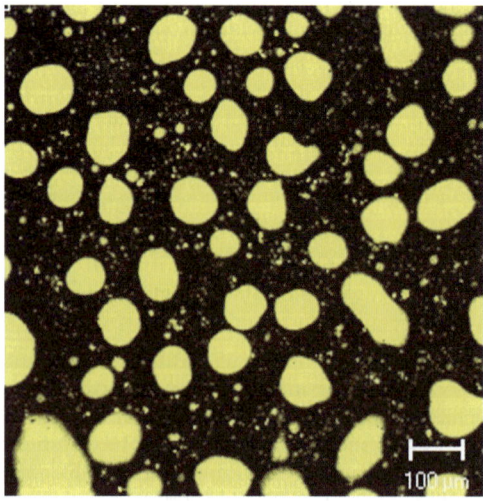

Fig. 6 Confocal image of a dewetted dye loaded film prior to rehydration in water.

Fig. 7 Vesicles in the process of rehydrating in water from a surface imaged using confocal microscopy.

(PEO$_{16}$PBO$_{22}$). In this work, rather than relying on dewetting from a uniformly hydrophobic and oleophobic surface, the authors patterned a gold surface and used thiol coupling chemistry to selectively deposit regions of vesicle forming polymer in regimented squares to control the vesicle diameter.

In this method, the vesicle formation process is slow, with vesicles remaining attached to the surface for some time, unless the system is stirred or otherwise agitated. Fig. 7 shows multilamellar vesicles and vesicles within vesicles forming from the surface. They have a distribution in the vesicle diameter of 20 μm and below.

(a) **In 125 μM dye solution.** Rehydrating a polymer film into a solution containing the active compound was 2% efficient at encapsulating the dye within the vesicles. This demonstrates that high loading concentrations could be achieved by using substantially higher concentrations of the original aqueous loading solution than the one required for the application. The remaining unencapsulated material could then be recycled in subsequent rehydrations.

(b) **Surface rehydration in a pre-loaded film.** For the three different loading concentrations of 7%, 16% and 23% by mass, the loading efficiencies were 3.7%, 5.4% and 0.5%. This shows that somewhere in between 16% and 23% dye by mass the ability of the diblock to encapsulate is disrupted due to the higher loading.

(c) **In a pre-loaded film (left over 48 hours at ambient temperature).** This route gave an efficiency of 2.9%. This is lower than when the film is rehydrated and columned immediately.

(d) **In a pre-loaded (left over 48 hours at 60 °C).** The efficiency of this route was 0.1% and showed that far from aiding encapsulation the temperature gave one of the lowest values amongst the surface rehydration methods.

(e) **In a pre-loaded film annealed in a water saturated environment (extruded).** The annealing of the dye loaded film gave an efficiency of 13.6%. This was the most efficient of all the various methods used to encapsulate Rhodamine B. By annealing the layer in a humid environment we hoped to induce some water saturation within the film. This would improve the mobility in the film and allow ordering of the layer prior to the actual rehydration of the layer.

(f) **In a pre-loaded film (sonicated upon rehydration).** The motivation behind this route was to rehydrate a dye-loaded vesicle forming film in water as normal but

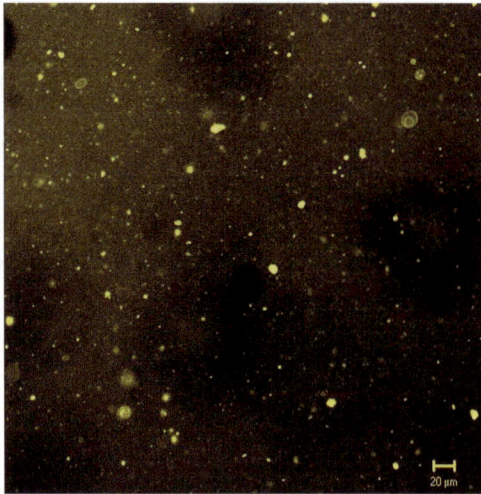

Fig. 8 A confocal microscopy image of vesicles formed using sonication upon rehydration from a surface.

perform the rehydration in the sonicator to provide the energy to drive vesicle formation from the surface. This would therefore reduce the leakage from the dye-loaded film and increase the loading efficiency in the final vesicle dispersion.

In the confocal image in Fig. 8, there are clearly vesicle structures. This formation technique gave an efficiency of 10% for the unextruded vesicles.

(g) **In a pre-loaded film (sonicated upon rehydration and extruded).** This method gave an efficiency value of 10.8% for the extruded vesicles, which is only 0.8% more than the unextruded case.

3.4 Encapsulation efficiency of the dehydration-rehydration method

The dehydration-rehydration encapsulation route involves surface hydration of a dye-loaded polymer film, followed by dehydration of the vesicle solution (*via* lypholization) and subsequent rehydration of the dried polymer. After rehydration, the solution was passed through the PD-10 columns to separate the vesicles from the unencapsulated dye. This technique aims to encapsulate the dye or active molecule in the lamellar layers when the vesicles are dehydrated, as they will tend to fuse to form much larger multilamellar vesicles. In lipid systems this technique has been used to encapsulate enzyme molecules that would be destroyed by other methods using organic solvent.

The results from this method were low with only 0.9% of the total dye encapsulated. It is, however, possible that there was some loss of materials during the rather involved experimental procedure, with its two freeze drying steps.

3.5 Summary of results

Table 1 summarises the results we have obtained for encapsulation efficiencies for rhodamine in PEO/PBO polymersomes prepared by a number of different routes.

4 Discussion

Routes based on the rehydration of a thin film of vesicle-forming block copolymer on a surface can be used to encapsulate an active molecule. Changes to the details of

Table 1 Encapsulation efficiencies and dye/polymer weight ratios for rhodamine in PEO/PBO polymersomes prepared by a variety of encapsulation routes

Rehydration method	Encapsulation efficiency	Mass ratio of dye to polymer
Surface rehydration	5.4%	1:7.1
Surface rehydration in 125 μM dye solution	2.0%	1:12.5
Surface rehydration in a pre-loaded film at different loadings		
7% dye by mass	3.7%	1 : 14.3
16% dye by mass	5.4%	1 : 6.3
23% dye by mass	0.5%	1 : 4.3
Surface rehydration in a pre-loaded film (left over 48 hours at ambient)	2.9%	1 : 7.2
Surface rehydration in a pre-loaded film (left over 48 hours at 60 °C)	0.1%	1 : 7.1
Surface rehydration in a pre-loaded film annealed in a water saturated environment (extruded)	13.6%	1 : 6.1
Surface rehydration in a pre-loaded film (sonicated upon rehydration)	10.0%	1 : 7.1
Surface rehydration in a pre-loaded film (sonicated upon rehydration and extruded)	10.8%	1 : 7.1
Dehydration/rehydration	0.9%	1 : 4.3

the process can lead to significant improvements in the encapsulation efficiency, with the highest value of encapsulation efficiency, at 13.6%, being obtained for a film, which was treated with water vapour before being rehydrated.

The majority of the literature so far has been concerned with demonstrating the release of an active molecule from polymersome systems.[22,32] The level of loading compared to that predicted from the encapsulation of a volume of water is low at either single percent figures or less.[18,22] However, there are very few studies that explicitly mention the loading efficiency.

One of the few studies on the encapsulation efficiency of polymer vesicles was performed by Adams et al.[12] In this work a pH change triggered the amphiphilic copolymers to aggregate and produce vesicles, encapsulating one of the two hydrophilic dyes, riboflavin and Rhodamine B. The loading was quantified using a fluorescence assay similar the one used in this work, although without the final disruption of the vesicles. The main finding of that work is that the pH switch route produced much lower encapsulation efficiencies than previously expected, assuming the vesicles enclosed dye at the concentration of the free solutions (*i.e.* the dye concentration inside vesicles is equal to the dye concentration in continuous phase). This assumption in itself limits the overall encapsulation efficiency (dye encapsulated *vs.* all dye in the system) to the highest attainable volume fraction of the solution inside the vesicles. This maximal internal phase volume is probably somewhere in the range below 10%.

Of that amount predicted from the outlined geometric arguments and the known dye concentration, the polymer concentration and the vesicle aggregation number, only 0.8% and 10% for Rhodamine B and riboflavin, respectively, could be encapsulated. Indeed, driven by the fact that the dye concentrations inside the vesicles were lower than in the bulk, significant dye diffusion *into* the vesicles took place.[22]

We hypothesise that the poor encapsulation efficiency of the solvent switch route can be ascribed to the details of the non-equilibrium process by which the vesicles are formed. The various observed intermediate stages of the self-assembly process (worms, tadpoles, octopi or ill defined "blobs")[33,34] suggest that the aggregation starts by the formation of a non-ergodic[35] compact object, which subsequently forms an internal compartment. This internal volume then swells by inward diffusion of water. If the molecule to be encapsulated has a significantly lower mobility through the aggregate than the water molecules (as should be expected for most hydrophilic actives), this route is expected to lead to very low encapsulation efficiencies.

One set of studies that did successfully encapsulate molecules with high efficiency within polymersomes using a solvent switch route was carried out by Lomas et al.[36,37] Their work encapsulated DNA within vesicles of a pH sensitive poly-(2-(methacryloyloxy)ethyl-phosphorylcholine)-co-poly(2-(diisopropylamino)ethyl methacrylate) (PMPC–PDPA) diblock copolymer. However, there are special aspects to this system that arise from the specific interactions between the amphiphile and the DNA, which allow vesicle loading efficiencies of up to 25% of the total DNA in solution. This has recently been improved to an efficiency of 55% for plasmid DNA.[38]

4.1 Rehydration efficiency

Turning now to the surface rehydration route, where we were able to achieve encapsulation efficiencies of around 10%, we see that it is possible to make substantial changes to the encapsulation efficiency by different pre-treatments of the amphiphile film. Once again, this stresses the likely importance of the details of the non-equilibrium process that results in the formation of vesicles.

At present we have not optimised the chemistry to actively encapsulate a specific molecule, as would be the ideal case for maximum encapsulation efficiency. At this stage we are primarily interested in a generic technology that would be universal for active molecules. The rehydration of vesicle forming droplets could be scaled up to a continuous process by having a recirculating sheet of PTFE with a doctorblade system to spread the polymer solution. This would provide a way to scale-up the process to an industrial scale.

Fig. 9 schematically shows one possible route for the evolution of a dewetted film of amphiphilic polymer as it comes into contact with water. The reason for using the dewetted route was to speed up vesicle formation as well as to template the size distribution of vesicles formed, as both of these would be required for a scale up of the process. The use of small layered droplets will make vesicle formation quicker as the edges of the lamellae are exposed, which is the main driving force in the formation of vesicles. Also as the droplets are separate from one another it is easier for them to detach compared to a continuous polymer film. After spin coating at point (i) the film has dewetted from the surface and the droplet contains a mixture of dye and vesicle forming block copolymer. At this point, the important factors are likely to be the degree and character of the ordering of the block copolymer, and the degree to which the dye is dispersed through the film, as opposed to being phase separated

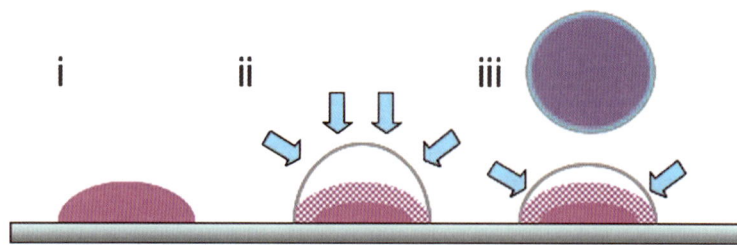

Fig. 9 Schematic representation of the formation of vesicles from dewetted droplets (i) is the initial dewetted droplet, being a mix of the block copolymer and the active molecule. Part (ii) is the initial budding off of the block copolymer as the droplet is rehydrated. Part (iii) is the detachment and enclosure of the polymer bilayer to form a vesicle.

or aggregated. It is to be expected that the degree of ordering would be affected by annealing of the film, either by exposure to elevated temperatures or by swelling with solvent vapour. On the other hand, if there is a significant driving force for aggregation of the dye, as might be anticipated for high dye loadings, the extra mobility induced by this pre-treatment might be expected to lead to poorer encapsulation efficiencies.

After the introduction of liquid water to the system at point (ii), a swollen copolymer phase develops whose water content varies both with proximity to the surface and with time. Since the equilibrium morphology of the swollen copolymer film depends on water concentration, this can lead to the formation of a sequence of different phases near the surface. In a similar system to ours, previous work has shown an initial microphase separation into a hexagonal rod phase followed by the formation of surface lamellae as well as the removal of grain boundaries.[21] It is the expansion and detachment of these surface lamellae that leads to the formation and separation of vesicles in stage (iii).

This detachment or budding off process proceeds *via* the unbinding of lamellae[39] as the layer is hydrated, this process is shown in Fig. 10. When the layer spacing of the lamellar layer u is greater than the equilibrium separation d unbinding takes place. The unbinding process of the lamellae layers is driven by osmotic pressure as well as undulation forces between neighbouring lamellae. This mechanism does not proceed homogeneously as there are always defects in the lamellar structure of the layer.[21] The point at which the lamellae curve and form a vesicle is determined by the Rayleigh instability of a cylinder under tension, which minimizes the surface area. This is shown in (ii) where the vesicle tube has a height L greater than length of the lamellae D. There will exist a point attachment where the layer will be incomplete prior to detachment, this defect will allow transit of the active molecules out of the vesicle interior by the process of effusion. The osmotic pressure difference as the layer is hydrated also has a negative effect on the encapsulation efficiency in the system as it acts to force the active molecule (in our case Rhodamine B) out of any defect in the pre-vesicle aggregate. An as yet untested route would be to match the osmotic pressure inside and outside of the vesicle as it forms from the surface. This would reduce the osmotic pressure that is forcing the active molecules out of the point defects by effusion.

One can speculate that the most efficient encapsulation route would be achieved if the polymer film was already in a well-ordered microphase separated, lamellar state, with the dye molecules uniformly dispersed in the domains containing hydrophilic

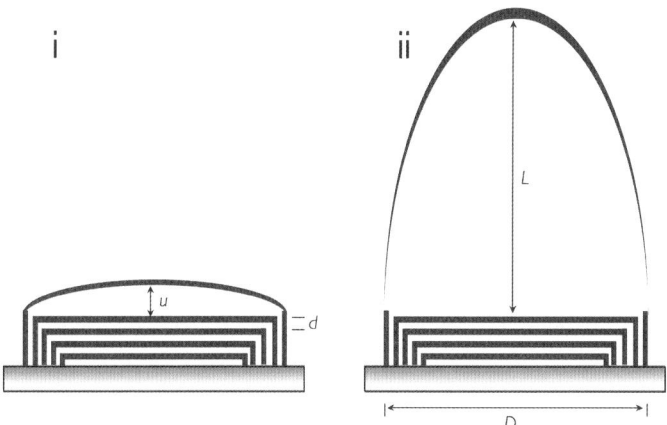

Fig. 10 Schematic representation of the unbinding transition (i) and the onset of vesicle formation due to a Rayleigh instability (ii).

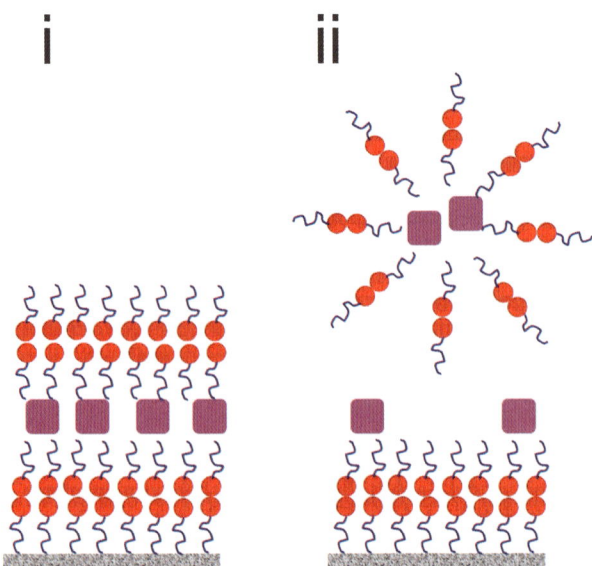

Fig. 11 Schematic representation of the theoretical maximum encapsulation possible for surface rehydration. Part (i) is the ideal vesicle forming situation prior to rehydration and (ii) is after vesicle formation and encapsulation of the active molecules. The red circles represent the hydrophobic part of the diblock and the blue chains the hydrophilic units.

polymer. This ideal layering to produce vesicles is shown in Fig. 11 with the red circles being the hydrophobic part of the diblock and the blue chains the hydrophilic regions; the pink squares are the active molecules. In Fig. 11 (i) as the layer proceeds from a film after rehydration in (ii) the unencapsulated external dye will be free in solution whist the dye enclosed by the polymer layer will be contained. In these circumstances one can anticipate an ideal upper limit of encapsulation efficiency of 50% (since we expect half of the hydrophilic domains to end up on the outside of a vesicle, and half on the inside), with actual values somewhat less due to the leakage of dye from the hydrophilic domains into the aqueous phase before the vesicle formation process is complete. The ideal system in Fig. 11 is not the case for our experimental system as we will have a much more mixed system, which will influence the maximum encapsulation efficiency. We can go further and say that the best case would be if there was some preferential ordering or interaction of the dye molecules with the hydrophilic regions that formed the inside of the vesicle, then we would expect a much higher loading efficiency.

In fact, the highest encapsulation efficiency we have achieved is somewhat less than a third of this upper limit, at 13.6%. We find that encapsulation efficiency can indeed be increased by pre-treatment of the film by exposure to water vapour (section 2.6(e)), presumably due to an increase in ordering of the film before exposure to water. Annealing at elevated temperature decreases the efficiency, however, as does using very high values of dye loading, perhaps in both cases due to an increase in the degree of aggregation of the dye. It seems likely that further work to optimise film ordering and dye aggregation state, informed by microstructural studies of the vesicle formation process, would lead to further increases in encapsulation efficiency.

5. Summary

Polymersomes have considerable potential as molecular delivery devices, with a number of important advantages over liposomes. However, this potential will only be realised when better control is achieved over the process by which molecules are encapsulated. In our work, we have shown that encapsulation can be achieved by a simple method of hydrating a surface-deposited thin film. We related these results to the non-equilibrium mechanisms that underlie vesicle formation and discuss how an understanding of these mechanisms can help optimize encapsulation efficiencies. Using this understanding we could finally improve this efficiency by annealing and pre-hydration steps. Our conclusion is that, even considering the good encapsulation efficiency, surface methods are still unsuitable for the massive scale-up needed when applied to commercial "mass market" molecular delivery scenarios. However, targeting more specialized applications for high value ingredients (like pharmaceuticals) might be more feasible.

Acknowledgements

We would like to thank Drs P. Schuetz, J. Bent and M. F. Butler (Unilever R&D Colworth) for their invaluable discussions and help during the preparation of this publication. This work was jointly funded by the Technology Strategy Board (TSB), Unilever and AkzoNobel (TSB project Natural Nanotechnology TSB 64).

References

1 D. E. Discher and A. Eisenberg, *Science*, 2002, **297**, 967–973.
2 D. E. Discher and F. Ahmed, *Annu. Rev. Biomed. Eng.*, 2006, **8**, 323–341.
3 P. Ghoroghchian, P. Frail, K. Susumu, D. Blessington, A. Brannan, F. Bates, B. Chance, D. Hammer and M. Therien, *Proc. Natl. Acad. Sci. U. S. A.*, 2005, **102**, 2922.
4 P. Ghoroghchian, J. Lin, A. Brannan, P. Frail, F. Bates, M. Therien and D. Hammer, *Soft Matter*, 2006, **2**, 973.
5 B. M. Discher, Y. Y. Won, D. S. Ege, J. C. M. Lee, F. S. Bates, D. E. Discher and D. A. Hammer, *Science*, 1999, **284**, 1143–1146.
6 K. P. Davis, T. P. Lodge and F. S. Bates, *Macromolecules*, 2008, **41**, 8289–8291.
7 T. Smart, H. Lomas, M. Massignani, M. V. Flores-Merino, L. R. Pérez and G. Battaglia, *Nano Today*, 2008, **3**, 38–46.
8 G. H. Fredrickson and F. S. Bates, *Annu. Rev. Mater. Sci.*, 1996, **26**, 501–550.
9 P. Walde and S. Ichikawa, *Biomol. Eng.*, 2001, **18**, 143–177.
10 D. J. Adams, D. Atkins, A. I. Cooper, S. Furzeland, A. Trewin and I. Young, *Biomacromolecules*, 2008, **9**, 2997–3003.
11 S. Jain and F. S. Bates, *Macromolecules*, 2004, **37**, 1511–1523.
12 D. J. Adams, S. Adams, D. Atkins, M. F. Butler and S. Furzeland, *J. Controlled Release*, 2008, **128**, 165–170.
13 M. Krack, H. Hohenberg, A. Kornowski, P. Lindner, H. Weller and S. Förster, *J. Am. Chem. Soc.*, 2008, **130**, 7315–7320.
14 B. Sun and D. T. Chiu, *Anal. Chem.*, 2005, **77**, 2770–2776.
15 D. D. Lasic, *Biochem. J.*, 1988, **256**, 1–11.
16 A. S. Utada, E. Lorenceau, D. R. Link, P. D. Kaplan, H. A. Stone and D. A. Weitz, *Science*, 2005, **308**, 537–541.
17 S. Hauschild, U. Lipprandt, A. Rumplecker, U. Borchert, A. Rank, R. Schubert and S. Förster, *Small*, 2005, **1**, 1177–1180.
18 G. Battaglia, A. J. Ryan and S. Tomas, *Langmuir*, 2006, **22**, 4910–4913.
19 N. Napoli, M. J. Boerakker, N. Tirelli, R. J. M. Nolte, N. A. J. M. Sommerdijk and J. A. Hubbell, *Langmuir*, 2004, **20**, 3487–3491.
20 L. M. Dominak and C. D. Keating, *Langmuir*, 2007, **23**, 7148–7154.
21 G. Battaglia and A. J. Ryan, *Nat. Mater.*, 2005, **4**, 869–876.
22 F. Meng, G. Engbers and J. Feijen, *J. Controlled Release*, 2005, **101**, 187–198.
23 S. Rameez, H. Alosta and A. F. Palmer, *Bioconjugate Chem.*, 2008, **19**, 1025–1032.
24 G. Battaglia and A. J. Ryan, *J. Am. Chem. Soc.*, 2005, **127**, 8757–8764.
25 G. Battaglia and A. J. Ryan, *J. Phys. Chem. B*, 2006, **110**, 10272–10279.
26 S. M. Mai, J. P. A. Fairclough, I. W. Hamley, M. W. Matsen, R. C. Denny, B. X. Liao, C. Booth and A. J. Ryan, *Macromolecules*, 1996, **29**, 6212–6221.
27 A. Napoli, M. Valentini, N. Tirelli, M. Müller and J. A. Hubbell, *Nat. Mater.*, 2004, **3**, 183–189.
28 S. W. Provencher, *Comput. Phys. Commun.*, 1982, **27**, 213–227.

29 S. Förster, B. Berton, H. Hentze, E. Krämer, M. Antonietti and P. M. Linder, *Macromolecules*, 2001, **34**, 4610–4623.
30 M. Geoghegan and G. Krausch, *Prog. Polym. Sci.*, 2003, **28**, 261–302.
31 J. R. Howse, R. A. L. Jones, G. Battaglia, R. E. Ducker, G. J. Leggett and A. J. Ryan, *Nat. Mater.*, 2009, **8**, 507–511.
32 U. Borchert, U. Lipprandt, M. Bilang, A. Kimpfler, A. Rank, R. Peschka-Süss, R. Schubert, P. Linder and S. Förster, *Langmuir*, 2006, **22**, 5843–5847.
33 A. Rank, S. Hauschild, S. Förster and R. Schubert, *Langmuir*, 2009, **25**, 1337–1344.
34 L. Chen, H. Shen and A. Eisenberg, *J. Phys. Chem. B*, 1999, **103**, 9488–9497.
35 Y. Y. Won, H. T. Davis and F. S. Bates, *Macromolecules*, 2003, **36**, 953–955.
36 H. Lomas, I. Canton, S. MacNeil, J. Du, S. P. Armes, A. J. Ryan, A. L. Lewis and G. Battaglia, *Adv. Mater.*, 2007, **19**, 4238–4243.
37 H. Lomas, M. Massignani, K. A. Abdullah, I. Canton, C. L. Presti, S. MacNeil, J. Du, A. Blanazs, J. Madsen, S. P. Armes, A. L. Lewis and G. Battaglia, *Faraday Discuss.*, 2008, **139**, 143–159.
38 H. Lomas, personal communication, 2008.
39 R. Lipowsky, *Nature*, 1991, **349**, 475–482.

Molecular control of ionic conduction in polymer nanopores†

Eduardo R. Cruz-Chu,[ab] Thorsten Ritz,[c] Zuzanna S. Siwy[c] and Klaus Schulten[*ab]

Received 30th March 2009, Accepted 14th April 2009
First published as an Advance Article on the web 27th July 2009
DOI: 10.1039/b906279n

Polymeric nanopores show unique transport properties and have attracted a great deal of scientific interest as a test system to study ionic and molecular transport at the nanoscale. By means of all-atom molecular dynamics, we simulated the ion dynamics inside polymeric polyethylene terephthalate nanopores. For this purpose, we established a protocol to assemble atomic models of polymeric material into which we sculpted a nanopore model with the key features of experimental devices, namely a conical geometry and a negative surface charge density. Molecular dynamics simulations of ion currents through the pore show that the protonation state of the carboxyl group of exposed residues have a considerable effect on ion selectivity, by affecting ionic densities and electrostatic potentials inside the nanopores. The role of high concentrations of Ca^{2+} ions was investigated in detail.

1 Introduction

Solid-state nanopores are small pores with radii of a few nanometres that are built in synthetic materials, such as silicon nitride,[1,2] silicon oxide,[3] silicon[4] and polymer films.[5–8] An atomic model of a polymer nanopore immersed in electrolyte solution is shown in Fig. 1. Due to their tunable geometry, resistance to harsh solvent conditions, and selectivity for surface modifications, solid-state nanopores have become promising tools in nanotechnology. To date, they have been used to unravel the fine-tuned interactions between DNA and proteins,[9,10] to build proteinaceous channels that mimic trafficking in cells[11] and to evaluate the feasibility of a fast-sequencing DNA technique.[12–14] Besides this wide range of applications, solid-state nanopores have also been used as models to test fundamental theories about ion dynamics in nanoscale confinements.[15,16] Studying ionic transport on the nanoscale is crucial not only for controlling the flow of ions across the nanopore,[15–20] but also for understanding how biological channels function. In this regard, theoretical and computational modeling have been of critical importance, since experimental

---

[a]*Beckman Institute for Advanced Science and Technology, University of Illinois, Urbana-Champaign, IL, USA. E-mail: kschulte@ks.uiuc.edu*
[b]*Center for Biophysics and Computational Biology, University of Illinois, Urbana-Champaign, IL, USA*
[c]*Department of Physics and Astronomy, University of California, Irvine, CA, USA*

† Electronic supplementary information (ESI) available: Atomic coordinates for PET monomers as well as topology and parameter files needed to perform MD simulations of PET polymers. See DOI: 10.1039/b906279n

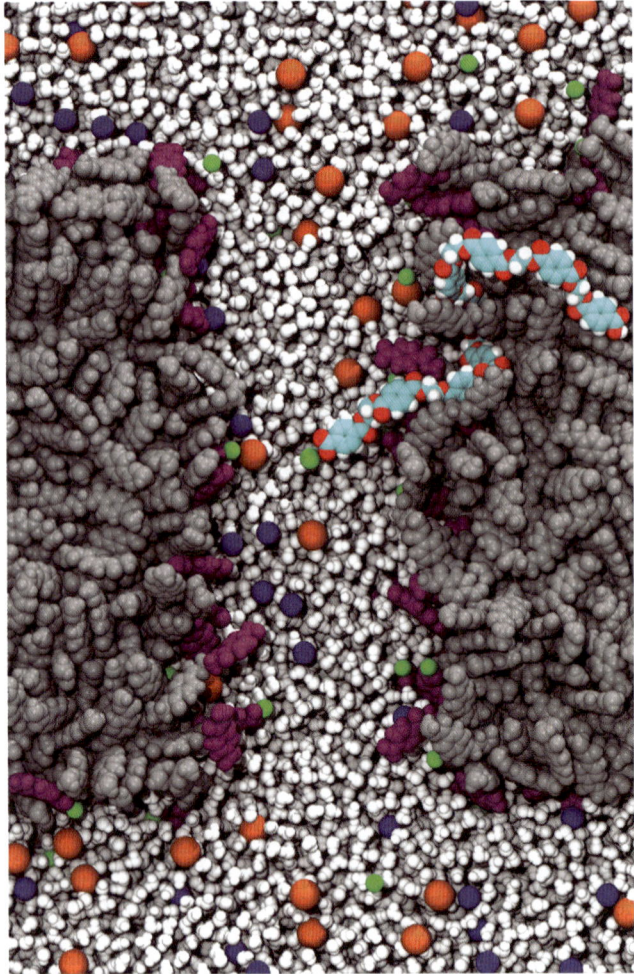

Fig. 1 Atomic model of a polymer nanopore. The image shows a polyethylene terephthalate (PET) nanopore immersed in electrolyte solution. The nanopore bulk material is shown in gray. PET residues with deprotonated carboxyl groups are colored in purple. K^+, Cl^- and Ca^{2+} ions are pictured as blue, red and green beads respectively. Water molecules are shown in white. A single PET polymer chain is highlighted, colored in cyan (carbon atoms), red (oxygen atoms), and white (hydrogen atoms).

techniques are not yet able to resolve the dynamics of the few ions contained in the nanopore volume.

Modeling based on the Poisson–Nernst–Planck (PNP) equations, Monte Carlo (MC) as well as molecular dynamics (MD) simulations have accompanied experimental efforts. The continuum approach of PNP successfully described the effect of ion current rectification observed in asymmetric pores[21–23] as well as ionic selectivity, *i.e.* a pore preference for transporting one type of ion, cations or anions.[24,25] However, continuum modeling cannot provide a molecular understanding of how ions compete for space and surface charges in the restricted volume of a nanopore. Monte Carlo simulations, which take into account the finite sizes of ions, offer more detail on the behavior of ions in nanopores. Applied to calcium-selective channels, these simulations successfully described channel selectivity properties.[26] MD approaches reveal most details about ionic trajectories and their interactions with other ions, with water and with the nanopore surface in equilibrium as well as non-equilibrium conditions. Increases in computational power have extended the duration of MD simulations, today approaching a time scale that is accessible in ion current measurements. Indeed, MD simulations successfully described transport properties of proteinaceous nanopores such as α-hemolysin[27–29] and solid-state silicon-based nanopores.[30–33] The MD methodology had not been extended initially to polymer nanopores due to the lack of an accurate model for the chemical features of the pore surface, a key factor that determines the transport properties.

Here, we report all-atom simulations of ion transport through a polyethylene terephthalate (PET) nanopore. PET nanopores have emerged as a nano-scale model system of interest, because of the experimental observation of multiple effects allowing for the control of ion transport properties, *i.e.*, ion current rectification[8,34] and selectivity,[22,24,25,35] as well as the realization of ionic diodes[19,20] and ionic transistors.[36]

Recently, it has been shown that local charge inversion can be a powerful tool to control ionic transport properties in PET nanopores.[37] Charge inversion means that the total positive charge brought by the cations close to the surface becomes larger than the total negative surface charge on the walls.[38–41] Charge inversion can be observed through reversal of the sign of the electrophoretic mobility of charged colloids.[42,43] In a nanopore with negative walls, charge inversion results in switching the selectivity properties of the nanopore. While a negatively charged nanopore in contact with *e.g.* KCl, would be cation selective, the same nanopore in contact with multivalent cations, such as Ca^{2+}, will become selective for anions. Such a selectivity switch induced through charge inversion has been observed in protein pore OmpF[44] as well as in PET nanopores.[37] Therefore, studying the interactions of multivalent and monovalent ions with the pore structure is crucial for gaining a better understanding of induced charge inversion.

In this article, we present a procedure to build all-atom models of PET nanopores and use it to accurately replicate experimental devices. After validation of the model, we study the dynamics of K^+, Cl^- and Ca^{2+} ions inside the nanopore using the MD approach. The following questions are discussed: how does the PET nanopore surface affect the ratio of the current carried by K^+, Cl^- and Ca^{2+} ions? How does Ca^{2+} interact with the charged groups of the pore walls?

2 Computational methods

In this section, the protocol to build PET nanopores is presented, along with the needed procedures for solvation/ionization, and setup of ionic conduction simulations.

The MD simulations were performed using the program NAMD,[45] the post-processing analysis of the MD trajectories was performed with custom VMD[46] and MatLab[47] scripts. The protocols for the MD simulations have been described in detail before[33,45] and are briefly stated here. For all simulations, an integration time step of 1 fs was used. For periodic boundary conditions, full electrostatic forces were computed using the particle-mesh Ewald method with a grid density of 1 Å^{-3} or finer. For van der Waals and electrostatic interactions in non-periodic boundary conditions, a cut-off of 12 Å with a switching function starting at 10 Å was assumed. Langevin dynamics was utilized to control system temperature. For MD simulations in the NpT ensemble, the pressure (p) was kept at 1 atm using a hybrid Nosé–Hoover–Langevin piston. Force field parameters for K^+, Cl^-, and Ca^{+2} ions were taken from the CHARMM force field;[48] the TIP3P[49] model was used for water.

2.1 PET model

A PET monomer can exist as a *trans* or a *cis* isomer, depicted in Fig. 2a and 2b, respectively. Monomers connect to form polymers and terminate in carboxyl groups

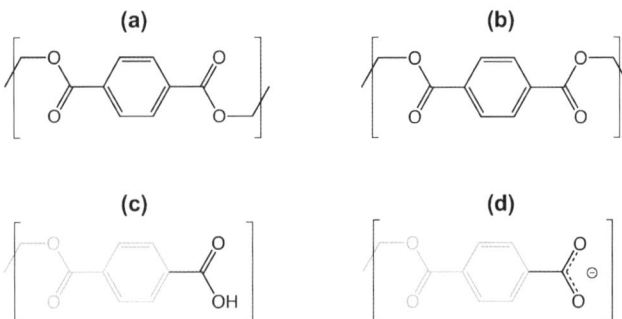

Fig. 2 PET chemical structure. Shown are the chemical structure of a PET monomer in (a) *trans* and (b) *cis* isomeric forms. Also shown are the terminal PET monomers with (c) protonated and (d) deprotonated carboxyl groups.

that are protonated at pH 4 and deprotonated at pH 7 (*cf.* Fig. 2c and 2d). Several MD models of PET polymers had already been developed[50–52] and successfully used to study the atomic origin of PET mechanical properties.[53–55] However, these models had been designed to describe properties of the bulk material and used force fields incompatible with explicit water models. To develop a PET model suitable for simulation of ion currents in a nanopore system, we built the PET moiety with the widely used CHARMM force field.[48] This force field was employed extensively to study a variety of biomolecular channels,[29,56–58] the limitations of the model in representing interactions with water and ions being well-known. The force field parameters for PET were obtained in analogy to previously parameterized model compounds. Topology and parameter files as well as atomic models of the *cis* and *trans* isomers are provided in the ESI.†

In previous studies,[55,59] PET bulk models were built using a stepwise procedure,[60] where monomers are added to a growing chain confined within a periodic cell. A disadvantage of such a method is that the accessible space is reduced as the polymer grows, which can lead to residues crossing through the center of aromatic rings. To avoid this problem, we built a periodic PET bulk using a collapsing–annealing MD procedure, presented schematically in Fig. 3.

First, ten PET polymers were generated. The number of subunits for each polymer was set to 9, corresponding to the most abundant polymer length observed in amorphous PET materials.[55] The initial conformation for each 9-mer was linear, the isomeric form (*cis* or *trans*) for each residue was assigned randomly, and carboxyl groups at both ends were protonated. To randomize the structure, each 9-mer was minimized for 200 steps and equilibrated for 1 ns at 1000 K. A single PET 9-mer obtained by this method is shown in Fig. 3a. Subsequently, one of the ten PET 9-mers was randomly selected, reoriented in space, and assigned to a point on a 7 × 7 × 5 grid with 75 Å spacing between grid points. After that, the grid was collapsed into a 90 Å × 90 Å × 60 Å rectangular box by applying forces of 5 pN towards the center of the box to each aromatic carbon atom located outside of the box. This simulation was performed at 1000 K to avoid folding of the 9-mers when they were moving toward the rectangular region. The collapsing grid at 0, 0.4, 0.7 and 2 ns is shown in Fig. 3b, c, d and e, respectively. The collapsed structure was placed within a periodic cell of dimensions 120 Å × 120 Å × 90 Å, *i.e.*, sufficiently large to avoid contacts among periodic images, as shown in Fig. 3g. To obtain homogeneous density, the PET structure was annealed in the NpT ensemble. First, the PET cell was simulated for 5 ns at 1000 K. At this temperature, the PET structure is a fluid and fills the entire cell volume. Snapshots of that simulation after 1 and 5 ns are shown in Fig. 3h and i, respectively. After that, the temperature was gradually reduced to 400 K, with a cooling step of 0.2 ns per −100 K. The annealing cycle

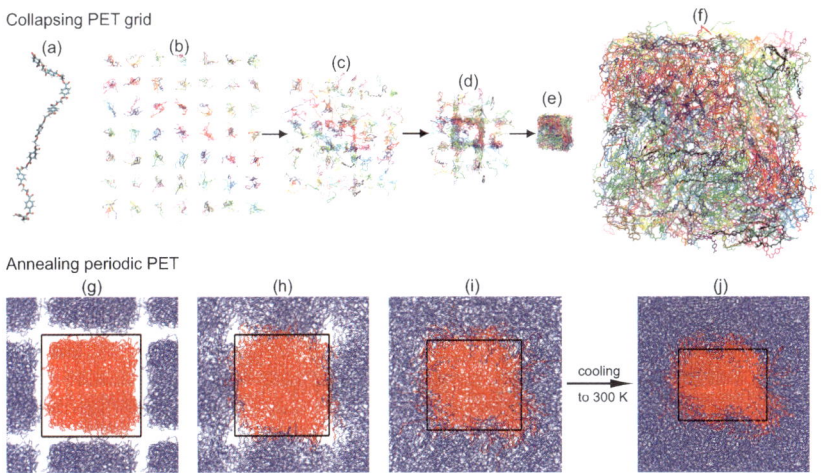

Fig. 3 Building bulk PET. The image illustrates the collapsing–annealing procedure used to build amorphous PET. The first stage of the procedure involved collapsing a grid of PET 9-mers. (a) Model of a PET 9-mer. (b) Initial conformation of the PET grid. Ten different 9-mers were randomly located on a grid of 7 × 7 × 5 points. Each 9-mer is presented in a different color. Subsequently, the grid was collapsed into a rectangular box. The snapshots at (c), (d) and (e) present the collapsing simulation after 0.4, 0.7 and 2 ns, respectively. (f) Enlarged view of the collapsed PET structure. The second stage of the procedure involved annealing the PET structure, in order to convert it into a periodic, amorphous box of homogeneous density. (g) Collapsed PET structure, colored in red, placed in a periodic box. The periodic images are colored in blue. The first step of the annealing cycle consisted of equilibration at 1000 K. Snapshots (h) and (i) show the simulation after 1 and 5 ns, respectively. The system was then cooled until reaching room temperature. The last frame of the annealing cycle is shown in (j); the PET polymers are seen to be tightly braided across the periodic cells.

finished with 5 ns of equilibration at 300 K. The resulting amorphous PET is shown in Fig. 3j, the periodic cell having a dimension of 107 Å × 84 Å × 64 Å.

A nanopore was molded into the PET material by defining a conical pore geometry,[8] according to eqn (1):

$$x^2 + y^2 = \left[\left(z + \frac{h}{2}\right)\tan\theta + R_{\min}\right]^2, \quad -\frac{h}{2} \leq z \leq \frac{h}{2} \qquad (1)$$

Here, θ is the angle between nanopore wall and pore axis, h is the height of the pore and $R_{\min}$ is the minimum radius of the cone. Through deletion of residues, a pore was created with $R_{\min} = 16$ Å, $h = 105$ Å, $\theta = 3°$. The height of the pore was achieved through replication of the bulk structure unit cell, followed by deletion of residues above $z = h/2$ and below $z = -h/2$. Protonated carboxyl groups were added to all polymer terminals.

Deletion of residues cut some of the 9-mers into smaller segments. To prevent small segments from being released from the nanopore surface, polymers with a length of three subunits or less were removed. To relax the surface while preserving the optimal geometry, a force of 1 pN guided non-terminal residues entering the pore volume back into the PET bulk during 2000 steps of minimization and 1 ns of simulation. After that, the system was equilibrated in a constant volume simulation without guiding forces for 4 ns. Fig. 4b shows the radius of the pore along the z-axis after the initial cut (circles) and after surface rearrangement (squares). As it can be observed, the conical geometry is preserved during the entire procedure. The narrowest pore region has a radius of 1 nm, resembling the smallest nanopores that can be manufactured by track-etching.[62]

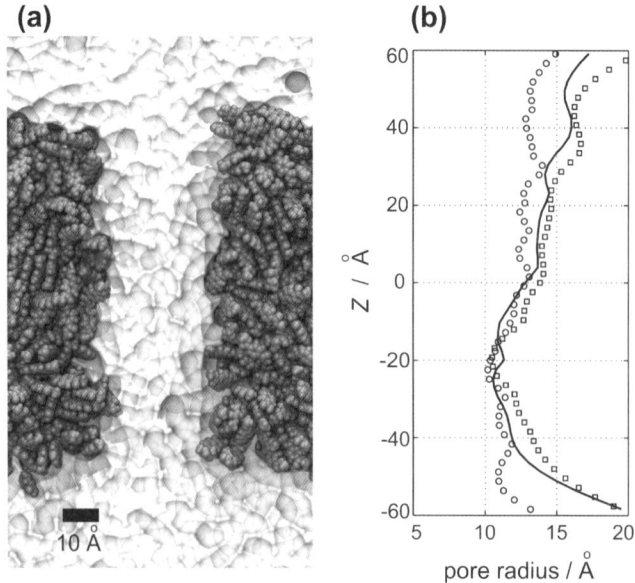

Fig. 4 Nanopore radius. Panel (a) shows a snapshot of a solvated PET nanopore. The PET structure is pictured as gray beads and water as a white surface. Panel (b) shows the radius of the PET nanopore along the z-axis at two different stages of creation and after solvation. Circles denote the radius after the initial cut and patching of terminal residues. Squares denote the radius after rearrangement of the surface in vacuum. The solid line denotes the radius after solvation. The radius values were computed using the program HOLE[61]. Solid line and square symbols represent the average values over the last 2 ns of MD simulation at each stage.

2.2 Ionic current simulations

Simulations of ionic currents closely follow protocols used in earlier MD studies.[29,33,45,56–58] The PET nanopore was solvated in a rectangular water box using the *Solvate* plugin of VMD.[46] To allow the solvent molecules fill all small voids at the PET surface, the system was simulated for 5 ns at 300 K in the NpT ensemble, with the PET structure restrained using harmonic forces with a spring constant of 5 kcal mol^{-1} Å^{-2}. Subsequently, the restraints were removed, and the system simulated further for 7 ns at 300 K in the NpT ensemble. The radius of the pore along the z-axis after solvation is shown in Fig. 4b (solid line).

The solvated nanopore was used to study ionic conduction at pH 4 and pH 7. To mimic pH 7 conditions, terminal PET residues in contact with water were deprotonated, resulting in a surface charge of -1.03 e nm^{-2}, which agrees well with experimental values.[63] The system was neutralized by adding counter ions, either K$^+$ or Ca^{2+}, using the program *cionize*[64] that located the ions according to the electrostatic potential of the PET structure. To mimic pH 4 conditions, all carboxyl groups remained protonated. Subsequently, K$^+$ and Cl$^-$ ions were randomly placed in solvent compartments, corresponding to 1 M KCl concentration. After adding ions, the systems were equilibrated at 300 K for 1 ns in the NpT ensemble, followed by 1 ns equilibration in the NVT ensemble. The final systems have periodic lattice vectors of 107 Å × 84 Å × 265 Å.

To simulate ionic currents, a uniform electrostatic field E_z was applied to all atoms of the system along the z-axis. The voltage difference ϕ across the simulated cell is

$$\phi = -L_z E_z \qquad (2)$$

Table 1 Ionic current simulations of PET nanopores. The table shows the conditions for the various simulations. The number of K$^+$ ions was 994 for all simulations. The number of Ca^{2+} ions in simulations 5–8 is sufficient to saturate half of the carboxyl groups, and sufficient to saturate all carboxyl groups in simulations 9 and 10. Cl$^-$ ions were added in order to obtain charge neutrality in each of the simulations. MD simulations were performed under constant volume conditions. For pH 4, the simulations were performed for 10 ns (simulations 1, 2, 5 and 6); for pH 7, the simulations were performed for 20 ns (simulations 3–4 and 7–10). Averaged ionic currents and their standard error were calculated using eqn (4) for the last 5 ns for each simulation

Index	pH	ϕ/V	Ca^{2+} (# ions)	Cl$^-$ (# ions)	K$^+$ current/ nA	Cl$^-$ current/ nA	Ca^{+2} current/ nA	Total current/ nA
1	4	+1	0	994	2.02 ± 0.19	1.79 ± 0.18	0	3.81
2	4	−1	0	994	−1.60 ± 0.18	−1.85 ± 0.18	0	−3.45
3	7	+1	0	703	3.09 ± 0.17	0.83 ± 0.15	0	3.92
4	7	−1	0	703	−3.09 ± 0.17	−0.80 ± 0.15	0	−3.89
5	4	+1	145	1285	1.33 ± 0.18	2.24 ± 0.21	0.50 ± 0.08	4.07
6	4	−1	145	1285	−1.34 ± 0.18	−2.27 ± 0.21	−0.42 ± 0.08	−4.03
7	7	+1	145	995	1.62 ± 0.18	1.44 ± 0.18	0.06 ± 0.07	3.12
8	7	−1	145	995	−1.70 ± 0.18	−1.56 ± 0.18	−0.05 ± 0.07	−3.31
9	7	+1	291	1286	1.24 ± 0.18	1.98 ± 0.21	0.12 ± 0.10	3.34
10	7	−1	291	1286	−1.04 ± 0.18	−1.71 ± 0.21	−0.22 ± 0.10	−2.97

where L_z is the dimension of the system in the z-direction. Table 1 presents a summary of the ion conduction simulations performed. The voltage signs are defined relative to the lower solvent compartment which is in contact with the narrowest pore opening. A +1 V bias induces cations to move from the lower to the upper solvent compartment. To prevent the PET membranes from drifting, the benzene carbons located in a belt region of 40 Å height and 30 Å away from the pore axis were restrained by harmonic forces with a spring constant value of 1 kcal mol^{-1} Å^{-2}.

3 Results and discussion

In this section we present and discuss the results of our simulations. First, we compare our PET bulk model with previously reported PET simulations and the expected values from experimental PET materials. Then, we report the ionic current for different pH conditions and electrolyte solution and relate the observed ionic currents to the individual ionic behavior and surface properties of the pore.

3.1 PET structure

Experimentally, single nanopores in PET are prepared by the track-etching technique described previously.[7] Briefly, PET films are irradiated with single swift heavy ions, which were accelerated to a total kinetic energy of 2.2 GeV (UNILAC, GSI Darmstadt, Germany). This irradiation process causes the formation of a single damage track through the films. It is expected that the impact of the heavy ion modifies the biaxially-oriented PET into an isotropic amorphous structure.[65] Conically shaped nanopores are obtained by asymmetric etching of the irradiated foils in 9 M NaOH while the other side of the membrane is in contact with an acidic stopping medium.[6] The heavy ion irradiation and chemical etching cause formation of 1 carboxyl group per nm^2.[63]

An atomic-model of a PET nanopore should be built from amorphous bulk material. The procedure for building an amorphous bulk cube has been described in the computational methods section (see Fig. 3). The final step of this procedure involves

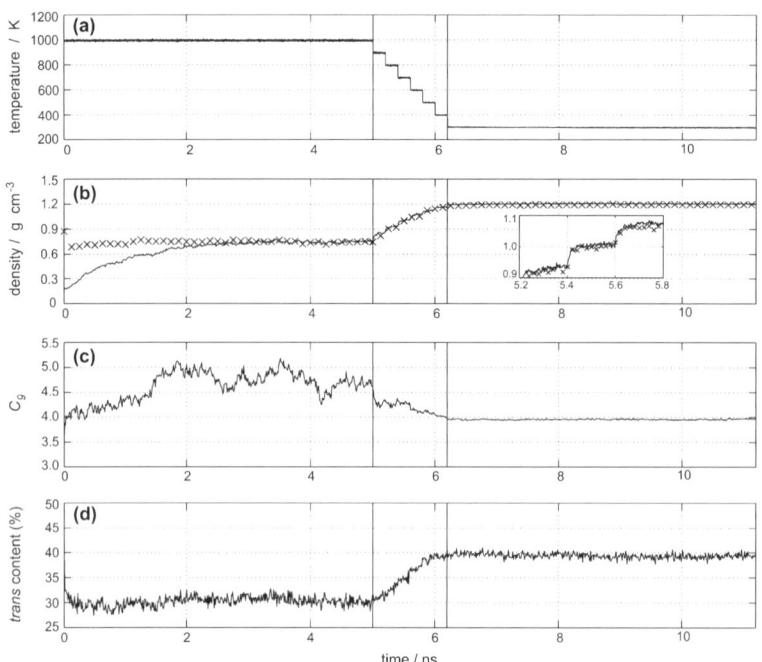

Fig. 5 Analysis of the annealing cycle for periodic bulk PET. Panel (a) shows the variation of temperature during the annealing cycle and the respective variation in density (b), characteristic ratio (c) and percentage of glycol dihedral in *trans* conformation (d). Vertical black guidelines divide each panel into three regions: the first region corresponds to 5 ns of equilibration at 1000 K, the second region to 1.2 ns when cooling from 900 K to 400 K, and the third region to 5 ns of equilibration at 300 K. In panel (b), the density values shown were computed for two cubic regions of equal volume (see text), an inner cube (crosses) and an outer cube (solid line). The inset in panel (b) shows a closer view of the density variation during cooling.

an annealing cycle, shown in Fig. 5a, with 5 ns equilibration at 1000 K, cooling in 100 K steps with 0.2 ns equilibration per step, and 5 ns equilibration at room temperature (300 K).

To compare our bulk material with other PET models and experimental PET material, we analyzed three measures commonly used to characterize PET bulk properties, namely volume density, the so-called characteristic ratio for a 9-mer, C_9, and the glycol dihedral angle between two joint PET residues.

The density of our PET bulk model during the annealing cycle is shown in Fig. 5b. At the initiation of the annealing cycle, densities in the center (crosses) and periphery (solid line) of the bulk cube differed significantly, but both converged to a constant value, demonstrating the homogeneity of the bulk material. The equilibrium density of 1.20 g cm^{-3} is in good agreement with previously reported PET models[51,52,55] with densities of 1.28 g cm^{-3} and 1.13 g cm^{-3}, but still slightly lower than the experimental value[53] of 1.34 g cm^{-3}.

The characteristic ratio is defined as the ratio between the mean-square end-to-end length R of a polymer and the length l_i of each rigid bond within the polymer.[51] We calculate the characteristic ratio C_9 according to

$$C_9 = \frac{\langle R^2 \rangle}{\sum_i l_i^2} \quad (3)$$

At 1000 K, the PET bulk behaves as a fluid and C_9 fluctuates, reaching values up to 5. As the system becomes more dense, C_9 decreases, obtaining a final value of

3.96, which is close to the experimental value of 4 and in good agreement with previously reported simulations.[51]

Fig. 5d shows the percentage of glycol dihedral in the *trans* conformation, the *trans* angle being defined as 180° ± 20°. At the end of the annealing cycle, we obtained 39.4% of *trans* content. Here we found our largest discrepancy with previous atomistic models and experimental values that report 25% *trans*.[55] The percentage of *trans* content is strongly influenced by the values of the O–C–C–O dihedral parameters. Cail and Stepto[54] have reported that for dihedrals with barrier heights of 1.40, 1.00 and 0.24 kcal mol^{-1}, the *trans* content at 541 K is 19%, 24% and 39%, respectively. The value we used was taken from the propane-1,2-diyl diacetate compound model present in CHARMM, which has a barrier height of 0.2 kcal mol^{-1} (dihedral parameters available in ESI†). In order to obtain 25% *trans* content, a further refinement of this dihedral parameter would be needed. The discrepancy in the percentage of the glycol dihedral angle does not alter the properties of the pore surface, *i.e.*, charge density and conical geometry; therefore, that discrepancy does not affect the ionic current measurements.

Finally, the accuracy of the bulk model was evaluated using the pair distribution function $g(r)$. Fig. 6 shows $g(r)$ for all carbon–carbon pairs. The overall shape of the curve compares well with previous studies.[51] The peaks observed between 1 and 4 Å describe the carbon–carbon distances defined by the connectivity within each polymer.[60] The inset in Fig. 6 shows the $g(r)$ computed for carbon–carbon pairs, where each carbon belongs to a different polymer chain. The shoulder around 5 Å is related to the van der Waals contact between two carbons that are not covalently bonded.

From the PET bulk structure we built a conical PET nanopore with a minimum radius of 10 Å and 100 Å height and with a surface charge density as observed in experiments.[63] The detailed procedure is described in the computational methods section.

3.2 Ionic currents

Previous sections presented the results from building an amorphous form of the PET polymer material and from modeling a nanopore with conical shape and surface charge density closely matching the experimentally studied systems. Following the validation of material characteristics, we used the nanopore system to study its transport properties. Table 1 lists the ionic conduction simulations that were performed under various electrolyte conditions.

As mentioned in the introduction, we employed MD simulations to obtain information about how the surface charge of the pore walls affects the ion current passing

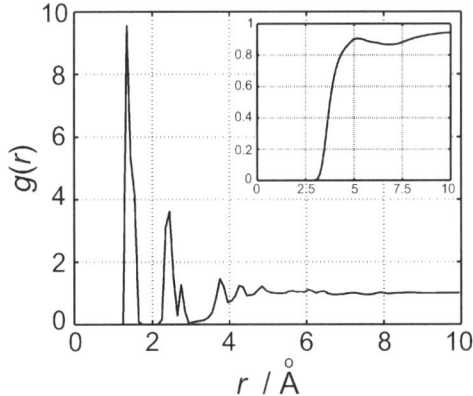

Fig. 6 Pair distribution function, $g(r)$, of carbon–carbon pairs separated by a distance r. $g(r)$ was computed over the last 2 ns of a 5 ns MD equilibration at 300 K. The inset shows the $g(r)$ for carbon–carbon pairs, each carbon belonging to a different PET 9-mer.

through the pore, and about how the presence of divalent cations such as Ca^{2+} changes the ionic selectivity of the pore.

We performed ten simulations, varying pH values between pH 4, where the carboxyl groups are protonated, and pH 7, where the carboxyl groups are deprotonated, also varying voltage biases between +1 and −1 V, and choosing different Ca^{2+} concentrations. For each simulation, the instantaneous ionic currents were computed using eqn (4):

$$I(t) = \frac{1}{\Delta t L_z} \sum_{i=1}^{N} q_i [z_i(t+\Delta t) - z_i(t)] \qquad (4)$$

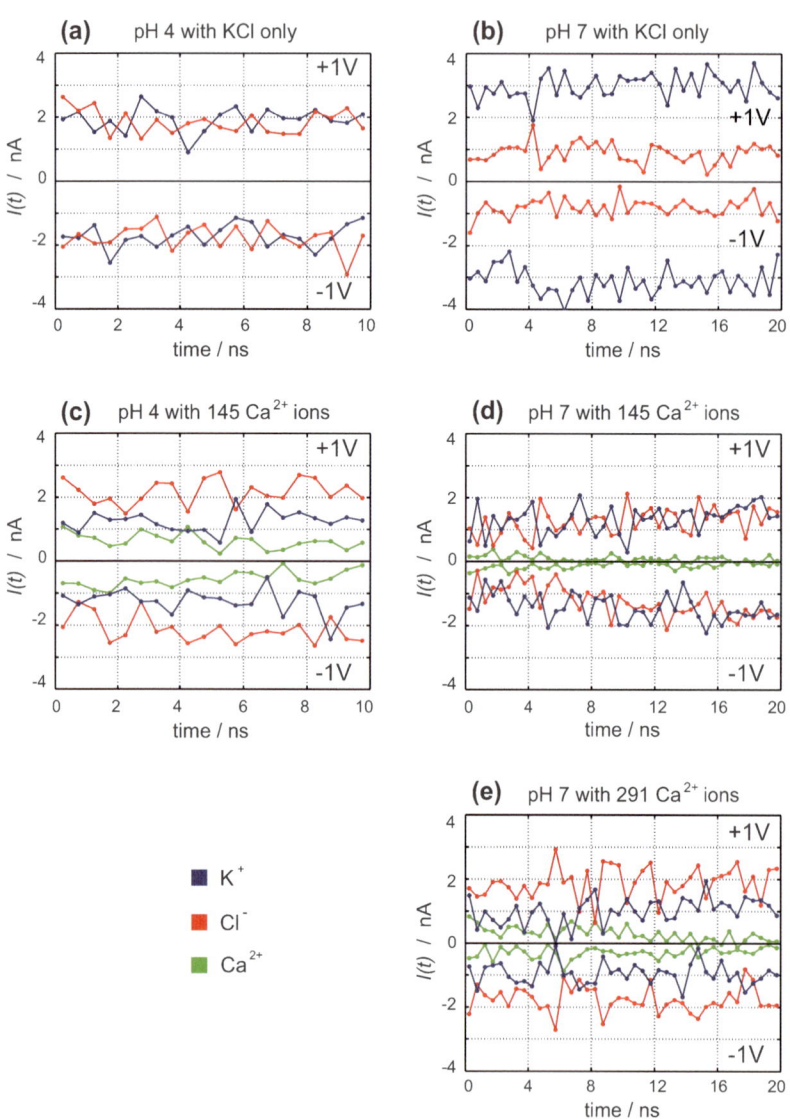

Fig. 7 Instantaneous ionic currents through PET nanopores. The Figure shows the ionic currents computed using eqn (4), each point representing an average over 0.5 ns. The ionic currents were calculated for each ionic species, K^+ (blue), Cl^- (red) and Ca^{2+} (green). Two different voltages are used: +1 V (positive currents) and −1 V (negative currents): (a) shows simulations 1 and 2, (b) shows simulations 3 and 4, (c) shows simulation 5 and 6, (d) shows simulations 7 and 8, (e) shows simulations 9 and 10. Simulation conditions are listed in Table 1.

where q_i is the charge of ion i, z_i its z-coordinate as a function of time t, L_z is the length of the system along the z-axis, and Δt is 1 ps. The averaged ionic currents over the last 5 ns of each MD simulation are presented in Table 1. Individual ionic currents for each MD simulation are presented in Fig. 7.

Our simulations suggest a strong dependence of ion selectivity properties of PET nanopores on the nanopore surface charge. At pH 4, when the nanopore surface is not charged, the anionic and cationic currents are of about the same magnitude (simulations 1 and 2), as shown in Fig. 7a and Table 1. As expected, there is no ion current selectivity in a neutral nanopore. At pH 7 however, where the carboxyl groups are negatively charged, the ion current carried by K^+ is about three times larger than the current carried by Cl^- (simulations 3 and 4), as shown in Fig. 7b and Table 1. Ion selectivity is typically expressed as the ratio of the current carried by cations and the total current carried by cations and anions. The ion selectivity value of 0.79, as predicted by our simulations, agrees well with the values measured experimentally at elevated KCl concentrations.[37,66] Due to the comparatively small volume accessible in all-atom simulations, the ion concentration of 1 M KCl is chosen larger than the concentrations typically used in experimental measurements. However, experimental values of ion currents and ion selectivity have been determined at a very wide range of KCl concentrations including 1 M KCl.[37,66] One should note that due to the required electroneutrality of the system, the number of Cl^- ions changes from 995 in simulations 1 and 2 (pH 4) to 703 in simulations 3 and 4 (pH 7). This change can account for some of the reduction in anionic current. However, the reduction in anionic current by a factor of two is larger than would be expected solely due to the changed concentration.

The pH dependence of the ion selectivity properties of the PET pore can also be seen by analyzing the concentration profiles of different ion species as a function of distance from the center of the nanopore (Fig. 8a and 8b). At pH 4, the concentrations of K^+ and Cl^- ions are practically identical, while at pH 7 there is a huge increase of the K^+ concentration, which agrees well with the observed increase of the cationic current (Table 1).

We also examined ion currents in the presence of both KCl and $CaCl_2$. In a control condition at pH 4 (simulations 5 and 6), with an uncharged surface, addition of $CaCl_2$ had little effect on the total current and the ion selectivity when compared with a pure KCl solution (simulations 1 and 2). The cationic current was simply split between K^+ and Ca^{2+} ions with a ratio roughly corresponding to their relative concentrations. The increase of the anionic current compared to the results of simulations 1 and 2 is due to the increase of the Cl^- concentration (Fig. 8c).

At pH 7, with negatively charged surfaces due to the deprotonated carboxyl groups in the system, the addition of $CaCl_2$ resulted in a more complex and interesting behavior. In simulations 7 and 8, 145 Ca^{2+} ions were added to neutralize the surface charge. In simulations 9 and 10, 291 Ca^{2+} were added, which could, in principle, saturate every carboxyl group, thereby possibly resulting in overcharging of the system walls. In all of the simulations 7–10, Ca^{2+} competed with K^+ ions for locations close to the negative carboxyl groups (see below).[67] During the simulations, most Ca^{2+} ions remained practically immobile, in contact with carboxyl groups. For the smaller number of 145 Ca^{2+} ions, indeed only a very small fraction of Ca^{2+} ions was mobile and the contribution to the total ion current was negligible, as shown in Fig. 7c and Table 1. With Ca^{2+} ions linked to carboxyl groups, the system is effectively neutralized (simulations 7 and 8). Differences between cationic and anionic currents are within the standard error of the simulations, indicating that there is no selectivity of ion current in this case.

Addition of 291 Ca^{2+} ions (simulations 9 and 10) resulted in a further change of the ionic conduction behavior. The currents carried by Cl^- are now larger than the combined K^+ and Ca^{2+} currents (Table 1 and Fig. 7d). This suggests that the nanopore in the latter system has changed selectivity, possibly due to charge inversion in the nanopores. To analyze the reason underlying the changed ionic currents,

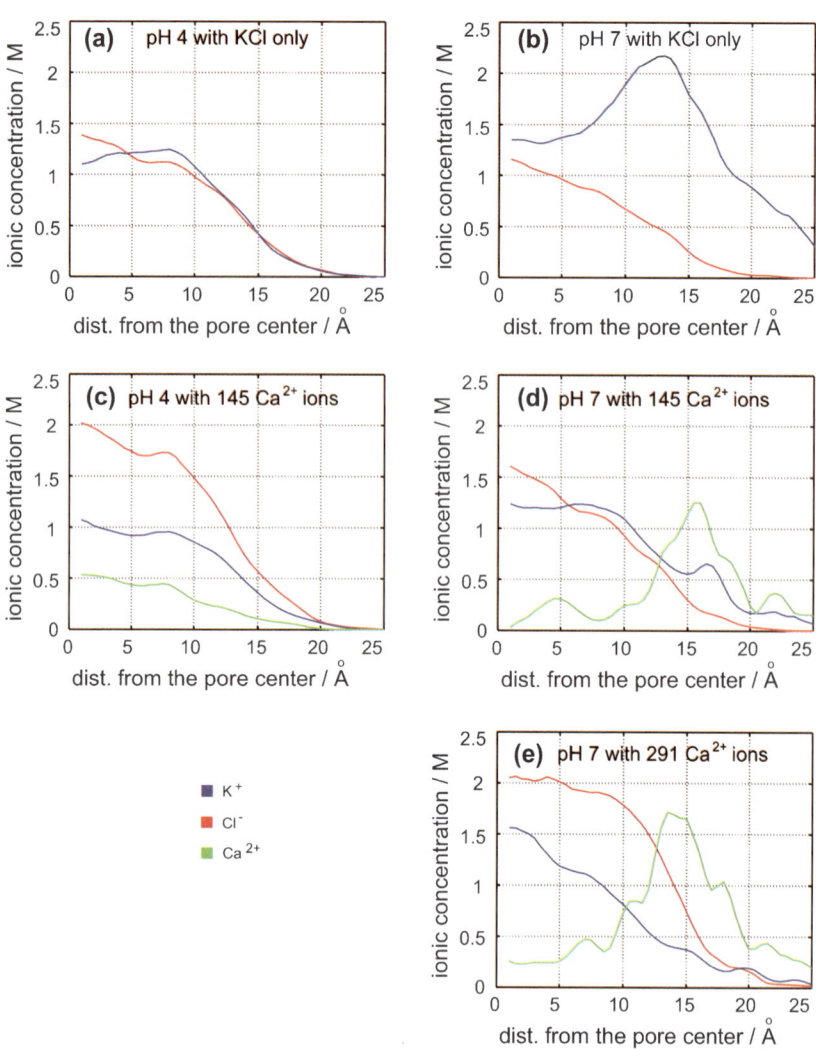

Fig. 8 Ionic concentrations as a function of radial distance from the nanopore center. Concentrations are averaged over an 8 nm [−40; 40 Å] compartment along the z-axis. The Figures show the ion concentration inside the pores for each MD simulation at +1 V bias: (a) simulation 1, (b) simulation 3, (c) simulation 5, (d) simulation 7, (e) simulation 9. K^+, Cl^- and Ca^{2+} concentrations are depicted in blue, red and green, respectively. Simulation conditions are listed in Table 1.

we determined the ionic concentration profile in the pore as well as the electrostatic potential along the z-axis of the nanopore.

An analysis of the ion densities in the nanopore shows that Ca^{2+} outcompeted K^+ in the region close to the pore walls, the K^+ concentration being significantly less than in the situation without Ca^{2+} (compare Fig. 8b with Fig. 8d and e). However, the Ca^{2+} concentration at the pore walls increased only slightly when the overall concentration was changed from 145 to 291 ions. This indicates that the fraction of the carboxyl groups with an adsorbed Ca^{2+} did not change much for the higher Ca^{2+} concentrations, *i.e.* the pore walls remained roughly electroneutral. But there arose an increase of the total Ca^{2+} concentration in the pore, accompanied by a corresponding increase in the amount of chloride ions. The distribution of the Cl⁻ ions, however, does not show any maximum in the vicinity of the position with the increased Ca^{2+} concentration. Such a maximum of Cl⁻ concentration near the pore wall is typically indicative of charge inversion.[68] Since one does not see such a maximum here, one can rule out overcharging of the nanopore walls as the cause for increased anionic currents (see Table 1). We suggest then that the slight increase of the adsorbed Ca^{2+} causes changes in the electrostatic potentials. These changes, together with the increased concentration of Cl⁻ ions, can explain the increased anionic current. Our results are in agreement with experimental data[40,67] and previous modeling,[69] which showed that the charge inversion and ion selectivity switch do not occur if a system contains a high concentration of monovalent cations. This effect, called *salting-out*, is explained by the competition of K⁺ and Ca^{2+} ions for

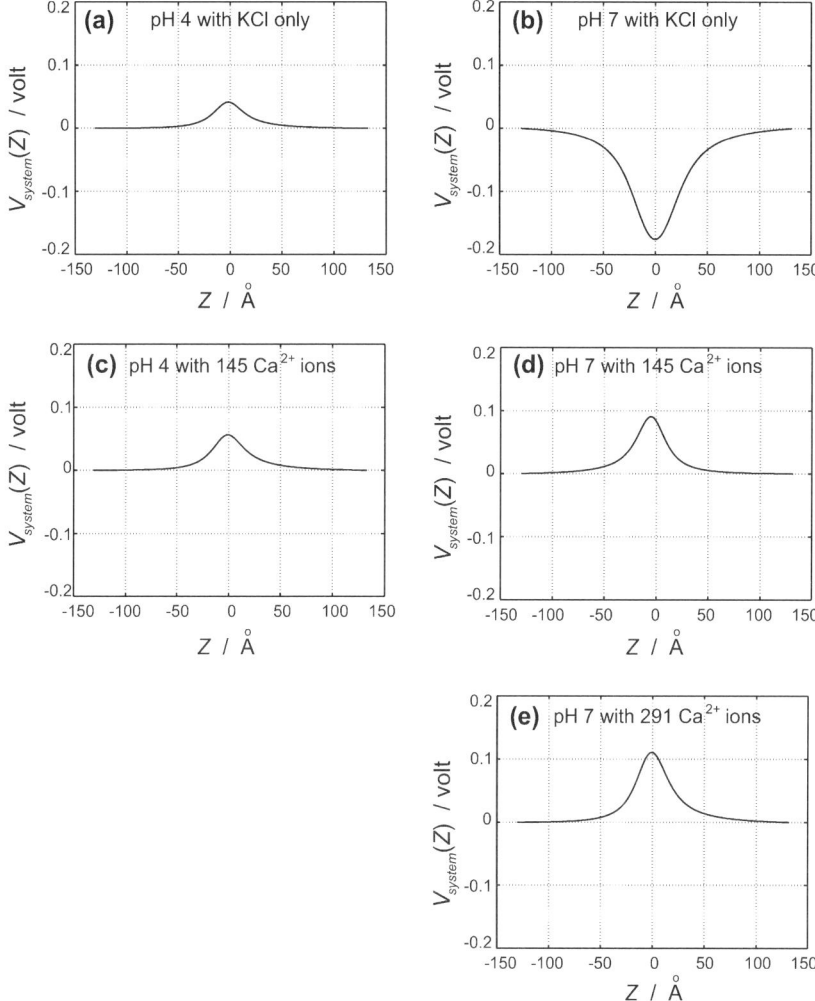

Fig. 9 Electrostatic potential of the system $V_{system}(z)$. $V_{system}(z)$ is defined in eqn (5). Figures show $V_{system}(z)$ for the systems at: (a) pH 4 and 1 M KCl, (b) pH 7 and 1 M KCl, (c) pH 4 and 1 M KCl and 145 Ca^{2+} ions, (d) pH 7 and 1 M KCl and 145 Ca^{2+} ions, (e) pH 7 and 1 M KCl and 291 Ca^{2+} ions. Lower solvent compartment, nanopore region, and upper solvent compartment are located in the intervals [−150; 50 Å], [−50; 50 Å], and [50; 150 Å] along the z-axis, respectively.

the negative charges at the walls. The K$^+$ ion concentration at the wall remains high enough so that surface overcharging does not occur.

To study how the adsorbed ions affect the electrostatic properties of the nanopore channel, we computed the electrostatic potential along the pore z-axis, $V_{system}(Z)$. The procedure used has been described in detail before.[33] Briefly, the averaged three-dimensional electrostatic potential was computed for the entire system using the *PMEpot* plugin[29] available in VMD[46] for the last 5 ns of each trajectory. From the three-dimensional electrostatic potentials, we extracted the electrostatic potential across the pore z-axis at +1 V and −1 V biases, denoted as $V_{+1V}(z)$ and $V_{-1V}(z)$, respectively. Then, $V_{system}(z)$ was calculated using eqn (5):

$$V_{system}(z) = \frac{V_{+1V}(z) + V_{-1V}(z)}{2} \quad (5)$$

Adding $V_{+1V}(z)$ and $V_{-1V}(z)$ cancels out the influence of the external applied voltage biases +1 V and −1 V, leaving only the electrostatic contributions from all charged species in the system. Therefore, V_{system} describes the electrostatic potential of the system itself acting on a line coincident with the pore z-axis.

Fig. 9 shows V_{system} for all systems. There is a significant change in V_{system} at different pH conditions. At pH 4, the PET surface is not charged, there is no ion adsorption, and the resulting V_{system} is small and positive (Fig. 9a and c). However, at pH 7, the PET surface becomes negatively charged and in the presence of 1 M KCl, V_{system} becomes negative (Fig. 9b), which implies cation selectivity of the pore. After adding Ca^{2+}, the negatively charged surface binds the Ca^{2+} ions, V_{system} becomes positive again as shown in Fig. 9d and e. The reversal of the sign of V_{system} after Ca^{2+} binding suggests that charge inversion has occurred; however, we do not see selectivity for Cl$^-$ ions, a characteristic feature of charge inversion. The absence of Cl$^-$ selectivity in our simulations is due to the high ionic concentration assumed; only at a sufficiently low ionic concentration would the effect of charge inversion and Cl$^-$ selectivity arise.

4 Conclusions

In this article, we have reported the first all-atom MD simulations of PET polymer nanopores. The study presents several methodological and scientific accomplishments and opens an avenue for future research. First, we have presented a collapsing–annealing procedure to produce polymeric PET bulk structures. This procedure overcomes the difficulties identified for the growing-chain methodology[60] and can be easily used to build other polymeric bulk structures. Secondly, we have built a PET nanopore model that accurately reproduces geometry and surface charge density of the experimental systems. Third, we have built an MD model based on the CHARMM force field, which permits extension for future simulations, for instance, to study functionalized PET nanopores.[20,70] Our all-atom MD simulations give insight into transport properties of these pores. We have described the ion dynamics inside PET nanopores immersed in KCl solution at different pH conditions, which correspond to different surface charge properties of the pore walls. Our results reproduced the cation selectivity observed at neutral pH conditions, at which the PET surface is negatively charged. Furthermore, we have calculated the ionic current carried by each ion species separately as well as the ionic distribution inside the pore. We have shown a very high affinity of Ca^{2+} ions to deprotonated carboxyl groups, which is important for understanding the effect of charge inversion. Our simulations showed that mobility of Ca^{2+} ions in the pore is very low and that Ca^{2+} ions close to the carboxyl groups are practically immobile. Since our simulations were performed in the presence of a high concentration of KCl, charge inversion could not be observed.

Acknowledgements

We thank the members of the Theoretical and Computational Biophysics Group for helpful discussions. Z. S. is an Alfred P. Sloan Fellow. This work was supported by grants from NIH (P41-RR05969) and NSF (CCR-02-10843, CHE-0747237). We acknowledge supercomputer time provided at the National Center for Supercomputing Applications through Large Resources Allocation Committee grant MCA93S028 and provided on the Turing Xserve Cluster at the University of Illinois at Urbana-Champaign.

References

1 J. Li, D. Stein, C. McMullan, D. Branton, M. J. Aziz and J. A. Golovchenko, *Nature*, 2001, **412**, 166–169.
2 J. B. Heng, C. Ho, T. Kim, R. Timp, A. Aksimentiev, Y. V. Grinkova, S. Sligar, K. Schulten and G. Timp, *Biophys. J.*, 2004, **87**, 2905–2911.
3 A. J. Storm, J. H. Chen, X. S. Ling, H. W. Zandbergen and C. Dekker, *Nat. Mater.*, 2003, **2**, 537–540.
4 S. R. Park, H. Peng and X. S. Ling, *Small*, 2007, **3**, 116–119.
5 R. L. Fleischer, P. B. Price and R. M. Walker, *Nuclear tracks in solids. Principles and applications*, University of California Press, Berkeley, 1975.
6 P. Y. Apel, Y. E. Korchev, Z. Siwy, R. Spohr and M. Yoshida, *Nucl. Instrum. Methods Phys. Res., Sect. B*, 2001, **184**, 337–346.
7 R. Spohr, German Patent DE 2951376 C2. U. S. Patent 4369370, 1983.
8 Z. Siwy, *Adv. Funct. Mater.*, 2006, **16**, 735–746.
9 Q. Zhao, G. Sigalov, V. Dimitrov, B. Dorvel, U. Mirsaidov, S. Sligar, A. Aksimentiev and G. Timp, *Nano Lett.*, 2007, **7**, 1680–1685.
10 R. M. M. Smeets, S. W. Kowalczyk, A. R. Hall, N. H. Dekker and C. Dekker, *Nano Lett.*, DOI: 10.1021/nl803189k.
11 T. Jovanovic-Talisman, J. Tetenbaum-Novatt, A. S. McKenney, A. Zilman, R. Peters, M. P. Rout and B. T. Chait, *Nature*, 2009, **457**, 1023–1027.
12 K. Healy, *Nanomedicine*, 2007, **2**, 459–481.
13 D. Branton, D. W. Deamer, A. Marziali, H. Bayley, S. A. Benner, T. Butler, M. D. Ventra, S. Garaj, A. Hibbs, X. Huang, S. B. Jovanovich, P. S. Krstic, S. Lindsay, X. S. Ling, C. H. Mastrangelo, A. Meller, J. S. Oliver, Y. V. Pershin, J. M. Ramsey, R. Riehn, G. V. Soni, V. Tabard-Cossa, M. Wanunu, M. Wiggin and J. A. Schloss, *Nat. Biotechnol.*, 2008, **26**, 1146–1153.
14 G. Sigalov, J. Comer, G. Timp and A. Aksimentiev, *Nano Lett.*, 2008, **8**, 56–63.
15 A. Lev, Y. Korchev, T. Rostovtseva, C. Bashford, D. Edmonds and C. Pasternak, *Proc. R. Soc. London, Ser. B*, 1993, **252**, 187–192.
16 D. Gillespie, D. Boda, Y. He, P. Apel and Z. S. Siwy, *Biophys. J.*, 2008, **95**, 609–619.
17 C. Dekker, *Nat. Nanotechnol.*, 2007, **2**, 209–215.
18 H. Daiguji, Y. Oka and K. Shirono, *Nano Lett.*, 2005, **5**, 2274–2280.
19 R. Karnik, C. Duan, K. Castelino, H. Daiguji and A. Majumdar, *Nano Lett.*, 2007, **7**, 547–551.
20 I. Vlassiouk and Z. Siwy, *Nano Lett.*, 2007, **7**, 552–556.
21 J. Cervera, B. Schiedt and P. Ramírez, *Europhys. Lett.*, 2005, **71**, 35–41.
22 J. Cervera, B. Schiedt, R. Neumann, S. Mafé and P. Ramírez, *J. Chem. Phys.*, 2006, **124**, 104706.
23 H. S. White and A. Bund, *Langmuir*, 2008, **24**, 2212–2218.
24 A. Plecis, R. B. Schoch and P. Renaud, *Nano Lett.*, 2005, **5**, 1147–1155.
25 I. Vlassiouk, S. Smirnov and Z. Siwy, *Nano Lett.*, 2008, **8**, 1978–1985.
26 D. Boda, M. Valiskó, B. Eisenberg, W. Nonner, D. J. Henderson and D. Gillespie, *Phys. Rev. Lett.*, 2007, **98**, 168102.
27 D. Wells, V. Abramkina and A. Aksimentiev, *J. Chem. Phys.*, 2007, **127**, 125101.
28 J. Mathé, A. Aksimentiev, D. R. Nelson, K. Schulten and A. Meller, *Proc. Natl. Acad. Sci. U. S. A.*, 2005, **102**, 12377–12382.
29 A. Aksimentiev and K. Schulten, *Biophys. J.*, 2005, **88**, 3745–3761.
30 J. B. Heng, A. Aksimentiev, C. Ho, P. Marks, Y. V. Grinkova, S. Sligar, K. Schulten and G. Timp, *Nano Lett.*, 2005, **5**, 1883–1888.
31 J. B. Heng, A. Aksimentiev, C. Ho, P. Marks, Y. V. Grinkova, S. Sligar, K. Schulten and G. Timp, *Biophys. J.*, 2006, **90**, 1098–1106.
32 Q. Zhao, J. Comer, V. Dimitrov, S. Yemenicioglu, A. Aksimentiev and G. Timp, *Nucleic Acids Res.*, 2008, **36**, 1532–1541.

33 E. R. Cruz-Chu, A. Aksimentiev and K. Schulten, *J. Phys. Chem. C*, 2009, **113**, 1850–1862.
34 Z. Siwy and A. Fuliński, *Phys. Rev. Lett.*, 2002, **89**, 198103.
35 Z. Siwy, Y. Gu, H. A. Spohr, D. Baur, A. Wolf-Reber, R. Spohr, P. Apel and Y. E. Korchev, *Europhys. Lett.*, 2002, **60**, 349–355.
36 R. Fan, M. Yue, R. Karnik, A. Majumdar and P. Yang, *Phys. Rev. Lett.*, 2005, **95**, 086607.
37 Y. He, D. Gillespie, D. Boda, I. Vlassiouk, R. S. Eisenberg and Z. S. Siwy, *J. Am. Chem. Soc.*, 2009, **131**, 5194–5202.
38 K. Besteman, M. A. G. Zevenbergen, H. A. Heering and S. G. Lemay, *Phys. Rev. Lett.*, 2004, **93**, 170802.
39 K. Besteman, M. A. G. Zevenbergen and S. G. Lemay, *Phys. Rev. E: Stat., Nonlinear, Soft Matter Phys.*, 2005, **72**, 061501.
40 F. H. J. van der Heyden, D. Stein, K. Besteman, S. G. Lemay and C. Dekker, *Phys. Rev. Lett.*, 2006, **96**, 224502.
41 M. Valiskó, D. Boda and D. Gillespie, *J. Phys. Chem. C*, 2007, **111**, 15575–15585.
42 R. James and T. Healy, *J. Colloid Interface Sci.*, 1972, **40**, 53–64.
43 A. Martín-Molina, M. Quesada-Pérez, F. Galisteo-González and R. Hidalgo-Álvarez, *J. Chem. Phys.*, 2003, **118**, 4183–4189.
44 A. Alcaraz, E. M. Nestorovich, M. L. López, E. García-Giménez, S. M. Bezrukov and V. M. Aguilella, *Biophys. J.*, 2009, **96**, 56–66.
45 J. C. Phillips, R. Braun, W. Wang, J. Gumbart, E. Tajkhorshid, E. Villa, C. Chipot, R. D. Skeel, L. Kale and K. Schulten, *J. Comput. Chem.*, 2005, **26**, 1781–1802.
46 W. Humphrey, A. Dalke and K. Schulten, *J. Mol. Graphics*, 1996, **14**, 33–38.
47 *Matlab v.6*, The MathWorks Inc., Natic, Massachusetts, USA, 2002.
48 A. MacKerell, Jr., D. Bashford, M. Bellott, R. L. Dunbrack, Jr., J. Evanseck, M. J. Field, S. Fischer, J. Gao, H. Guo, S. Ha, D. Joseph, L. Kuchnir, K. Kuczera, F. T. K. Lau, C. Mattos, S. Michnick, T. Ngo, D. T. Nguyen, B. Prodhom, I. W. E. Reiher, B. Roux, M. Schlenkrich, J. Smith, R. Stote, J. Straub, M. Watanabe, J. Wiorkiewicz-Kuczera, D. Yin and M. Karplus, *J. Phys. Chem. B*, 1998, **102**, 3586–3616.
49 W. L. Jorgensen, J. Chandrasekhar, J. D. Madura, R. W. Impey and M. L. Klein, *J. Chem. Phys.*, 1983, **79**, 926–935.
50 H. Sun, *J. Comput. Chem.*, 1994, **15**, 752–768.
51 M. S. Hedenqvist, R. Bharadwaj and R. H. Boyd, *Macromolecules*, 1998, **31**, 1556–1564.
52 S. U. Boyd and R. H. Boyd, *Macromolecules*, 2001, **34**, 7219–7229.
53 J. Zhou, T. M. Nicholson, G. R. Davies and I. M. Ward, *Comput. Theor. Polym. Sci.*, 2000, **10**, 43–51.
54 J. I. Cail and R. F. T. Stepto, *Polymer*, 2003, **44**, 6077–6087.
55 M. Roberge, R. E. Prud'homme and J. Brisson, *Polymer*, 2004, **45**, 1401–1411.
56 F. Khalili-Araghi, E. Tajkhorshid and K. Schulten, *Biophys. J.*, 2006, **91**, L72–L74.
57 M. Sotomayor, V. Vasquez, E. Perozo and K. Schulten, *Biophys. J.*, 2007, **92**, 886–902.
58 J. Gumbart and K. Schulten, *J. Gen. Physiol.*, 2008, **132**, 709–719.
59 P. Gestoso and J. Brisson, *Polymer*, 2003, **44**, 7765–7776.
60 D. N. Theodorou and U. W. Suter, *Macromolecules*, 1985, **18**, 1467–1478.
61 O. S. Smart, J. G. Neduvelil, X. Wang, B. A. Wallace and M. S. P. Sansom, *J. Mol. Graphics*, 1996, **14**, 354–360.
62 Z. Siwy, P. Apel, D. Baur, D. D. Dobrev, Y. E. Korchev, R. Neumann, R. Spohr, C. Trautmann and K. Voss, *Surf. Sci.*, 2003, **532–535**, 1061–1066.
63 A. Wolf-Reber, PhD Thesis, University of Frankfurt, Germany, 2002.
64 J. E. Stone, J. C. Phillips, P. L. Freddolino, D. J. Hardy, L. G. Trabuco and K. Schulten, *J. Comput. Chem.*, 2007, **28**, 2618–2640.
65 V. Singh, T. Singh, A. Chandra, S. K. Bandyopadhyay, P. Sen, K. Witte, U. W. Scherer and A. Srivastava, *Nucl. Instrum. Methods Phys. Res., Sect. B*, 2006, **244**, 243–247.
66 P. Ramírez, V. Gómez and J. Cervera, *J. Chem. Phys.*, 2007, **126**, 194703.
67 F. H. J. van der Heyden, D. Stein and C. Dekker, *Phys. Rev. Lett.*, 2005, **95**, 116104.
68 G. M. Torrie and J. P. Valleau, *J. Phys. Chem.*, 1982, **86**, 3251–3257.
69 Y. Chen, Z. Ni, G. Wang, D. Xu and D. Li, *Nano Lett.*, 2008, **8**, 42–48.
70 F. Xia, W. Guo, Y. Mao, X. Hou, J. Xue, H. Xia, L. Wang, Y. Song, H. Ji, Q. Ouyang, Y. Wang and L. Jiang, *J. Am. Chem. Soc.*, 2008, **130**, 8345–8350.

PAPER

Nanomechanics of organic/inorganic interfaces: a theoretical insight

Maria L. Sushko†*

Received 15th January 2009, Accepted 25th March 2009
First published as an Advance Article on the web 23rd July 2009
DOI: 10.1039/b900861f

Microfabricated arrays of cantilevers coated with active layers represent ultrasensitive devices for the label-free detection of chemical and biochemical reactions. The development of these sensors for practical applications requires an understanding of the mechanism of transduction of chemical or physical changes in the active layer of the cantilever into its mechanical bending. In order to eliminate non-specific effects, differential detection with respect to reference cantilevers with an inert coating is used. However, the convolution of different specific effects leading to cantilever bending does not allow their direct decoupling based on experiments alone. We propose a quantitative mesoscopic model showing that there are two competing components to the differential deflection: the component associated with specific chemical or physical reaction on the active cantilever and the component due to a difference in elastic properties of the active and reference coatings. We apply the model to study the origin of the chemomechanical response in cantilever arrays for experimentally studied reactions, including deprotonation of pH sensitive self-assembled monolayers, DNA hybridization and swelling of polyelectrolyte brushes. We show that for all these diverse systems the theoretical model gives good quantitative agreement with the experimental data and provides a guide for designing cantilever sensors with significantly improved sensitivity.

Introduction

Mechanochemistry is a unique technique for altering the reaction landscape through mechanical forces. This technique is utilized in synthetic chemistry, film growth for MEMS, electronic and optoelectronic devises, photochemical and electrochemical processes in solar cells.[1–4] It is widely used by Nature for governing the most fundamental biochemical reactions and processes, such as transcription with RNA polymerase, which utilizes chemical catalysis to move along DNA,[5] signal transduction in cell membranes,[6] and the operation of myosin motors in muscles.[7]

The mechanical stress can be either a driving force or a consequence of chemical transformations, and in the latter case can serve as a measurable quantity for the characterization of these processes. Curvature measurement is one of the most common techniques for measuring stresses in films. With miniaturization of the devices to micron and nanoscale, the sensitivity of the curvature measurements has become high enough to follow chemical reactions and subtle physical transformations in

London Centre for Nanotechnology, University College London, Gower Street, London, UK WC1H 0AH

† Present address: Pacific Northwest National Laboratory, PO Box 999, MS K2-12, Richland, WA 99354, USA. E-mail: maria.sushko@pnl.gov; Fax: +1 (1)509 371 6498; Tel: +1 (1)509 371 7286

the active layer of the sensor.[8,9] This stipulated application of cantilever sensors to a variety of chemical and biological processes such as deprotonation reactions,[10,11] swelling of polymer brushes,[12–14] DNA hybridisation,[15–19] protein recognition[20,21] and cell adhesion.[22]

Cantilever sensors are coated on one side with the active layer, such as self-assembled monolayers, monolayers of biomolecules, polymer films, *etc.* The creation of an organic/inorganic interface changes the surface free energy and induces stress at the interface. This change in the surface stress difference on the upper and lower surfaces of the cantilever leads to its bending, which is measured. The cantilever mechanical bending is largely determined by the elastic properties of the material. The pioneering work of Stoney on the theory of the elastic response in thin long beams,[23] widely used in practical applications, is based on several assumptions: (1) the bending of the cantilever is very small; (2) the length of the cantilever is large compared to its width, which itself is large compared to its thickness; (3) the thickness of the active layer is negligible compared to the cantilever thickness (so called thin film approximation). This model links the observed bending signal with the stress generated due to physical or chemical transformations in the active layer of the cantilever sensor and is routinely used for interpretation of the experimental results.[10–22] However, there are two problems. Firstly, the key assumption of Stoney's theory, the thin film approximation, justified for macroscopic devices, may not be applicable to micro- and nanodevices with a generally higher ratio of the thicknesses of the active layer and the cantilever. This deficiency of Stoney's theory was first highlighted with respect to residual stress measurements in MEMS devices, having typical substrate thicknesses of 300–400 μm and films several microns thick.[24] In our previous combined experimental and theoretical study[25] we have shown that this problem is also relevant for nanodevices for surface stress measurements. Using a model system for the detection of mechanochemical transformations we have shown that for cantilever sensors (typical cantilever and active layer thicknesses of 1 μm and 30–50 nm, respectively) Stoney's theory can lead to qualitatively incorrect results. Here we will discuss the theoretical model in more detail and show that in typical cantilever sensors the bending signal is determined not only by the elastic properties of the cantilever, as proposed by Stoney's theory, but also by the elasticity of the active layer. Secondly, all these mechanical models, linking the observed curvature to the stress, do not provide a direct correlation between the nature of the processes in the active layer with the observed nanomechanical signal. The thermodynamic model of the relation between the surface free energy per unit area (γ) and the surface stress (σ) has been developed by Shuttleworth.[26] It gives the following expression for the surface stress:

$$\sigma = \gamma + A\frac{\partial \gamma}{\partial A} \quad (1)$$

where A is the area of the cantilever. The methodology for calculation of the surface free energy change with curvature and, therefore, the stress, based on molecular dynamics has been proposed.[27,28] Although molecular dynamics simulations can potentially pinpoint the main interactions in the film responsible for cantilever bending, these studies have been limited to calculations of stress in inorganic materials.[27] Partially this is due to the complexity of bioactive coatings used in experiments. For these systems the mapping of the configurational space requires long time-scale dynamics and significant computational overheads.

An alternative approach is based on more approximate mesoscopic models, which do not require detailed atomic-level knowledge of the structure of the inorganic surface and of the conformations of the organic molecules in the film. The advantage of mesoscopic theories is their intrinsic ability to separate different types of interactions, and, therefore, identify the main contributions to the measured cantilever bending. Although mesoscopic models describing a wide variety of polymer and colloidal systems have been developed (see for example ref. 29), the application of these models to study surface stress in soft systems on solid substrates is very limited.[30,31]

This lack of theoretical understanding of the origin of surface stress hinders the development and calibration of cantilever sensors. In this paper we present a mesoscopic model for the differential signal, which goes beyond the approximations used in Stoney's formalism. Our model shows that the measured differential deflection is determined by two contributions: the elasticity, caused by the difference in elastic moduli of the active and reference coatings, and the contribution due to the specific reaction in the active coating. We present a general model for the calculation of stress induced by specific chemical reactions in the active layer, associated with changes in its electrostatic properties. We apply the model to reveal the fundamental physics of the cantilever response to partial charging of pH sensitive self-assembled monolayers, DNA hybridization and conformational changes in polyelectrolyte brushes and perform a direct comparison of the theoretical results with the available experimental data.

Theory

Structure of the cantilever sensor

Typical stress sensors are based on Si or SiN cantilevers approximately 1.0 μm thick. The active/reference coating is either directly grafted onto a cantilever surface[12,13] or self-assembled on a gold surface evaporated onto an adhesive layer of Cr or Ti (ref. 10, 11, 15–22 and Fig. 1). The environment for the operation of the sensor is usually an aqueous buffer solution, in which the sensor is fully immersed.

In experiments, in order to eliminate non-specific effects, such as ion binding to the passive side of the cantilever, the differential signal is usually measured.[10,11] The coating of the reference cantilever should be inert with respect to the reaction of interest. Therefore, the observed differential signal is usually attributed to the stress generated in the active cantilever upon chemical or physical transformations in its coating.[10–15,17–22]

Differential signal: general theory

A cantilever sensor can be approximated as a classical bilayered plate system consisting of elastically isotropic substrate (Si) and a coating (organic layer on gold). The curvature of the cantilever can be calculated from the experimentally measured cantilever deflection as

$$K = \frac{1}{R} = \frac{2\Delta z}{L^2} \qquad (2)$$

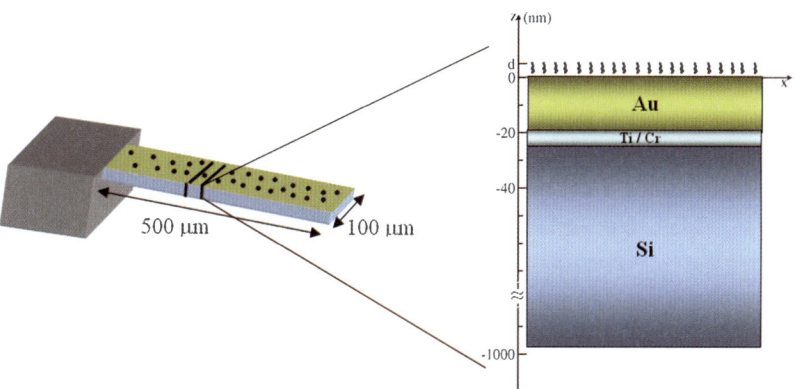

Fig. 1 Structure of a typical cantilever sensor functionalized with an organic monolayer.

where R is the radius of curvature, Δz is the deflection and L is the length of the cantilever.

The curvature of a bimaterial strip can be expressed as[32]

$$K = \frac{6E_c E_s (t_c + t_s) t_c t_s}{E_c^2 t_c^4 + 4E_c E_s t_c^3 t_s + 6E_c E_s t_c^2 t_s^2 + 4E_c E_s t_c t_s^3 + E_s^2 t_s^4} \Delta\eta_0 \equiv M_{\text{elast}} \Delta\eta_0 \quad (3)$$

where E_c and E_s are the biaxial elastic moduli ($E = Y/(1 - \nu)$, Y is Young's modulus and ν is the Poisson's ratio), and t_c and t_s are the thicknesses of the coating and substrate, respectively. $\Delta\eta_0 = \eta_{c,0} - \eta_{s,0}$ is the mismatch strain between the coating and the substrate prior to mechanical relaxation (bending) of the cantilever (see Fig. 2). In a thin-film approximation ($t_c \ll t_s$) eqn (3) reduces to

$$K = \frac{6 E_c t_c \Delta\eta_0}{E_s t_s^2} \quad (4)$$

By making a thin-film approximation for the macrostress acting on the film as $\sigma_c = E_c t_c \Delta\eta_0$, we recover Stoney's equation.[32] However, it will mask the physics underlying the cantilever bending. The expression (4) suggests that there are two contributions to the differential deflection: the differences in elastic moduli for the reference and active coatings and the difference in the strain mismatch. The strain mismatch can have different origins, but it is mainly due to the mismatch in the lattice constants between two solids. We can write the mismatch strain for the reference coating as $\Delta\eta_{\text{ref},0} = \eta_{\text{Au}} - \eta_{\text{Si}}$, where we define the strain in all layers lower than gold by η_{Si} (Fig. 2). Using a linear approximation valid for small deformations we can write the mismatch strain in the active cantilever as $\Delta\eta_{\text{active},0} = \eta_{\text{chem}} + \eta_{\text{Au}} - \eta_{\text{Si}}$, where the strain component η_{chem} is due to chemical or physical reactions in the active layer of the sensor. Therefore, the differential deflection or the differential curvature is

$$\Delta K = K_{\text{active}} - K_{\text{ref}} = \frac{6}{E_s t_s^2} \left(E_{\text{active}} t_{\text{active}} [\eta_{\text{chem}} + \eta_{\text{Au}} - \eta_{\text{Si}}] - E_{\text{ref}} t_{\text{ref}} [\eta_{\text{Au}} - \eta_{\text{Si}}] \right) \quad (5)$$

Regrouping the terms we can separate the chemical and the elastic contributions to the differential deflection:

$$\Delta K = \frac{6 E_{\text{active}} t_{\text{active}}}{E_s t_s^2} \eta_{\text{chem}} + \frac{6 (E_{\text{active}} t_{\text{active}} - E_{\text{ref}} t_{\text{ref}})}{E_s t_s^2} [\eta_{\text{Au}} - \eta_{\text{Si}}] \quad (6)$$

Converting to stress using Atkinson's formula,[33] valid for $E_c/E_s \leq 3.5$ and $t_c/t_s < 0.1$, we obtain

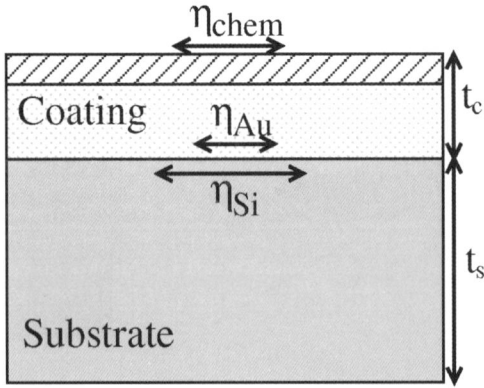

Fig. 2 Schematic of a cantilever sensor (not to scale).

$$\Delta\sigma = \frac{E_s t_s^3}{6(t_s + t_c)} \Delta K = \sigma_{\text{chem}} + \sigma_{\text{elast}} \qquad (7)$$

The chemical strain and stress can be calculated using the Shuttleworth equation (see below), while the mismatch strain at the gold/substrate interface can be calculated from the absolute deflection of the reference cantilever:

$$\eta_{\text{Au}} - \eta_{\text{Si}} = \frac{\sigma_{\text{ref}}}{E_s t_s} = \frac{1}{3} \frac{t_s \Delta z_{\text{ref}}}{L^2} \qquad (8)$$

Elastic effects

The main aim of introducing the *in situ* reference is to filter out non-specific effects. However, eqn (6) and (7) suggest that the differential signal apart from the specific cantilever response on the physical or chemical transformations in the active layer also contains a contribution from the difference in elastic properties of the active and reference coatings. The elastic properties of the coatings, in turn, depend on the nature of the chemical bonding between the organic molecules and the substrate and the strength of intermolecular interactions in the organic layer. For example, adsorption of thiols on the gold (111) surface results in asymmetric relief of the intrinsic tensile stress on the gold surface.[34] The stress relief is due to the redistribution of the electron density and the relaxation of the metal surface, induced by gold–sulfur binding. Interactions between the adsorbates are also of major importance. We have shown previously that the elastic properties of self-assembled monolayers on the gold surface depend on the strength of the van der Waals and dipole interactions between the molecules in the SAM.[25]

The sign and magnitude of the elastic term depend on the difference in the biaxial elastic moduli of the active and reference coatings (eqn (6)). Therefore, the elastic properties of the reference coating can be used to maximize the sensitivity of the cantilever sensor. At least two strategies can be used: (i) the elasticity of the reference coating can be chosen so that $E_{\text{ref}} = E_{\text{active}} t_{\text{active}}/t_{\text{ref}}$, then the differential signal will be equal to the specific chemical contribution; (ii) E_{ref} can be chosen so that the sign of the elastic term will be the same as the sign of the chemical term, which will provide a constant shift for the differential signal. The latter may be beneficial for the detection of weak chemical signals.

Clearly, if the chemical and elastic contributions are of opposite signs, the magnitude and the sign of the total differential signal will depend on the competition of these two contributions and the interpretation of the measured data becomes complicated.[25] The sensitivity also appears reduced in this regime.

To choose the best reference for differential measurements it is essential to know the elastic properties of organic/inorganic interfaces. Cantilever measurements can be used for this purpose. Indeed, in the regime of zero chemical stress ($\sigma_{\text{chem}} = 0$) differential measurements can be used to determine the biaxial elastic modulus of the "active" coating with respect to the reference with known E_{ref}. For example, self-assembled monolayers of alkanethiols on the gold (111) surface, for which the biaxial elastic modulus has been calculated using quantum mechanical and classical simulations,[25] can serve as a reference coating.

In the following we will assume an ideal reference, for which $\sigma_{\text{elast}} = 0$ and we will focus on the calculation of σ_{chem} for the most common biochemical reactions in the active layers of cantilever sensors.

Chemical stress

To utilize biorecognition reactions in cantilever sensing the surface of the active cantilever is usually coated with a monolayer of biomolecules. Thiol chemistry is often used to self-assemble biomolecules on metal surfaces to produce well-defined reproducible active layers[35] (see also Fig. 1). Since most biological molecules, such as

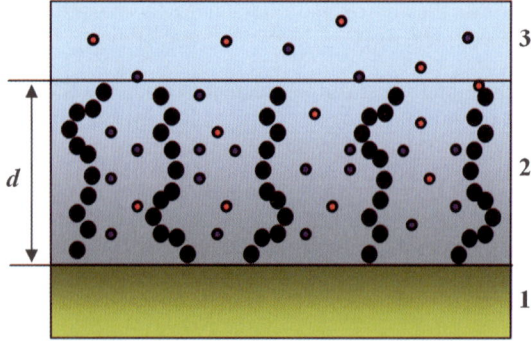

Fig. 3 Slab geometry of the system: region 1 is semi-infinite metal, region 2 is the monolayer of organic molecules of thickness d and region 3 is semi-infinite bulk aqueous salt solution.

DNA, RNA and proteins, are charged under physiological conditions, the corresponding chemical contribution to the stress is determined by chemical or physical transformations in the monolayer of charged molecules on the metal surface immersed in aqueous salt solutions (Fig. 3). In general, these transformations in the cantilever active layer may involve changes in the density of fixed charges, the distribution of mobile charges in the monolayer and in solution and conformation of the molecules in the monolayer. Our model depends parametrically on the conformational properties of the monolayer, which determine the monolayer thickness (d) and its volume charge density. These parameters of the active layer will be calculated separately using classical Density Functional Theory (see below). First we will derive the expression for chemical stress in the active layer, induced by the changes in its electrostatic properties.

As mentioned above, the surface stress relates to the change in surface free energy of the active layer *via* the Shuttleworth eqn (1). While the electrostatic free energy of the partially charged organic layer can be calculated as

$$F = \tfrac{1}{2}\int dr \rho(r)\psi(r) \qquad (9)$$

where $\rho(r)$ is the volume charge density of the monolayer and $\psi(r)$ is the electrostatic potential from the surface, which, in turn, can be calculated on the basis of the Debye–Hückel theory for slab geometries.[36] In cantilever sensors we are dealing with the three layer system: conducting gold slab/low dielectric constant organic layer/aqueous salt solution with a dielectric constant of 78.

Here we will address two most typical distributions of charges in the active layer: monolayers of molecules with neutral main chains and charged tail-groups (surface charge) and monolayers of charged chains (volume charge). In both cases we will assume that the charge is evenly smeared over the surface or volume, respectively. Another approximation made is the mean-field approximation, which implies that the aqueous solution is a dielectric medium characterized by the dielectric constant and the Debye screening length κ^{-1}. Finally we approximate the cantilever with an ideal surface covered with a single domain monolayer.

Surface charge density

Typical examples of systems with surface charge are pH-sensitive self-assembled monolayers[10,11,25] and lipid bilayers.[37] In these systems the pH-induced change in surface free energy is equal to the electrostatic free energy of the charged surface. Since only the surface of the monolayer carries a non-zero charge the volume charge density of the organic layer $\rho(r) = \rho(x,y)\delta(z - d)$, where $\delta(z - d)$ is the Dirac delta function ($\delta(z - d) = 1$ for $z = d$ and $\delta(z - d) = 0$ for $z \neq d$) and $z = 0$ at the gold surface (Fig. 1). Within the approximation of an evenly charged surface $\rho(x,y) = \rho = $ const and the surface free energy is

$$F(d) = \tfrac{1}{2}\rho\psi(d)A \tag{10}$$

In order to calculate the surface stress in this model we use the expression for the potential above the surface of the monolayer ($z \geq d$) obtained by Netz[36]

$$\psi(z) = \frac{4\pi l_B e^{-(z-d)\kappa_3}}{\kappa_3}\left[\frac{\rho(1+e^{-2d\kappa_2})}{1+e^{-2d\kappa_2}+\frac{\varepsilon_2\kappa_2}{\varepsilon_3\kappa_3}(1-e^{-2d\kappa_2})}\right] \tag{11}$$

where l_B is the Bjerrum length ($l_B = e^2/(4\pi\varepsilon_3 k_B T)$ with $k_B T$ being the thermal energy and e the elementary charge), κ_3, κ_2 are the inverse screening lengths, and ε_3, ε_2 are the static dielectric constants of the solvent and the monolayer, respectively. In the case when there are no ions inside the monolayer (corresponding to $\kappa_2 = 0$), the expression for the potential reduces to

$$\psi(z) = \frac{4\pi l_B e^{-(z-d)\kappa_3}}{\kappa_3}\frac{Q}{A} \tag{12}$$

and $\psi(d) = \dfrac{4\pi l_B}{\kappa_3}\dfrac{Q}{A}$ at the surface of the monolayer (Q is the total charge on the cantilever surface, thus, Q/A is the surface charge density). Combined with eqn (10) this gives

$$F = \frac{1}{2}\frac{4\pi l_B}{\kappa_3 d}\frac{Q^2}{A} = \gamma A \tag{13}$$

and substitution into (1) leads to

$$\sigma = \frac{F}{A} + A\frac{\partial(F/A)}{\partial A} = -\frac{1}{2}\frac{4\pi l_B}{\kappa_3 d}\frac{Q^2}{A^2} \tag{14}$$

This result shows that the first term, which has the physical meaning of the energy of formation of a unit surface area or surface tension, is essentially positive and two times smaller by the absolute value than the second negative term. The second term represents the change in energy upon extension. In the case of charged monolayers on the metal surface eqn (14) shows that the surface free energy decreases upon cantilever extension, resulting in a compressive stress. This result agrees with the experimental observation of an always compressive *absolute* signal.[10,11] Eqn (14) shows that charged monolayers on the metal surface exhibit a behavior of typical elastic solids, for which stretching results in a decrease in the surface density and, therefore, in a negative differential term in the Shuttleworth equation.[38] Note that in general for solid surfaces both terms in the Shuttleworth equation (eqn (1) and (14)) are of the same order of magnitude and the sign of the total surface stress is determined by the sign and the magnitude of the differential term (see for example ref. 38 and 39). In contrast, for equilibrium liquid interfaces the differential term is zero. Therefore, the surface stress of the equilibrium liquid interface is equal to the surface tension and is essentially positive (tensile).

Volume charge density

Polyelectrolyte brushes and DNA monolayers are typical examples of active coatings with charged chains immersed in aqueous salt solution. Calculation of stress in these systems involves calculation of total free energy of these monolayers (see previous section and eqn (1)). Extensive theoretical work on the theory of polyelectrolyte brushes based on mean-field and scaling concepts provides analytical models for the energetic and conformational properties of these monolayers.[40–43] Application of mean-field models to the calculation of surface stress in DNA monolayers predicted an exponential dependence of the stress on DNA grafting density[30] and revealed the leading role of ion osmotic pressure and hydration forces in stress generation. However, several fitting parameters had to be introduced into the model to obtain a quantitative agreement with experiment.[30]

The main oversimplification of the mesoscopic models of polyelectrolyte brushes with respect to their applicability to cantilever sensors is the approximation of the substrate as an inert flat surface, which provides the anchoring points for the molecules, but otherwise does not interact with them. However, in cantilever sensors the interactions of the organic film with the underlying metal are of utmost importance for the quantitative determination of the chemical stress in the active layer.[25] In our model image interactions with the metal are included, which allows obtaining quantitative results for the chemical stress without introducing any fitting parameters.

Similarly to the case of the surface charge density, we make the approximation of an evenly charged monolayer and consider $\rho(r) = \rho = \text{const}$. It should be noted that for very dense monolayers the lateral distributions of mobile ions between the molecules is independent of position within the monolayer (r), while for monolayers with lower densities the lateral concentration profile of the ions will have a sinusoidal character. Nevertheless, these lateral variations in the volume charge density on the Angstrom length scale introduce just a small correction to the macroscopic stress on cantilevers tens of microns long.

The solution of the Poisson–Boltzmann equation for this three-layer system gives the following expression for the electrostatic potential:[36]

$$\psi(z) = \frac{4\pi l_B \rho e^{-(z-d)\kappa_3}}{\kappa_2 \kappa_3} \frac{(1 - e^{-2d\kappa_2})}{1 - e^{-2d\kappa_2} + \frac{\varepsilon_2 \kappa_2}{\varepsilon_3 \kappa_3}(1 + e^{-2d\kappa_2})} \quad (15)$$

The surface free energy of the cantilever coated with gold film and the DNA monolayer can then be found as

$$F(d) = \frac{1}{2}\rho \int_0^d \psi(r)dr = \frac{2\pi l_B \rho^2}{\kappa_2 \kappa_3^2} \frac{(1 - e^{-2d\kappa_2})}{1 - e^{-2d\kappa_2} + \frac{\varepsilon_2 \kappa_2}{\varepsilon_3 \kappa_3}(1 + e^{-2d\kappa_2})} \quad (16)$$

Noting that $F(d) = \gamma A$, where A is the area of the cantilever and substituting (16) into (1) gives

$$\sigma(\rho, d) = -\frac{4\pi l_B}{\kappa_2 \kappa_3^2} \frac{(1 - e^{-2d\kappa_2})}{1 - e^{-2d\kappa_2} + \frac{\varepsilon_2 \kappa_2}{\varepsilon_3 \kappa_3}(1 + e^{-2d\kappa_2})} \frac{\rho^2}{A} \quad (17)$$

Therefore, the absolute stress induced by the adsorption of polyelectrolyte layers is compressive and it strongly depends on the volume charge density of the monolayer and its thickness. Note that for very thick monolayers, i.e. for $d > 100$ nm, the dependence on d disappears and expression (17) simplifies to

$$\sigma(\rho) = -\frac{4\pi l_B}{\kappa_2 \kappa_3^2} \frac{\varepsilon_3 \kappa_3}{\varepsilon_2 \kappa_2 + \varepsilon_3 \kappa_3} \frac{\rho^2}{A} \quad (18)$$

The volume charge density of polyelectrolyte monolayers is determined by the linear charge density of macromolecules, the concentration of bound counter-ions and the concentrations of non-bound co- and counter-ions inside the monolayer. The latter depends on the chemical potential difference between cations and anions ($\Delta\mu$) and the chemical potential difference between the salt ions in the aqueous half-space and in the monolayer, which can be described by the Born energy $\Delta W = \frac{e^2}{8\pi k_B T R \varepsilon_2} - \frac{e^2}{8\pi k_B T R \varepsilon_3}$, where R is the radius of the ion. The concentration of ions in the monolayer can be then expressed as $C_{monolayer} = C_{bulk} e^{-\Delta\mu - \Delta W}$, where C_{bulk} is their bulk concentration. These properties can be determined more precisely using classical Density Functional Theory of polyelectrorlyte brushes, described in the next section.

Conformational changes

To determine the structure and electrostatic properties of monolayers of charged chains we used a classical Density Functional Theory.[44] In this approach polyelectrolyte molecules are approximated as chains of charged spheres (Fig. 3). Each sphere represents a monomer, for example DNA base for single-stranded DNA (ssDNA) or DNA base-pair for double-stranded DNA (dsDNA). Mobile ions are also represented as charged spheres with the radii equal to the corresponding ionic radii. The total free energy of the solvated polyelectrolyte layer can be written as

$$F_{tot} = F_{id} + F_{ex} \qquad (19)$$

where F_{id} is the ideal part of the free energy of the isolated polyelectrolyte chain free from all non-bonded interactions and F_{ex} is the excess free energy accounting for non-bonded, intra- and intermolecular interactions.[45] The latter includes contributions from the hard-sphere repulsion, the chain connectivity, Coulomb energy and electrostatic correlations.

The total free energy of the system is minimized with respect to the densities of polymer segments, co- and counter-ions to obtain the equilibrium distributions of the densities of all components of the system. A typical plot of the densities of ssDNA bases, co- and counter-ions is shown in Fig. 4. The monolayer thickness (d) was calculated using the distribution of polyelectrolyte segments, $\rho_p(z)$, as

$$d = \frac{2\int_0^\infty z\rho_p(z)dz}{\int_0^\infty \rho_p(z)dz} \qquad (20)$$

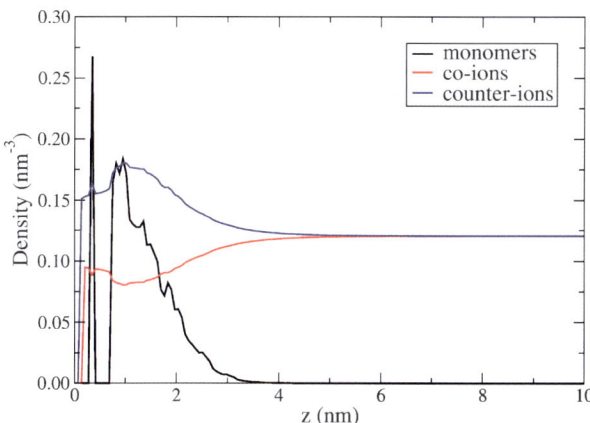

Fig. 4 Densities of DNA monomers, co- and counter-ions in a monolayer of 10-mer ssDNA as a function of the distance from the gold surface. Ionic strength of 1:1 electrolyte is 200 mM and DNA grafting density is 0.03 nm^{-2}. Monolayer thickness, calculated using (20), is equal to 2.65 nm.

The main sources of error in the calculation of the stress using eqn (17) are the uncertainty in the concentration of ions in the polyelectrolyte monolayer and the limitations in the applicability of the linear Poisson–Boltzmann theory. The first factor dominates the error for low-density monolayers, while the second is dominant for dense monolayers of highly charged polyelectrolytes, such as dsDNA.

We have estimated that the 10% variation in κ_2^{-1}, which corresponds to a 2% variation in the concentration of ions, in low density DNA monolayers leads to 10% variation in the calculated stress. For dense monolayers the concentration of ions inside the DNA array is largely independent of the bulk ionic strength[46,47] and is determined by the volume charge density of fixed ions ($\rho_{ss(ds)}$) and the uncertainty in κ_2^{-1} is smaller.

The second source of error is due to the breakdown of the linear Poisson–Boltzmann theory for highly charged systems. The linear theory is valid if the electrostatic potential is less than approximately 25 mV everywhere in the system. This condition is satisfied for ssDNA monolayers and polyelectrolyte brushes for all ionic strengths and grafting densities studied here. The electrostatic potential, however, peaks at 58 mV near the gold surface exponentially decreasing further away from the metal (eqn (15)) for the dsDNA monolayer in pure water. This introduces a small error to our calculated value of the stress. However, it has been shown that non-linear effects become significant only for long polymer brushes.[48] Therefore, these can be ignored for 10- to 30-mer DNA layers used in experiments.

Results

Surface charging effect: pH-sensitive SAMs

In this section we will focus on the nanomechanical response of cantilevers coated with pH-sensitive COOH-terminated SAMs. Since partial charging of the monolayer takes place with the increase in pH, the experimental results were interpreted based on electrostatic interactions between the molecules in the SAM.[10,11] Although this qualitative model does not contradict the experimental data, it remains unclear whether electrostatic interactions alone are responsible for the observed bending of the cantilever.

In order to compare the predictions of the theory with the available experimental data[10,11] we considered the pH dependence of the chemical component of the stress in 16-mercaptohexadecanoic acid ($S(CH_2)_{15}COOH$) monolayers. In experiments, phosphate buffer with a constant ionic strength of 0.1 was used. We took into account in the calculation of the Debye length of the solvent ($1/\kappa_3$) that the buffer solution is a mixture of monobasic (NaH_2PO_4) and dibasic (Na_2HPO_4) sodium phosphates with the ratio of the monobasic and dibasic salts changing with pH. The experimental data for the degree of deprotonation of the monolayer[49,50] has been used for the calculation of the surface charge density of the SAM. Both experimental groups used 500 μm × 100 μm cantilevers, which gives $A = 5 \times 10^4$ μm^2.

Table 1 pH dependence of stress in cantilevers functionalized with $S(CH_2)_{15}COOH$ SAMs

		$\Delta\sigma_{meas}$ (mN m^{-1})	
pH	σ_{chem} (mN m^{-1})	ref. 10	ref. 11[a]
6	−0.2 ± −4.8	+0.9 ± 0.3	−2.2
7	−7.8 ± 1.1	−2.4 ± 0.6	−4.8
8	−12.2 ± 1.3	−7.7 ± 0.6	−10.2
9	−24.9 ± 1.8	−14.5 ± 0.5	−18.1

[a] The error bars are not given in ref. 11

According to our molecular mechanics calculations[50] the thickness of the SAM is 1.84 nm. The comparison of the theoretical results calculated using eqn (14) with the above parameters and the experimental data are summarized in Table 1.

It follows from Table 1 that the proposed theory gives a quantitative agreement with the experimental data, suggesting that the strength of electrostatic interactions in partially charged SAMs is sufficient to induce the observed bending of the cantilever. This result is, to our knowledge, the first direct proof that chemical transformations in the active layer are able to induce the observed mechanical response in cantilever sensors.

The accuracy of the calculated data is determined by the variance of the parameters in eqn (14). One of the main sources of the uncertainty of the calculated stress is the variance in the fraction of the deprotonated molecules in the SAM at a given pH. The experimental titration curves suggest that the transition from the fully protonated to fully deprotonated SAM takes place in the region of pH $(5.5–12.0) \pm 0.2$ (see ref. 49). Since pH 6 corresponds to an almost fully protonated monolayer small fluctuations in pH induce the largest fluctuations in the fraction of deprotonated molecules (θ) in the SAM. At this pH, θ varies from 0.5 to 2.5%. The surface charge density of the SAM at higher pH is better defined with $\theta = (8 \pm 2)\%$, $(10 \pm 2)\%$, $(30 \pm 3)\%$ at pH 7, 8 and 9, respectively, leading to a smaller uncertainty in the calculated data. The uncertainty in the thickness of the SAM also contributes to the uncertainty of the calculated data. Since in the presented model the thickness of the SAM is determined by the position of the charges with respect to the metal surface, d is determined by two factors: the tilt angle of the SAM and the localization of the negative charge in the deprotonated molecule. Experimentally the tilt angle is determined up to $0.8°$,[35] which results in a 0.02 nm uncertainty in the value of d and 0.2 mN m^{-1} uncertainty in the calculated stress. It has been shown that the negative charge of the deprotonated molecule is distributed over the oxygens of the carboxylic group.[52] For $S(CH_2)_{15}COO(H)$ monolayers the position of the oxygens with respect to the gold surface differs by 0.18 nm,[53] which gives 5% uncertainty for the stress. The variances for the calculated stress reported in Table 1 are the sums of these three contributions. Other parameters of the system, such as the bulk concentration of salts in solution and the dielectric constant of the solvent are defined up to 1% and give at least an order of magnitude smaller contribution to the uncertainty of the calculated stress.

According to eqn (14) the chemical component of the stress is always compressive, which correlates with the experimental data for the *absolute* signal.[10,11] In experiments CH_3-terminated monolayers with the same chain-length as corresponding COOH-SAMs on gold surfaces were used as reference coatings. We have shown previously that in this case the elastic stress is always tensile.[25] At low pH the compressive electrostatic and tensile elastic contributions to the differential deflection become comparable, which can lead to either a compressive or tensile total differential signal. In particular, for reference and active cantilevers functionalized with $S(CH_2)_{15}CH_3$ and $S(CH_2)_{15}COOH$ SAMs, respectively, the chemical component of the differential stress at pH 6 varies between -0.2 and -4.8 mN m^{-1} (mainly due to fluctuations in the surface charge density of the COOH-SAM), while $\sigma_{elast} = 0.3$ mN m^{-1}, giving rise to a total differential stress of $+0.1$ to -4.5 mN m^{-1}.

Experimentally both groups obtained negative *differential* signals in medium and high pH regimes (pH $\geq$ 7), which corresponds to 8–30% deprotonation of the COOH-terminated SAM[49] and, therefore, a large chemical contribution to the stress (Table 1). On the other hand, at lower pH (pH 6, corresponding to 2.5% deprotonation of the SAM) the average differential stress measured by Watari *et al.*[10] is tensile, while it is compressive according to the results of Fritz *et al.*[11] This qualitative difference arises from a subtle balance of tensile (non-electrostatic) and compressive (electrostatic) contributions to the differential signal, which is sensitive to the properties of the active layer and depends on the method of deposition of the gold film and the quality of the SAMs.

DNA monolayers

Fast and reliable label-free detection of specific DNA sequences is very important for medical diagnostics. Not surprisingly considerable effort has been devoted to elucidation of sensitivity of cantilever devices coated with single-stranded DNA molecules of specific sequence to the detection of small concentrations of complementary strands in solution.[15–22] Systematic experimental study of the dependence of the cantilever response on ssDNA grafting density and electrolyte concentration revealed important trends.[54] However, interpretation of these experimental data is complicated since changes in one of the parameters of the system leads to simultaneous changes in several other parameters. For example, the change in bulk concentration of ions in solution leads to changes in Debye screening lengths in regions 3 and 2, in the conformation of DNA molecules in the monolayer and, therefore, in monolayer thickness (see Fig. 3 and eqn (17)). Upon hybridization reaction on the cantilever surface both κ_2 and d change again due to changes in the volume charge density of fixed ions, which induces further conformational changes in the DNA monolayer and redistribution of mobile ions in the system. Therefore, it is not possible to attribute the changes in the observed cantilever signal to the change of one particular parameter of the system.

According to eqn (17) the absolute stress induced by the adsorption of DNA monolayers is compressive and it strongly depends on the volume charge density of the monolayer and its thickness. Both parameters change upon hybridization and the corresponding change in the stress can be calculated as

$$\Delta\sigma = \sigma_{ds}(\rho_{ds}, d_{ds}) - \sigma_{ss}(\rho_{ss}, d_{ss}) \qquad (21)$$

where indices "ss" and "ds" stand for single- and double-stranded DNA, respectively.

Here we will discuss in detail the dependence of surface stress in cantilever sensors coated with DNA monolayers on gold surfaces on DNA grafting density (ρ_{gr}) and compare theoretical results with experimental data reported by Stachowiak et. al.[54] Before discussing the difference signal $\Delta\sigma$ it is instructive to follow the changes in the structure and surface stress of ss- and dsDNA monolayers as a function of grafting density.

DFT calculations of the distributions of monomers and mobile ions in DNA monolayers revealed linear dependencies of the thicknesses of ssDNA and dsDNA monolayers on their grafting densities (Fig. 5a). The increase in monolayer thicknesses is, however, slower than the increase in the corresponding grafting densities resulting in the increase in the volume densities of the monomers with ρ_{gr}. This leads to a decrease in the effective dielectric constant of the monolayer and an increase in the Born energy of mobile counter-ions. Hence the decrease in their concentration in the monolayers. Therefore, the degree of counter-ion compensation of the density of fixed ions decreases and the total absolute volume charge density in the DNA monolayers increases with the grafting density (Fig. 5b). Since the stress is proportional to the square of the volume charge density in DNA monolayers (see eqn (17)), cantilever bending increases with grafting density. Note that this increase in stress is weaker than quadratic for short DNA molecules due to the exponential dependence on the thickness of the monolayer, which also changes with the grafting density. As mentioned above this dependence of the chemical stress on the monolayer thickness becomes negligible for very long molecules and the stress should quadratically depend on the volume charge density of these monolayers.

These calculations also showed that for all grafting densities the volume charge density of dsDNA monolayers is approximately an order of magnitude larger than that of ssDNA monolayers (Fig. 5b). This is the major factor of the differences in the stress in ss- and dsDNA monolayers (Fig. 5c). This large difference between

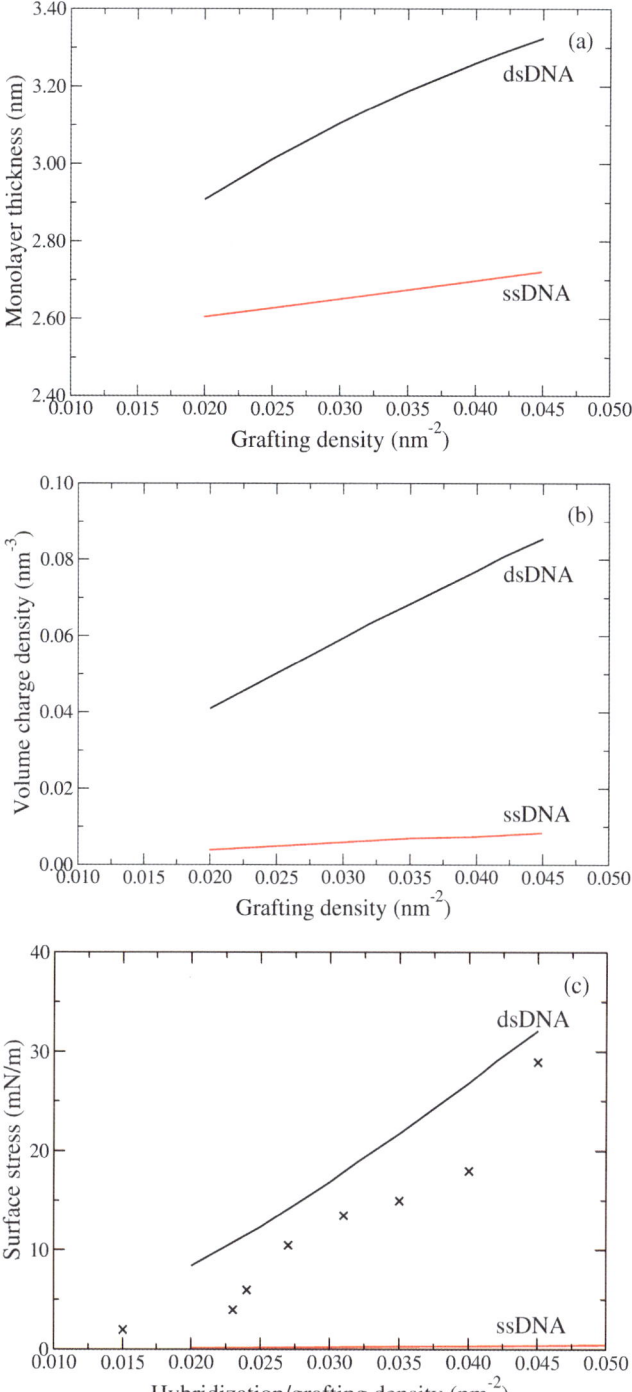

Fig. 5 Grafting density dependence of (a) thickness; (b) absolute volume charge density, and (c) surface stress of single- and double-stranded 10-mer DNA monolayers on gold surfaces: Experimental data for surface stress from ref. 54 are shown in (c) with crosses. The uncertainty of the experimental data varies between 4 to 12 mN m^{-1}. The bulk concentration of the 1:1 electrolyte is 200 mM.

the stresses in ss- and dsDNA monolayers justifies the direct comparison of the calculated results for $\Delta\sigma$ at a certain grafting density with the experimental results for equal *hybridization* density. Experimentally not all the ssDNA molecules on the cantilever surface take part in the hybridization reaction.[54] Therefore, in the experimental system the cantilever surface is coated with a mixed ss/dsDNA monolayer. However, according to our results the contribution to the total stress from ssDNAs is negligibly small at all grafting densities compared to the stress induced by the dsDNA part of the active layer. Within this approximation the agreement of the calculated data with experiment is satisfactory (Fig. 5c), suggesting that the presented model captures the physics of the hybridization induced cantilever bending.

Another important effect, relevant to denser DNA monolayers, which is not considered in our model, is the influence of hydration forces on the surface stress in the active layer. Experimental studies of dense DNA monolayers revealed a strong dependence of cantilever bending on humidity.[55] This result is likely to be a manifestation of two effects: (i) lateral pressure of water molecules in the DNA monolayer and (ii) partial deprotonation of the phosphate groups of DNA molecules. Both effects increase with hydration and lead to an increase in the total stress. In these partially hydrated DNA monolayers lateral interactions strongly depend on the structure of water inside the monolayer. Although several phenomenological mesoscopic models of hydration forces exist,[56] to determine the relative contribution of hydration and osmotic effects accurate atomistic free energy calculations are required.

Polyelectrolyte brushes

Electrolyte-induced swelling and collapse of polyelectrolyte brushes on cantilever surface mimics the mechanical response of natural organisms on fluctuations in ion concentration[57,58] and represents a good model system for studying cantilever response on conformational changes in its active layer.

To elucidate the accuracy of the theoretical model in predicting surface stress, induced by conformational changes in polyelectrolyte brushes, we have compared theoretical results with experimental data reported by Zhou *et. al.*[14] In particular, we have performed calculations of surface stress in polyelectrolyte brushes with fixed grafting density and charge of the monomers as a function of bulk concentration of 1:1 electrolyte.

To perform a direct comparison with experimental data for the bending of cantilevers coated with polymethacryloyl ethylene phosphate (PMEP) brushes on gold surfaces as a function of KCl concentration,[14] we have first determined the grafting density of the brush used in the experiment.

Our DFT calculations suggest that the thickness of the highly charged polyelectrolyte brushes at fixed bulk concentration of salt depends linearly on the number of monomers (N) of polyelectrolytes. This result agrees with the predictions of the scaling theory and mesoscopic models of polyelectrolyte brushes.[41,42] Fitting these two parameters to the experimental values for brush thicknesses at two concentrations of KCl gives $\rho_{gr} = 0.2$ nm^{-2} and $N = 210$, which leads to $d = 62.10$ and 58.16 nm at 0 and 100 mM KCl, respectively. These values of d compare well with the AFM data on the brush thicknesses of 62 and 58 nm at 0 and 100 mM KCl, respectively.[14]

Using the determined value for the brush grafting density we have performed DFT calculations of the properties of the brush in the range of KCl concentrations of 0–100mM. These calculations showed that the brush thickness decreases monotonically with bulk concentration of salt (C_{bulk}), following the well established scaling laws in the osmotic and salted brush regimes.[41,42] The decrease in the excess volume charge density is essentially non-linear (Fig. 6). The excess charge density is equal to the sum of the total densities of fixed charges, co- and counter-ions:

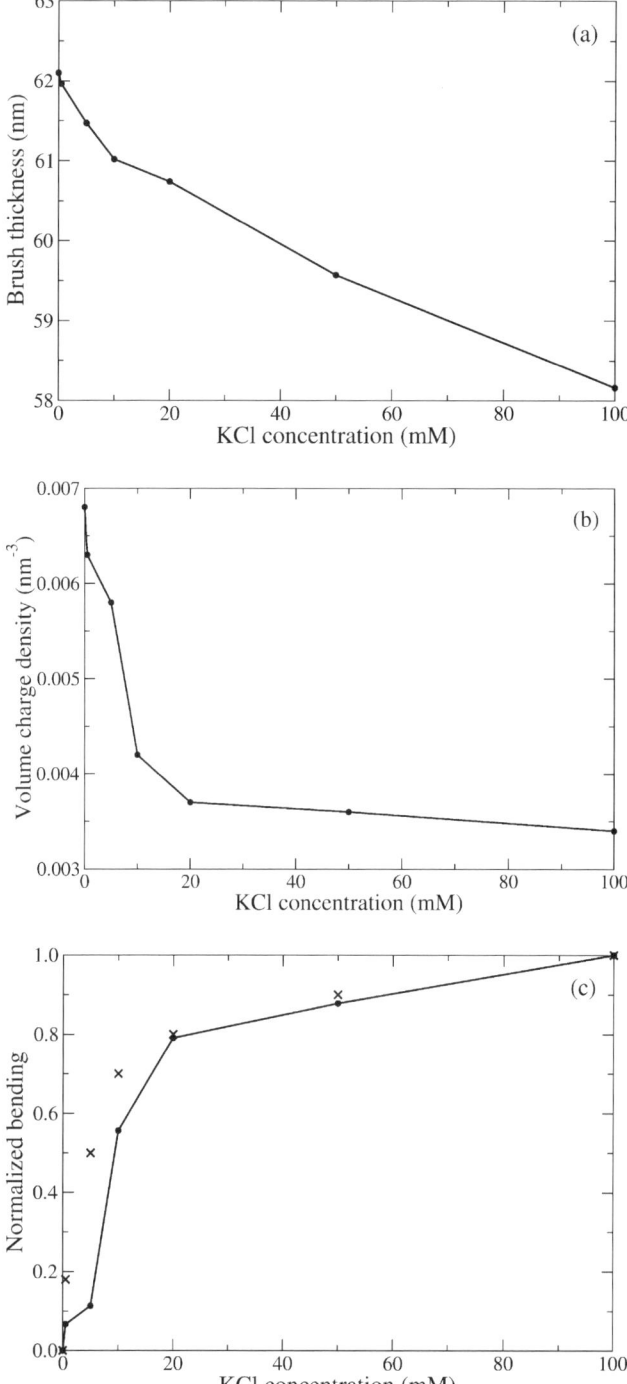

Fig. 6 Dependence of the (a) thickness and (b) volume charge density of polyelectrolyte brushes on gold surfaces on the bulk concentration of KCl: Calculated normalized cantilever deflection, induced by swelling of the polyelectrolyte brush, are shown in (c). Experimental data for z_n from ref. 14 are shown in (c) with crosses. Brush grafting density is 0.2 nm^{-2}.

$\rho = \rho_p + \rho_{co} - \rho_{ctr}$. Therefore, its non-linear dependence on C_{bulk} is due to a combination of three competing trends: (i) the increase in C_{bulk} leading to monotonic decrease in d, and, therefore, monotonic increase in the density of charged monomers, ρ_p. The concentration of mobile ions is (ii) proportional to C_{bulk}; and (iii) an exponential function of Born energy: $C_{monolayer} = C_{bulk}e^{-\Delta\mu - \Delta W}$ (see Theory section). The Born energy, in turn, increases with the increase in the monomer density in the brush, and therefore it increases with C_{bulk}. Hence the exponential term decreases with C_{bulk}, which leads to a slower than linear overall increase in the concentrations (or densities) of mobile ions in the brush (Fig. 6b).

This non-linearity in the dependence of ρ on C_{bulk} is also reflected in the calculated values for the stress, which depend quadratically on ρ (see eqn (17)). Our model predicts a monotonic decrease in the absolute value of compressive stress in the brush with the increase in bulk concentration of salt. To perform a quantitative comparison with the experimental data of Zhou et. al.[14] we have converted the stress into absolute deflection, $\Delta z(C_{bulk})$, using Stoney's equation and then into normalised bending using

$$z_n(C_{bulk}) = \frac{\Delta z(C_{bulk}) - \Delta z(0)}{\Delta z(100) - \Delta z(0)} \quad (22)$$

where bulk concentrations of salt are in mM.

In these calculations the experimental geometrical parameters of the cantilever[14] were used. Note that studied conformational changes in polyelectrolyte brushes lead to chemical stress larger than 0.05 N m^{-1} by absolute value and elastic effects can be ignored here. Excellent quantitative agreement of the calculated values of z_n with experimental data (Fig. 6c) confirms the validity of the model to study conformational changes in the active layer of cantilever sensors.

Discussion

One of the main approximations we have made considering the chemical component of surface stress is the approximation of ideal monolayers on atomically flat gold surfaces. In cantilever experiments, however, the gold surface is usually not a single crystal, but has a mean square roughness of the order of 0.3–0.9 nm with a grain size of about 10 nm.[59] These non-ideal monolayers can partially relieve stress via relaxation of the molecules at the boundaries. We will discuss this effect on the example of self-assembled monolayers, though it is equally relevant to other systems.

The non-ideal character of SAMs is manifested in a possible non-uniformity of monolayer parameters, such as the tilt angle and the orientation of the molecules with respect to the surface. These structural variations across the cantilever are more pronounced for SAMs with short molecules. For these monolayers the tilt angle is ill defined[35] and a characteristic variation of the tilt angle around the average is 5–7° at room temperature for chain-lengths lower or equal to 8 carbons.[51] Therefore, some stress can be relieved via the disorder in SAMs with short chains. SAMs with longer chains ($n \geq 12$) do not have this degree of freedom to relieve stress and their properties are closer to the properties of ideal monolayers. For these monolayers the tilt angle remains uniform at room temperature decreasing by several degrees compared with the zero-temperature tilt, while the monolayer may have some temperature-induced rotational defects, which do not change the height of the charged groups with respect to the rest of the monolayer.[60,53]

The quality of the gold film and the SAM become crucial when the chemical and elastic components of the differential signal become comparable. Hence the difference in the sign for the differential stress in the low pH regime observed by different groups. In this regime of weakly charged monolayer with the delicate balance of the compressive and tensile effects it becomes important to take into account all specific structural properties of the film, such as gold grain size, distribution of charged groups on the surface of the film, distribution of co- and counter-ions of the solvent in the vicinity of the monolayer.

Conclusions

Our results suggest that mesoscopic models are able to capture the physics underlying the observed cantilever bending induced by chemical transformations in its active layer. The developed mesoscopic models for chemical and elastic components of the stress give quantitative results for the stress generated in the cantilevers coated with self-assembled monolayers, DNA arrays and polyelectrolyte brushes. Therefore, these models can be used to screen the parameters of the system and develop sensors with significantly improved sensitivity.

We have shown that the sign of the differential signal is determined by the competition of the contribution from chemical reactions in the active layer and the elasticity mismatch between the active and reference coatings. This is a very general result for differential signals, independent of the specific nature of reactions taking place in the active layer of the cantilever. It suggests that the choice of coating for the reference cantilever should be governed not only by its inertness with respect to the reactions of interest, but also by the similarity of its elastic properties to the elastic properties of the active coating.

Acknowledgements

The author thanks Rachel McKendry, Alex Shluger, John Harding, Peter Grutter, Wilhelm Huck and Gabriel Aeppli for stimulating discussions. The work has been funded by EPSRC, "Materials Modeling Initiative" programme, grant GR/S80103/01.

References

1 B. M. Rosen and V. Percec, *Nature*, 2007, **446**, 381.
2 R. A. Cross, *Nature*, 1997, **385**, 18.
3 O. Azzaroni, B. Trappmann, P. van Rijn, F. Zhou, B. Kong and W. T. S. Huck, *Angew. Chem., Int. Ed.*, 2006, **45**, 7440.
4 I. Frank, *Angew. Chem., Int. Ed.*, 2006, **45**, 852.
5 L. Bai, T. J. Santangelo and M. D. Wang, *Annu. Rev. Biophys. Biomol. Struct.*, 2006, **35**, 343.
6 J. H. White, R. A. McIllhinney, A. Wise, F. Ciruela, W. Y. Chan, P. C. Emson, A. Billinton and F. H. Marshall, *Proc. Natl. Acad. Sci. U. S. A.*, 2000, **97**, 13967.
7 T. Sakamoto, M. R. Webb, E. Forgacs, H. D. White and J. R. Sellers, *Nature*, 2008, **455**, 128.
8 J. K. Gimzewski, C. Gerber, E. Meyer and R. R. Schlittler, *Chem. Phys. Lett.*, 1994, **217**, 589.
9 H. P. Lang and C. Gerber, *STM and AFM Studies on (Bio)Molecular Systems: Unravelling the Nanoworld*, Springer, New York, 2008, vol. 285, p. 1.
10 M. Watari, J. Galbraith, H. P. Lang, M. Sousa, M. Hegner, C. Gerber, M. A. Horton and R. A. McKendry, *J. Am. Chem. Soc.*, 2007, **129**, 601.
11 J. Fritz, M. K. Baller, H. P. Lang, T. Strunz, E. Meyer, H. J. Guntherodt, E. Delamarche, C. Gerber and J. K. Gimzewski, *Langmuir*, 2000, **16**, 9694.
12 G. G. Bumbu, G. Kircher, M. Wolkenhauer, R. Berger and J. S. Gutmann, *Macromol. Chem. Phys.*, 2004, **205**, 1713.
13 G. G. Bumbu, M. Wolkenhauer, G. Kircher, J. S. Gutmann and D. Berger, *Langmuir*, 2007, **23**, 2203.
14 F. Zhou, W. Shu, M. E. Welland and W. T. S. Huck, *J. Am. Chem. Soc.*, 2006, **128**, 5326.
15 J. Fritz, E. B. Cooper, S. Gaudet, P. K. Sorger and S. R. Manalis, *Proc. Natl. Acad. Sci. U. S. A.*, 2002, **99**, 14142.
16 K. M. Hansen, H. F. Ji, G. H. Wu, R. Datar, R. Cote, A. Majumdar and T. Thundat, *Anal. Chem.*, 2001, **73**, 1567.

17 R. McKendry, J. Y. Zhang, Y. Arntz, T. Strunz, M. Hegner, H. P. Lang, M. K. Baller, U. Certa, E. Meyer, H. J. Guntherodt and C. Gerber, *Proc. Natl. Acad. Sci. U. S. A.*, 2002, **99**, 9783.
18 R. Mukhopadhyay, M. Lorentzen, J. Kjems and F. Besenbacher, *Langmuir*, 2005, **21**, 8400.
19 W. M. Shu, D. S. Liu, M. Watari, C. K. Riener, T. Strunz, M. E. Welland, S. Balasubramanian and R. A. McKendry, *J. Am. Chem. Soc.*, 2005, **127**, 17054.
20 J. Fritz, M. K. Baller, H. P. Lang, H. Rothuizen, P. Vettiger, E. Meyer, H. J. Guntherodt, C. Gerber and J. K. Gimzewski, *Science*, 2000, **288**, 316.
21 H. P. Lang, M. Hegner, E. Meyer and C. Gerber, *Nanotechnology*, 2002, **13**, R29.
22 A. Ikai, R. Afrin, H. Sekiguchi, T. Okajima, M. T. Alam and S. Nishida, *Curr. Protein Pept. Sci.*, 2003, **4**, 181.
23 G. G. Stoney, *Proc. R. Soc. London, Ser. A*, 1909, **82**, 172.
24 K. S. Chen and K. S. Ou, *J. Micromech. Microeng.*, 2002, **12**, 917.
25 M. L. Sushko, J. H. Harding, A. L. Shluger, R. A. McKendry and M. Watari, *Adv. Mater.*, 2008, **20**, 3848.
26 R. Shuttleworth, *Proc. Phys. Soc., London, Sect. A*, 1950, **63**, 444.
27 U. Tartaglino, D. Passerone, E. Tosatti and F. Di Tolla, *Surf. Sci.*, 2001, **482–485**, 1331.
28 D. Passerone, E. Tosatti, G. L. Chiarotti and F. Ercolessi, *Phys. Rev. B: Condens. Matter Mater. Phys.*, 1999, **59**, 7687.
29 J. N. Israelachvili, *Intermolecular and Surface Forces*, Academic Press, London, 1992.
30 M. F. Hagan, A. Majumdar and A. K. Chakraborty, *J. Phys. Chem. B*, 2002, **106**, 10163.
31 A. M. Jonas, Z. J. Hu, K. Glinel and W. T. S. Huck, *Nano Lett.*, 2008, **8**, 3819.
32 C. A. Klein, *J. Appl. Phys.*, 2000, **88**, 5487.
33 A. Atkinson, *Br. Ceram. Proc.*, 1995, **54**, 1.
34 V. Srinivasan, G. Cicero and J. C. Grossman, *Phys. Rev. Lett.*, 2008, **101**, 185504.
35 F. Schreiber, *Prog. Surf. Sci.*, 2000, **65**, 151.
36 R. R. Netz, *Eur. Phys. J. E*, 2000, **3**, 131.
37 I. Pera and J. Fritz, *Langmuir*, 2007, **23**, 1543.
38 I. A. Rusanov, *Surf. Sci. Rep.*, 1996, **23**, 173.
39 I. A. Rusanov, *Surf. Sci. Rep.*, 2005, **58**, 111.
40 R. Toomey and M. Tirrell, *Annu. Rev. Phys. Chem.*, 2008, **59**, 493.
41 P. Pincus, *Macromolecules*, 1991, **24**, 2912.
42 O. V. Borisov, T. M. Birshtein and E. B. Zhulina, *J. Phys. II*, 1991, **1**, 521.
43 O. V. Borisov, E. B. Zhulina and T. M. Birshtein, *Macromolecules*, 1994, **27**, 4795.
44 J. Wu and Z. Li, *Annu. Rev. Phys. Chem.*, 2007, **58**, 85.
45 T. Jiang, Z. Li and J. Wu, *Macromolecules*, 2007, **40**, 334.
46 P. Gong, T. Wu, J. Genzer and I. Szleifer, *Macromolecules*, 2007, **40**, 8765.
47 T. Wu, P. Gong, I. Szleifer, P. Vlcek, V. Subr and J. Genzer, *Macromolecules*, 2007, **40**, 8756.
48 A. Naji, R. R. Netz and C. Seidel, *Eur. Phys. J. E*, 2003, **12**, 223.
49 K. Aoki and T. Kakiuchi, *J. Electroanal. Chem.*, 1999, **478**, 101.
50 S. E. Creager and J. Clarke, *Langmuir*, 1994, **10**, 3675.
51 M. L. Sushko and A. L. Shluger, *J. Phys. Chem. B*, 2007, **111**, 4019.
52 R. Schweiss, P. B. Welzel, C. Werner and W. Knoll, *Langmuir*, 2001, **17**, 4304.
53 M. L. Sushko and A. L. Shluger, *Adv. Mater.*, 2009, **21**, 1111.
54 J. C. Stachowiak, M. Yue, K. Castelino, A. Chakraborty and A. Majumdar, *Langmuir*, 2006, **22**, 263.
55 J. Mertens, C. Rogero, M. Calleja, D. Ramos, J. A. Martin-Gago, C. Briones and J. Tamayo, *Nat. Nanotechnol.*, 2008, **3**, 301.
56 S. Leikin, V. A. Parsegian, D. C. Rau and R. P. Rand, *Annu. Rev. Phys. Chem.*, 1993, **44**, 369.
57 L. M. Routledge, *J. Cell Biol.*, 1978, **77**, 358.
58 W. B. Amos, *Nature*, 1971, **229**, 127.
59 M. Godin, P. J. Williams, V. Tabard-Cossa, O. Laroche, L. Y. Beaulieu, R. B. Lennox and P. Grutter, *Langmuir*, 2004, **20**, 7090.
60 W. Mar and M. L. Klein, *Langmuir*, 1994, **10**, 188.

General discussion

Professor Steiner opened the discussion of the introductory lecture by Professor Jones: My question concerns the motility of the object involving the actuation of macromolecules. The relaxation rates that are found (*i.e.* that we have found in experiments in collaboration with Professor Huck) are very slow, often on the order of many minutes to hours. The reason for this is that macromolecules which are taken far from equilibrium may be trapped in conformations from which they recover only very slowly. This affects the actuation by polymer brushes you have shown in your lecture, as well as the motion of gels. Because the trapped conformations are properties of single macromolecules, reducing the size of the system does not necessarily speed up the kinetics in a significant fashion.

Can you comment on this? Are there useful strategies for the use of *e.g.* polyelectrolytes as molecular actuators? Rather than switching the system between two equilibrium states, which require the complete equilibration of the molecules, should we think of strategies to shuttle the system between two non-equilibrium states, the transition between which is not kinetically hindered?

Professor Jones responded: Your comment is very pertinent—like you, we too have found that the kinetics of response of polymer brushes to those changes in environment that lead to a change in conformation can be surprisingly slow. This is both interesting—in that it points to the existence of the sort of trapped configurations that you refer to—and frustrating, in that I think it does suggest real limitations to the use of systems like polyelectrolyte brushes as actuators. I suspect that some of the limitations arise from entanglement issues (as, in fact, pointed out in related circumstances by O'Connor and McLeish[1] some years ago), which in principle could be decreased by reducing the grafting density. In any case, as you suggest, more consideration not just of the equilibrium end-states but of the pathways between the end states is probably required.

1 K. O'Connor and T. McLeish, *Faraday Discuss.*, 1994, **98**, 67–78.

Professor Chan remarked: The notion of an osmotic pressure stems from the fact that the presence of solute species on one side of the semi-permeable membrane alters the chemical potential of the permeable solvent species because of solute–solvent interactions. This change in chemical potential of the solvent species can be viewed as a pure solvent being held at a higher pressure. The osmotic pressure is the difference between this higher pressure and the pressure of the pure solvent.

Your explanation of the origin of the osmotic pressure as being attributed to an excluded volume effect of the solvent near the semi-permeable membrane is novel. How does this explanation relate to the above conventional explanation where the size of the solute does not enter at all?

Professor Jones responded: My explanation, rather than being novel, was based on some rather old ideas due to Derjaguin and others (the original reference is, I believe, ref. 1). I agree entirely with your statement of the classical explanation of osmotic pressure, but it is important to distinguish between the thermodynamics of the situation and the kinetics. What I was trying to explain was not the thermodynamic origin of osmotic pressure, but the kinetics of the process by which osmotic pressure becomes equalised by the flow of fluid through a semi-permeable membrane.

1 B. V. Derjaguin, G. P. Sidorenkov, E. A. Zubashchenkov, E. V. Kiseleva, *Kolloidn. Zh.*, 1947, **9**, 335.

Professor Kornyshev opened the discussion of Professor Sen's paper: What is rate determining in this motility generation process—the elementary act of redox reaction or proton diffusion? With this comes a second question: have you considered any other kind of redox reactions, including the light assisted ones?

Professor Sen answered: In general, proton diffusion over the length of the rod is fast compared to the turnover-limited catalytic activity of the particles. We have experimented with a number of different metal pairs for redox reactions.[1] Other redox-based fuel systems have also been explored including hydrazine fuels[2] and glucose.[3]

1 Y. Wang, R. M. Hernandex, D. J. Bartlett, J. M. Bingham, T. R. Kline, A. Sen and T. E. Mallouk, *Langmuir*, 2006, **22**, 10451–10456.
2 M. E. Ibele, Y. Wang, T. R. Kline, T. E. Mallouk and A. Sen, *J. Am. Chem. Soc.*, 2007, **129**, 7762–7773.
3 N. Mano and A. Heller, *J. Am. Chem. Soc.*, 2005, **127**, 11574–11575.

Dr Titmuss opened the discussion of the paper by Dr Parnell: Have you tried to change the architecture of your polymer in such a way as to impose a curvature on the system, to encourage the formation of vesicles? For example, an A_xBA_y triblock in which $x > y$ should tend to form vesicles with the longer block on the outside and the shorter block inside.

Are there any biological systems that encapsulate in the manner described in your paper? If not, could you follow the path of bio-inspiration outlined in Professor Jones' introductory lecture to look to biological systems that do encapsulate efficiently, such as viruses? In such systems, self-assembly of the vessel and loading of the vessel are often separate processes.

Dr Parnell responded: This effect of curvature has been designed into the current design of the diblock, by tuning the ratio of hydrophobic to hydrophilic units, but not to control which of the blocks is on the interior of the polymersome. Work on a triblock copolymer system by Blanazs *et al.*[1] has varied the A and C block lengths of an ABC triblock copolymer to control the block that decorates either the interior or exterior of the vesicle. This system uses a solvent switch to form the polymer vesicle from the molecularly dissolved unimers. The decision to use the surface rehydration mechanism in our work stems from ideas on the origins of life on the prebiotic earth. Phospholipids might have been present at this early time and a series of rehydration and dehydrations could have been used to encapsulate and contain molecules within the membranes. The evolution of membrane structures has been seen as one of the great leaps in biological evolution as it allowed compartmentalization and specificity within the membrane bound structures. The more complex routes that you suggest which are now used by biology are highly specific and adapted to their task. We aimed in our work to develop a simple generic technology that would enable the encapsulation of a series of molecules at high loading efficiency. Maybe we will have to examine multiple stages in the processing as a more efficient strategy for encapsulating functional molecules.

1 A. Blanazs, M. Massignani, G. Battaglia, S. P. Armes and A. J. Ryan, *Adv. Funct. Mater.*, 2009, DOI:10.1002/adfm.200900201

Professor Jones also responded: The relationship between molecular architecture and propensity to form vesicles is an important one that needs more exploration. I do not know whether there are direct analogies between encapsulation in biological systems and our method of surface rehydration. However, it is worth mentioning that a very similar idea has been suggested in the context of discussions of the origin of life, as a mechanism by which prebiotic "protocells" might have been formed from the rehydration of films of mixtures of amphiphiles and other organic molecules and macromolecules.[1] Of course, there are many much more sophisticated encapsulation mechanisms in use in biology, many of which require energy inputs, and these offer tempting, but distant, targets for synthetic emulation.

1 D. W. Deamer and G. L. Barchfeld, *J. Mol. Evol.*, 1982, **18**, 203–206.

Mr Ahangar continued the discussion of the paper by Professor Sen: Is it possible to compartmentalize the nanobots and attach marker molecules to their ends which can then be driven towards cancer cells and used to kill them?

Professor Sen responded: This is a long term goal of the project. The main difficulties are (i) getting the motors to move in the high electrolyte concentrations present in living systems and (ii) functionalizing the motors without losing catalytic efficiency. As discussed in this paper, however, we have made great strides in determining that these nanobots can sense chemical gradients and travel directionally in response to them. This may aid in tracking down cancer cells. We have also enabled the nanorods to carry colloidal particle cargos, which may be replaced with therapeutic agents.

Dr Howse said: In your paper you talk about the autonomous motion of silver chloride particles when exposed to light. However, these particles, as represented in your Fig. 6 are symmetrical. One of the key features of propulsion at the nanoscale is asymmetry, yet it does not appear to be present in this system. Do you have any SEM or TEM images of these particles which might assist in the understanding of the mechanism for propulsion, and which might explain and provide some evidence for asymmetry in these systems?

Professor Sen replied: There are several potential sources for asymmetry in this system:
(a) Fresh silver chloride particles are not completely symmetric.
(b) The silver metal produced by photolysis deposits asymmetrically onto the particles over time (see ref. 17 in our paper for FESEM images of fresh and UV-exposed colloids).
(c) The UV-light strikes the particles from below. As the particle rotates *via* Brownian rotation, different faces of the particle get exposed differently.
(d) See the recent theory paper on the diffusion of symmetric and asymmetric particles.[1]

1 R. Golestanian, *Phys. Rev. Lett.*, 2009, **102**, 188305(1–4).

Mr Ahangar said: While it is important to streamline the motion of silver chloride particles for achieving efficiency, surely it is also important to know how it can be done?

Professor Sen replied: This is part of our ongoing research. Our modeling work is guiding this process, since it provides a way to think analytically about the particle motion and design for alternatives. At this stage, however, our primary goal is to achieve designed transport functionality, above obtaining efficiency.

Professor Sen returned to the discussion of the paper by Professor Jones: Do your Pt-based Janus particles slow down at higher ionic concentrations? If so, that would suggest a diffusiophoretic mechanism.

Professor Jones responded: We have not tried such an experiment, but I agree entirely that this approach could give us useful information about the mechanisms at work in our observations.

Professor Sen asked: Based on a given reaction rate, is it possible to predict an approximate lower size limit for particles capable of directed osmophoretic motion in the face of Brownian perturbations?

Dr Howse answered: There will be some point at which the average step size for the ballistic (propelled) step is equivalent to that of the step size for Brownian motion. A very important consideration, however, is the effect of rotation diffusion. Rotational diffusion scales with r^3 and so as the size is reduced the rotational diffusion time (the time taken for a singularly ordered system to become isotropic) is reduced, thus diminishing the effect of propulsion, resulting more in enhanced diffusion.

Professor Whitesides remarked: What's unique about nanoscale? There are a number of parameters around which one might build a gradient: chemical concentration, temperature (and "pressure" which is part of the same thing) number of particles, orientation of molecules (and *e.g.* local dipole moment), configuration of molecules. How do we distinguish between these mechanisms?

Professor Jones responded: The nanoscale world has a few general features which make the engineering that cell biology does very different from the sort of engineering we do at the macroscopic scale. These include the dominating importance of Brownian motion, the nature of fluid flow at very low Reynolds number and the strength of surface forces between objects at nanoscale separations. There are, of course, other unique features of the nanoscale world which arise from essentially quantum effects; these, though, are perhaps less important in the context of nanotechnology that is done in a warm, wet environment such as that in which cell biology operates (with a few exceptions, of which the mechanism of photosynthesis is probably the most important).

In the specific example I talked about, of using self-diffusiophoresis to propel microscale particles, the phenomenon depends on flows generated within distances set by a molecular scale of the surface by concentration gradients. You are quite right to suggest that gradients of other quantities (temperature, for example) can generate similar flows, and corresponding phenomena such as thermophoresis are closely analogous to diffusiophoresis and are controlled by equations of very similar functional form. This means that it is difficult to rule out the possibility that such other mechanisms do contribute in part to the phenomena we see; our identification of diffusiophoresis as the dominating mechanism is supported by order-of-magnitude estimates.

Professor Kornyshev continued the discussion of the paper by Professor Sen: The result must depend on the rate of reaction on the two sides of the particles. If you roughen the platinum surface (*i.e.* increase its reaction-active surface area) and gradually reduce the amount of gold on the other end (or block it by covering it with some inhibitor), will you see a difference in the velocity of the particle?

Professor Sen answered: We have done a detailed study of the reaction rate in an analogous system where two metals (silver and gold) were evaporated onto a surface and the fluid pumping over them was observed with tracer particles.[1] We have also performed mixed potential studies (ref. 14 in our paper and ref. 2 here). In the first system the slow step appears to involve gold. The fluid pumping rate increases with increasing area of the gold surface until a plateau is reached.

1 T. R. Kline, J. Iwata, P. E. Lammert, T. E. Mallouk, A. Sen and D. Velegol, *J. Phys. Chem. B*, 2006, **110**, 24513–24521.
2 Y. Wang, R. M. Hernandez, D. J. Bartlett, J. M. Bingham, T. R. Kline, A. Sen and T. Mallouk, *Langmuir*, 2006, **22**, 10451–10456.

Professor Chau inquired: What is the major limitation of using electrokinetic propulsion to direct the particle movement in bio-related applications such as drug delivery?

Professor Sen responded: The ionic strength of the solution is a key parameter to monitor for all electrokinetically propelled particles. The electric field that drives the particles is inversely proportional to the ionic strength (conductivity) of the solution. In biological systems, the ionic strength is relatively high so the motion would be quite slow. However, there are other mechanisms, such as osmophoresis, that can operate by non-ionic mechanisms.

Mr Ahangar continued the discussion of the paper by Dr Parnell: I think that the answer lies in having a container for the killing of cancer cells.

Dr Parnell replied: The ability to encapsulate effective treatments for a number of diseases/conditions using polymer vesicles would be extremely useful. These could be silencing-RNA treatments and other non-conventional treatments that cannot exist for long in the environment of the cell due to their degradation by enzymes and other biological conditions. Work by Lomas *et al.*[1] has shown the low cytoxicity of these materials and their targeted nature that can be incorporated into the design of polymersomes for drug delivery applications.

1 H. Lomas, I. Canton, S. MacNeil, J. Du, S. P. Armes, A. J. Ryan, A. L. Lewis and G. Battaglia, *Adv. Mater.*, 2007, **19**, 4238–4243.

Professor Jones then responded: Drug delivery is an application that clearly suggests itself for polymersomes, and previous experience with liposomal systems makes it clear that there is value in using these approaches for anticancer therapeutic agents.

Professor Chau enquired of Professor Sen: Currently, is there any attempt to incorporate temporal sensing into self-propelling particles?

Professor Sen answered: We have begun to consider the idea of creating particles which "remember" their history, for instance, using porous particles to absorb small amounts of fuel or inhibitors from the areas of solution they have previously encountered. Therefore, their current speed would depend on their history.

Professor Matile queried Dr Parnell: Compared to biological liposomes, I wonder how polymersome membranes can be disrupted, which detergents are used, and how the process is detected in solution. Secondly, I wonder how polymersomes respond to osmotic pressure, if they shrink and swell (*i.e.*, if they are water permeable), and whether shape changes in response to osmotic stress have been used to generate motion.

Dr Parnell responded: The polymer vesicles used in this study can be disrupted with sodium dodecyl sulfate (SDS), dioxane or Triton X (a surfactant). The disruption of the vesicles can be observed using either dynamic light scattering or confocal microscopy. I have not subjected the polymersomes to extreme changes in osmotic pressure as you suggest. The membranes used in our polymer system are permeable to water and small molecules and under certain conditions they are as permeable as phosphatidylcholine membranes. Work by Battaglia *et al.*[1] has measured the permeability of these membranes to a series of charged species and demonstrated the tuneability of the membrane. The motility of these systems in response to a gradient in osmotic pressure might possibly provide some asymmetry. Another interesting approach would be to incorporate a single pore or opening into the vesicle wall and encapsulate a fuel that would react with the outside medium and propel it along.

1 G. Battaglia, A. J. Ryan and S. Tomas, *Langmuir*, 2006, **22**, 4910–4913.

Professor Huskens asked: Can rehydration be induced by topographical rather than chemical surface patterns? This could give easy access to spatially and size-controlled vesicle formation.

Dr Parnell responded: This is an intriguing possibility which we have not considered as yet. However, we will attempt it in the future as this would give us greater flexibility in the array of sizes and shapes of polymer vesicles that we can form.

Professor Huck said: Your results on the formation of polymersomes from hydrated and patterned polymer films suggest that you can control the overall shape and dimensions of the resulting structures in solution. Could you discuss the mechanism by which the 2D sheet converts into a closed 3D structure and have you seen different structures arising from different surface patches?

Dr Parnell replied: In the paper we discuss the physical mechanism by which vesicles form from the surface of a polymer film by the change in film morphology upon rehydration.[1] The initially disordered polymer layer proceeds *via* a rod phase to a lamellar phase. The lamellae become swollen and unbind from one another by detaching and budding off after the ingress of water through the membrane. At some point, due to a Rayleigh instability and depending on the interfacial tension of the layer, it will pinch and seal to form a closed structure. Nearly all the structures observed have been spherical, although myelin-type tube structures have been seen in some cases as filament structures coming off the surface. We have recently been awarded time at the ESRF to study the formation of vesicles from individual patches using a microfocus X-ray beamline. This will allow us to study the formation from individual templated regions. By further understanding the formation and phase transition mechanisms we hope to design encapsulation strategies informed by this research.

1 G. Battaglia and A. J. Ryan, *J. Phys. Chem. B*, 2006, **110**, 10272–10279.

Professor Huskens addressed Dr Parnell and Professor Jones: Can one control vesicle shape during the rehydration process by controlling the internal vesicle volume (in the same way as osmosis can lead to decrease of the volume and concomitant shape deformations)? It would probably require a faster closure of the vesicle structure upon rehydration compared to deformation of the intermediate partially open form, and also a slow rate of solvent transfer across the vesicle wall.

Dr Parnell replied: The polymersome membranes are robust and therefore not as deformable as conventional phospholipid liposomes. Also the free energy of the structure, even if it were temporarily deformed, would probably drive the structure to return to a spherical structure.

Professor Jones replied: This is something we have not thought about, though I can see that in principle it should work if one has sufficient control over the rehydration process.

Dr Bittner continued the discussion of the paper by Professor Sen: During photolyis of AgCl particles the ionic strength increases due to production of Cl^- and H^+. This should shield the electric field which acts on surrounding particles, hence particles close to others should slow down. Is that effect observed, *e.g.* do particles slow down inside a "school" of particles?

Professor Sen replied: One would expect this to be the case. Experimentally, however, this is a difficult phenomenon to observe. The "powered speed" of the particles within the schools is a somewhat ill-defined term. If the displacement is viewed over long times, it would approach zero as the particle in general remains confined within the school. If the displacement is viewed over short times, it is difficult to distinguish powered motion from ordinary Brownian "kicks"—not to mention the fact that the uncertainty in the distance measurements would increase as the distances in question approach the microscope's limit of resolution.

Mr Barbero asked: How do the electric fields formed by the separation of H and Cl ions evolve with time? Is there a way to measure the strength of these fields? Also what is the driving force for the separation of these ions? One would expect oppositely charged ions to attract each other.

Professor Sen answered: (a) See ref. 17 of our paper for a rearranged version of eqn (4) which specifically identifies the electric field component. This electric field depends on the natural log of the reaction rate. It is difficult to measure the magnitude of these fields experimentally owing to the small length scales and particle heterogeneity.

(b) The driving force for the separation of these ions is purely the difference in their diffusion constants. An electric field set up in solution to prevent the H and Cl ions from separating on the macroscale is counteracting this. It is this electric field that also acts on the particles in the system. It may be helpful to look at the schematic in our Fig. 6 not as a discrete separation of H and Cl ions on the microscale, but as a series of H and Cl ion pairs in which the H ion has statistically diffused out slightly farther than its Cl ion counterpart while still remaining paired. (See our ref. 19 for a more in depth discussion.)

Dr Channon remarked: It seems that the dipolar nature of the electrostatic field generated by the catalytic action of the micro-swimmers could cause an antiparallel assembly of the swimmers if they collide in solution. This would probably manifest itself primarily as dimerisation, since such a dimer would should simply spin, remaining in one place. My question is: do you ever see such an effect? If so, is it controllable by some means, for example, reducing the concentration? If not, would you like to speculate as to why not?

Professor Sen answered: This is an interesting idea. However, if such localized rod–rod interactions do occur, it is a subtle effect. Even if a rod does get close enough to another to feel the other's "tug", Brownian motion (and the powered motion) quickly pulls them apart again. The flow fields for electrophoresis and diffusiophoresis decay as $1/r^3$, and so are much shorter range than usual force-produced hydrodynamic flows, which decay as $1/r$.

Dr Howse communicated: Are the velocities you quote for your particles obtained from the root mean squared displacements or by another method?

Professor Sen communicated in reply: We do not use root mean squared speed. Generally, we track particles manually with time steps of roughly 1 s. This time interval is long enough for the powered motion to greatly exceed the Brownian motion, but short enough so that the particles move in roughly straight lines, before Brownian rotation changes their directions significantly (in time $1/D_r$, where D_r is the rotational diffusion constant of the particle in question).

Mr Carew opened the discussion of the paper by Professor Schulten: How much force can be exerted on a DNA molecule as it moves through your pore?

Professor Schulten responded: For a nanopore with a diameter as small as the DNA diameter, *i.e.* 2 nm, the minimum force value for translocation is 60 pN. Such force is required to stretch and narrow the helix structure, allowing the DNA molecule to pass through the pore (see ref. 30 and 31 in our paper). For pores with larger diameters, stretching of the DNA is not needed. In these cases, the force required for DNA passage decreases as the nanopore diameter increases as shown in ref. 1.

1 S. van Dorp, U. F. Keyser, N. H. Dekker, C. Dekker and S. G. Lemay, *Nat. Phys.*, 2009, **5**, 347–351.

Mr Carew then asked: If you were to use the pore to pull through a DNA molecule connected to a larger object, how large might that object be? Or, how much pulling force can be exerted by the pore on the DNA molecule?

Professor Schulten answered: This is an interesting question. In order for a particle adhering to DNA to pass through the nanopore, its size cannot exceed the space between DNA and the pore wall. In case the particle is larger, as for a DNA-binding protein, a sufficiently strong electrostatic potential pulling the DNA through the pore will strip the protein off the DNA. This effect has been explained in ref. 1.

1 Q. Zhao, G. Sigalov, V. Dimitrov, B. Dorvel, U. Mirsaidov, S. Sligar, A. Aksimentiev and G. Timp, *Nano. Lett.*, 2007, **7**, 1680–1685.

Mr Hopkinson communicated: In your presentation you showed a simulation indicating that the DNA would have to be highly extended to pass through the nanopore. As DNA is capable of forming structures other than the double helix—I am thinking specifically of the quadruplex in ref. 1—would you be able to detect the unfolding, and hence examine the stability of this structure?

1 J. L. Huppert, *Chem. Soc. Rev.*, 2008, **37**, 1375–1384.

Professor Schulten communicated in reply: Yes, nanopores can detect the unfolding and stability of a DNA quadruplex. In a recent publication,[1] the proteinaceous pore α-hemolysin was used to capture a DNA quadruplex and the unfolding of the quadruplex structure was monitored by observing changes in the ionic conductance.

In general, nanopores produce different ionic current signals depending on the structure of the DNA molecule present in the pore. For instance, solid-state nanopores have been used to differentiate between single-stranded and double-stranded DNA (see ref. 30 and 31 in our paper), as well as to observe the unfolding from double-stranded to single-stranded DNA in hairpins (see our ref. 32).

1 J. W. Shim and L. Q. Gu, *J. Phys. Chem. B*, 2008, **112**, 8354–8360.

Dr Titmuss opened the discussion of the paper by Dr Sushko: Given that you are trying to look at the release of a lateral stress, how appropriate is it that your calculations assume a laterally homogeneous density distribution through the brush region? This is particularly relevant as there are plenty of experimental and theoretical observations of lateral inhomogeneity and domain formation (length-scale 4–100 nm) at interfaces decorated by brushes. How difficult would it be to allow the density to vary laterally in your calculations?

In your paper you mention the difference in the Born energy for ions inside and outside the brush region: do you have a numerical value in terms of thermal energy $k_B T$ for this difference? We have indirect measurements for the difference in chemical potential for a surfactant molecule inside and outside a brush bearing an opposite charge to the surfactant ($\Delta \mu \sim 2\text{–}4\, k_B T$), and would be interested to compare the order of magnitude.

Dr Sushko answered: The question of inhomogeneity of molecular layers on gold surface is addressed in the "Discussion" section of my paper. Indeed, the discrepancy between the calculated and experimental data for the stress increases with the increase in disorder in the monolayers (see Fig. 3A in ref. 1). However, one should bear in mind that in-plane forces are averaged over the whole surface of the cantilever, *i.e.* typically over the surface of 5×10^4 μm². On this length-scale the domain structure of the monolayer with the grain sizes of 4–100nm does not lead to significant deviations of the measured stress with the calculated one, providing that each domain is well ordered. If apart from the domain structure of the sample, the domains themselves are not well ordered, then the idealized model would not be appropriate. The latter case corresponds to a high degree of global disorder in the sample, which should inevitably be reflected in the surface stress.

The classical DFT model for polyelectrolyte brushes presented in the paper can be extended to 2D and 3D cases to study disordered brushes. This extension of the model will increase the computational costs, but these will remain manageable.

The Born energy of mobile ions strongly depends on the grafting density of the polyelectrolyte brush. For example, for a polyelectrolyte brush with the grafting density $\rho_{gr} = 0.03$ nm^{-2} and other parameters as in Fig. 4 of the paper, the Born energy of sodium ions is approximately equal to 0.10 $k_B T$. For denser brushes with $\rho_{gr} = 0.2$ nm^{-2} and other parameters as in ref. 2, the Born energy is approximately equal to 0.84 $k_B T$.

In these calculations the radius of the Na$^+$ ion was equal to its van der Waals radius, $R(Na^+) = 0.116$ nm.

1 M. L. Sushko, J. H. Harding, A. L. Shluger, R. A. McKendry and M. Watari, *Adv. Mater.*, 2008, **20**, 3848.
2 F. Zhou, W. Shu, M. E. Welland and W. T. S. Huck, *J. Am. Chem. Soc.*, 2006, **128**, 5326.

Professor Kornyshev continued the discussion of the paper by Professor Schulten: Have you facilities to stimulate the effect of inverse osmosis? Your pores are very narrow, and because of that if you apply pressure across the membrane you expect to have a filtering effect. Water will go through, but ions will not be that eager to enter the pore because the membrane is nonpolar and electrostatic energy of an ion in a pore will be higher than in the bulk, so there will be an additional barrier for ions to pass through the pore.

Professor Schulten replied: A computational method for generating a hydrostatic pressure gradient across a nanopore, thereby inducing water flow, has been described and applied in ref. 1.

Your suggestion of studying an inverse osmosis effect using nanopores with charged surfaces is ideal for molecular dynamics simulations, which provide an atomic-detail description of the solvent, ions and pore surface. In fact, in a collaboration with an external group my coworkers are investigating the feasibility of your suggestion already.

1 F. Zhu, E. Tajkhorshid and K. Schulten, *Biophys. J.*, 2002, **83**, 154–160.

Professor Huskens commented: You observe charge neutralization by calcium at the channel wall, and strong adsorption of these Ca^{2+} ions. A goal was to see charge reversal, but this was not observed. What are the reasons for these observations? Strong binding of calcium is common only for chelating ligands, so here by two carboxylate groups, but apparently this is not the case here. I regard competition by K$^+$ as highly unlikely, since the divalent Ca^{2+} has much higher binding constants by orders of magnitude. Another option could be a lack of counter-anions, which would mean that overcharging could actually occur at higher (rather than lower) KCl concentrations. I would welcome your opinion on this.

Professor Schulten answered: Evidence of charge inversion arises as a maximum in the Cl$^-$ concentration profile. The Cl$^-$ maximum should be located near the pore wall, indicating that the pore surface has "inverted" its charge, from negative to positive, such that the PET surface recruits anions instead of cations (see Fig. 8b in our paper).

As discussed in our paper, we attributed the lack of a Cl$^-$ peak to the high concentration of K$^+$ ions. This mechanism has been proposed by van del Heyden *et al.* (see our ref. 40), where the KCl concentration is high enough to allow the K$^+$ ions to occupy spaces near the wall, hiding the Cl$^-$ peak.

Your suggestion that the absence of charge inversion is due to the lack of counter-anions cannot explain our results. In our simulations, we used a high concentration of Cl⁻ ions. Table 1 lists the number of Cl⁻ ions for each simulation at pH 7 (index 7 to 10), which correspond to 1 M (see Methods section). Furthermore, the Cl⁻ ionic current is similar or slightly higher than the K⁺ ionic current (Table 1, 6th and 7th columns), revealing that a considerable number of Cl⁻ ions cross the pore.

Professor Reinhoudt remarked: You mentioned the difference in transport of native DNA and methylated DNA through the nanopores. Could you use this for detecting differently methylated DNAs in the presence of native DNA?

Professor Schulten responded: Yes. Indeed, in a recent joint publication with an experimental group, we proposed that nanopores can be used to detect methylated DNA.[1]

When the diameter of a nanopore is as small as the diameter of double-stranded DNA, a minimum voltage of 3.5 V is required to translocate the DNA. We found that if the double-stranded DNA is methylated, the flexibility of the structure is affected, and the minimum voltage drops to 2.5 V.

1 U. M. Mirsaidov, W. Timp, X. Zou, V. Dimitrov, K. Schulten, A. P. Feinberg and G. Timp, *Biophys. J.*, 2009, **96**, L32–L34.

Professor Whitesides continued the discussion of Dr Sushko's paper: Can this technique examine hydration of SAMs, especially SAMs that are resistant to adsorption of protein?

Dr Sushko responded: The classical density functional theory usually describes the solvent, *e.g.* water, as a uniform dielectric medium.[1] This approximation along with the approximation of mobile ions and polymer segment with hard spheres of certain radii and charges allows calculations of the equilibrium distribution of densities of these species in the system. These are the key parameters of organic layers for the calculation of in-plane forces and stress in DNA monolayers and polyelectrolyte brushes. Note that atomistic molecular dynamics calculations for these systems in a wide range of concentrations of salt in solution are prohibitively expensive.

Hydration is essentially a quantum phenomenon (see a brief discussion of the hydration effects at the end of the "DNA monolayers" section of the paper). The enthalpy of hydration is determined by the mutual polarization of water molecules and SAMs. Therefore, mesoscopic methods and even molecular dynamics with standard non-polarizable force-fields are not appropriate. The reliable theoretical description of hydration requires quantum mechanical free energy calculations since the entropic contribution is likely to be comparable to the enthalpy. In a periodic model these calculations can be performed using, for example, the metadynamics method implemented in a CP2K code.[2]

The adsorption of proteins onto SAMs is yet another theoretical challenge. Isolated macromolecules at surfaces represent a system too large for periodic QM calculations, which requires truly multiscale approaches. I am involved in the development of such Meso/QM/MM methods[3] and would like to invite you to review the current status of the field.[4]

1 J. Z. Wu and Z. D. Li, *Annu. Rev. Phys. Chem.*, 2007, **58**, 85.
2 C. Michel, A. Laio, F. Mohamed, M. Krack, M. Parinello and A. Millet, *Organometallics*, 2007, **26**, 1241.
3 M. L. Sushko, P. V. Sushko, I. V. Abarenkov and A. L. Shluger, Embedding cluster approach for metal/organic interfaces, in preparation.
4 J. H. Harding, D. M. Duffy, M. L. Sushko, P. M. Rodger, D. Quigley and J. A. Elliott, *Chem. Rev.*, 2008, **108**, 4823.

Dr Steinke addressed Professor Schulten: To what extent is the modelling software presented in this talk capable of predicting DNA sequencing events and selectivity (at nulceotide base level) for systems that use biological nanopore systems?

Professor Schulten answered: Inorganic nanopores are presently being evaluated for so-called fourth-generation sequencing techniques, *i.e.* for very fast and low cost sequencing of human DNA. The software employed in the present study, NAMD and VMD (ref. 45 and 46 in our paper) are used to interpret and guide the development of new nanopore-based sequencing. A review of the type of simulations done is ref. 1.

Biological nanopores are also being used in experiments and simulations to study selective DNA translocation, for example, in case of an α-hemolysin pore and hairpin DNA (see ref. 28 in our paper). For both inorganic and biological nanopores, molecular modeling with VMD and NAMD complements experimental efforts to develop new sequencing devices.

1 A. Aksimentiev, R. Brunner, J. Cohen, J. Comer, E. R. Cruz-Chu, D. Hardy, A. Rajan, A. Shih, G. Sigalov, Y. Yin and K. Schulten, *Methods Mol. Biol.*, Humana Press, NJ-USA, 2008, pp. 181–234.

Professor Huck continued the discussion of the paper by Dr Sushko: It has already been mentioned several times during this meeting, including in Professor Jones' opening lecture, that the response time of polymer brushes to external stimuli such as pH or salt concentration is rather slow. With your current understanding of the structure of polymer brushes and the presence of different forces, would you be able to comment or give your opinion on possible reasons for the slow conformational changes in polymer brushes?

Dr Sushko replied: In general, the relaxation time of polymers in solutions and at surfaces scales with the number of degrees of freedom of the macromolecule. For example, this effect has been predicted theoretically and observed experimentally for the adsorption of DNA molecules onto a mica surface.[1] It has been found that the relaxation time for DNA adsorption scales with the length of the molecule. The measured relaxation time of 5994 base pairs of DNA was as long as 5 min.[1] Similarly, one can expect that the relaxation time should increase with the number of polyelectrolyte molecules in the brush.

Therefore, the slow response of polyelectrolyte brushes on changes in pH or salt content of the solvent is likely to be a manifestation of the large size of the brushes studied experimentally. In particular, in typical cantilever experiments the brush covers the surface of 500 µm × 100 µm. Considering that the grafting density is 0.2 nm^{-2}, the number of polymer molecules in the brush is $N = 10^{10}$. The number of degrees of freedom of the brush would then be equal to $n_{brush} = Nn_p - n_{conf}$, where n_p is the number of degrees of freedom for an isolated polyelectrolyte molecule tethered to the surface and n_{conf} is the number of degrees of freedom lost due to confinement of polyelectrolytes in the brush. This suggests that the denser the brush, the stronger the confinement effects and the cooperativity of relaxation of polyelectrolytes in the brush. Therefore, one can expect shorter relaxation times for dense brushes than for less dense brushes with *the same number of molecules*.

One of the main measurable responses of polyelectrolyte brushes on the changes in the properties of the solvent is the change in the *average* brush thickness, which, in turn, is characterised by the distribution of the monomers in the brush. Weak cooperativity in molecule relaxation leads to a long tail for the distribution of the densities of the monomers in the brush, stabilized by fast mobile ion exchange between the brush and the bulk solution. In these conditions the global relaxation of the brush has to proceed through a number of local relaxations of several molecules and yielding events for the conformational changes of neighboring regions in the brush.

To obtain quantitative data for brush relaxation I am currently developing a dynamic classical DFT model *via* coupling the static DFT with the Poisson–Nernst–Planck formalism for drift diffusion.

1 C. Rivetti, M. Guthold and C. Bustamante, *J. Mol. Biol.*, 1996, **264**, 919.

Dr Bittner continued the discussion of the paper by Professor Schulten: How "normal" is the water in the nanoscale pores? For example, capillaries fill up with Lennard-Jones fluids obeying the standard $\sqrt{t}$ law down to some molecular diameters,[1,2] but carbon nanotubes much faster.[3] Is the water self-diffusion constant in the pore the same as in the bulk?

1 D. I. Dimitrov, A. Milchev and K. Binder, *Phys. Rev. Lett.*, 2007, **99**, 054501.
2 P. Huber, S. Grüner, C. Schäfer, K. Knorr and A. V. Kityk, *Eur. Phys. J. Special Topics*, 2007, **141**, 101.
3 M. Whitby and N. Quirke, *Nat. Nanotechnol.*, 2007, **2**, 87.

Professor Schulten responded: We have not determined water self diffusion from our present simulations. Water diffusion inside sufficiently narrow pores differs from diffusion in bulk. For instance, in hydrophobic pores, water diffuses faster as the pore radius decreases, due to the lack of hydrogen-bond partners among water molecules in the pore.[1]

Water diffusion can also be influenced by interactions with a particular, *i.e.* not inert, pore surface. A molecular dynamics study of silica slits showed water ordering and low water diffusion within 1.5 nm from the wall.[2]

A similar behavior should result in PET nanopores described in our paper. Both silica and PET surface models are negatively charged, and the ionic concentration profiles in ref. 2 and in our paper in this issue are similar.

1 M. L. Brewera, U. W. Schmitt and G. A. Voth, *Biophys. J.*, 2001, **80**, 1691–1702.
2 S. Joseph and N. R. Aluru, *Langmuir*, 2006, **22**, 9041–9051.

Dr Titmuss asked: You mentioned that you can currently use your nanopores to count bases but not to identify the bases, so you acknowledge that to make a gene sequencer you still have much to do. At the same time, more conventional methods of gene sequencing are becoming ever cheaper and more efficient. Will you get there in time or will conventional methods become so cheap that it is not worth following your approach?

Professor Schulten replied: Only time will tell. The improvement in sequencing speed and decrease in price, goals of the fourth generation DNA sequencing, involves advances in technology beyond improving "conventional" methods. For instance, see ref. 1.

1 R. F. Service, *Science*, 2006, **311**, 1544–1546.

Professor Chan addressed Professor Sen: If you were able to change the hydrodynamic boundary condition from the normal no-slip condition on solid surfaces to say a partial- or full-slip (zero tangential stress) condition by modifying the surface of your swimming microbots, would you expect them to travel faster or slower under otherwise identical solution conditions and concentration gradients? I think unlike self-propelled swimmers they may move slower with partial-slip or full-slip boundaries.

Professor Sen answered: We believe that the larger the hydrodynamic slip allowed at the particle surface, the faster will be the speed of the particles. In the cases discussed in our paper, there is an electric field set up by the reaction, and the motion of the particles results from the electric field acting in one direction on the particles themselves and in the opposite direction on the counterions in the double layer of the particles. Since the velocity of the fluid in the double layer is in the opposite direction to the particle motion, a no-slip condition will cause the fluid to exert a drag force along the particle surface. Without this extra drag (*i.e.* with partial or full slip) the particle speed would be faster. A recent paper by Khair and Squires[1] specifically explores this relationship between hydrodynamic slip at the surface (note: this is different from the "slip velocity" sometimes used for hydrodynamic calculations) and electrophoretic mobility, showing that until one reaches very high ζ-potentials, perhaps greater than 150 mV in magnitude, hydrodynamic slip enhances the particle speed.

1 A. S. Khair and T. M. Squires, *Phys. Fluid*, 2009, **21**, 042001.

PAPER

Solid state nanofibers based on self-assemblies: from cleaving from self-assemblies to multilevel hierarchical constructs

Olli Ikkala,[*a] Robin H. A. Ras,[a] Nikolay Houbenov,[a] Janne Ruokolainen,[a] Marjo Pääkkö,[a] Janne Laine,[b] Markku Leskelä,[c] Lars A. Berglund,[d] Tom Lindström,[e] Gerrit ten Brinke,[f] Hermis Iatrou,[g] Nikos Hadjichristidis[g] and Charl F. J. Faul[h]

Received 13th March 2009, Accepted 31st March 2009
First published as an Advance Article on the web 6th August 2009
DOI: 10.1039/b905204f

Self-assemblies and their hierarchies are useful to construct soft materials with structures at different length scales and to tune the materials properties for various functions. Here we address routes for solid nanofibers based on different forms of self-assemblies. On the other hand, we discuss rational "bottom-up" routes for multi-level hierarchical self-assembled constructs, with the aim of learning more about design principles for competing interactions and packing frustrations. Here we use the triblock copolypeptide poly(L-lysine)-b-poly(γ-benzyl-L-glutamate)-b-poly(L-lysine) complexed with 2′-deoxyguanosine 5′-monophosphate. Supramolecular disks (G-quartets) stabilized by metal cations are formed and their columnar assembly leads to a packing frustration with the cylindrical packing of helical poly(γ-benzyl-L-glutamate), which we suggest is important in controlling the lateral dimensions of the nanofibers. We foresee routes for functionalities by selecting different metal cations within the G-quartets. On the other hand, we discuss nanofibers that are cleaved from bulk self-assemblies in a "top-down" manner. After a short introduction based on cleaving nanofibers from diblock copolymeric self-assemblies, we focus on native cellulose nanofibers, as cleaved from plant cell wall fibers, which are expected to have feasible mechanical properties and to be templates for functional nanomaterials. Long nanofibers with 5–20 nm lateral dimensions can be cleaved within an aqueous medium to allow hydrogels and water can be removed to allow highly porous, lightweight, and flexible aerogels. We further describe inorganic/

[a]Department of Applied Physics, Helsinki University of Technology, FIN-02015 TKK Espoo, Finland. E-mail: Olli.Ikkala@tkk.fi
[b]Department of Forest Products Technology, Helsinki University of Technology, FIN-02015 TKK Espoo, Finland
[c]Department of Chemistry, University of Helsinki, FIN-00014 Helsinki, Finland
[d]Department of Fiber and Polymer Technology, Royal Institute of Technology, SE 100 44 Stockholm, Sweden
[e]Innventia AB, P.O. Box 5604, SE-114 86 Stockholm, Sweden
[f]Zernike Institute for Advanced Materials, University of Groningen, 9747, AGGroningen, The Netherlands
[g]Department of Chemistry, University of Athens, Panepistimiopolis Zografou, 157 71 Athens, Greece
[h]School of Chemistry, University of Bristol, Bristol, UK BS8 1TS

organic hybrids as prepared by chemical vapour deposition and atomic layer deposition of the different nanofibers. We foresee functional materials by selecting inorganic coatings. Finally we briefly discuss how the organic template can be removed *e.g.*, by thermal treatments to allow completely inorganic hollow nanofibrillar structures.

Introduction

There exists extensive research towards self-assembled synthetic, biological, and bioinspired soft materials to explore concepts for structural control, increasing complexity, and to achieve various functionalities relevant in applications.[1–7] Self-assemblies can be achieved based on competing repulsive and attractive interactions, where the latter ones can be permanent covalent or weaker physical interactions.[8] Hierarchical structure formation takes place if the different constituent mechanisms act simultaneously at different length scales.[9,10] Nature provides a wealth of examples on self-assemblies allowing tailored materials properties, for example, based on proteins, tough and strong inorganic–organic hybrid structures, plant cell wall cellulosic structures, and multilevel inorganic/organic hierarchical fibrillar constructs.[5,11,12] The last two examples are particularly inspiring for the present paper, in our efforts to investigate different aspects of self-assemblies for rational solid nanofiber construction.

We will first discuss examples of "top-down" methods to cleave solid nanofibers from bulk host self-assemblies, which act as templates for their formation. An extensively used method to prepare nanofibers especially for bioapplications is provided by electrospinning, where even smaller fibers become cleaved due to splaying.[13,14] However, towards the rational use of self-assemblies to construct nanofibers, conceptually perhaps the simplest model material is provided by diblock copolymers[15,16] with hexagonally self-assembled cylindrical cores which can be cleaved by selective solvent processes after shear alignment.[17–19] Such fibers can be post-modified by various ways[20] and we will describe inorganic modification using atomic layer deposition[21] which is a self-limiting sequential chemical vapour deposition concept.[22–24] This allows high precision both in the organic and inorganic parts of the hybrids. On the other, plant cell wall cellulose can be a feasible starting material as it contains within its hierarchical structure mechanically strong native cellulose nanofibers with lateral dimensions of down to a few nm.[12] This in combination with its sustainability has spurred interest for different forms of cellulose in nanoscience and functional materials.[12,25–39] We discuss the cleavage of the native nanofibers (also denoted as microfibrils) to form hydrogels[40] and aerogels,[41] and as an example of their post-functionalization we discuss chemical vapour deposition with TiO_2. In more general terms, widely different functional materials are expected based on different post-modifications.

On the other hand, solid nanofibers can be constructed directly based on self-assemblies "in a bottom-up manner". Nanofiber or nanoribbon formation in aqueous, biological, and solvent environments have received extensive interest, for example by self-assembling helix-coil block copolypeptides, oligopeptide-containing amphiphiles, and even amyloids.[42–47] On the other hand, two-level self-assembled hierarchies and related functional properties have been demonstrated based on supramolecular combinations of block copolymers and surfactants, which allow combinations of structures on the 10–100 nm length scale of block copolymers with length scale structures an order of magnitude smaller than the latter ones.[4,7,9,10] Here we aim to generalize towards higher level hierarchical self-assemblies and structural control by combining several competing motifs, such as competition between disc-like and rod-like mesogens, polypeptides with α-helical, β-sheet and random conformations, additionally involving metal cation binding.[48] This is performed in the context of nanofibers.

Experimental

Sample preparation

Polymeric nanofibers were prepared based on polystyrene-*block*-poly(4-vinylpyridine) (PS-*b*-P4VP) diblock copolymer (M_{nPS} = 21.4 kg mol^{-1}, M_{nP4VP} = 20.7 kg mol^{-1}, M_w/M_n = 1.13, Polymer Source, Inc.), pentadecylphenol (PDP) or dodecylphenol (Aldrich), and poly(2,6-dimethyl-1,4- phenylene oxide) (PPE, M_w = 25.7 kg mol^{-1}, M_w/M_n = 1.37) as described in detail in ref. 17 and 18. The large amplitude shear alignment is made using a dynamic rheometer as also discussed in ref. 17 and 18.

The poly(L-lysine hydrochloride)-*b*-poly(γ-benzyl-d7-L-glutamate)-*b*-poly(L-lysine hydrochloride) (PLL-*b*-PBLG-*b*-PLL) triblock copolypeptide was synthesized from the corresponding precursor poly(ε-*tert*-butyloxycarbonyl-L-lysine)-*b*-poly(γ-benzyl-d7-L-glutamate)-*b*-poly(ε-*tert*-butyloxycarbonyl-L-lysine) by selective deprotection of the ε-amine group of ε-*tert*-butyloxycarbonyl-L-lysine. The precursor was synthesized by sequential ring opening polymerization of γ-benzyl-d7-L-glutamate *N*-carboxy anhydride and ε-*tert*-butyloxycarbonyl-L-lysine *N*-carboxy anhydride with the difunctional initiator 1,6-diaminohexane using high vacuum techniques.[49] The block lengths of PLL-*b*-PBLG-*b*-PLL are M_{nPBLG} = 31.0 kg mol^{-1} and $M_{nPLL}/2$ = 29.9 kg mol^{-1}, respectively, and the polydispersity index = 1.16.[50] 2'-Deoxyguanosine 5'-monophosphate (dGMP, sodium salt, Sigma–Aldrich) and PLL-*b*-PBLG-*b*-PLL in the HCl form were used to prepare complexes from 3mM KCl aq. solutions at pH 5.2 in a 1:1 (*vs* lysine residue) molar ratio. The concentration of the complex was 90 mM. The complexes precipitated immediately and were centrifuged in cold ethanol, rinsed three times with isopropyl alcohol/H$_2$O (30/70) mixture and three times with H$_2$O and dried at ambient conditions.

Native cellulose nanofibers were prepared from bleached sulfite softwood pulp (Domsjö ECO Bright; Domsjö Fabriker AB) consisting of 40% pine and 60% spruce with high hemicellulose content (13.8%) and low lignin content (1%). The fibrillation of the aqueous pulp was achieved by mechanical shearing, enzymatic hydrolysis by monocomponent endoglucanase (Novozym 476, Novozym A/S) followed by washing, mechanical shearing and high-pressure homogenization (Microfluidizer M-110EH, Microfluidics Corp.),[35] which leads to a hydrogel.[40] To prepare the aerogel, the aqueous gel was placed on a mould which was quickly plunged in liquid propane. Thereafter, the frozen sample in the mould was transferred into a vacuum oven and the sample was kept frozen during the drying by a massive cryogenically cooled copper plate underneath. The drying was finished when the pressure in the oven remained stable at *ca.* 10^{-2} mbar.

X-Ray scattering

The small angle X-ray measurements (SAXS) were performed using a Microstar microfocus X-ray source with a rotating anode (CuKα radiation, λ = 1.54Å) and Montel Optics. The magnitude of the scattering vector is given by $q = (4\pi/\lambda)\sin\theta$, where 2θ is the scattering angle. Wide angle X-ray (WAXS) measurements were performed at the HASYLAB at DESY (Hamburg), Beamline A2.

Chemical vapour deposition and atomic layer deposition

Atomic Layer Deposition (ALD): The Al$_2$O$_3$ films were deposited by ALD at 80 °C using trimethylaluminium and H$_2$O as reactants. The depositions were performed in a F-120 ALD reactor (ASM Microchemistry Ltd., Finland) under a pressure of 10 mbar with N$_2$ as the carrier and purging gas. One growth cycle consisted of a trimethylaluminium pulse (2 s), a N$_2$ purge (60 s), a H$_2$O pulse (0.5 s), and N$_2$ purge (150 s).

Chemical Vapor Deposition (CVD). TiO$_2$ films were deposited on dried cellulose nanofibers using an F-120 reactor (Microchemistry Ltd., Finland). The CVD process consisted of pre-heating of the support at 190 °C for 1.5 h; reaction of titanium isopropoxide, Ti(OC$_3$H$_7$)$_4$ at 190 °C and 1–5 kPa for 2 h, where the precursor was sublimated at 40 °C and carried with N$_2$ flow through the chamber; purging with N$_2$.

Electron microscopy

Bright-field TEM was performed using a Tecnai 12 microscope operating at an accelerating voltage of 120 kV. SEM was performed using Hitachi S-4700 FE-SEM and Leo Gemini DSM 982 microscopes.

Results and discussion

Solid state nanofibers by cleaving from synthetic and biological self-assemblies: "Top-down" nanofiber construction

We will now discuss preparation of solid state nanofibers by cleaving from bulk self-assemblies. Here the main emphasis is cleaving mechanically strong native nanocellulose fibers from macroscopic hierarchically ordered plant cell fibers, but as an introduction we will discuss rational constructs for fiber formation based on synthetic diblock copolymer templates (Fig. 1 Top Part). To this end, diblock copolymers are used which undergo cylindrical self-assembly in bulk. Herein, the cylindrical core is selected to be glassy polystyrene (PS) and it can be reinforced with poly(2,6-dimethyl-1,4-phenylene oxide) (PPE). PPE has a high glass transition temperature (T_g = 216 °C) and due to the molecular level miscibility with PS, its blending increases the glass transition temperature of the core material. Even more importantly, PPE promotes entanglements, and therefore addition of minor fractions of PPE leads to reinforcement, as discussed by van Zoelen et al.[18] In order to facilitate simple cleaving, the matrix phase is selected to be poly(4-vinylpyridine) (P4VP) as complexed with alkylphenol: It is well documented based on FTIR that phenols form hydrogen bonds with pyridines, see e.g. ref 51. Differential scanning calorimetry shows that alkylphenols, such as dodecylphenol (DDP) or pentadecylphenol (PDP) plasticize P4VP due to the supramolecular spacer-like side chains. If the nonpolar side chain is long enough, such as in PDP, lamellar self-assemblies of P4VP(PDP)$_{1.0}$ are obtained due to the long repulsive side chains.[51] In fact, in PS-b-P4VP(PDP)$_{1.0}$ a hierarchical self-assembly at the block copolymer length scale (10–100 nm) and surfactant length scale (ca. 3 nm) takes place.[52] For the present PS-b-P4VP and the nominally stoichiometric amount of DDP or PDP in comparison to P4VP repeat units, and optionally adding PPE (e.g. weight fraction 23% vs PS) one obtains cylindrical self-assemblies.[17,18,52] In order to have overall alignment, shear flow processing by either large amplitude dynamic rheometry or, more practically, by an extruder allows high overall alignment.[17,19] Finally, the nanofibers forming the self-assembled cylindrical cores can be cleaved by selective polar solvent treatment using ethanol, thus releasing long individualized nanofibers with a PS core and a P4VP corona, see Fig. 1.[17,18]

These nanofibers can be further functionalized. Here we emphasize that conformal inorganic/inorganic layers, mostly oxide, can be prepared in a well defined way with nanometer precision using ALD. The concept is a specific sequential form of CVD additionally incorporating self-limiting growth. In order to allow combination with the organic block copolymer template, materials and processes requiring low temperature ALD processes have to be selected. As a characteristic example, Al$_2$O$_3$ coating is prepared by exposing the nanofibers into a cycle of repeating trimethyl aluminium vapour and humidity, with an inert gas flushing in between. For example, after 200 cycles, Al$_2$O$_3$ with a thickness of ca. 15 nm is

Top-down nanofiber constructions, examples

1. Cleaving nanofibers from synthetic self-assemblies

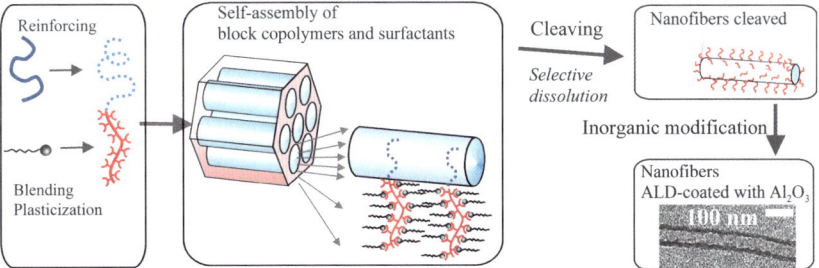

2. Cleaving nanofibers from biomatter hierarchical self-assemblies

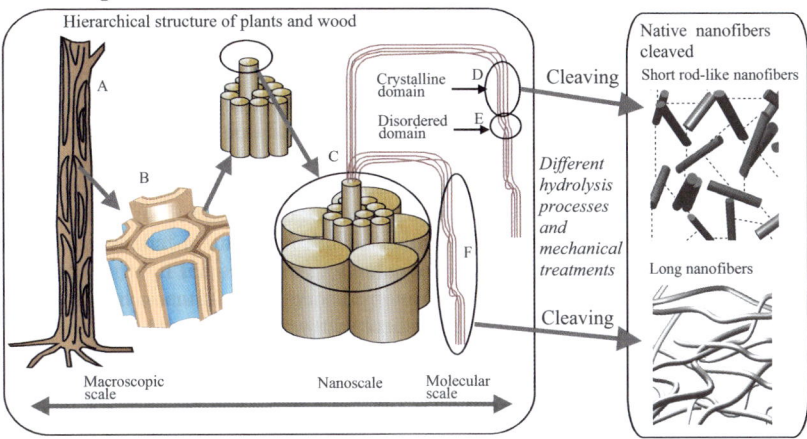

Fig. 1 Examples of cleaving nanofibers from bulk templates. 1: Schematics for a diblock copolymer self-assembled template consisting polystyrene-*b*-poly(4-vinylpyridine) where the polystyrene phase is reinforced with poly(2,6-dimethyl-1,4-phenylene oxide) and poly(4-vinylpyridine) is plasticized using dodecylphenol.[18] Atomic layer deposition allows conformal inorganic coatings where the polymer template can even be removed by heat treatment at the end.[21] 2: Cleaving native cellulose nanofibers from plant cell wall material. Strong acidic hydrolysis leads to rod-like highly crystalline cellulose I nanowhiskers, whereas milder hydrolysis and shearing allows long and entangled native nanofibers. The diameters of the fibers are in the nanometer range. Part of the latter scheme has been adapted from ref. 54.

achieved.[21] Importantly, the TEM micrograph suggests that the inorganic coating is continuous and non-granular (see Fig. 1).

The previous example indicates that combination of polymeric self-assembly and self-limiting inorganic vapour deposition techniques allows well defined inorganic/inorganic matter, where the prerequisite is existence of open surfaces available for the gas phase reactants of ALD, at least during some stage of the process. As ALD allows the construction of various well defined inorganic layers with dielectric and semiconducting properties, the concept paves the way towards functionalized nano-fibers and more generally functionalized inorganic/organic hybrids.

After the above model example, we next describe cleaving high-strength cellulose nanofibers from sustainable plant cell wall templates and their subsequent post-functionalization, of which inorganic CVD is taken as a specific example. We think that such concepts will have substantial importance as sustainable and bioinspired nanomaterials.

Plant cell walls incorporate macroscopic cellulosic fibers (Fig. 1 Bottom Part).[12] They are multicomponent complex materials with a hierarchical internal composition and structuring at different length scales takes place due to the naturally occurring self-assembly process. Even if the cellulose structure has been in many respects known for cellulose chemists for a long time,[53] the full structural subtleness and possibilities for nanomaterial science has only recently started to be appreciated. Cellulose consists of polysaccharide polymer chains consisting of β-(1→4)-D-glucose repeat units. Due to the biosynthesis, the cellulose chains pack in a specific crystalline I native form in parallel fashion and the chains are mutually interlocked by a substantial amount of hydrogen bonds based on the hydroxyl groups. Therefore, in a lose analogy, the hydrogen bonded structure between the chains in some respects resembles the β-sheets in proteins which act as reinforcements in proteinic materials. The mechanical properties resulting from this native cellulose I crystalline structure are not known in detail as yet, and have been evaluated so far indirectly. The stiffness, as specified by the Young's modulus, is expected to be in the range of 130 GPa.[55] This is high, compared with polymers which typically show 1–4 GPa, aromatic polyamides *ca.* 130 GPa, steel *ca.* 200 GPa, and carbon nanotubes and diamond near 1000 GPa. The strength is difficult to predict, but it is expected to be even up to the range of a few GPa,[39,56] which is comparable with steel *ca.* 0.5–2 GPa and carbon nanotubes, a few tens of GPa. For comparison it is also instructive to compare the predicted values to those of major ampullate silk which has a modulus of 10 GPa and a strength of 1.1 GPa. At the smallest length scale, the cellulose I crystals form nanoscale fibers, that are a few nanometers in the lateral dimension, depending on the cellulose source. They are connected to form long nanoscale fibers *via* disordered domains. Such long nanofibers, in turn, aggregate to form fiber bundles with disordered hemicellulose and lignins. These, in combination with disordered matter are combined at the largest length scale to form the macroscopic cellulose fibers (Fig. 1), which have commonly been used in *e.g.* in papermaking. In conclusion, the macroscopic cellulose fibers have a hierarchically self-assembled structure, which at the lowest level of hierarchy have nanometer fibers with a cellulose I native crystalline structure with expected feasible mechanical properties.

The problem is to prepare distinct cellulose I containing nanofibers. Common procedures to dissolve cellulose or to chemically modify the repeat units generally led to amorphous material or other non-native crystal structures with less than expected optimal mechanical properties. It had already been recognized early on[57,58] that a way could be to cleave the strong nanofibers from the macroscopic fibers by mechanical shearing. In the early days, this typically led to nonuniform materials and the processes were not of extensive practical importance. Recently several interesting methods have been developed based on cleaving the native crystalline cellulose nanofibers by controlled chemical, biochemical, or mechanical treatments that disintegrate the weaker constituents. Acidic hydrolysis leads to extensive hydrolysis of all disordered matter between the nanofibers and also within the disordered domains along the nanofibers, thus leaving only rod-like cellulose nanowhiskers, Fig. 1.[26] They are highly crystalline and have diameters of the order of a few nanometers. Due to their rod-like character and surface charges, they can form liquid crystalline solutions. Here we emphasize that longer, coiled, and entangled nanofibers (Fig. 1) can be obtained if only the interfibrillar disordered matter is disintegrated. Even if purely mechanical treatments were shown to cleave the nanofibers early on, it is only more recently that practical preparations have been developed, *e.g.* based on combination of enzymatic hydrolysis and shearing.[35,40] Here we will discuss efforts to construct nanostructured materials based on such techniques.

As described in the experimental part, following the enzymatic treatment and extensive shearing, the material exists as a dilute mixture in water, typically having a concentration of 0.1–6%wt. Fig. 2A depicts a cryo-TEM micrograph of the aqueous sample, showing a well defined nanofibrillar structure with a lateral dimension of down to *ca.* 5–6 nm, however, thicker occasional bundles are also detected.[40] The micrograph indicates long and entangled nanofibers. This manifests as gelation

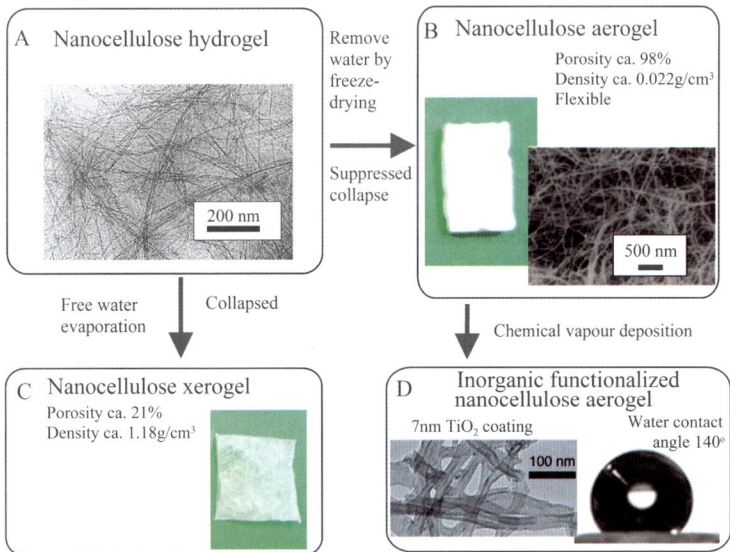

Fig. 2 Native cellulose nanofibers containing cellulose I crystal structure as cleaved from macroscopic cellulose fibers by a combination of enzymatic treatment and shearing. (A) Cryo-TEM of a (aqueous) hydrogel at a concentration of 2%wt, showing mostly 5–6 nm native cellulose nanofibers.[40] (B) Freeze drying to remove water allows lightweight aerogels.[41] Some aggregation of the nanofibers takes place during the water removal. (C) Free evaporation of water from hydrogels leads to a major collapse of the network structure. The samples even start showing some translucence. (D) The aerogels can be modified based on CVD. TEM illustrates TiO$_2$ coating which leads to a high contact angle by the combined effect of the surface topography and the coating.

in aqueous medium even by visual inspection. In dynamic rheology the storage and loss moduli are essentially independent of frequency for the whole investigated range of concentration from 0.125–6.5%wt and the loss modulus is more than an order of magnitude smaller than the storage modulus. The feasible mechanical properties of the nanofibers manifest themselves as a high value of the storage modulus of 2 MPa for the concentration of 2%wt. This is to be compared with a typical modulus of a rubber. In summary, the long and entangled cellulose nanofibers form a strong aqueous gel.[40]

Here we address the exploitation and functionalization of such nanofibers: for that end the water medium from the aqueous hydrogel is first removed. First, if the water of the hydrogel is just removed by evaporation from the liquid state, a compacted material results due to the collapse of the hydrogel network due to the high surface tension of the water surface receding throughout the sample in the process of drying. The resulting density and porosity depend on the exact method of evaporation, but typically it is $ca.$ 1.18 g cm^{-3} and the porosity is around 21% as measured using Hg porosimetry and the material can be denoted as a xerogel (Fig. 2C). After such a drying, the xerogel shows some degree of translucence and sheets thereof show a tensile modulus of $ca.$ 6 GPa, a tensile strength of $ca.$ 75–80 MPa, and a maximum strain of 2.5%. Such material properties start to be comparable with commodity polymers but obviously such a simple evaporation under laboratory conditions does not allow exploitation of the full mechanical potential of a cellulose nanofiber network. In fact, optimization of the materials and processes allows considerable increases in the values, as shown by Berglund *et al.*: the modulus can be increased to $ca.$ 13 GPa and the strength to in excess of 200 MPa by nanocellulose fibers with charged surface groups and using solvent exchange techniques.[36] Also, using various types of nanocellulose, highly transparent films have been obtained by infiltrating acrylate polymers within the nanocellulose followed by compaction by pressing.[37] Note that the diameter of the fiber is much smaller than the wavelength of light. On the other hand, if the water is removed in the frozen state by using freeze drying, the open hydrogel network structure is essentially preserved even in the dried state without a major collapse.[41] For example, if the hydrogel is simply plunged in liquid propane and the water is subsequently removed in a vacuum oven while keeping the sample frozen, highly porous materials with a very small density of 0.022 g cm^{-3} and high porosity of 98% are obtained (Fig. 2B). Such lightweight and highly porous materials are denoted as aerogels. We point out that aerogel-based inorganic materials, such as silica, as prepared by sol–gel methods form an important class of materials that are extensively used in applications.[59] A common problem therein is that they are very brittle, typically capable of only a fraction of % deformation. However, the present nanocellulose aerogels exhibit a very high deformability without breaking. For example 0.5 mm thick sheets can be reversibly bent back and forth and upon compression the maximum strain is more than 70%; such ultimate compressive strains, however, do not lead to fully reversible deformations.

The nanocellulose aerogels consist of a network of native cellulose nanofibers of diameter of *ca.* 30 nm, see Fig. 2B. Therefore some aggregation has taken place in the process of water removal. The nanofibers have a high density of hydroxyl groups on their surfaces due to the β-(1→4)-D-glucose repeat units. TiO_2 is an interesting oxide material to investigate for functionalization, *e.g.* due to its photocatalytic properties, potential to control wettability, and its possibilities for device making. In an effort to functionalize the native cellulose nanofibers using inorganic matter, CVD of TiO_2 was performed. Dried aerogel was coated by CVD at 190 °C using titanium isopropoxide, $Ti(OC_3H_7)_4$. The detailed analysis is presently in progress based on *e.g.* XPS and spectroscopies and they preliminarily point towards TiO_2 layers. TEM shows a well defined layer of *ca.* 7 nm deposited material on the surface of the native cellulose nanofibers (Fig. 2D). The coating manifests in the wetting behaviour, as expected for TiO_2: A pristine aerogel absorbs a water droplet instantly, so that one can even classify the nanocellulose aerogel as superhydrophilic and superabsorbent. On the other hand, the aerogel with deposited inorganic coating leads to a high contact angle of 130–140°. This is quite high, indicating that the nanocellulose aerogel surface has become extremely hydrophobic. Such high contact angles, approaching those of superhydrophobic materials, are manifestations of a combined effect of surface chemical functionalization and an additional nanoscopic and potentially hierarchical surface topography.

We have shown two routes for organic nanofiber cleaving, based on synthetic block copolymeric bulk self-assembled templating as well as biomatter plant cell fiber templating. We showed that these nanofibers, in turn, are useful templates to grow inorganic coatings. In particular, the ALD is useful for allowing self-limiting well controlled conformal growth. Finally, the organic nanofibrillar template can also be removed, thus allowing us to construct purely inorganic nanoscale hollow objects. We expect that such sequential template routes open up new possibilities for *e.g.* inorganic/organic hybrids, highly porous material and materials for semiconductor devices.

Solid state nanofibers by rational hierarchical construction: "Bottom-up nanofiber construction"

Here we address solid state nanofibrillar constructs containing competing supramolecular discotic and helical rod-like mesogenic groups, which open up systematic ways for more general multilevel hierarchical self-assemblies.[48] In the aqueous and organic solvent medium, there exist extensive efforts to construct nanofibers and nanoribbons based on self-assemblies: One example is provided by diblock copolypeptides with rod-like α-helical blocks and coil-like blocks. Therein, the lateral dimension of the fibers is controlled by the packing frustration. The α-helical chains have a strong tendency to pack with a small lateral dimension whereas the coiled blocks take more lateral space, depending on the stretching and related entropy.[46] Another approach consists of amphiphiles with oligopeptidic blocks and alkyl blocks, also leading to fiber formation.[42] Rod–coil self-assemblies allow ribbon formation in organic solvents,[43] and nanofibers of rod–coil block copolymers containing conjugated blocks have been investigated *e.g.* for electroactive materials on substrates.[60]

Previously, concepts have been developed for two level self-assemblies by supramolecular combination of block copolymers and surfactants, all based on flexible chains, where the side chains are bonded using *e.g.* hydrogen bonding or ionic bonding.[4,7,52,61] The self-assembly is based on the "chemical" contrast between the three types of constituents. The side-chains can also be rod-like mesogenic, thus incorporating conformation contrast between the constituents to drive towards self-assembling hierarchies.[8] For generalization, we have started to investigate higher level hierarchies by making use of the packing frustration of disks, rods, and coils, in combination with the polypeptide conformational control.[48]

Fig. 3 shows one form which uses packing frustration between rods and disks to allow nanofiber formation. The block copolypeptide (Fig. 3A) consisting of a α-helical poly(γ-benzyl-L-glutamate) (PBLG)[62] central block and two poly(L-lysine) (PLL) hydrochloride endblocks, with molecular weights of 29.9–31.0–29.9 kg mol^{-1}. It has been synthesized using high vacuum techniques using 1,6-diaminohexane initiator as described in the experimental part. Due to the difunctional initiator, the central PBLG block, in fact, consists of two PBLG blocks with a short flexible hexyl spacer in between, see Fig. 3A. For shorthand notation, the block polypeptide is denoted as PLL-*b*-PBLG-*b*-PLL as the overall effect of the short spacer on the central block conformation is expected to be minor, as typically the PBLG helices can have "kinks" anyway within the α-helical conformation.[63] The poly(L-lysine hydrochloride) endblocks are ionically complexed with 2′-deoxyguanosine 5′-monophosphates (dGMP) (Fig. 3B) in aqueous solution containing 3 mM KCl, whereupon segregation takes place after their combination. After centrifuging, rinsing and drying (see experimental part), fibrillation in the solid state was observed using TEM (see Fig. 3G, to be

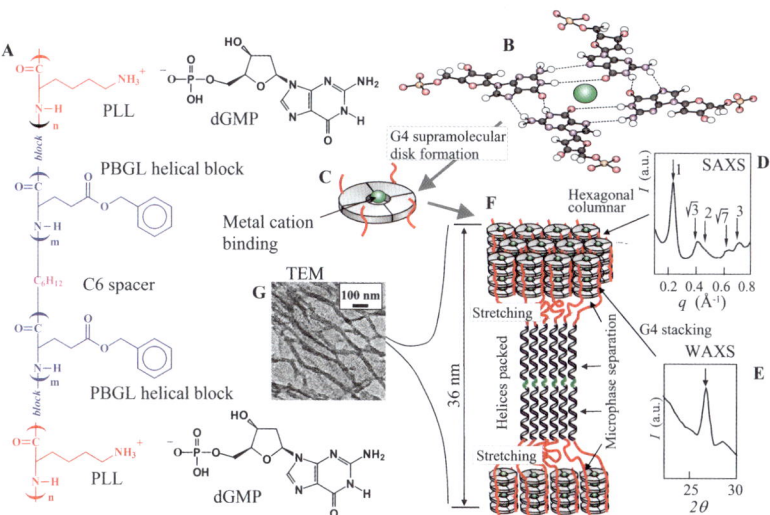

Fig. 3 Hierarchically self-assembled constructs for nanofiber formation,[48] based on competing supramolecular discotic motifs (G-quartets) and helical rod-like motifs (PBLG), block copolypeptide PLL-*b*-PBLG-*b*-PLL microphase separation and metal ion binding.

discussed in more detail later) and also investigated using SAXS and FTIR. For short-hand notation, the resulting ionically complexed adduct is denoted as PLL(dGMP)-b-PBLG-b-PLL(dGMP).

The structural hierarchy will next be followed step-by-step from the smallest structures up to the largest ones. First, existence of ionic complexation between the phosphoric acid head groups of dGMP and basic lysines of PLL (Fig. 3A) is suggested in FTIR showing distinct absorption peaks at 1059 cm^{-1} due to asymmetric stretching and at 970 cm^{-1} due to the symmetric stretch vibration characteristic for phosphates.[48,64] Elemental analysis shows that in the complex the degree of complexation is 85% vs the number of lysine groups. Later we suggest a possible reason for the less than nominally complete complexation, as there has to be an uncomplexed interface region between the block copolypeptide domains, see Fig. 3F later. In general, it is well known that four guanosines tend to form supramolecular discs, denoted as G-quartets or G4, based on 8 hydrogen bonds incorporating so-called Hoogsteen pairings (Fig. 3B and 3C).[65] Such supramolecular discs have a diameter of $ca.$ 2.5 nm. Direct spectroscopic evidence is not straightforward to achieve in the present complicated material. However, indirectly the G-quartet formation is unambiguously shown by SAXS (see Fig. 3D), which indicates characteristic reflections at $q_1^* = 0.24$ Å^{-1}, $\sqrt{3}q_1^*$, $2q_1^*$, $\sqrt{7}q_1$, and $3q_1^*$, which can be assigned to hexagonal cylindrical packing with a cylinder-to-cylinder distance of $ca.$ 3.0 nm. That the cylinder diameter is slightly larger than the diameter of the G-quartets, can be explained due to the PLL complexed with the dGMP within the intercolumnar regions (Fig. 3F). One aspect has to be emphasized: added metal salts are needed to stabilize complexes. In the simplest form, the salt selected is KCl, later we suggest that other salts can also be used to allow functionalities. For PLL(dGMP)-b-PBLG-b-PLL(dGMP) as prepared in the presence of KCl, WAXS shows a relatively narrow peak at $ca.$ 27° (Fig. 3E), which is typical for π-stacking of dimensions 0.33 nm. This indicates that the supramolecular G4 disks stack to form columns (Fig. 3F), as promoted by the cation–dipole interaction between the K$^+$ cations within the G-quartets. Based on the FTIR evidence on the PLL/dGMP ionic interaction, these columnar assemblies are located in the domains also containing the PLL blocks.

On the other hand, FTIR shows that there are α-helices (1650 cm^{-1} and 1544 cm^{-1}), β-sheets (1693 cm^{-1}, 1631 cm^{-1}, and 1529 cm^{-1}), and random coil (1654 cm^{-1} and 1535 cm^{-1}) conformations within the complex.[48] It is most natural to assess the α-helical conformation to the PBLG, as PBLG is a prototypical coil-forming polypeptide.[62,66] The hexagonal packing of the PBLG helices is supported by X-ray scattering: WAXS shows small but clear reflections that could be assigned as $\sqrt{3}q_2^*$, $2q_2^*$, and $\sqrt{7}q_2$, where the main peak $q_2^* = ca.$ 0.45 Å^{-1} is within the SAXS regime and is masked within a composite reflection peak and cannot be resolved separately. This indicates hexagonal order at the periodicity of 1.3 nm that is close to the value expected for PBLG cylinders.[67] Taken these aspects into account, the central PBLG block has a smaller lateral periodicity due to packed rod-like helical chains whereas the end blocks have a larger periodicity due to the supramolecular disk-like entities. This leads to packing frustration between the PLL(dGMP) end blocks and PBLG central block, while the central and end blocks must microphase separate due to their vastly different natures. TEM (Fig. 3G) gives a hint how the self-assembly is achieved: It indicates fiber formation where the maximum lateral dimension is $ca.$ 36 nm. This is closely what would be expected for the PLL-b-PBLG-b-PLL of the present molecular weight and α-helical central blocks. We also point out that in spite of various staining protocols, the internal structure of the fibers could not be resolved. Neither is SAXS useful for structural assessment due to the small number of block copolypeptide layers within the fibers. Therefore, the exact details leading from packing shown in Fig. 3F to the fibers shown in Fig. 3G are not yet fully elucidated. Finally, we recall that the degree of complexation of dGMP was less than nominal based on the elemental analysis (85%). Taking Fig. 3F, this would be completely expected, as part of the PLL chain should remain uncomplexed to facilitate connection between the discotic and α-helical domains.

In summary, we suggest that the fiber formation is a manifestation of multilevel hierarchical self-assembly. Recently the G4-assemblies have attracted interest as functional units in assemblies beyond biochemistry.[68] We foresee that the metal cations bound in the cores of the G4-disks obviously due to cation–dipole interactions open up a platform for functionalities: Preliminary results indicate that besides K^+, e.g. Fe^{3+} and Tb^{3+} can be loaded within the self-assemblies and investigations are in progress to explore the potential functionalities. We think that judicious selection of the metal cation can open up interesting applications, for example in redox controlled fibers.

Conclusion

We have discussed two routes for nanoscopic solid state fibers based on self-assemblies. The first one can be denoted as a "top down concept", where constituent smaller scale nanofibers are cleaved from macroscopic synthetic or biological bulk self-assemblies. The first example deals with shear aligned diblock copolymeric self-assemblies where the cylindrical cores are cleaved to form distinct fibers. There is more emphasis here on native cellulose nanofibers upon cleaving from macroscopic fibers. Such fibers are expected to have extraordinary mechanical properties and they can allow a useful template for functional nanomaterials. The nanofibers can be post-modified for functionalities using several methods. From the many possibilities, we concentrate here on chemical vapour deposition and atomic layer deposition. In particular, the latter is very feasible in connection with the soft matter and block copolymer self-assemblies, atomic layer deposition is a self-limiting controlled synthesis of inorganic matter. We also describe "bottom-up" solid state nanofibers, based on several length scale self-assemblies. We describe how combining several competing motifs in a rational way, higher level hierarchical self-assemblies are obtained for nanofibrillar constructs, for example by competing discotic and rod-like mesogens, various polypeptide conformations, and microphase separations. We expect interesting new developments for functional materials.

Acknowledgements

We thank P. Hiekkataipale, S. Hanski, S. Funari, M. Kemell, J. de Wit, M. Ritala, W. de Zoelen, G.A. Alberda van Ekenstein, E. Polushkin, H. Nijland, M. Ankerfors, H. Kosonen, A. Nykänen, S. Ahola, M. Österberg, P.T. Larsson, R. Silvennoinen, J. Vapaavuori, V. Pore, L. Wågberg, and J. Sainio for discussions and assistance on the various aspects of the present work. We acknowledge funding from Marie Curie Network 'BioPolySurf', Vinnova, Finnish National Agency for Technology and Innovation, Academy of Finland, UPM, and NOKIA Research Center.

References

1 G. M. Whitesides and B. Grzybowski, *Science*, 2002, **295**, 2418–2421.
2 I. W. Hamley, *Angew. Chem., Int. Ed.*, 2003, **42**, 1692–1712.
3 F. J. M. Hoeben, P. Jonkheijm, E. W. Meijer and A. P. H. J. Schenning, *Chem. Rev.*, 2005, **105**, 1491–1546.
4 O. Ikkala and G. ten Brinke, *Chem. Commun.*, 2004, 2131–2137.
5 M. A. Meyers, P.-Y. Chen, A. Y.-M. Lin and Y. Seki, *Prog. Mater. Sci.*, 2008, **53**, 1–206.
6 C. Park, J. Yoon and E. L. Thomas, *Polymer*, 2003, **44**, 6725–6760.
7 G. ten Brinke, J. Ruokolainen and O. Ikkala, *Adv. Polym. Sci.*, 2007, **207**, 113–177.
8 M. Muthukumar, C. K. Ober and E. L. Thomas, *Science*, 1997, **277**, 1225–1232.
9 O. Ikkala and G. ten Brinke, *Science*, 2002, **295**, 2407–2409.
10 J. Ruokolainen, R. Mäkinen, M. Torkkeli, T. Mäkelä, R. Serimaa, G. ten Brinke and O. Ikkala, *Science*, 1998, **280**, 557–560.

11 J. Aizenberg, J. C. Weaver, M. S. Thanawala, V. C. Sundar, D. E. Morse and P. Fratzl, *Science*, 2005, **309**, 275–278.
12 D. Klemm, B. Heublein, H.-P. Fink and A. Bohn, *Angew. Chem., Int. Ed.*, 2005, **44**, 3358–3393.
13 D. H. Reneker and A. L. Yarin, *Polymer*, 2008, **49**, 2387–2425.
14 G. C. Rutledge and S. V. Fridrikh, *Adv. Drug Delivery Rev.*, 2007, **59**, 1384–1391.
15 F. S. Bates and G. H. Fredrickson, *Phys. Today*, 1999, **52**, 32–38.
16 I. W. Hamley, *The Physics of Block Copolymers*, Oxford University Press, Oxford, 1998.
17 G. Alberda van Ekenstein, E. Polushkin, H. Nijland, O. Ikkala and G. ten Brinke, *Macromolecules*, 2003, **36**, 3684–3688.
18 W. van Zoelen, G. Alberda van Ekenstein, E. Polushkin, O. Ikkala and G. ten Brinke, *Soft Matter*, 2005, **1**, 280–283.
19 A. W. Fahmi, H. Bruenig, R. Weidisch and M. Stamm, *Macromol. Mater. Eng.*, 2005, **290**, 136–142.
20 A. W. Fahmi, H.-G. Braun and M. Stamm, *Adv. Mater.*, 2003, **15**, 1201–1204.
21 R. H. A. Ras, M. Kemell, J. de Wit, M. Ritala, G. ten Brinke, M. Leskelä and O. Ikkala, *Adv. Mater.*, 2007, **19**, 102–106.
22 M. Leskela and M. Ritala, *Angew. Chem., Int. Ed.*, 2003, **42**, 5548–5554.
23 R. L. Puurunen, *J. Appl. Phys.*, 2005, **97**, 121301.
24 M. Ritala and M. Leskelä, in *Handbook of Thin Film Materials*, ed. H. S. Nalwa, Academic Press, San Diego, 2002, pp. 103–159.
25 J.-F. Revol, L. Godbout, X. M. Dong, D. G. Gray, H. Chanzy and G. Maret, *Liq. Cryst.*, 1994, **16**, 127–134.
26 K. Fleming, D. G. Gray and S. Matthews, *Chem.–Eur. J.*, 2001, **7**, 1831–1835.
27 A. N. Nakagaito and H. Yano, *Appl. Phys. A: Mater. Sci. Process.*, 2004, **78**, 547–552.
28 M. A. S. A. Samir, F. Alloin and A. Dufresne, *Biomacromolecules*, 2005, **6**, 612–626.
29 T. Saito, S. Kimura, Y. Nishiyama and A. Isogai, *Biomacromolecules*, 2007, **8**, 2485–2491.
30 O. van den Berg, J. R. Capadona and C. Weder, *Biomacromolecules*, 2007, **8**, 1353–1357.
31 J. R. Capadona, O. van den Berg, L. A. Capadona, M. Schroeter, S. J. Rowan, D. J. Tyler and C. Weder, *Nat. Nanotechnol.*, 2007, **2**, 765–768.
32 A. J. Svagan, M. A. S. A. Samir and L. A. Berglund, *Biomacromolecules*, 2007, **8**, 2556–2563.
33 J. R. Capadona, K. Shanmuganathan, D. J. Tyler, S. J. Rowan and C. Weder, *Science*, 2008, **319**, 1370–1374.
34 S. Ahola, M. Österberg and J. Laine, *Cellulose*, 2008, **15**, 303–314.
35 M. Henriksson, G. Henriksson, L. A. Berglund and T. Lindström, *Eur. Polym. J.*, 2007, **43**, 3434–3441.
36 M. Henriksson, L. A. Berglund, P. Isaksson, T. Lindström and T. Nishino, *Biomacromolecules*, 2008, **9**, 1579–1585.
37 M. Nogi and H. Yano, *Adv. Mater.*, 2008, **20**, 1849–1852.
38 L. Wågberg, G. Decher, M. Norgren, T. Lindström, M. Ankerfors and K. Axnäs, *Langmuir*, 2008, **24**, 784–795.
39 M. Nogi, S. Iwamoto, A. N. Nakagaito and H. Yano, *Adv. Mater.*, 2009, **20**, 1–4.
40 M. Pääkkö, M. Ankerfors, H. Kosonen, A. Nykänen, S. Ahola, M. Österberg, J. Ruokolainen, J. Laine, P. T. Larsson, O. Ikkala and T. Lindström, *Biomacromolecules*, 2007, **8**, 1934–1941.
41 M. Pääkkö, J. Vapaavuori, R. Silvennoinen, H. Kosonen, M. Ankerfors, T. Lindström, L. A. Berglund and O. Ikkala, *Soft Matter*, 2008, **4**, 2492–2499.
42 J. D. Hartgerink, E. Beniash and S. I. Stupp, *Proc. Natl. Acad. Sci. U. S. A.*, 2002, **99**, 5133.
43 L. C. Palmer and S. I. Stupp, *Acc. Chem. Res.*, 2008, **41**, 1674–1684.
44 T. J. Deming, *Adv. Drug Delivery Rev.*, 2002, **54**, 1145–1155.
45 A. P. Nowak, V. Sato, V. Breedveld and T. J. Deming, *Supramol. Chem.*, 2006, **18**, 423–427.
46 T. J. Deming, *Prog. Polym. Sci.*, 2007, **32**, 858–875.
47 T. P. Knowles, A. W. Fitzpatrick, S. Meehan, H. R. Mott, M. Vendruscolo, C. M. Dobson and M. E. Welland, *Science*, 2007, **318**, 1900–1903.
48 N. Houbenov, A. Nykänen, H. Iatrou, J. Ruokolainen, N. Hadjichristidis, C. F. J. Faul and O. Ikkala, *Adv. Funct. Mater.*, 2008, **18**, 2041–2047.
49 T. Aliferis, H. Iatrou and N. Hadjichristidis, *Biomacromolecules*, 2004, **5**, 1653–1656.
50 H. Iatrou, H. Frielinghaus, S. Hanski, N. Ferderigos, J. Ruokolainen, O. Ikkala, D. Richter, J. Mays and N. Hadjichristidis, *Biomacromolecules*, 2007, **8**, 2173–2181.
51 J. Ruokolainen, G. ten Brinke, O. Ikkala, M. Torkkeli and R. Serimaa, *Macromolecules*, 1996, **29**, 3409–3415.
52 J. Ruokolainen, M. Saariaho, O. Ikkala, G. ten Brinke, E. L. Thomas, M. Torkkeli and R. Serimaa, *Macromolecules*, 1999, **32**, 1152–1158.
53 A. C. O'Sullivan, *Cellulose*, 1997, **4**, 173–207.

54 T. Zimmerman, E. Pöhler and T. Geiger, *Adv. Eng. Mater.*, 2004, **6**, 754–761.
55 T. Nishino, K. Takano and K. Nakamae, *J. Polym. Sci., Part B: Polym. Phys.*, 1995, **33**, 1647–1651.
56 D. H. Page and F. EL-Hosseny, *J. Pulp Paper Sci.*, 1983, **9**, 99–100.
57 F. W. Herrick, R. L. Casebier, J. K. Hamilton and K. R. Sandberg, *J. Appl. Polym. Sci.: Appl. Polym. Symp.*, 1983, **37**, 797–813.
58 A. F. Turbak, F. W. Snyder and K. R. Sandberg, *J. Appl. Polym. Sci.: Appl. Polym. Symp.*, 1983, **37**, 815–827.
59 A. C. Pierre and G. M. Pajonk, *Chem. Rev.*, 2002, **102**, 4243–4265.
60 P. Leclere, E. Hennebicq, A. Calderone, P. Brocorens, A. C. Grimsdale, K. Müllen, J. L. Bredas and R. Lazzaroni, *Prog. Polym. Sci.*, 2003, **28**, 55–81.
61 S. Hanski, N. Houbenov, J. Ruokolainen, D. Chondronicola, H. Iatrou, N. Hadjichristidis and O. Ikkala, *Biomacromolecules*, 2006, **7**, 3379–3384.
62 See *e.g.* G. Kess and R. S. Porter, in *Mechanical and Thermophysical Properties of Polymer Liquid Crystals*, ed. W. Brostow, Chapman & Hall, London, 1998, pp. 342–406.
63 P. Papadopoulos, G. Floudas, I. Schnell, T. Aliferis, H. Iatrou and N. Hadjichristidis, *Biomacromolecules*, 2005, **6**, 2352–2361.
64 B. Ozer, C. F. J. Faul, B. Smarsly and M. Antonietti, *Soft Matter*, 2006, **2**, 329–336.
65 J. T. Davis, *Angew. Chem., Int. Ed.*, 2004, **43**, 668–698.
66 P. Papadopoulos, G. Floudas, H.-A. Klok, I. Schnell and T. Pakula, *Biomacromolecules*, 2004, **5**, 81–91.
67 E. A. Minich, A. P. Nowak, T. J. Deming and D. J. Pochan, *Polymer*, 2004, **45**, 1951–1957.
68 H. Cohen, T. Sapir, N. Borovok, T. Molotsky, R. Di Felice, A. B. Kotlyar and D. Porath, *Nano Lett.*, 2007, **7**, 981–986.

Nanoparticles at electrified liquid–liquid interfaces: new options for electro-optics

M. E. Flatté,[a] A. A. Kornyshev[b] and M. Urbakh[c]

Received 20th January 2009, Accepted 6th April 2009
First published as an Advance Article on the web 24th July 2009
DOI: 10.1039/b901253m

We describe the results of a theoretical analysis of the localization of functionalized metal or semiconductor nanoparticles at the interface of two immiscible electrolytic solutions and discuss various options that this interface may offer for a new kind of self-assembled, electro-optic devices.

Functionalization of liquid interfaces is driven by aspirations to build self-assembled soft nanostructures with unique properties, accessible to light from both sides of the interface. Adsorption of nanoparticles, which may change the optical properties of the interface, is an example of such functionalization. Interesting new developments take place in electrochemical liquid–liquid systems, consisting of two immiscible electrolytic solutions that form an interface impermeable to ions unless a large voltage (>1 V) is applied across the interface. Functionalizing interfaces between two immiscible electrolytic solutions (ITIES) with nano-objects may lead to novel optical, sensing, or catalytic applications, or even to creating objects of scientific curiosity, such as optical molecular machines.[1,2,3,4]

When two immiscible liquids come in contact, one aqueous and one organic, and these liquids are mixed with salts such that there are hydrophilic ions in the water and hydrophobic ions in the oil, a sharp interface is formed.[5,6] When a voltage difference is applied across the bulks of the two liquids, two "back-to-back" electrical double layers are formed on the two sides of the interface (see Fig.1). This ITIES can be very thin ($\sim$1 nm, depending on salt concentrations) and robust to voltage (supporting a voltage up to $\sim$0.8 V without ion flow across the interface).[5,6] The resulting electric fields exceed typical breakdown voltages in semiconductors (usually $\sim$0.02 V nm^{-1}) and are similar to those of good oxides (such as silicon dioxide or aluminum oxide, $\sim$1 V nm^{-1}), but with much lower leakage currents below breakdown.

Nanoparticles can be dissolved in either the water or oil phase depending on the functionalization of their surfaces. If they are functionalized by hydrophobic ligands they will tend to reside in oil. For us, more interesting will be the case when the ligands have hydrophilic acidic terminal groups that could dissociate in water. The number of charges per particle in equilibrium depends on the pH and acid ionization constants. The charge of these groups prevents agglomeration of nanoparticles and it gives a unique opportunity for the coupling to electric field, which will allow manipulating their localization at the interface with variation of the bias voltage. The electric field at the interface, along with the other forces acting on the nanoparticles, cause nanoparticles to be drawn to and localized at the interface.[7,8,9] Thus, the nanoparticle surface concentration adjusts subject to the applied voltage,[8,9] permitting the reversible assembly and disassembly of nanoparticles at the ITIES.

---

[a]*Department of Physics & Astronomy, University of Iowa, Iowa City, IA, 52242, USA*
[b]*Department of Chemistry, Imperial College, London, United Kingdom SW7 2AZ*
[c]*School of Chemistry, Tel Aviv University, Ramat Aviv, 69978, Israel*

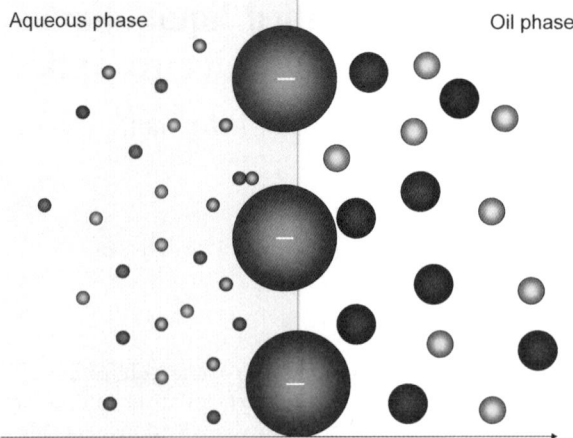

Fig. 1 A cartoon of ITIES with nanoparticles adsorbed at the interface (in fact nanoparticles are much larger than the ions).

At normal pH the net charge of the nanoparticles is typically much smaller than the number of ligands. The potential well at the ITIES for small nanoparticles (of the order of one or several nm) is formed mainly by the balance between solvation (which drives nanoparticles towards the aqueous phase) and the electric field (which, for the appropriate sign, will drive them towards the organic phase). The well is deepened by surface tension and nanoparticle polarizability.[10]

The potential well for particles of ten and several tens of nanometres is dominated by the effect of surface tension. A particle localized at the interface replaces a section of water–oil interface by two interfaces, those of particle–water and particle–oil. The sum of the surface tensions of the latter for silver and gold, and some other metals, is smaller than that of the replaced oil–water interfaces plus its surface with water. This effect gives a very strong incentive for the particle to localize at the interface without any bias fields. In order to move the particle away from the surface, one must apply a strong field of the corresponding sign. Since sustainable voltages are limited at the ITIES, this will be possible if most of the ligand's acidic groups are dissociated. Estimates show that even 50 nm particles can be removed from the surface to water bulk by sustainable fields at complete dissociation of ligands[11] which is possible at high pH.[12] In this paper we will discuss mainly small nanoparticles, the field-induced localization of which at ITIES has been already experimentally proven.[8,9] Consideration of localization of large nanoparticles will be reported in a forthcoming paper,[11] which will also show several astounding optical effects that could be achieved manipulating with large nanoparticles.

As just mentioned, the potential well for small nanoparticles at the interface is much less deep than that for large nano- and colloidal microparticles, as the surface tension energy gain from particle adsorption scales as the square of the particle radius.[13] Thus, multiply-charged small nanoparticles prefer to stay in the water phase. Polarizing water negatively with respect to oil will result in an electric field across the interface that will force the nanoparticles towards the oil and localizes them at the interface. Our estimates[10] show that the electric fields available for ITIES are sufficient to permit a stable, substantial concentration of nanoparticles at the interface, and that switching of this electric field can reversibly control the well depth and therefore the concentration at the interface of nanoparticles.

Altogether, the confining potential for the charged nanoparticles at ITIES consists of: competitive wetting, line tension, the solvation energy, the potential contribution from the applied electric field, and the nanoparticle polarizability in an external field. The last effect is usually negligible. Fig. 2 shows estimates of the other contributions

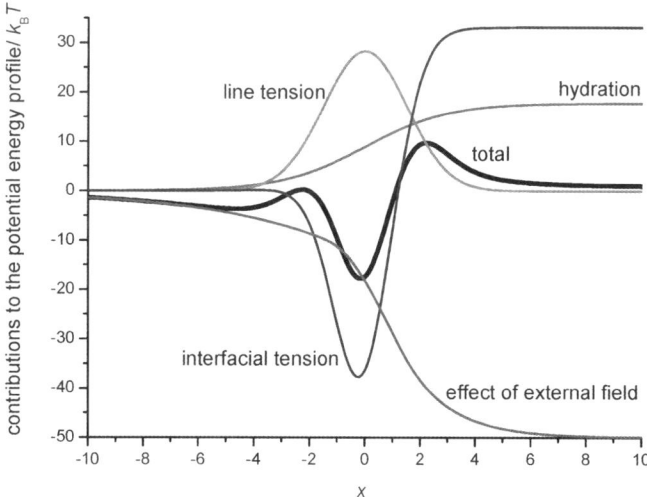

Fig. 2 Different contributions to the nanoparticle energy profile at ITIES, compared to the total profile. Parameters: potential drop across the interface is −250 mV; all others as in Fig. 3. x is the distance from the interface (at $x = 0$), given in nm.

to the confining potential of nanoparticles at ITIES, using methods described in ref. 10. The much higher dielectric constant of water relative to oil (78.8 for water vs. 10.7 for, e.g., 1,2-dichloroethane) leads to a very large electrostatic re-solvation energy if a nanoparticle is transferred from water to oil. This energy is also quadratic in nanoparticle charge and roughly linear in radius (the charge is proportional to the square of the radius but the hydration energy of unit charge is inversely proportional

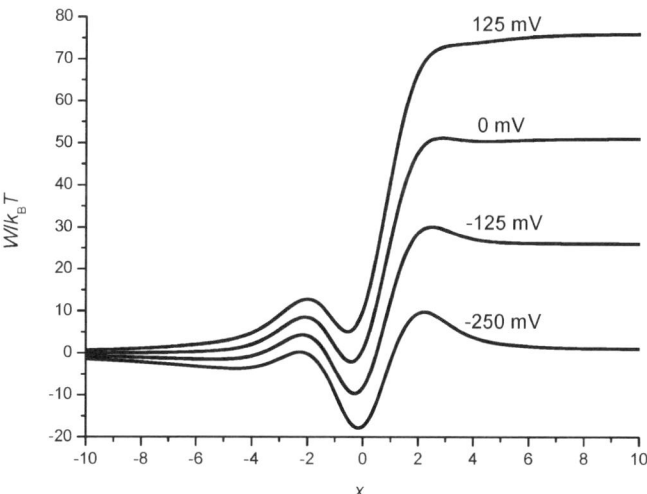

Fig. 3 Nanoparticle energy profile at ITIES: the effect of potential drop across the interface between the aqueous and organic phases. For details of calculation see ref. 10. Parameters: radius of the nanoparticle = 1.5 nm; dielectric constants $\varepsilon_w = 78.8$ (water), $\varepsilon_o = 10.7$ (1,2-dichloroethane); charge of the nanoparticles, $z = -5$; concentrations of electrolytes in water and 1.2 dicloroethane, $c = 5 \times 10^{-3}$ M (which corresponds to Debye lengths 4.31 nm in water and 1.6 nm in oil); interfacial tension between water and oil = 30 mN m^{-1}; line tension = 10^{-11} N; three phase contact angle between the particle surface and water–oil interface (cf. Fig.2) $\theta = 0.55\pi$.

to the radius). This effect, as shown in Fig. 2, makes nanoparticle localization at ITIES unlikely unless an electric field is applied.

The effect of applied potential bias is shown in Fig.3. The negatively charged nanoparticles tend to remain in the aqueous phase when no external voltage is applied (or for a positive voltage). This differs from the behavior of uncharged particles, which will adsorb at the ITIES to reduce the interfacial energy. Our calculations also demonstrate that the electric fields sustainable at an ITIES can provide a deep enough potential well to localize nanoparticles at the interface. For sufficient electric field the potential energy minimum for charged nanoparticles can even be deeper than that for uncharged particles. For example, uncharged nanoparticles smaller than 2 nm will not be localized at the ITIES because the surface tension contribution provides only $1-2k_BT$ energy of adsorption.

The barriers surrounding the well are the result of the positive line tension. The value of the latter is, in fact, poorly known, and to make the things even more complicated, the line tension may depend on the degree of ionization of the ligands. It also strongly affects the depth of the main well. For substantially lower line tension the barriers will not be remarkable and the well much deeper. Consequently for much greater nanoparticles, say of 50 nm radius, the relative importance of the line tension will be much smaller, and the barriers practically vanish.

Varying the applied voltage permits reversible control of the energy minimum depth, as shown in Fig. 3. The change in the depth of the potential well leads directly to a change in the coverage of the interface by nanoparticles. The change in voltage also slightly shifts the position of the energy minima relative to the interface, which could affect nanoparticle-nanoparticle interactions.

Fig. 4 shows the dependence of the energy profile to the nanoparticle size (at a constant charge of the coated nanoparticle). As indicated before, the potential minimum deepens for larger particles (surface tension). However, in practice the charge of the nanoparticle is proportional to its surface area, and thus depends on the square of the nanoparticle diameter. As the effect of the voltage is proportional to the nanoparticle charge, the influence of the electric field will increase with nanoparticle size as well. However, this does not change the trend of deeper potential minima for larger particles shown in Fig. 4.

A possibility to control a concentration of nanoparticles at ITIES by variation of electric field opens the door to novel electrically-tunable optical devices. But if the

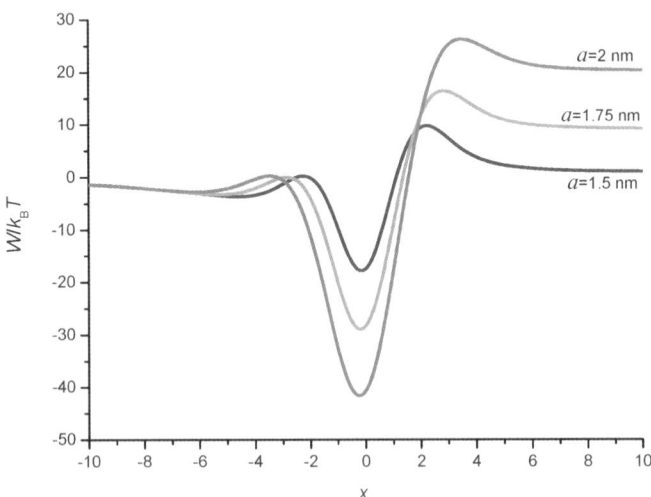

Fig. 4 Nanoparticle energy profile at ITIES: effect of particle radius (at constant charge on its surface). Calculations performed within the same model as in Fig.3. Parameters: potential drop across the interface is -250 mV; all others as in Fig.3.

particles stably reside at the surface for a substantial and sustainable range of electric fields, other kind of devices may be considered, where the varying field can affect the optical properties of the particles themselves. For instance, for quantum dots localized at ITIES, this may lead, to tunable optical filters. Principles of operation of such filters follow from our recent theoretical prediction of a giant Stark effect in quantum dots localized at ITIES.[3] In order for the Stark shifts of the optical absorption features of colloidal quantum dots to exceed the room-temperature linewidth and the local fluctuating electric fields in the environment of these dots, 10^7 V cm^{-1} stable electric fields are required; these are possible in the ITIES systems. Thus Stark-shift filters may be a possibility. In previous work, Stark shifts of the optical absorption features were difficult or impossible to see due to the large room-temperature line width and to local fluctuating electric fields from the movement of charges on the ligands.[14]

A Stark shift of the electron and hole energies occurs when an electric field is applied to a nanoparticle. This shift then modifies the absorption energies for optical transitions. As the nanoparticles would consist of semiconductor nanocrystals with dielectric constants of ~10, at high concentrations of electrolyte the dominant capacitance of the interface will come from the double layers in the aqueous and organic phases. Thus more than 90% of the applied electric potential should drop across the layer of semiconductor nanocrystals. We approximate the potential drop as linear. The electric field in the dot will shift carriers of opposite charge to opposite sides of the dot, and in the process lower the transition energy.

In spherical quantum dots the electric field mixes the lowest-order S state and one of the lowest-order P states with a Hamiltonian term of magnitude eEa. The P state mixed in has its quantization axis parallel to the applied electric field. The mixing energy that results is one-sixth of the potential energy drop across the nanoparticles, $eV/6$.[3] Fig. 4 shows the peak energy of the lowest-energy absorption feature as a function of electric field for three sizes of CdSe/ZnS dots. For CdSe/ZnS quantum dots whose lowest-energy absorption line is at 560 nm in the absence of an electric field, a 1 V applied potential shifts the absorption peak from 560 nm to 620 nm. This shift, as shown in Fig. 5, corresponds to a change from green to red, and is twice as large as typical room-temperature linewidth of 30 nm for a quantum dot size distribution of 5%. Thus the Stark shift in the optical absorption should survive to room temperature.

We also find that nanocrystals of different shape do not have substantially different Stark shifts, so long as the potential energy drop across the nanoparticles remains the same. For example, one-dimensional nanorods with long axes oriented parallel to the applied electric field have Stark shifts that are equal (within 10%) to the Stark shifts for spherical nanocrystals with the same potential energy drop.

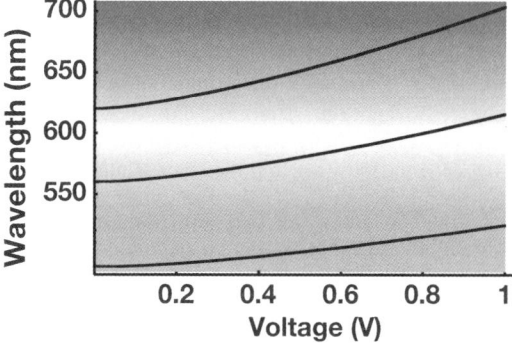

Fig. 5 Calculated peak wavelength of the low-energy optical absorption as a function of voltage for three sizes of quantum dots.[3] The three curves correspond to zero-field absorption peaks of 490, 560, and 620 nm.

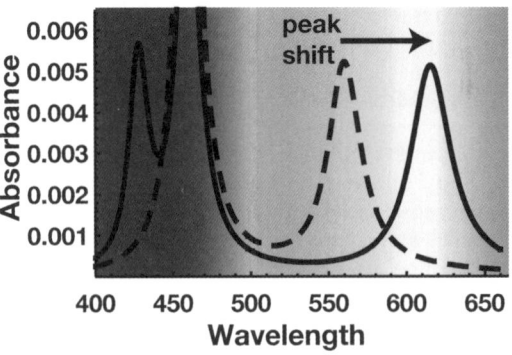

Fig. 6 Calculated optical absorbance spectrum for a CdSe/ZnS quantum dot with a peak absorption at 560 nm for zero field (dashed line).[3] In the ITIES region an applied voltage of 1 V shifts the peak to 620 nm (solid line).

We calculated, according to the method of ref. 7, the optical absorbance of a two-dimensional layer of nanoparticles with an optical linewidth of 100 meV, and plot in Fig. 6. Optical transitions proceed from S valence states to S conduction states, and from P valence states to P conduction states. The two P states whose quantitation axes are perpendicular to the applied electric field will not shift energies in the applied electric field (to first order). Thus the absorption feature near 450 nm in Fig. 6 does not shift in energy. Without an electric field that feature consists of transitions involving 3 P states, and with a field it only involves 2 P states, so the peak absorption is reduced with electric field. The third P state has shifted up in energy and forms the new feature near 425 nm. The S states have shifted to lower energy, from green to red.

The response time of the optical properties will be determined by the response time of the electrical double layers, τ_m, and the response time of the re-localization of quantum dots in the modified electric field, τ_d. Whichever time is longer will dominate. A change of the potential at the interface requires changing the charge in the "back-to-back" double layers at the interface. This will require the so-called *migration time* to equilibrate, which is estimated as $\tau_m = LL_D/D_i$, where L is the macroscopic dimension of the cell, L_D is the Debye screening length or its equivalent in a double layer nonlinearly responding to the overall voltage drop,[15,16] and D_i is the diffusion coefficient of the ions. Migration of ions through the oil will provide the limiting time, primarily because of the higher oil viscosity relative to water. Considering a microfluidic (thin) cell, *e.g.* $L = 100$ μm, $L_D = 1$ nm, and $D_i = 10^{-6}$ cm^2 s^{-1}, we get $\tau_m = 10^{-3}$ s.

The relaxation time τ_d may dominate if the surface concentration of the nanoparticles changes significantly with voltage. This relaxation time will depend on the average concentration and the Stokes radii of the nanoparticles. The Stokes radius influences the timescale through the nanoparticle diffusion coefficient. We estimate $\tau_d = l^2/D_p$, where $l = (4\pi c_p/3)^{-1/3}$ is the average distance between nanoparticles and, correspondingly, between the interface and the layer of nanoparticles closest to it (c_p is their number concentration), and D_p is the nanoparticle diffusion constant. For $l = 1000$ nm and $D_p = 10^{-7}$ cm^2 s^{-1} the relaxation time $\tau_d = 10^{-1}$ s. If the particle stays at the interface for the full range of electric fields the field variation will lead to nanoparticle transport distances smaller than 1 nm, yielding an irrelevant τ_d on the order of 10^{-7} s.

We find the room-temperature extinction coefficient of a monolayer of quantum dots with surface density 10^{17} m^{-2} to be 0.5% (Fig. 6). Multipass geometries can be constructed to bring the extinction coefficient near to unity. Applications of the proposed tunable optical filters may include fast, high-extinction, angle-independent energy filters for time-resolved optical probes of biological systems, including two-photon fluorescence.

As for metal nanoparticles, sustainable electric fields would be ineffective at changing their properties. However, by linking metal nanoparticles with molecules or supramolecular complexes capable of intramolecular electron transfer (ET), one can expect surface-plasmon-enhanced ET and giant Raman effects.[17] This may provide a whole new area of research: plasmon-enhanced ET Raman signals could be very strong and observable without multiple passage geometries. Furthermore, they may exhibit strong electrochromic effects, *via* the electric field effect on the free energy of elementary act of ET. The voltage variation, moving particles from the surface to the bulk, or *vice versa* should give very different Raman signals. It is not yet clear how this can be used, but the prospect may become more visible after we describe and experimentally characterize this phenomenon.

As for large metal nanoparticles, they comprise a special story. Dramatic effects on the optical properties of ITIES can be achieved with their help, which can be put into the heart of the principles of new electro-optic devices,[11] which we may not yet discuss here.

This work was supported by the interchange grant of the Levehulme Trust (UK), F/07058/P, the Israel Science Foundation (grant 773/05), and an US Army Research Office MURI grant.

References

1. Y. Cheng and D. J. Schiffrin, *J. Chem. Soc., Faraday Trans.*, 1996, **92**, 3865; C. Johans, R. Lahtinen, K. Kontturi and D. J. Schiffrin, *J. Electroanal. Chem.*, 2000, **488**, 99.
2. H. H. Girault, *Nat. Mater.*, 2006, **5**, 851.
3. M. E. Flatté, A. A. Kornyshev and M. Urbakh, *Proc. Natl. Acad. Sci. U. S. A.*, 2008, **105**, 18212.
4. A. A. Kornyshev, M. Kuimova, A. M. Kuznetsov, J. Ulstrup and M. Urbakh, *J. Phys.: Condens. Matter*, 2008, **20**, 073102.
5. H. H. Girault and D. H. Schiffrin, *Electroanalytical Chemistry*, ed. A. J. Bard, Marcel Dekker, New York, 1989, vol. 15, p. 1.
6. H. H. Girault, in *Modern Aspects of Electrochemistry*, ed. J. O. Bockris, B. E. Conway and R. E. White, Plenum, New York, 1993, vol. 25, p. 1.
7. W. H. Binder, *Angew. Chem., Int. Ed.*, 2005, **44**, 5172.
8. B. Su, J.-P. Abid, D. J. Fermin, H. H. Girault, H. Hoffmannova, P. Krtil and Z. Samec, *J. Am. Chem. Soc.*, 2004, **126**, 915.
9. B. Su, D. J. Fermin, J.-P. Abid, N. Eugster and H. H. Girault, *J. Electroanal. Chem.*, 2005, **583**, 241.
10. M. E. Flatté, A. A. Kornyshev and M. Urbakh, *J. Phys.: Condens. Matter*, 2008, **20**, 073102.
11. M. Flatte, A. A. Kornyshev and M. Urbakh, 2009, to be published.
12. R. Murray, *Chem. Rev.*, 2008, **108**, 2688.
13. P. Pieranski, *Phys. Rev. Lett.*, 1980, **45**, 569.
14. S. A. Empedocles and M. G. Bawendi, *Science*, 1997, **278**, 2114.
15. M. S. Kilic, M. Z. Bazant and A. Ajdari, *Phys. Rev. E: Stat., Nonlinear, Soft Matter Phys.*, 2007, **75**, 021502.
16. C. W. Monroe, M. Urbakh and A. A. Kornyshev, *J. Electroanal. Chem.*, 2005, **582**, 28.
17. T. Shegai, Z. P. Lee, T. Dadosh, Z. Y. Zhang, H. X. Xu and G. Haran, *Proc. Natl. Acad. Sci. U. S. A.*, 2008, **105**, 16448.

Free-standing porous supramolecular assemblies of nanoparticles made using a double-templating strategy

Xing Yi Ling,[a] In Yee Phang,[b] David N. Reinhoudt,[a] G. Julius Vancso[b] and Jurriaan Huskens*[a]

Received 10th December 2008, Accepted 9th March 2009
First published as an Advance Article on the web 30th June 2009
DOI: 10.1039/b822156a

The formation of stable and ordered free-standing porous supramolecular assemblies of nanoparticles with sizes and geometries controlled at different length scales is demonstrated by a double-templating strategy. Our technique combines the directed assembly of particles, templating using nanoimprint lithography (NIL), and supramolecular layer-by-layer (LbL) assembly. First, 500-nm β-cyclodextrin (CD)-functionalized polystyrene (PS) particles were assembled by convective assembly onto a sacrificial polymer template patterned with a predefined geometry and size using NIL, forming a 3D crystal architecture of particles. LbL assembly of alternating supramolecular host- and guest-functionalized glues of CD-functionalized Au (Au-CD) nanoparticles and adamantyl (Ad) dendrimers, sized between 3–5 nm, within the preformed PS-particle crystal effectively bound the particles together into a particle composite. These particle composites were released from the substrate together with the polymer template, and transferred onto a target substrate. The particle crystal integrity, order and functionality were preserved. Rinsing the structure with dichloromethane removed the PS core material together with the polymer template, resulting in interconnected porous capsules, the sizes and shapes of which are fully determined by the PS core size and the polymer template definition. Again, integrity and shape were preserved in the rinsing step. These capsules were capable of storing organic fluorescent molecules using specific interactions.

1 Introduction

The fabrication of well-defined free-standing micro- or nano-objects is important for specific applications in membranes,[1,2] sensors,[3] nanomechanical films and nanoreactors.[4,5] Nanoparticles that exhibit interesting optical, electronic and catalytic properties are potential candidates for the fabrication of new types of functional materials. In particular, the assembly of well-defined nanoparticle-based structures is promising in the current trend of miniaturizing devices. Considerable progress has been achieved in forming self-assembled 1D or 2D nanoparticle structures in solution by means of specific chemical interactions, including linear structures of nanoparticles,[6] nanowires,[7] and spherical and network aggregates of nanoparticles.[8]

[a]*Molecular Nanofabrication Group, MESA+ Institute for Nanotechnology, University of Twente, P.O. Box 217, 7500AE Enschede, The Netherlands. E-mail: j.huskens@utwente.nl; Tel: +31 (0)53 489 2980*
[b]*Materials Science and Technology of Polymers, MESA+ Institute for Nanotechnology, University of Twente, P.O. Box 217, 7500AE Enschede, The Netherlands*

Most solution strategies to form nanoparticle structures are based on intrinsic interparticle interactions and/or the control over kinetics of nanoparticle assembly in solution. Generally, these strategies suffer from a limited complexity of the assemblies. Moreover, the assembly of free-standing structures in solution is limited to the nanometre length scale, whereas the bottom-up assembly of the nanoparticles into nano- or microobjects with defined composition and geometry remains experimentally challenging. Recently, the layer-by-layer (LbL) assembly technique[9] has been extended to the formation of flexible free-standing (mostly unpatterned) thin films in air, at the liquid/liquid interface, or at the air/liquid interface.[3] It generally involves the LbL growth of a polymer or polyelectrolyte[10] film with a well-defined thickness and composition on a substrate and its removal from the substrate by using a sacrificial layer, such as cellulose acetate.[11] The incorporation of fillers such as metallic nanoparticles,[2,12] clays,[13,14] and nanotubes[15] into the LbL films are essential to enhance the mechanical robustness[1] and/or to introduce additional optical properties[5] into the thin films. However, the macroscopic robustness of the free-standing structures and their potential functionalities in optics, chemical sensing, or nano- or microelectromechanical systems have to be further explored to lead to technical applications.

The self-assembly of nanoparticles can be combined with top-down nanofabrication techniques to achieve 3D nanoparticle-based structures on surfaces. Nanofabrication techniques such as soft lithography,[16,17] nanoimprint lithography (NIL),[18,19] microcontact printing,[17,20–22] photolithography[23,24] and dip-pen nanolithography[25,26] enable the design and manipulation of the shapes and sizes of the nano- and microstructures. For instance, Santhanam and Andres have prepared monolayers of hexagonally ordered thiolate-functionalized Au nanoparticles on a water surface and transferred them onto a poly(dimethylsiloxane) (PDMS) stamp for the printing of a single layer of particles on a Au substrate.[27] With the combination of LbL assembly, multilayers of nanocrystals and/or polyelectrolytes have been formed on (patterned) self-assembled monolayers (SAMs).[12,27–29] Our group employs the specific inclusion complexation of hydrophobic guest molecules by β-cyclodextrin (CD) to direct the supramolecular assembly of complementary (bio)molecules and nanoparticles.[30] By using ferrocenyl- or adamantyl-functionalized dendrimers[31,32] or nanoparticles[33] and CD-functionalized nanoparticles[34,35] as basic building blocks, 3D nanoparticle assemblies have been formed on molecular printboards patterned by μCP and NIL in a LbL fashion.[19,36–38]

Here, a versatile approach towards the formation of stable and ordered free-standing particle and porous assemblies with controllable sizes and geometries is reported. The self-assembly of particles, NIL templating, and supramolecular chemistry are combined to obtain free-standing 3D particle structures. This approach extends the use of supramolecular chemistry in forming particle composite bridges with macroscopic robustness[39] to stand-alone and freely-suspended porous nanoparticle assemblies with specific functionality. Polystyrene-particle composite structures with predefined geometry and size were formed on an NIL-patterned polymer template by using self-assembly of particles and LbL assembly of supramolecular host- and guest-functionalized glues. The subsequent removal of the polystyrene cores of the hybrid particle structure resulted in the formation of interconnected porous capsules of macroscopic structures. The integrity, order and functionality of the porous nanoparticle composites are preserved. The potential of these structured capsules of supramolecular materials as carriers and storerooms of organic fluorescent molecules is also examined.

2 Experimental details

2.1 Materials

Adamantyl-terminated poly(propylene imine) dendrimer of generation 5 (G5-PPI-$(Ad)_{64}$)[32] was synthesized as described before. Poly(methyl methacrylate) (PMMA, M_w: 38 000), 8-anilino-1-naphthalenesulfonic acid (ANS), lissamine rhodamine

B sulfonylchloride and 6-hydroxy-2-naphthoic acid were obtained from Sigma Aldrich, Germany. Carboxylate-functionalized polystyrene particles of 500 nm were purchased from Polysciences Inc. CD-Functionalized polystyrene particles (PS-CD), prepared from the carboxylate-functionalized particles, and gold nanoparticles (Au-CD, $d \sim 3$ nm) were prepared as described before.[34,40] Milli-Q water with a resistivity higher than 18 MΩ cm was used in all experiments.

2.2 Substrate and monolayer preparation

Flat silicon substrates were cleaned by immersion in piranha solution (conc. H_2SO_4 and 33% H_2O_2 in a 3 : 1 volume ratio) for 15 min to form a SiO_2 layer on the surface. **Warning!** Piranha should be handled with caution; it is a highly corrosive oxidizing agent. The substrates were then sonicated in Milli-Q water and ethanol for 1 min, and dried with N_2.

Nanoimprint lithography (NIL) was performed by putting a silicon stamp (3 μm lines at 8 μm period with a height of 800 nm, and 8 μm holes at 3 μm spacing with a height of 500 nm) in contact with a 1–1.5 μm thick layer of PMMA on a SiO_2 substrate. A pressure of 40 bar was applied at a temperature of 180 °C using a hydraulic press (Specac). The residual PMMA layer was not removed after the hot embossing process. The PMMA templates were oxidized in an O_2 plasma etcher (Tepla 300E) for 30 s to render them hydrophilic.

2.3 Assembly of PS-CD particles

On oxidized PMMA templates, 500 nm CD-functionalized polystyrene (PS-CD) particles were convectively assembled into the grooves of the NIL substrates to form particle crystals by using a capillary-assisted deposition setup.[19,37] The preformed particle array was then gently dipped in a 1 mM aqueous solution of G5-PPI-(Ad)$_{64}$ for 30 min, rinsed with water, and blown dry with N_2.

2.4 Formation of hybrid particle crystals

The nanoparticle composites were prepared by layer-by-layer (LbL) assembly of Au-CD nanoparticles ($d \sim 3$ nm) and G5-PPI-(Ad)$_{64}$ on the pre-assembled PS-CD particle layers, according to a published procedure.[34,39] Up to 30 LbL cycles were performed. After each adsorption step, the substrate was blown dry with N_2.

2.5 Infiltration of the particle composites with fluorescent molecules

Infiltration of fluorescent molecules into the particle composites was performed prior to the release-and-transfer process. The particle composites were immersed in an aqueous solution of 1 mM ANS, lissamine rhodamine, or naphthoic acid for 10 min. The nanoparticle composites were subsequently gently rinsed with Milli-Q water and dried with N_2.

2.6 Release-and-transfer of particle composites

Oxidized NIL-patterned PMMA on a SiO_2 substrate, onto which particle composites were deposited, was gently immersed in water at 50 °C. The entire PMMA layer, attached to the particle composites, was slowly peeled off from the SiO_2 substrate in warm water, resulting in a floating PMMA sheet attached with particle composites at the water/air interface. A target substrate was used to pick up the entire layer which was subsequently blown dry with N_2. Removal of the PMMA layer and PS cores was performed by gently immersing the entire substrate in dichloromethane for 1 min, followed by drying with N_2.

2.7 Scanning and transmission electron microscopy (SEM, TEM)

All SEM images were taken with a HR-LEO 1550 FEF SEM. TEM was performed on a Philips CM 30 Twin STEM fitted Kevex delta and Gatan model 666 PEELs operating at 300 kV.

2.8 Atomic force microscopy (AFM)

AFM measurements were carried out using a Dimension D3100 atomic force microscope equipped with a NanoScope IVa controller and a hybrid scanner (H-153) with x-, y- and z- feedbacks from Veeco (Veeco/Digital Instruments (DI), Santa Barbara, CA) in ambient condition. Silicon cantilevers, PointProbe®Plus non-contact high resonance frequency (PPP-NCH) from Nanosensors (Nanosensors, Wetzlar, Germany) were used for intermittent contact (tapping) mode operation to obtain high-resolution images of the samples. Scan rates were varied from 0.35–0.5 Hz and free amplitude (A_o) set-point values were around 1.6 V. Prior to nanoindentation, the nanostructures were imaged, and the 'point and shoot' function was used to pre-position the location of the cantilever. The AFM tip was pressed upon the sample at a preset force. The indentation marks on the nanostructure due to the high loading force were observed and confirmed that the force was applied at correct location. The force applied on the sample was determined by the deflection of the cantilever multiple by the value of the spring constant of the cantilever. The spring constants of the cantilevers were assessed according to Sader's method.[41] The spring constant values were in the range of 24.7–31.1 N m^{-1} with a mean value of 28.3 N m^{-1}. Data conversion was carried out with Nanoscope® software version 613b.

2.9 Fluorescence microscopy

Fluorescence microscopy was performed using an Olympus inverted research microscope IX71 equipped with a mercury burner U-RFL-T as the light source and a digital camera Olympus DP70 (12.5 million pixel cooled digital color camera) for image acquisition. The fluorescence imaging of the ANS- and rhodamine-infiltrated structures was performed at blue excitation (450 $\leq \lambda_{ex} \leq$ 480 nm, $\lambda_{em} \geq$ 515 nm). The images of the naphthoic acid-infiltrated structures were taken at UV excitation (300 $\leq \lambda_{ex} \leq$ 400 nm, $\lambda_{em} \geq$ 400 nm), which were filtered using a U-MWG Olympus filter cube.

3 Results and discussion

The preparation procedure of free-standing hybrid particle crystals and porous assemblies is shown in Scheme 1. A PMMA template with a pattern made by NIL

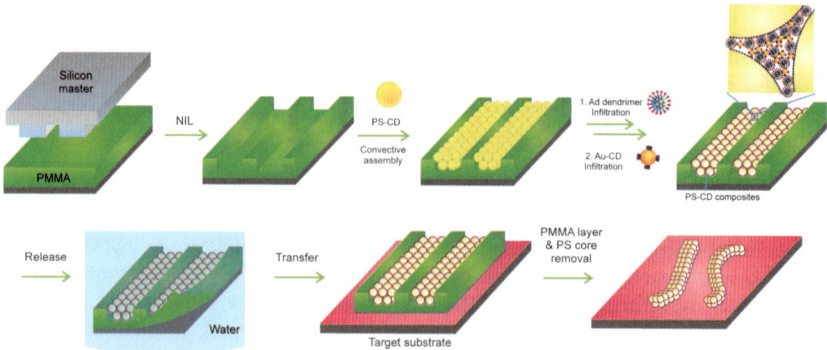

Scheme 1 The preparation of a PMMA template and the release-and-transfer of free-standing hydrid particle composites, followed by template and PS core removal.

was used (Scheme 1). The PMMA layer was mildly oxidized in O_2 plasma to render its surface hydrophilic. Subsequently, the CD-functionalized polystyrene particles (PS-CD, $d \sim 500$ nm), prepared as described before,[40] were assembled onto the template *via* capillary-assisted assembly,[19,37] with no specific affinity between the nanoparticles and the substrate. By controlling the assembly parameters,[40] the particles were directed into the grooves of the PMMA template. The resulting particle crystals inherited the overall shape from the underlying template, thus allowing the assembly of the particles into highly organized structures of desired sizes and shapes.

Adamantyl-functionalized poly(propylene imine) dendrimer of generation 5 (Ad dendrimer)[32] and CD-functionalized Au nanoparticles (Au-CD, $d \sim 3$ nm),[19] respectively, were used as the supramolecular guest- and host-functionalized glues to adhere the PS-CD particles within the PS-CD crystal to each other and to the substrate *via* supramolecular LbL assembly.[19] The PS-CD crystal was first infiltrated with Ad dendrimers, followed by the complementary Au-CD nanoparticles. These steps were repeated in a LbL fashion to reach the desired 30 number of bilayers to form hybrid particle composite.[34] By using the PMMA template as a carrier, the entire particle composites were released from the silicon substrate into water, resulting in a floating PMMA template (to which the hybrid nanoparticle composites were attached) at the water/air interface. A target substrate was subsequently used to pick up the PMMA template with the particle composites for visualization.

Fig. 1A shows a photograph of a hybrid particle composite embedded in a PMMA template after transfer onto a target SiO_2 substrate. As a result of its thickness of <1 μm, the PMMA–particle structure appeared flexible. SEM images (Fig. 1B and C) reveal that the hybrid particle crystals, consisting of two layers of PS-CD particles in 3 μm line structures, remained adhered to the PMMA template, probably as a result of van der Waals forces[42] and electrostatic interactions between

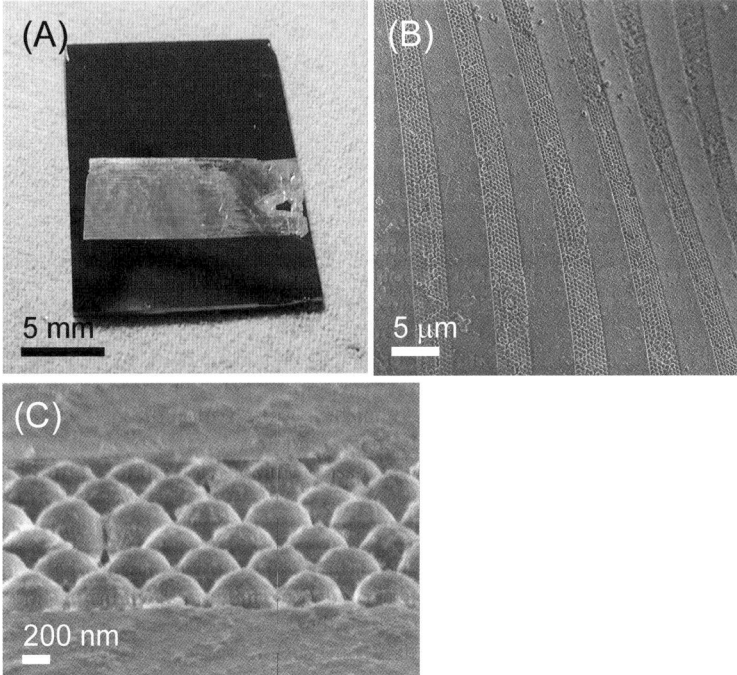

Fig. 1 A photograph (A) and SEM images (B, C) of infiltrated particle crystals embedded in the PMMA template, after release-and-transfer onto a target SiO_2 substrate.

the positively charged Ad dendrimers and the negatively charged oxidized PMMA surface. The polymer layer was slightly stretched during the transfer process. However, no obvious changes in the order and integrity of the particle crystals were observed, indicating the robustness and flexibility of the hybrid particle crystals during the release-and-transfer process. Only incidentally, particle vacancies were observed in the hybrid particle crystal. In contrast, when the supramolecular LbL assembly within the particle crystal was omitted, most of the particle crystals disintegrated, and the PS particles were removed from the polymer layer during the layer transfer process (data not shown). Thus, the stability and integrity of the hybrid particle crystals can be attributed to the strong multivalent interactions[30] within the crystal.

TEM images (Fig. 2A and B) reveal the infiltrated particle crystal after transfer onto a copper grid. The hybrid composite after 30 LbL cycles maintained a high quality hcp structure, indicating that all particles were effectively bound together during the release-and-transfer process. At the periphery of each PS particle and at the interface between the adjacent particles (Fig. 2B), material of high contrast is visible, indicating individual Au nanoparticles at the shell of the hybrid PS particle crystal. Energy-dispersive X-ray spectroscopy (EDX) performed on a PS-CD particle and at a void between particles confirmed the presence of a large amount of Au on the PS-CD particle as compared to the void area (Fig. 2C).

After transferring the hybrid particle crystal onto a target substrate, the substrate was rinsed with dichloromethane to remove the PS core and the PMMA template simultaneously. As shown in Fig. 3, ribbons of interconnected hollow hybrid

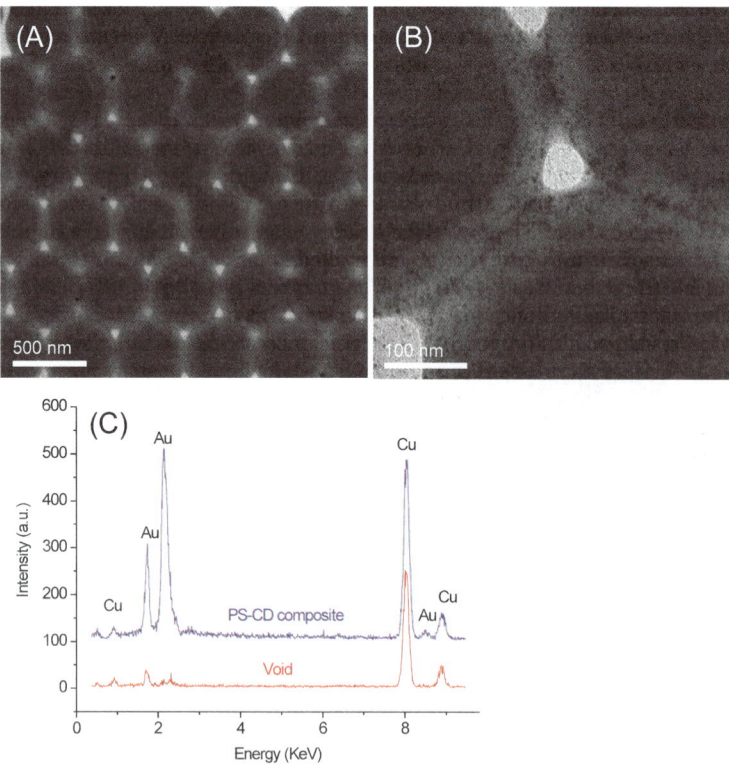

Fig. 2 TEM images of the hybrid particle crystal after 30 LbL cycles and release-and-transfer onto a copper grid (A, B). EDX spectra (C) taken on top of a PS particle within the hybrid particle crystal (top) and at a void between three neighboring particles (bottom).

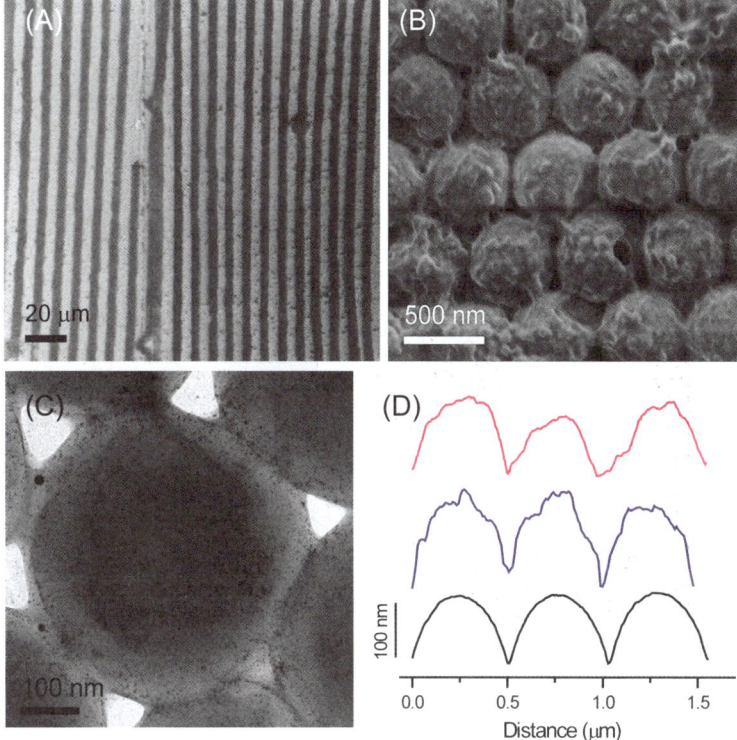

Fig. 3 (A–C) SEM images of the ribbons of hybrid hollow capsules, after transfer of the hybrid particle composite onto a target substrate and removal of the PS core and the PMMA template. (D) AFM height profiles of the as-prepared particle crystal (bottom), the particle composite (center), and the hollow capsules (top).

capsules, approximately 500 μm in length were formed. The hollow ribbons exhibited slight curvature in appearance, as a result of the PMMA template removal. Zoomed-in SEM and TEM images (Fig. 3B and C) show that the internal cohesion by the supramolecular glues remained intact. The hollow nature of the capsules is confirmed by occasional defects in the LbL structure (Fig. 3B) and by AFM (see below). Even though some indented capsules were observed, the majority of the hollow capsules appeared to maintain their original spherical shape. A comparison of the AFM height profiles of the as-prepared particle crystal, the infiltrated particle composites, and the inverted ribbons revealed that their thicknesses are comparable (Fig. 3D), indicating that the structural integrity of the hollow capsules are preserved. Thus, the multilayered assemblies of Ad dendrimers and Au-CD nanoparticles have formed supramolecular capsules fused together into stable macroscopic entities, the shape of which is governed by the PS particles and the PMMA template, both of which have been removed completely.

The micromechanical properties and structural deformation of the hybrid particle structures and hollow capsule ribbons (infiltrated with 30 bilayers) were probed by AFM-based nanoindentation. Fig. 4A and D show typical force–penetration curves obtained by applying point loads onto a hybrid particle and a hollow capsule. A sigmoidal curve was obtained when the AFM tip struck the hybrid multilayer-coated PS particle. The hybrid particle experienced a relatively low-force penetration in the beginning of indentation, probably as a result of indentation of PS and/or Ad dendrimers and Au-CD nanoparticles. Upon 70–100 nm penetration, a drastic increase in force was observed. In the retracting part of the cycle, a large hysteresis

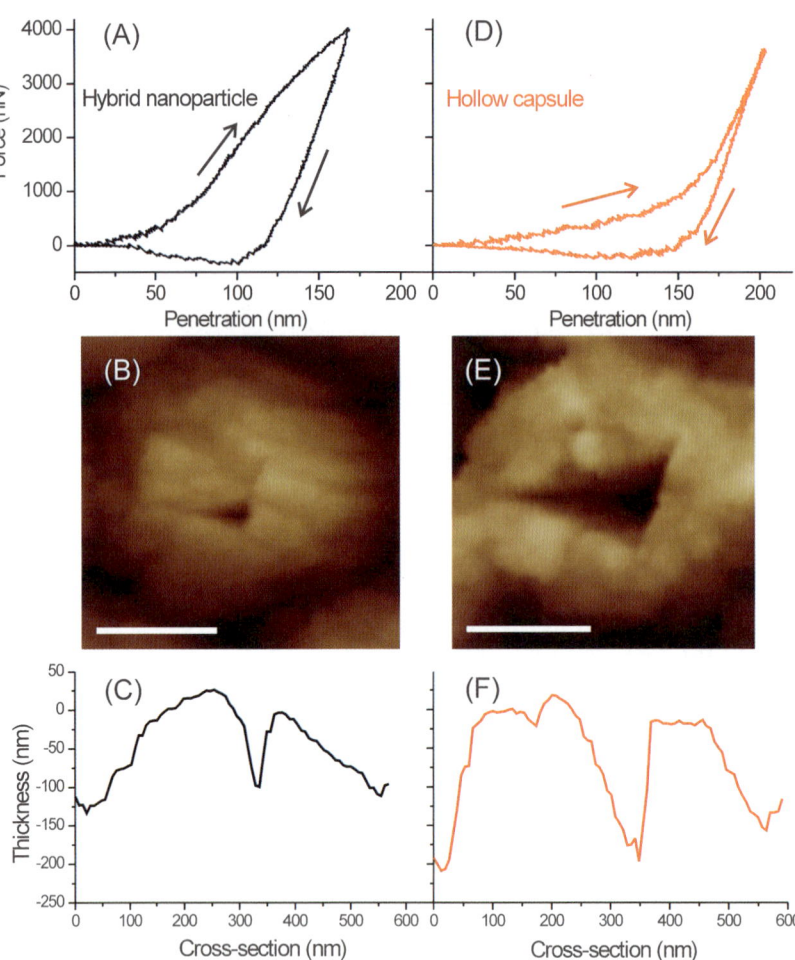

Fig. 4 Force–penetration curves (A, D), AFM images (B, E) and cross-section profiles after indentation (C, F) of a PS particle composite coated with 30 bilayers (A, B, C), and a hollow capsule of 30 bilayers (D, E, F). All scale bars indicate 200 nm.

was observed. The hollow capsules obtained after dichloromethane rinsing experienced a different mechanical behavior. A larger penetration was observed at the same 4 μN force load, suggesting a change in the material properties. The force–penetration curve was smooth with no discontinuities, indicating that no cracking occurred during the indentation.[43] Compared to the hybrid particle structure, only a low force is needed to penetrate the hollow capsule. Fig. 4B and E show the AFM images of the hybrid particle structure and the capsule after indentation. The indented area on the hybrid particle appears eight times smaller than on the capsule, which is attributed to the resistance of the PS particle towards indentation. The AFM cross-section profiles (Fig. 4C and F) show an indentation of ∼200 nm on the capsule, as compared to <100 nm on the PS particle, confirming the hollow nature of the capsule.

To examine the integrity of the sub-millimetre hollow capsule ribbons and the potential of these capsule ribbons in storing fluorescent molecules, the as-prepared particle composites, while still on the PMMA template, were dipped into an aqueous solution of 8-anilino-1-naphthalene sulfonic acid (ANS). ANS is a fluorescent probe that is highly sensitive toward its micro-environment.[44] The fluorescence properties of ANS depend on the polarity of the medium, and severe fluorescence quenching is observed in water. The negatively charged ANS is known to penetrate into the core of Ad dendrimers *via* electrostatic interactions to restore its fluorescence intensity.[32] Here, the recognition between ANS and the Ad dendrimers is employed to observe the presence of the Ad dendrimers and thus the complete capsule ribbons by fluorescence microscopy.

The ANS-infiltrated particle composites were subsequently released-and-transferred onto a target substrate. After removal of the PMMA layer and the PS cores by CH_2Cl_2, the ribbons were monitored by fluorescence microscopy. The fluorescence image of the composites (Fig. 5A) clearly shows the highly fluorescent micrometre-sized ribbons. This indicates that the ANS molecules were stored within the shells of the hollow capsules and remained in the structure upon release, transfer and rinsing. Besides the imaging, the results suggest the potential of the hybrid hollow capsule as an effective storage and carrier for molecular information. A wide variety of organic dyes can be stored within such dendrimers.[45,46] As shown in Fig. 5B, separately made capsule ribbons infiltrated with ANS and lissamine rhodamine, respectively, were released-and-transferred onto the same target substrate. Subsequent dichloromethane rinsing removed the PMMA templates and the PS cores of both particle structures simultaneously. Despite their similar structural appearance, the different fluorescent information embedded within the hybrid structures can be readily read out under the fluorescence microscope (upon blue excitation) to distinguish the different molecular information stored within them.

Fig. 5C shows the shape versatility of our approach in forming capsules in a network structure, simply by changing the design of the PMMA template. Hollow

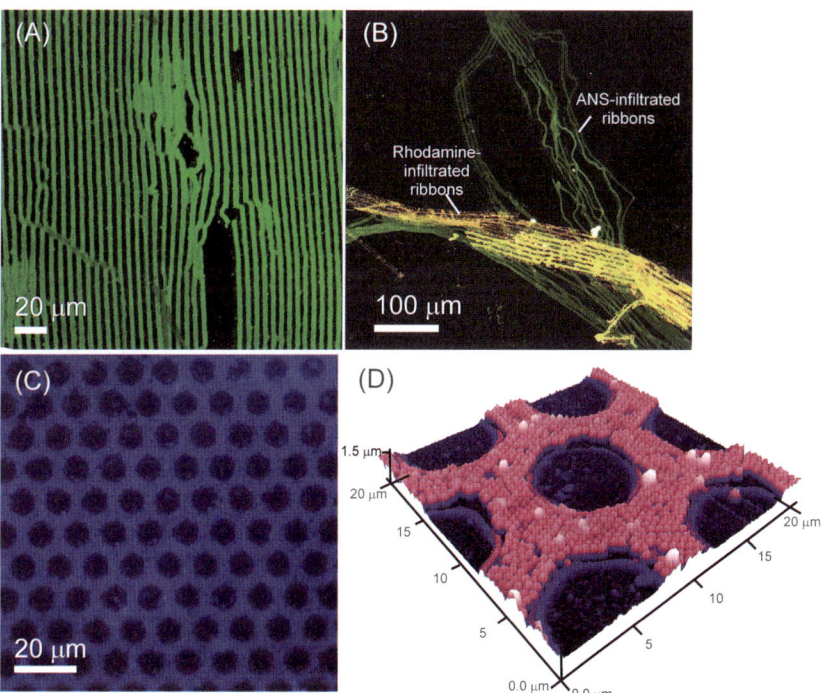

Fig. 5 Fluorescence microscopy images of the hybrid hollow capsule ribbons infiltrated with ANS (A), capsule ribbons overlapping structures infiltrated with lissamine rhodamine and ANS, deposited sequentially on the same target substrate (B), and a network capsule structure infiltrated with naphthoic acid (C). 3D AFM image (D) of the network structure shown in (C).

capsules of ring-shaped network structures (infiltrated with 30 bilayers) were thus formed by using a PMMA template with 8 μm dots with 3 μm spacing and a height of 500 nm. The height of the network structure is similar to that of the original template and corresponds to a single layer of PS-CD particles being assembled onto the PMMA template (Fig. 5D). The network was infiltrated with naphthoic acid, which is highly fluorescent under UV light, transferred and rinsed with dichloromethane, and imaged (Fig. 5C). The results indicate that the individual size of the hollow capsules can be manipulated by the choice of the size of the PS particles, while the overall shape and geometry of the entire capsule structures can be designed by the geometry and size of the PMMA template.

4 Conclusions

A versatile double-templating approach for the formation of stable and ordered free-standing porous assemblies with controllable size and geometry has been developed by combining the directed assembly of particles, supramolecular LbL assembly of Ad dendrimers and Au-CD nanoparticles, and NIL. This technique extends the LbL assembly of 2D flat films to 3D free-standing particle structures. The main component of the free-standing structure can be of arbitrary material provided the surface is engineered with supramolecular properties. The integrity of the particle crystals was induced by multivalent supramolecular interactions using the well-defined dendrimers and metallic nanoparticles (of 3–5 nm) *via* specific host–guest interactions within the particle crystal. The resulting particle composites can be tuned from micrometre to sub-millimetre objects with pre-designed shapes and geometries. After removal of the PS core and the underlying polymer template, the resulting supramolecular multilayered structures maintained the size and shape of the original 500 nm PS-CD particles to form free-standing interconnected hollow capsule ribbons with considerable mechanical stability. Such stable and ordered hollow capsule ribbons behave as three-dimensional receptor structures, which can potentially serve as molecular information carriers capable of storing and releasing molecular information.

Acknowledgements

Mr Iwan Heskamp is acknowledged for providing the lithographically patterned silicon masters. X. Y. L. and J. H. thank the Council for Chemical Sciences of the Netherlands Organization for Scientific Research (NWO-CW) for financial support (Vidi Vernieuwingsimpuls grant 700.52.423 to J. H.).

References

1 R. Vendamme, T. Ohzono, A. Nakao, M. Shimomura and T. Kunitake, *Langmuir*, 2007, **23**, 2792.
2 C. H. Lu, I. Donch, M. Nolte and A. Fery, *Chem. Mater.*, 2006, **18**, 6204.
3 C. Y. Jiang, S. Markutsya, Y. Pikus and V. V. Tsukruk, *Nat. Mater.*, 2004, **3**, 721.
4 Y. Lvov and F. Caruso, *Anal. Chem.*, 2001, **73**, 4212.
5 C. Y. Jiang and V. V. Tsukruk, *Adv. Mater.*, 2006, **18**, 829.
6 M. Li, H. Schnablegger and S. Mann, *Nature*, 1999, **402**, 393.
7 Z. Y. Tang, N. A. Kotov and M. Giersig, *Science*, 2002, **297**, 237.
8 A. K. Boal, F. Ilhan, J. E. DeRouchey, T. Thurn-Albrecht, T. P. Russell and V. M. Rotello, *Nature*, 2000, **404**, 746.
9 G. Decher, J. D. Hong and J. Schmitt, *Thin Solid Films*, 1992, **210–211**, 831.
10 W. T. S. Huck, A. D. Stroock and G. M. Whitesides, *Angew. Chem., Int. Ed.*, 2000, **39**, 1058.
11 X. Gao and L. Jiang, *Nature*, 2004, **432**, 36.
12 A. A. Mamedov and N. A. Kotov, *Langmuir*, 2000, **16**, 5530.
13 P. Podsiadlo, A. K. Kaushik, E. M. Arruda, A. M. Waas, B. S. Shim, J. D. Xu, H. Nandivada, B. G. Pumplin, J. Lahann, A. Ramamoorthy and N. A. Kotov, *Science*, 2007, **318**, 80.

14 F. Hua, T. Cui and Y. M. Lvov, *Nano Lett.*, 2004, **4**, 823.
15 Z. J. Liang, A. S. Susha, A. M. Yu and F. Caruso, *Adv. Mater.*, 2003, **15**, 1849.
16 Y. Xia, J. A. Rogers, K. E. Paul and G. M. Whitesides, *Chem. Rev.*, 1999, **99**, 1823.
17 P. C. Hidber, W. Helbig, E. Kim and G. M. Whitesides, *Langmuir*, 1996, **12**, 1375.
18 P. Maury, M. Escalante, D. N. Reinhoudt and J. Huskens, *Adv. Mater.*, 2005, **17**, 2718.
19 X. Y. Ling, I. Y. Phang, D. N. Reinhoudt, G. J. Vancso and J. Huskens, *Int. J. Mol. Sci.*, 2008, **9**, 486.
20 H. P. Zheng, I. Lee, M. F. Rubner and P. T. Hammond, *Adv. Mater.*, 2002, **14**, 569.
21 T. Kraus, L. Malaquin, H. Schmid, W. Riess, N. D. Spencer and H. Wolf, *Nat. Nanotechnol.*, 2007, **2**, 570.
22 H. X. He, H. Zhang, Q. G. Li, T. Zhu, S. F. Y. Li and Z. F. Liu, *Langmuir*, 2000, **16**, 3846.
23 T. Vossmeyer, S. Jia, E. DeIonno, M. R. Diehl, S. H. Kim, X. Peng, A. P. Alivisatos and J. R. Heath, *J. Appl. Phys.*, 1998, **84**, 3664.
24 Y. Masuda, T. Itoh and K. Koumoto, *Adv. Mater.*, 2005, **17**, 841.
25 X. G. Liu, L. Fu, S. H. Hong, V. P. Dravid and C. A. Mirkin, *Adv. Mater.*, 2002, **14**, 231.
26 J. W. Zheng, Z. H. Zhu, H. F. Chen and Z. F. Liu, *Langmuir*, 2000, **16**, 4409.
27 V. Santhanam and R. P. Andres, *Nano Lett.*, 2004, **4**, 41.
28 D. J. Zhou, A. Bruckbauer, C. Abell, D. Klenerman and D. J. Kang, *Adv. Mater.*, 2005, **17**, 1243.
29 Y. Lvov, G. Decher and H. Möhwald, *Langmuir*, 1993, **9**, 481.
30 M. J. W. Ludden, D. N. Reinhoudt and J. Huskens, *Chem. Soc. Rev.*, 2006, **35**, 1122.
31 C. A. Nijhuis, J. Huskens and D. N. Reinhoudt, *J. Am. Chem. Soc.*, 2004, **126**, 12266.
32 J. J. Michels, M. Baars, E. W. Meijer, J. Huskens and D. N. Reinhoudt, *J. Chem. Soc., Perkin Trans. 2*, 2000, 1914.
33 X. Y. Ling, D. N. Reinhoudt and J. Huskens, *Langmuir*, 2006, **22**, 8777.
34 O. Crespo-Biel, B. Dordi, D. N. Reinhoudt and J. Huskens, *J. Am. Chem. Soc.*, 2005, **127**, 7594.
35 V. Mahalingam, S. Onclin, M. Peter, B. J. Ravoo, J. Huskens and D. N. Reinhoudt, *Langmuir*, 2004, **20**, 11756.
36 X. Y. Ling, D. N. Reinhoudt and J. Huskens, *Chem. Mater.*, 2008, **20**, 3574.
37 P. Maury, M. Péter, O. Crespo-Biel, X. Y. Ling, D. N. Reinhoudt and J. Huskens, *Nanotechnology*, 2007, **18**, 044007.
38 O. Crespo-Biel, B. Dordi, P. Maury, M. Péter, D. N. Reinhoudt and J. Huskens, *Chem. Mater.*, 2006, **18**, 2545.
39 X. Y. Ling, I. Y. Phang, W. Maijenburg, H. Schönherr, D. N. Reinhoudt, G. J. Vancso and J. Huskens, *Angew. Chem., Int. Ed.*, 2009, **48**, 983.
40 X. Y. Ling, L. Malaquin, D. N. Reinhoudt, H. Wolf and J. Huskens, *Langmuir*, 2007, **23**, 9990.
41 J. E. Sader, I. Larson, P. Mulvaney and L. R. White, *Rev. Sci. Instrum.*, 1995, **66**, 3789.
42 M. A. Meitl, Z. T. Zhu, V. Kumar, K. J. Lee, X. Feng, Y. Y. Huang, I. Adesida, R. G. Nuzzo and J. A. Rogers, *Nat. Mater.*, 2006, **5**, 33.
43 X. Li, H. Gao, C. J. Murphy and L. Gou, *Nano Lett.*, 2004, **4**, 1903.
44 A. C. Saucier, S. Mariotti, S. A. Anderson and D. L. Purich, *Biochemistry*, 1985, **24**, 7581.
45 J. F. G. A. Jansen, E. M. M. de Brabander-van den Berg and E. W. Meijer, *Science*, 1994, **266**, 1226.
46 A. W. Bosman, H. M. Janssen and E. W. Meijer, *Chem. Rev.*, 1999, **99**, 1665.

Polymer crystallization under nano-confinement of droplets studied by molecular simulations

Wenbing Hu,[*a] Tao Cai,[a] Yu Ma,[a] Jamie K. Hobbs,[b] O. Farrance[b] and Günter Reiter[c]

Received 22nd January 2009, Accepted 9th March 2009
First published as an Advance Article on the web 21st July 2009
DOI: 10.1039/b901378d

Fabrication of polymer nano-crystals proceeds usually through hierarchical ordering of the different-scale structures. Nano-scale patterns are produced first, which serve as a spatial template for subsequent polymer crystallization under nano-confinement. We begin with a survey of the effects of nano-confinement on polymer crystallization, mainly on the basis of the knowledge obtained from molecular simulations. After that, we report dynamic Monte Carlo simulations of polymer crystallization confined in nano-droplets. We observed that the shape of droplets on a solid substrate appears as a pancake, and both initiation and development of crystallization are depressed with the decrease of droplet size. Surface-induced crystal nucleation guides the dominant edge-on crystal orientation at high temperatures; however, its contribution to nucleation rates is not much greater than crystal nucleation in the volume of the droplet. At low temperatures, edge-on crystals are frequent at both substrate/polymer and polymer/air interfaces. In conclusion, molecular simulations can shed light on the microscopic mechanisms of polymer crystallization under nano-confinement.

Introduction

Crystallization of polymers provides not only stable nano-scale structures, but also anisotropies of electrical conductivity, mechanical strength and optical dichroism *etc.* Therefore, understanding the crystallization behavior of polymers under nano-confinement is of essential importance in the fabrication of functional nano-structures involving crystalline polymers. Among various experimental, theoretical and simulation approaches reported, molecular simulations show particular advantages related to the ease to access nano-scale behavior. In practice, nano-confinement may result from restrictions imposed by geometries of different dimensionalities. Typical geometries include lamellae, cylinders and spheres, corresponding to spatial restriction in one, two and three dimensions respectively. In this report, we first present a brief survey of the effects of nano-confinement on polymer crystallization, knowledge obtained mainly through our dynamic Monte Carlo simulations of simple lattice model polymers. Then, we focus our attention on polymer crystallization under nano-confinement of droplets which results, for example, from the dewetting of a thin polymer film on a solid substrate.

---

[a]*Department of Polymer Science and Engineering, State Key Laboratory of Coordination Chemistry, School of Chemistry and Chemical Engineering, Nanjing University, Nanjing, 210093, China. E-mail: wbhu@nju.edu.cn*
[b]*Department of Chemistry, University of Sheffield, Sheffield, S3 7HF, United Kingdom*
[c]*Physikalisches Institut, Universität Freiburg, 79104 Freiburg, Germany*

In practice, such droplets of polymer melt can effectively isolate impurities that commonly initiate heterogeneous crystal nucleation under small supercoolings, and thus allow us to focus our observations on homogeneous crystal nucleation at large supercoolings.[1,2] The randomness of homogeneous crystal nucleation among isolated droplets has been convincingly demonstrated by the observations of atomic force microscopy (AFM) on nanometre-sized spherical microdomains of diblock copolymers.[3] Crystallization confined in nano-sized droplets appears to be fractionated, probably due to the distribution of droplet sizes and its influence on the nucleation probability; in addition, it produces very imperfect crystals of strongly reduced crystallinity.[4] Heating these imperfect crystals often leads to significant re-organization well before complete melting.[4,5]

Recently, Massa and Dalnoki-Veress found that the nucleation rate depends on the volume size of droplets that have been formed by the dewetting of polymers on a solid substrate.[6] Even the so-called self-nucleation is still randomly initiated among the droplets by a memory effect.[7] Moreover, the nucleation rates appear to be independent of chain lengths, similar to the behavior found in bulk polymer systems.[8] Kailas *et al.* have observed that the preferentially edge-on lamellar crystals were initiated in such nano-scale droplets resulting from dewetting of polymers on a solid substrate, with the nucleation rates mainly dependent on droplet thickness for a thickness of <5 nm.[9] Molecular dynamics simulations by Miura and Mikami observed that the nano-sized interface region of spherical domains induce crystal nucleation only at the early stage and depress crystal growth at later stages.[10] However, whether nucleation inside each nano-droplet crystal is preferentially initiated at interfaces or within the bulk phase requires further microscopic observations. We therefore employed dynamic Monte Carlo simulations to observe polymer crystallization confined in droplets of variable sizes. Our results will show that surface-induced crystal nucleation favors edge-on crystal orientation at high temperatures. However, its contribution to the overall nucleation rate does not seem to be significantly higher than the contribution of crystal nucleation in the bulk phase of the droplet, giving rise to nucleation rates proportional to both thickness and total volume of pancake-like droplets.

The content of this paper is organized as follows. We begin with an overview of the simulations of polymer crystallization confined in different nano-sized geometries. Then, we introduce the simulation techniques and sample preparation we used in the present study. After that, we present our simulation results together with a discussion on polymer crystallization confined in the geometry of nano-droplets. The paper ends with a summary of conclusions.

Overview: effects of nano-confinement on polymer crystallization as revealed by molecular simulations

In parallel to extensive experimental investigations, we have performed a series of dynamic Monte Carlo simulations on polymer crystallization confined in various forms such as ultrathin films,[11] nano-rods[12] and nano-droplets (presented here), all consisting of homopolymers, as well as within the lamellar,[13,14] cylindrical[15,16] and spherical[17] microdomains formed by diblock copolymers. There are two basic scenarios for nano-confinement imposed by the microdomains of diblock copolymers: hard confinement if the matrix constituted mainly by non-crystallizable blocks is glassy; and soft confinement if the matrix is deformable. Under soft confinement, crystallization often gives rise to breakout which destroys the nano-pattern generated by microdomains of diblock copolymers.

In a simple sketch, the key factors influencing polymer crystallization under nano-scale confinement can be divided into three categories: those related to the interface; to the geometry; and, in the case of block copolymers, to the block junction.

Flat interfaces impose both enthalpic and entropic effects on nano-confined polymers. The flatness of interfaces induces deformation in the contacting polymers, which are oriented preferentially parallel with the interfaces. This entropic effect will enhance crystal nucleation[18] and hence yields mainly edge-on oriented crystals at high temperatures if the flat walls do not have a high affinity for the polymer (*i.e.* are not sticky walls).[11,12] Such an effect also appears to be significant in nano-droplets as we report in this paper. On the other hand, a high affinity of polymers to interfaces decreases the mobility of polymers near interfaces. This enthalpic effect will slow down or even freeze polymer reorganization near interfaces. Under such conditions, the preferential crystal orientations switch from edge-on to flat-on for both lamellar[11] and cylindrical[12] confinement.

Spatial restrictions in low dimensions like in lamellae and cylinders cause anisotropic crystal growth. Only lateral crystal growth which advances in parallel with the extension axis (*i.e.* the unconfined direction) can proceed well and allows high crystallinity to be obtained.[12] In the droplets discussed below, polymer contacting at interfaces also plays an important role in the selection of crystal orientations at high temperatures.

Due to microphase separation of diblock copolymers, the junctions between blocks are oriented perpendicular to the interfaces. Such an orientation effect has important consequences at high temperatures when crystal nucleation is a rare event. It may guide the preference for perpendicular crystal orientations in lamellar[13] and cylindrical microdomains[15] of diblock copolymers. Molecular simulations have demonstrated that a single prealigned chain is already sufficient to induce crystal nucleation at high temperatures.[19] Under soft confinement, such preferred perpendicular crystal orientations may be responsible for "overcrowding" at the interfaces induced by crystal thickening, which leads first to thickness undulation and subsequently to the breakout of crystals from microdomains.[14,16] At high temperatures, when crystal thickness becomes larger than cylinder thickness, lateral crystal growth of lamellar crystals causes a breakdown of cylinders into a string of small crystallites aligned along the cylinder axis.[16] Under hard confinement of spherical microdomains, the block junctions are highly immobile and becomes responsible for low crystallinity especially when the sphere size is small.[17] On the other hand, the large curvature of spherical interfaces induces a slight chain extension of crystallizable blocks, which favors fast crystallization within spherical microdomains of diblock copolymers.[17]

Simulation techniques and sample preparation

Dynamic Monte Carlo simulations of lattice polymers have served as a powerful tool to improve our understanding of polymer crystallization behaviors.[20] Such simulations employ a micro-relaxation model to equilibrate chain conformations on the cubic lattice with polymer volume exclusion and periodic boundary conditions. Each step of micro-relaxation allows the monomer to jump to its vacant neighboring site with partial sliding diffusion along the chain, and thereby efficiently changes local conformations of polymer chains. Polymer bonds are not allowed to overlap or cross with each other due to their "hard" volume exclusion. Micro-relaxation allows bonds oriented either along the lattice axes or along the (face and body) diagonals, so the coordination number of the cubic lattice is as high as $6 + 12 + 8 = 26$, with 13 directions allowed for bond orientations.

In our study, we focus attention on the situation of a liquid droplet resulting from dewetting on a solid substrate, as previously used in experiments. To maintain stable droplet interfaces and to drive polymer crystallization, the well-known Metropolis importance-sampling algorithm has been employed to bias each step of micro-relaxation with the probability of acceptance as $\min\{1, \exp[-\Delta E/(kT)]\}$, where k is the Boltzmann constant, T is the temperature, and the energy barrier is defined as

$$\Delta E = cE_C + pE_P + b_1 B_1 + b_2 B_2 + gG = \left(c + p\frac{E_P}{E_C} + b_1\frac{B_1}{E_C} + b_2\frac{B_2}{E_C} + g\frac{G}{E_C} \right) E_C \quad (1)$$

Here, five interaction parameters were taken into consideration in accordance with the real situation in experiments. Referring to the fully ordered ground state, the first energy parameter E_C characterizes the bending energy for the non-collinear connection of two consecutive bonds along polymer chains, and c is the net amount of non-collinear connections; the second energy parameter E_P describes the energy penalty for non-parallel packing of two neighboring bonds, corresponding to the driving force for polymer crystallization,[20] and p is the net number of non-parallel pairs of polymer bonds; the third energy parameter B_1 is the interaction between the site of the substrate and the neighboring monomer, and b_1 is the net number of monomers contacting with the substrate; the fourth energy parameter B_2 is the interaction between a vacancy site (representing "air" surrounding the droplets) and the neighboring monomer, and b_2 measures the net number of "air"–monomer contacts; the last energy parameter G describes the energy penalty for lifting up the monomer away from the substrate (representing a weak 'gravitation' to collapse droplets), and g denotes the net height in those monomers moving up along the z direction.

We separately put 60, 40, 20, 5 and 1 polymer chains, each chain containing 2048 monomers, into a $x \times y \times 35$ lattice box, where $x \times y$ can be 64×64, 48×64, $48 \times$

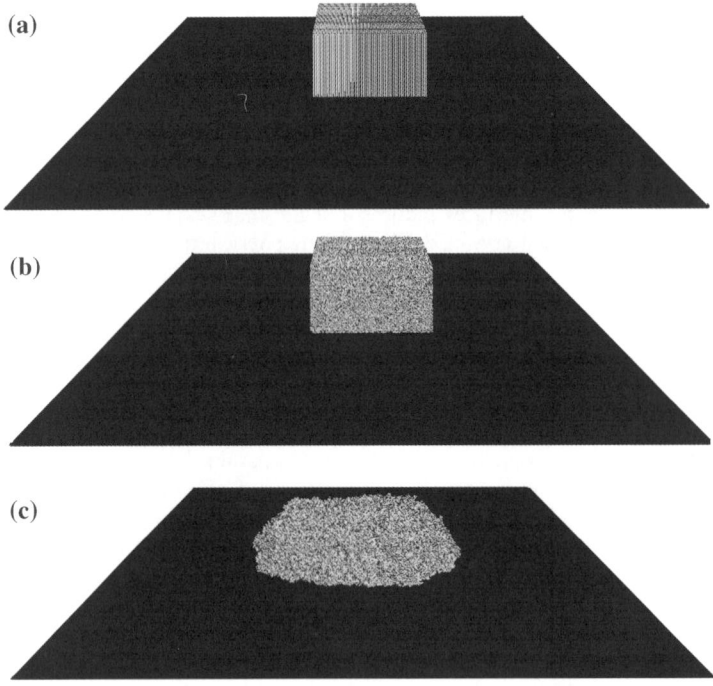

Fig. 1 Snapshots of the polymer sample system with 60 chains at different stages of droplet preparation on a solid substrate with the size 256×256 ($x \times y$). (a) The ordered initial state of the sample system in a box of $64 \times 64 \times 35$ ($x \times y \times z$); (b) after relaxation at $T = 6.0E_C/k$ for 2×10^6 MC cycles with hard walls as boundary conditions; (c) after relaxation at $T = 6.0E_C/k$ for 2×10^6 MC cycles with periodic boundary conditions along x- and y-axes. All the bonds are drawn as tiny cylinders with the substrate in black and the polymer chains paler.

32, 24 × 16 and 12 × 8 corresponding to the different numbers of polymer chains. The chains are regularly folded as in a single crystal (see, for example, Fig. 1a), and some vacancy sites are necessary in the box to allow polymer chains to relax. Then, we set the value of B_1/E_C to be constant at −0.05 to stand for weak attraction of the substrate to polymer chains in order to maintain a dewetted droplet, and the value of b_2/E_C fixed at 0.35 to represent the repulsion of the "air" to polymer chains, and the weak 'gravitation force' G/E_C to be constant at 0.01 to keep the droplet sticking at the substrate rather than floating away. For the sake of simplicity, we also set the value of E_P/E_C to be constant at one to represent a flexible chain. With hard walls as boundary conditions around the box, we relaxed polymer chains at a high temperature $T = 6.0 E_C/k$ for a period of 2×10^6 Monte Carlo (MC) cycles to reach the bulk random-coil states (see Fig. 1b), where each MC cycle was defined as the number of trial moves equal to the total number of monomers in the sample system. After that, we enlarged the box size to $256 \times 256 \times 35$ ($x \times y \times z$) and employed the periodic boundary conditions on x- and y-axes. In consequence, the sample systems collapsed down to form stable droplets after being annealed for another period of 2×10^6 MC cycles.

Fig. 2 shows the topographic pictures of the prepared droplets with various amounts of polymers from 60 to 5. One may recognize their roughly circular shapes

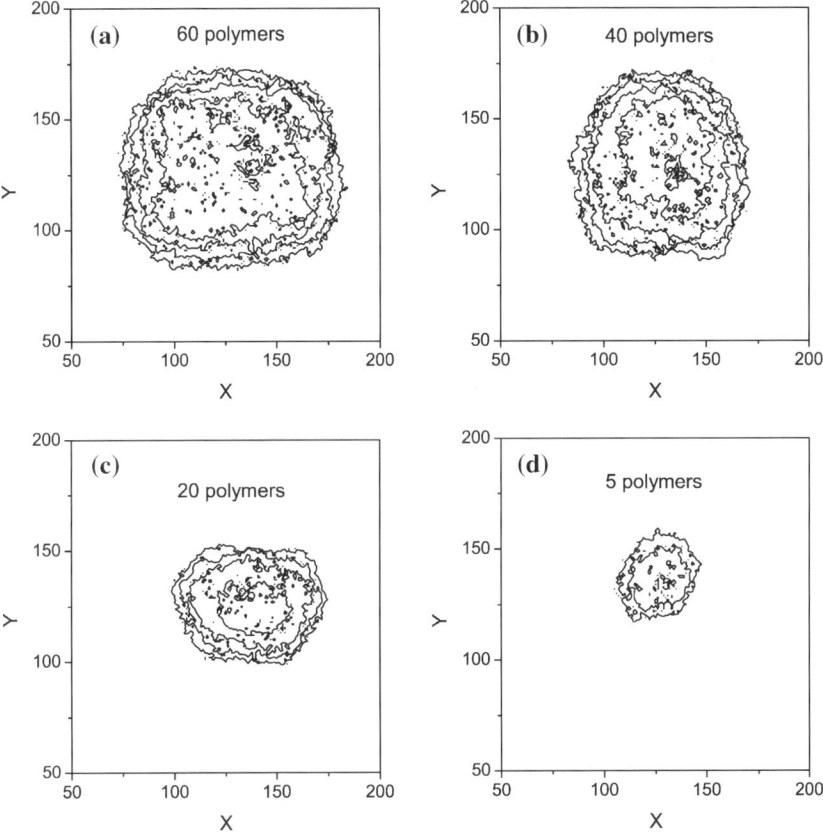

Fig. 2 Topographic pictures of the prepared droplets with different amount of polymer chains as denoted in the figures from (a) to (d). The lines are spaced with the height difference of five along the z-axis. The numbers printed in the highest domains indicate its lower height limit in lattice units.

with a flat center resembling pancakes. Such a shape is quite similar to the dewetting droplets observed under AFM.[9]

In the following, we will cool the sample systems down and observe their crystallization behaviors.

Results and discussion

We first changed the temperatures stepwise from 5 to 2 in steps of 0.002 separated by periods of 1000 MC cycles to mimic a slow cooling process. The crystallinity was monitored during the cooling process. Since in our simulations the number of parallel-packed bonds around each bond may vary from zero in the fully amorphous phase to 24 in the fully crystalline phase, we arbitrarily chose the number of 5 as the demarcation between the amorphous phase and the crystalline phase. Accordingly, the crystallinity was defined as the fraction of crystalline bonds in the total amount of bonds. Fig. 3 summarizes all the cooling curves. One may draw the conclusion that the smaller the amount of polymer, the slower the initiation of crystallization on cooling. Such a result is in agreement with expectation and with experimental observations. The nano-scale size of droplets represents a strong reduction in occurrence of the stochastic process of crystal nucleation.

We then quenched the droplets to various low temperatures in order to observe the isothermal crystallization. We performed the conventional Avrami analysis on these isothermal crystallization processes. The time (t) evolution of the relative crystallinity (X_c) can be treated by the Avrami equation

$$1 - X_c = \exp(-Kt^n) \qquad (2)$$

where the Avrami index n characterizes the situation of crystal nucleation and growth, and the kinetic constant K reflects the overall crystallization rate.[21] The results are summarized in Fig. 4a and 4b. One may recognize again that the smaller the droplet size, the slower the crystallization rate. The Avrami index increases from

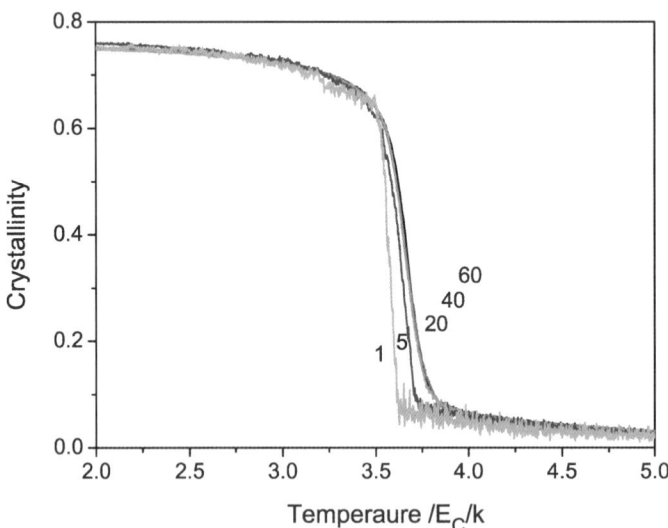

Fig. 3 Crystallinity upon cooling for the sample systems with variable numbers of polymer chains as denoted. A slow cooling process was mimicked by a stepwise change of the temperature in steps of 0.002 separated by periods of 1000 MC cycles. Only the segments connecting data points are drawn for clarity.

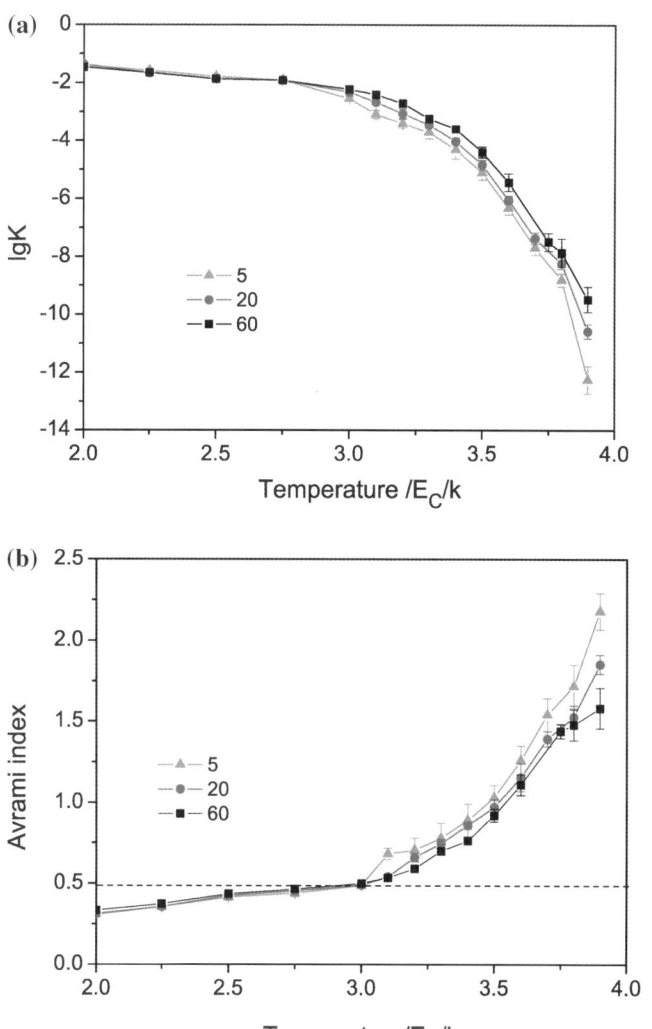

Fig. 4 Avrami analysis of isothermal crystallization of droplets at various temperatures with different sizes as labeled. (a) The kinetic constants; (b) the Avrami indexes. Each reported data was averaged over five individual processes of the sample systems with different seeds of random-number generation.

0.5 at low temperatures to more than two at high temperatures, implying that fewer nuclei generated at higher temperatures gain larger sizes due to crystal growth.

The saturated crystallinities obtained from long-term isothermal crystallization at various temperatures are summarized in Fig. 5. One can see that with the increase of temperatures, these saturated crystallinities increase, although all values are still far from 100% crystallinity. Meanwhile, we also estimated the degree of orientational order of those crystalline bonds by tracing the orientational order parameter F that was defined as

$$F = \frac{3\langle \cos^2 \theta \rangle - 1}{2} \qquad (3)$$

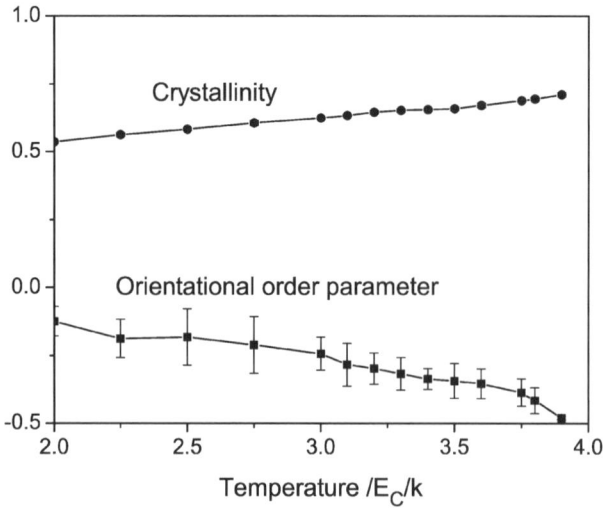

Fig. 5 Saturated crystallinities and orientational order parameters of crystalline bonds obtained from isothermal crystallization at various temperatures for the droplets containing 60 polymers. Each reported data point was averaged over five individual processes of the sample systems with different seeds of random-number generation.

where the angle θ was read with reference to the z-axis and $\langle ... \rangle$ means an assembled average over all the bonds containing more than 15 parallel neighbors. Here, we increased the criteria number of parallel neighbors to obtain a better resolution for the measurement. According to this definition, if all the concerned bonds are in parallel with the z-axis, F is equal to one; if they are perpendicular, F will be -0.5; and if they are random, F is close to zero. The results of orientational order parameters are also summarized in Fig. 5. One can see that the crystalline bonds have a strong tendency to orient perpendicular to z-axis, in other words, in parallel with the surface of the solid substrate, implying a preference of edge-on crystals on the substrate. This preference becomes even stronger with the increase of crystallization temperatures. The preferentially edge-on lamellar crystals are consistent with experimental observations.[8,9]

The preference for edge-on crystal orientations may be related to crystal nucleation induced by an entropic effect of the interfaces, as discussed in the above overview section. Therefore, it is worth having a close look at crystal nucleation at high temperatures. In the context of the well-known classical nucleation theory, each new crystallite has first to grow beyond a threshold size related to a free-energy barrier induced by the cost in surface free energy. Upon thermal fluctuations in the melt, the largest embryo crystallite has the highest probability to win the competition of crossing the threshold size. So upon thermal fluctuations at high temperatures, the properties of the largest crystallites represent the properties of crystal nuclei. One major advantage of this approach focusing only on the largest embryo crystallite is that we can calculate ensemble statistics with respect to the melt over a long time period during an isothermal process at a temperature high enough to avoid any irreversible crystal growth.

Fig. 6a shows the radial distribution of the probability of observing the largest crystallites which were generated across the droplet by thermal fluctuations at a high temperature. The very high probability around the central point of the droplet can be identified as a statistical artifact due to the discrete lattice space. There exists a plateau for the density of nucleation probability in the center of droplets. In

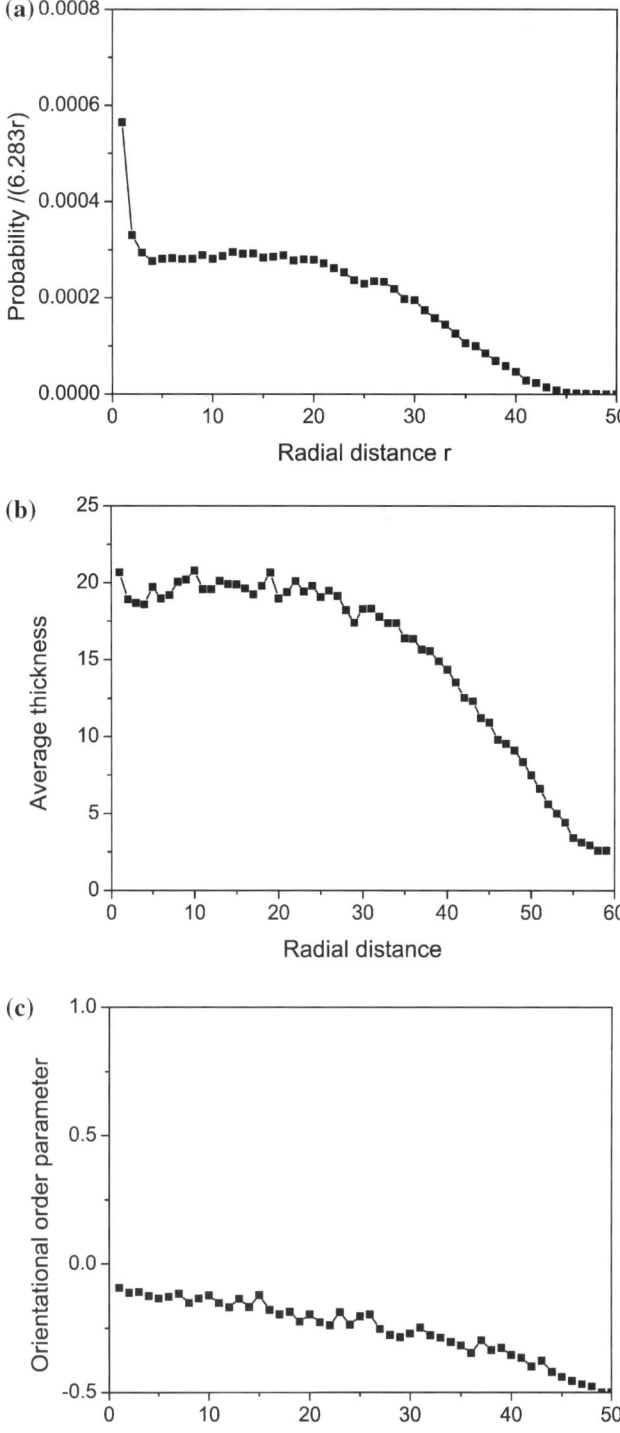

Fig. 6 Radial distribution of (a) the density of probability of observing the largest crystallites, (b) the droplet thickness, and (c) orientational order parameters of the largest crystallites calculated for the droplet containing 60 polymers at $T = 6.0E_C/k$. The largest crystallites were collected over 20 000 samples with 100 MC cycle intervals.

comparison with the radial distribution of droplet thickness shown in Fig. 6b, one may conclude that the probability of observing crystal nucleation appears to be proportional to the droplet thickness. This result implies that the surface-induced crystal nucleation cannot be much faster than homogenous crystal nucleation in the bulk phase of the droplet. The latter contributes to the total nucleation rates in a manner not very different from nucleation induced at the droplet surface.

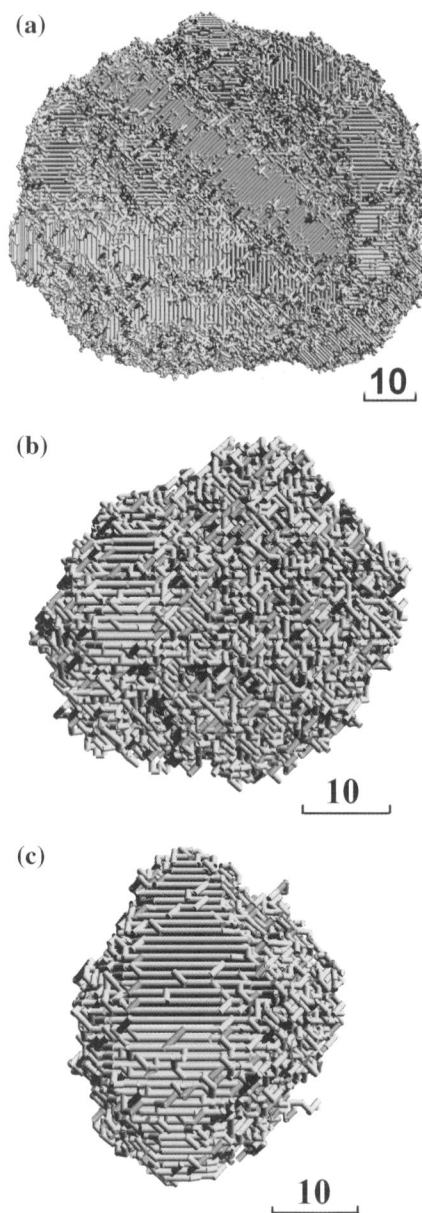

Fig. 7 Snapshots of the droplets after isothermal crystallization for various periods at $T = 3.9E_C/k$. (a) Containing 60 polymers after crystallization for 1.01×10^6 MC cycles; (b) containing 5 polymers after crystallization for 5.0×10^5 MC cycles; (c) containing 5 polymers after crystallization for 1.5×10^6 MC cycles. All the bonds are drawn as tiny cylinders. The snapshots represent views from the top of the droplets.

Indeed, the radial distribution of orientational order parameters of the largest crystallites shown in Fig. 6c demonstrates more homogeneous crystal nucleation with random crystal orientations in the center of the droplet. Hence, the surface-induced crystal nucleation dominates the crystal orientations at the surfaces of droplets, but its contribution to the total nucleation rates is not dominant over the volume crystal nucleation. So the rate of nucleation per unit of volume is closely similar to that in the bulk, but the orientation of nuclei 'edge-on' to the surface is strongly favored.

One major advantage of molecular simulations is that they allow visualization of crystallite structures with the resolution of molecular details. We therefore performed visual inspections on the snapshots obtained at both high and low temperatures. Fig. 7a shows the snapshot of the droplet containing 60 polymers after isothermal crystallization at a high temperature. The shape of the droplet is well preserved upon isothermal crystallization, with multiple crystallites being generated inside the droplet. The interesting point is that we only observed edge-on crystallites (see also Fig. 5) but with random orientations within the plane of the solid substrate. Fig. 7b and 7c demonstrate that with only one crystallite in the small droplet containing just five polymers, the shape of the droplet will be highly deformed as all polymers join into a single crystal.

Fig. 8a shows the snapshot of a droplet containing 60 polymers after isothermal crystallization at a low temperature. Many more small crystallites were generated in this case, but the top surface of the droplet shows a strong preference for crystal orientations with chains parallel to the top surface. The sectional view inside the droplet shown in Fig. 8b tells us that the chain-parallel preference only occurs in a limited region near the top surface. In the middle of the droplet, crystal

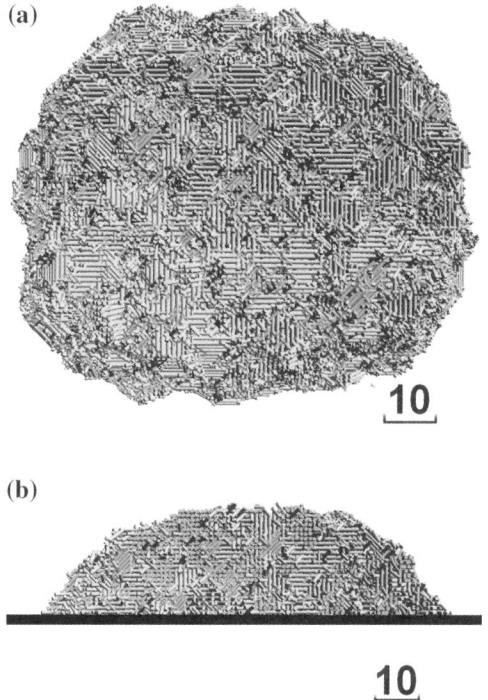

Fig. 8 Snapshot of the droplet containing 60 polymers after isothermal crystallization for 5.85×10^4 MC cycles at $T = 2.0E_C/k$. All the bonds are drawn as tiny cylinders. (a) The snapshot is viewed from the top side of the droplet; (b) sectional view of the inside structure of the droplet (cut at $y = 140$).

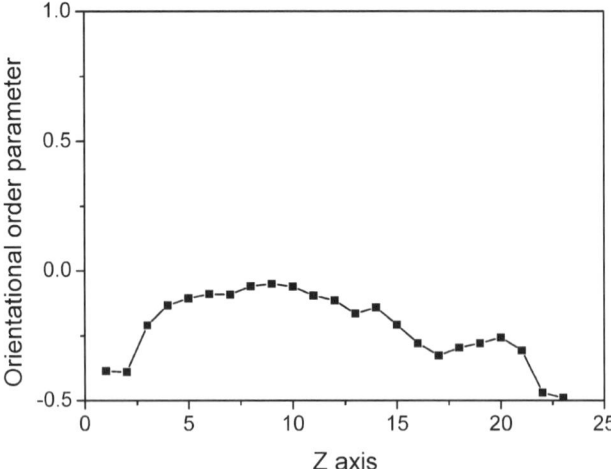

Fig. 9 Distribution of orientational order parameters of crystalline bonds along the z-axis for the droplet containing 60 polymers after isothermal crystallization for 5.85×10^4 MC cycles at $T = 2.0E_C/k$. The segments are drawn to guide the eyes.

orientations appear to be quite random. We made statistics of orientational order over the positions of crystalline bonds along the z-axis. The results are shown in Fig. 9. One can see that at low temperatures, the edge-on crystals are enriched at both top and bottom surfaces of the pancake-like droplets.

Conclusions

We performed dynamic Monte Carlo simulations of polymer crystallization confined in nano-droplets. The shape of droplets, which could have resulted from dewetting of polymer thinfilm on a solid substrate, resembles a pancake. With the decrease of droplet size, both initiation and evolution of crystallization inside droplets become slower. A strong edge-on preference of crystal orientations occurs at interfaces. However, the contribution of crystal nucleation induced at interfaces to the overall nucleation rate does not appear to be much more significant than the contribution from nucleation of random-oriented crystals inside the bulk phase of the droplet. Therefore, the nucleation rate for pancake-like droplets depends on both the thickness and the volume of these droplets.

Our results presented here provide another example which demonstrates the usefulness of molecular simulations for elucidating the microscopic mechanisms of polymer crystallization under nano-confinement. Such knowledge of physics on the nano-scale in the context of self-assembly of polymers will be beneficial for the nano-technology of fabricating 'soft-machines'.

Acknowledgements

The financial support from National Natural Science Foundation of China (NSFC Grant Nos. 20474027, 20674036, and 20825415) is appreciated. The joint funding from the NNSFC and the Royal Society of the United Kingdom is also appreciated.

References

1 D. Turnbull and R. L. Cormia, *J. Chem. Phys.*, 1961, **34**, 820–831.
2 R. L. Cormia, F. P. Price and D. Turnbull, *J. Chem. Phys.*, 1962, **37**, 1333–1340.

3 G. Reiter, G. Castelein, J.-U. Sommer, A. Röttele and T. Thurn-Albrecht, *Phys. Rev. Lett.*, 2001, **87**, 226101.
4 R. T. Tol, V. B. F. Mathot and G. Groeninckx, *Polymer*, 2005, **46**, 2955–2965.
5 A. Roettele, T. Thurn-Albrecht, J.-U. Sommer and G. Reiter, *Macromolecules*, 2003, **36**, 1257–1260.
6 M. V. Massa and K. Dalnoki-Veress, *Phys. Rev. Lett.*, 2004, **92**, 255509.
7 M. V. Massa, M. S. M. Lee and K. Dalnoki-Veress, *J. Polym. Sci., Part B: Polym. Phys.*, 2005, **43**, 3438–3443.
8 M. V. Massa, J. L. Carvalho and K. Dalnoki-Veress, *Phys. Rev. Lett.*, 2006, **97**, 247802.
9 L. Kailas, C. Vasilev, J.-N. Audinot, H.-N. Migeon and J. K. Hobbs, *Macromolecules*, 2007, **40**, 7223–7230.
10 T. Miura and M. Mikami, *Phys. Rev. E: Stat., Nonlinear, Soft Matter Phys.*, 2007, **75**, 031804.
11 Y. Ma, W.-B. Hu and G. Reiter, *Macromolecules*, 2006, **39**, 5159–5164.
12 Y. Ma, W.-B. Hu, J. Hobbs and G. Reiter, *Soft Matter*, 2008, **4**, 540–543.
13 W.-B. Hu and D. Frenkel, *Faraday Discuss.*, 2005, **128**, 253–260.
14 W.-B. Hu, *Macromolecules*, 2005, **38**, 3977–3983.
15 M.-X. Wang, W.-B. Hu, Y. Ma and Y.-Q. Ma, *J. Chem. Phys.*, 2006, **124**, 244901–244906.
16 Y. Qian, T. Cai and W.-B. Hu, *Macromolecules*, 2008, **41**, 7625–7629.
17 T. Cai, Y. Qian, Y. Ma and W.-B. Hu, *Macromolecules*, 2009, **42**, 3381–3385.
18 P. J. Flory, *J. Chem. Phys.*, 1947, **15**, 397–408.
19 W.-B. Hu, D. Frenkel and V. B. F. Mathot, *Macromolecules*, 2002, **35**, 7172–7174.
20 W.-B. Hu and D. Frenkel, *Adv. Polym. Sci.*, 2005, **191**, 1–35. See the Appendix.
21 M. Avrami, *J. Chem. Phys.*, 1939, **7**, 1103–1112.

PAPER

Nanostructured wrinkled surfaces for templating bionanoparticles—controlling and quantifying the degree of order

Anne Horn,[a] Heiko G. Schoberth,[a] Stephanie Hiltl,[a] Arnaud Chiche,[b] Qian Wang,[c] Alexandra Schweikart,[a] Andreas Fery*[a] and Alexander Böker*[ad]

Received 9th February 2009, Accepted 16th March 2009
First published as an Advance Article on the web 22nd July 2009
DOI: 10.1039/b902721a

We present a novel method to align the tobacco mosaic virus (TMV) on topographically structured surfaces. In order to gain defined patterns we use wrinkled polydimethlysiloxane (PDMS) sheets as templates. We aligned the virus with a simple spin-coating procedure on the PDMS sheet. The concentration of the virus solution and the spin speed are varied in order to identify ideal conditions for the arrangement of the viruses on the wrinkled templates. Here, we establish a simple analytical approach which allows quantifying the degree of order of the patterns, which is the basis for a quantitative discussion of templating efficiency. Furthermore, we discuss the role of dewetting processes for the particle assembly. TMVs can be used as reactive nanoparticles due to their well-defined surface chemistry. They can as well serve as a model system for alignment of anisotropic particles *via* spin coating from solution.

1 Introduction

The arrangement of colloidal particles is of common interest in technology and science as well. Potential applications in the field of data storage devices[1] and nano-electronics[2] were investigated by several groups. Therefore, the use of topographically structured substrates to align colloids gained increasing attention in the recent years.[3] Until now lithographic methods were the common techniques of obtaining those structures.[4,5] An alternative approach to generate patterned surfaces is controlled wrinkling or buckling of an elastomeric substrate with a hard top layer. Templated self-assembly of synthetic colloids like polystyrene, silica or gold particles has been explored recently.[3,6–8] However, anisotropic particles open a wider range of potential applications due to their unique electrical and optical properties. Rod-shaped biomolecules combine anisotropy and monodispersity along with chemical addressability and can serve as efficient templates for various reactions *e.g.* metallization or mineralization.[9–12] Hence, arrays of well-defined tobacco mosaic virus (TMV) on wrinkled substrates is an excellent system to study the self-assembly of anisotropic colloids on

[a] *Lehrstuhl für Physikalische Chemie II, Universität Bayreuth, D-95440 Bayreuth, Germany.*
E-mail: Andreas.Fery@uni-bayreuth.de
[b] *DSM-Material Science Centre, P.O. Box 18, 6160, MDGeleen, The Netherlands*
[c] *Department of Chemistry and Biochemistry and Nanocenter, University of South Carolina, Columbia, South Carolina, 29208, USA*
[d] *Lehrstuhl für Makromolekulare Materialien und Oberflächen and DWI an der RWTH Aachen e.V., RWTH Aachen University, D-52056 Aachen, Germany. E-mail: Boeker@dwi.rwth-aachen.de*

structured surfaces. In particular, we paid special attention to developing quantitative measures for the particle order, which allows optimizing assembly parameters.

In the work reported here, we developed a method yielding excellent control over TMV alignment on a large scale using wrinkled polydimethylsiloxane (PDMS) substrates. *Via* a simple spin-coating technique we are able to fill the grooves of the wrinkles with the virus and achieve patterns with high aspect ratios. The advantages of this approach are that it is not necessary to functionalize either the virus or the surface chemically and that there is no need of external physical forces other than those present during the spin-coating process in order to direct the self-assembly process. Additionally the PDMS substrates are produced in an easy, cheap and lithography-free way. Hence, we found a simple procedure to obtain highly aligned TMV strings with dimensions on the order of several microns in length but only a few nanometres in width on a cm^2 substrate.

2 Experimental

2.1 Production of wrinkled PDMS substrates

We prepared the PDMS by mixing Sylgard 184 Base (Dow Corning) and Sylgard 184 Curing Agent (Dow Corning) in a mass ratio of 10 : 1. The mixture was cast (3 mm in height) in a petri dish and heated to 80 °C for 24 h. Pieces of 6 mm width and 30 mm length were positioned in a custom-made stretching apparatus[13] and extended to 130% of the original length. These substrates were oxidized between 40 s and 120 s in an air plasma (1 mbar, 18 W, PDC-32 G, Harrick) in the stretched state. After relaxing the wrinkled samples could be used for further application. The application of wrinkles for templating has recently been reviewed.[14]

2.2 Alignment of TMV on wrinkled substrates

A TMV stock solution (10 mg ml^{-1}, 0.1 M potassium phosphate buffer, pH 7.8) was diluted with ultrapure water to the desired final concentration (between 0.2 mg ml^{-1} and 1.2 mg ml^{-1}). A droplet of 50 µl was spin-cast onto the wrinkled substrates (between 2000 rpm and 4000 rpm). After drying under N_2-flow the samples were characterized with scanning electron microscopy (SEM) and scanning force microscopy (SFM).

2.3 Characterization

SEM images were recorded by using FE-SEM (Zeiss 1530) operating at a voltage of 0.75 kV and with a working distance between 4 mm and 5 mm. To improve the measurements samples were sputtered with carbon (approximately 6 nm). SFM images were performed with a commercial SFM (Dimension 3100, Vecco Instruments Inc.) in Tapping Mode™. Si_3N_4 cantilevers were purchased from Olympus (spring constant ~40 N m^{-1}). Transmission electron microscopy (TEM) was performed on a Zeiss CEM902 microscope operated at 80 kV.

3 Results and discussion

The preparation of the wrinkled substrates was done using a previously reported method.[15–17] In Fig. 1a, a scanning force microscope (SFM) image is shown representing the topography of the obtained structures where the height profile illustrates the sinusoidal shape of the wrinkles. In order to verify this structure an epoxy resin replica of the wrinkled surface was made and imaged with transmission electron microscopy (TEM). The TEM cross section of the epoxy replica is shown in Fig. 1b. The desired wavelength λ can be tuned easily by the plasma exposure dose (in the range between 130 nm and 1 µm, see ref. 18) as it is dependent on the thickness of the hard oxide layer generated by the plasma. The amplitude A is

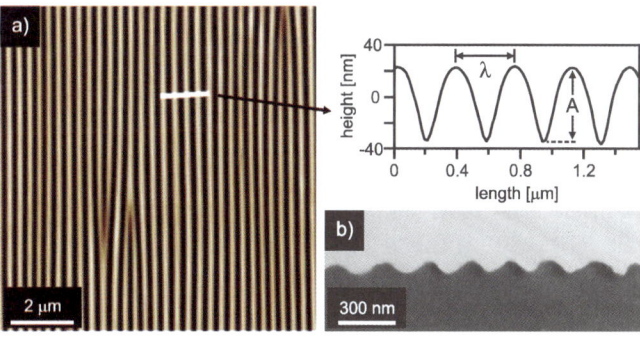

Fig. 1 (a) SFM height image (z-range = 0–60 nm) of wrinkled PDMS substrate ($\lambda = 324 \pm 22$ nm; $A = 58 \pm 2$ nm). The white line represents the position of the height profile shown on the right side. (b) TEM image showing a cross section of an epoxy replica of wrinkled PDMS ($\lambda = 232 \pm 24$ nm; $A = 59 \pm 8$ nm).

dependent on the applied strain and the present wavelength but the aspect ratio is limited to 0.1 for the conditions used here. Therefore it is possible to generate defined substrates with variable dimensions.

The aim of our study is to find optimum conditions to arrange the TMV in the grooves of the wrinkles. The rod-shaped virus is 300 nm long and 18 nm in diameter. To ensure comparable results we always used wrinkled templates with a wavelength λ of 300 ± 15 nm and an amplitude A of 30 ± 3 nm. Fig. 2 shows SFM images of aligned TMV on the wrinkled substrates. The viruses are mainly found in the grooves of the substrate. Previous studies have shown that, due to discontinuous dewetting, TMV particles selectively place at the base of relief structures.[10] The distances between the single virus strings are predetermined by the wavelength of the wrinkles. Therefore, the spacing of the strings can be controlled by adjusting the wavelength of the substrate. Furthermore it is possible to generate structural defects on the wrinkled supports. By regulating the release rate of the stretched PDMS the defect density can be controlled[15] and more complex geometries can be obtained.[19,20] Indeed, TMV is flexible enough to bend which means that it can be placed in e.g. a Y-junction (see Fig. 2b). As shown in Fig. 2c we can distinguish between one-fold and multiple adsorption by an SFM height profile. The height itself is the same in both cases but the shape of the topography is different. In the case with only one virus string lying in the groove the peak is much sharper than for multiple adsorption.

In the following, we focused on exploring the parameter range of the alignment conditions. Therefore we used SEM as our measuring technique, which allows one to examine larger regions to obtain statistically relevant information. We varied the concentration of the TMV solution (0.2 mg ml^{-1}, 0.4 mg ml^{-1}, 0.9 mg ml^{-1}, 1.2 mg ml^{-1}) while keeping the spin speed, the wavelength of the substrate and the amplitude constant at 3000 rpm, 300 nm and 30 nm, respectively. For every sample four images (2 × SFM, 2 × SEM) at different sample spots (a representative selection of SEM images is shown in Fig. 3a) were evaluated to calculate the virus occupancy parameter Ω (Here, we note that the following parameters V_k, W_k and O_k were extracted from the images by manually measuring the respective lengths):

$$\Omega = \frac{1}{\mathscr{A}_{\text{tot}}} \sum_{k=1}^{4} \mathscr{A}_k \frac{V_k}{W_k} \qquad (1)$$

where $\mathscr{A}_k$ is the evaluated area of the image k, $\mathscr{A}_{\text{tot}} = \sum_{k=1}^{4} \mathscr{A}_k$ is the total area of all images ($\mathscr{A}_{\text{tot}} \geq 770$ μm^2), V_k represents the total length of virus-filled grooves of

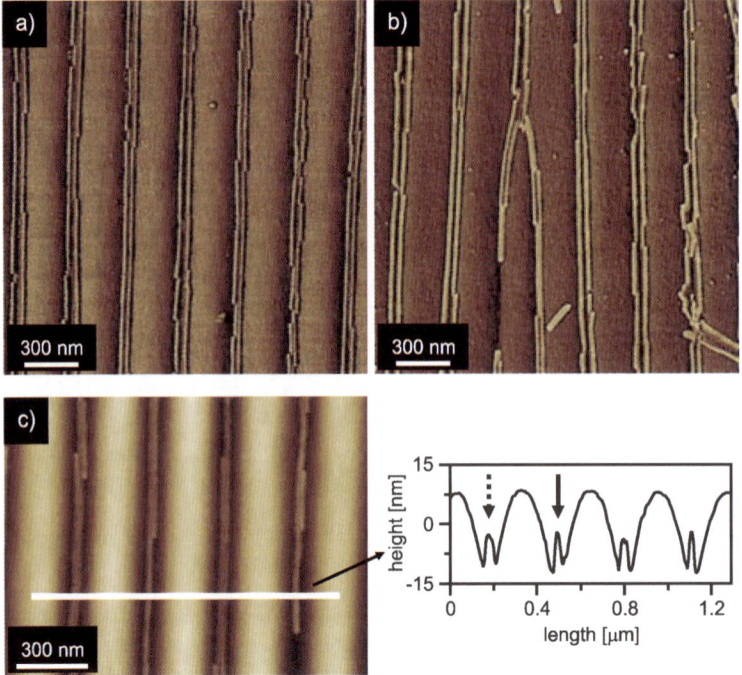

Fig. 2 (a) SFM phase images (z-range = 0–40°) of aligned TMV on wrinkled PDMS substrate showing multiple adsorption of the virus in the grooves ($\lambda = 294 \pm 21$ nm; $A = 32 \pm 1$ nm). (b) SFM phase image (z-range = 0–100°) of aligned TMV on a wrinkled PDMS substrate showing a Y-junction ($\lambda = 286 \pm 20$ nm; $A = 29 \pm 1$ nm). (c) SFM height image (z-range = 0–50 nm) with corresponding height profile showing one-fold (solid arrow) and multiple adsorption (dashed arrow) of the virus in the grooves ($\lambda = 299 \pm 18$ nm; $A = 27 \pm 1$ nm).

the image k and W_k is the total length of the grooves of the image k. The virus occupancy parameter characterizes how many viruses are in the grooves of the wrinkles ranging from $0 \rightarrow 1$ (where 1 represents 100% occupied grooves) but it does not take into account the viruses found outside the grooves. Therefore another parameter is necessary, namely the virus deviator parameter Φ:

$$\Phi = \frac{1}{\mathcal{A}_{tot}} \sum_{k=1}^{4} \mathcal{A}_k \frac{O_k}{f_k V_k + O_k} \qquad (2)$$

where O_k is the total length of viruses outside the grooves of image k and f_k is the multiple adsorption factor of image k. We extracted f_k by evaluating the averaged number of adsorption (one-fold, two-fold...) of the viruses from SFM images with higher magnification (e.g. 2 μm × 2 μm). The results of varying the concentration of the virus solution by keeping all other parameters constant are shown in Fig. 3b. The virus occupancy parameter increases with higher concentration. It reaches almost full groove occupancy ($\Omega = 1$) at 1.2 mg ml^{-1} which means that all the wrinkles are filled with the virus. As the virus deviator parameter is a relative value which compares how many viruses are out of the grooves to the total number of viruses on the sample it is obvious that the error bar at 0.2 mg ml^{-1} is very high. The number of viruses in the grooves at this concentration is rather low but there are always some viruses which stick directly to the surface. Relative to the total number of viruses the number of viruses outside the grooves decreases with increasing concentration. Hence, the virus deviator parameter decreases with

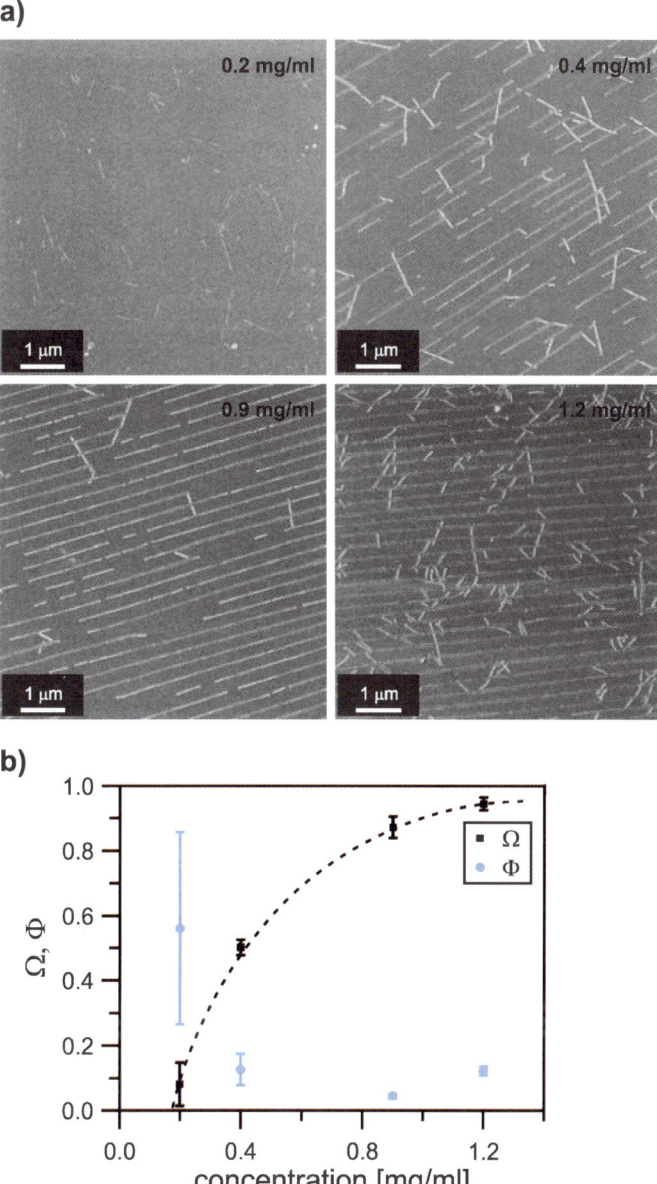

Fig. 3 (a) SEM images of TMVs aligned on wrinkled PDMS substrates prepared from virus solutions of different concentrations. (b) Plot of virus occupancy parameter (Ω) and virus deviator parameter (Φ) versus concentration (dashed line is a guideline for the eyes).

increasing concentration till reaching its minimum at 0.9 mg ml^{-1}. When the virus concentration reaches a certain value where nearly all the wrinkles are filled, the virus deviator parameter increases again. Obviously, in this case, the viruses have no space in the grooves any more and adsorb outside. Therefore, we find an optimum concentration of 0.9 mg ml^{-1} where the virus deviator has its minimum while the virus occupancy parameter has a high value. Nearly 90% of the wrinkles are filled with TMV and only a few adsorb unaligned on the surface.

Another important parameter which determines the TMV alignment is the spin speed. Therefore, we varied the spin speed (2000 rpm, 3000 rpm, 3500 rpm, 4000 rpm) by keeping the concentration constant at 0.9 mg ml^{-1}, the wavelength of the substrate at 300 nm and the amplitude at 30 nm. We evaluated the data in the same way as for the concentration dependency. Fig. 4a shows the dependence of the virus occupancy (Ω) and virus deviator (Φ) parameters on the spin speed. At a spin speed of 2000 rpm the value of Ω is around 0.5. By increasing the spin speed to 3000 rpm the value increases to approximately 0.9 which means that 90% of the wrinkles are filled with TMV. Further raising the spin speed leads to a decrease of the occupancy parameter again. This behaviour may be explained by the thickness of the water film. When starting from a continuous film it starts to dewet at certain areas of the substrate. At low spin speeds (*e.g.* 2000 rpm) the water film is rather thick and randomly formed droplets are generated. This leads to a discontinuous distribution of TMV on the substrate. There are areas with no virus (holes of the water film) and areas with a rather high concentration of virus (droplets). When the water is fully evaporated we have, on the one hand, regimes where no virus is found at all, and on the other hand regimes where the virus

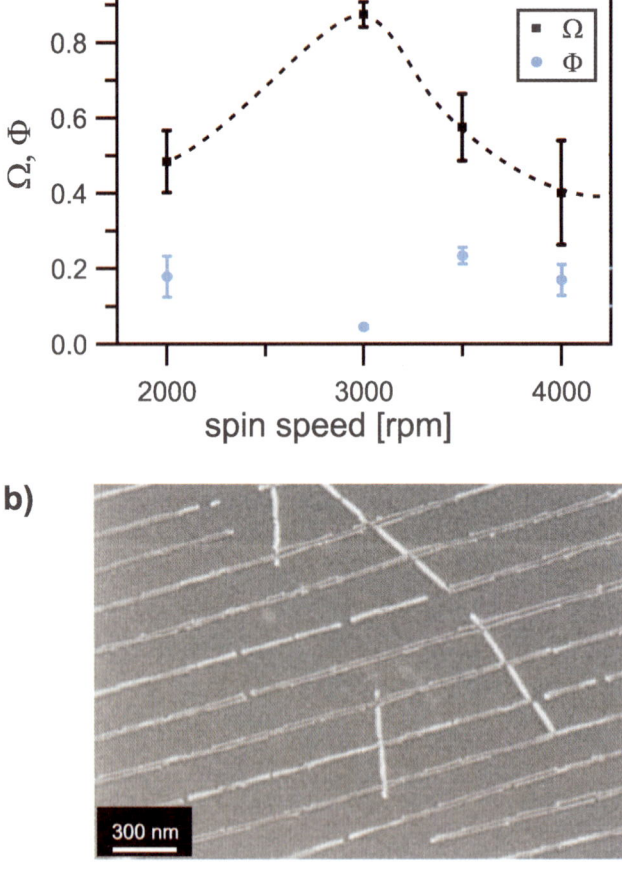

Fig. 4 (a) Plot of the virus occupancy parameter (Ω) and virus deviator parameter (Φ) *versus* spin speed (dashed line is a guideline for the eyes). (b) SEM image of TMVs aligned on wrinkled PDMS ($\lambda = 324 \pm 22$ nm; $A = 58 \pm 2$ nm) with optimum conditions (spin speed = 3000 rpm, $c = 0.9$ mg ml^{-1}).

concentration is so high that TMV not only fills the grooves but is also a part of the surface. This leads to a virus deviator parameter of around 0.2. The maximum of the virus occupancy parameter is at 3000 rpm and was found to be the optimum spin speed for the alignment of TMV. Here, the water film is thinner than in the case described before. Hence, it starts dewetting on the top of the wrinkles which means that the virus concentration is locally raised in the grooves and lowered to nearly zero on the top. Therefore, TMV is forced to align in the grooves and only a few viruses are found outside. For spin speeds faster than 3000 rpm the water film is too thin to result in a sufficient solvent evaporation time allowing for controlled filling of the grooves. The film thickness may be thinner than the amplitude of the wrinkles. Consequently the top of the wrinkles does not lead to a controlled "breaking" of the water film any more. Thus holes occur everywhere in the water film. As a consequence, with increasing spin speed the viruses directly stick to the surface widely spread over the sample. This effect can be seen in the graph in Fig. 4a where the virus occupancy parameter decreases from 3000 rpm to 4000 rpm while the virus deviator parameter—which is a relative value—first increases significantly and then remains constant within the error range. The SEM image in Fig. 4b shows a sample prepared at optimum conditions (spin speed = 3000 rpm, c(TMV) = 0.9 mg ml^{-1}). The TMVs are ordered parallel to each other with a regular distance determined by the wavelength of the substrate. Some of the viruses are disordered which is an effect of the above-mentioned phenomenon.

In summary, we present an effective technique for aligning anisotropic particles over large areas on predefined substrates. Using wrinkled PDMS substrates to direct the self-assembly process we create highly uniform TMV arrays. The control of alignment is achieved using a simple spin-coating technique. An important aspect, and not only for this particular case of particle templating, is the development of quantitative measures to reveal the order of the templated particles or templating efficiency. For this purpose, we introduce simple parameters and use them for a quantitative discussion of the influence of process conditions. We varied the concentration of the virus solution and the spin speed to determine the optimum conditions for the virus assembly. We found a concentration limit above which additional TMVs are increasingly deposited outside rather than in the grooves. Our results identify the spin speed as a critical parameter for structure formation. This can be explained by a dewetting-assisted structure-formation mechanism. While the particular concentration and speed range is characteristic for the particular combination of solvent, particle geometry and wetting properties of particle and substrates, we believe that this mechanism can be of interest for alignment of particles from solution and further theoretical work for a quantitative modelling of the effect is currently pursued. Until now, several groups used the chemical addressability of viruses to build silica-coated nanoparticles,[21] metal cages[22,23] or metal wires.[24,25] With these modified viruses, and by using the approach presented in this paper, it becomes feasible to develop metal wires where the distances are predefined by the wavelength of the wrinkled substrate. Furthermore, induced structural defects of the substrate open the possibility of producing branched virus assemblies and therefore branched metal wires. To conclude, we generated well-defined TMV arrays which may serve as metallisation or mineralisation templates for future applications and as a model system for the alignment of anisotropic nanoparticles.

Acknowledgements

The authors thank Stephan Herminghaus and Markus Hund for helpful discussions, and Benjamin Goßler and Carmen Kunert for help with the SEM and TEM pictures. This work is supported by the Lichtenberg-Program of the Volkswagen-Stiftung and the Sonderforschungsbereich 481 funded by the German Science Foundation (DFG). QW is partially supported by US NSF, US DoD and the W. M. Keck Foundation.

References

1. M. Spasova, U. Wiedwald, R. Ramchal, M. Farle, M. Hilgendorff and M. Giersig, *J. Magn. Magn. Mater.*, 2002, **240**, 40–43.
2. Y. Huang, X. Duan, Q. Wei and C. Lieber, *Science*, 2001, **291**, 630–633.
3. F. Juillerat, H. H. Solak, P. Bowen and H. Hofmann, *Nanotechnology*, 2005, **16**, 1311–1316.
4. H. H. Solak, C. David, J. Gobrecht, V. Golovkina, F. Cerrina, S. O. Kim and P. F. Nealey, *Microelectron. Eng.*, 2003, **67–68**, 56–62.
5. S. Y. Chou, P. R. Krauss and P. J. Renstrom, *Science*, 1996, **272**, 85–87.
6. B. Varghese, F. C. Cheong, S. Sindhu, T. Yu, C.-T. Lim, S. Valiyaveettil and C.-H. Sow, *Langmuir*, 2006, **22**, 8248–8252.
7. C. Lu, H. Möhwald and A. Fery, *Soft Matter*, 2007, **3**, 1530–1536.
8. Y. Xia, Y. Yin, Y. Lu and J. McLellan, *Adv. Funct. Mater.*, 2003, **13**, 907–918.
9. S. P. Wargacki, B. Pate and R. A. Vaia, *Langmuir*, 2008, **24**, 5439–5444.
10. S. Balci, D. M. Leinberger, M. Knez, A. M. Bittner, F. Boes, A. Kadri, C. Wege, H. Jeske and K. Kern, *Adv. Mater.*, 2008, **20**, 2195–2200.
11. D. M. Kuncicky, R. R. Naik and O. D. Velev, *Small*, 2006, **2**, 1462–1466.
12. P. J. Yoo, K. T. Nam, A. M. Belcher and P. T. Hammond, *Nano Lett.*, 2008, **8**, 1081–1089.
13. C. M. Stafford, S. Guo, C. Harrison and M. Y. M. Chiang, *Rev. Sci. Instrum.*, 2005, **76**, 062207.
14. A. Schweikart and A. Fery, *Microchim. Acta*, 2009, **165**, 249–263.
15. K. Efimenko, M. Rackaitis, E. Manias, A. Vaziri, L. Mahadevan and J. Genzer, *Nat. Mater.*, 2005, **4**, 293–297.
16. N. Bowden, W. T. S. Huck, K. E. Paul and G. M. Whitesides, *Appl. Phys. Lett.*, 1999, **75**, 2557–2559.
17. C. M. Stafford, C. Harrison, K. L. Beers, A. Karmin, E. J. Amis, M. R. Vanlandingham, H.-C. Kim, W. Volksen, R. D. Miller and E. E. Simonyi, *Nat. Mater.*, 2004, **3**, 545–550.
18. M. Pretzl, A. Schweikart, C. Hanske, A. Chiche, U. Zettl, A. Horn, A. Böker and A. Fery, *Langmuir*, 2008, **24**, 12748–12753.
19. J. Genzer and J. Groenewold, *Soft Matter*, 2006, **2**, 310–323.
20. A. Chiche, C. M. Stafford and J. T. Cabral, *Soft Matter*, 2008, **4**, 2360–2364.
21. E. Royston, S. Y. Lee, C.J.N. and M. Harris, *J. Colloid Interface Sci.*, 2006, **298**, 706–712.
22. C. Radloff, R. A. Vaia, J. Brunton, G. T. Bouwer and V. K. Ward, *Nano Lett.*, 2005, **5**, 1187–1191.
23. C. M. Soto, A. S. Blum, C. D. Wilson, J. Lazorcik, M. Kim, B. Gnade and B. R. Ratna, *Electrophoresis*, 2004, **25**, 2901–2906.
24. M. Knez, M. Sumser, A. M. Bittner, C. Wege, H. Jeske, T. P. Martin and K. Kern, *Adv. Funct. Mater.*, 2004, **14**, 116–124.
25. E. Dujardin, C. Peet, G. Stubbs, J. N. Culver and S. Mann, *Nano Lett.*, 2003, **3**, 413–417.

PAPER

Silica nano-particle super-hydrophobic surfaces: the effects of surface morphology and trapped air pockets on hydrodynamic drainage forces

Derek Y. C. Chan,[*a] Md. Hemayet Uddin,[bc] Kwun L. Cho,[b] Irving I. Liaw,[b] Robert N. Lamb,[b] Geoffrey W. Stevens,[c] Franz Grieser[b] and Raymond R. Dagastine[*c]

Received 19th January 2009, Accepted 26th March 2009
First published as an Advance Article on the web 23rd July 2009
DOI: 10.1039/b901134j

We used atomic force microscopy to study dynamic forces between a rigid silica sphere (radius ~45 μm) and a silica nano-particle super-hydrophobic surface (SNP-SHS) in aqueous electrolyte, in the presence and absence of surfactant. Characterization of the SNP-SHS surface in air showed a surface roughness of up to two microns. When in contact with an aqueous phase, the SNP-SHS traps large, soft and stable air pockets in the surface interstices. The inherent roughness of the SNP-SHS together with the trapped air pockets are responsible for the superior hydrophobic properties of SNP-SHS such as high equilibrium contact angle (>140°) of water sessile drops on these surfaces and low hydrodynamic friction as observed in force measurements. We also observed that added surfactants adsorbed at the surface of air pockets magnified hydrodynamic interactions involving the SNP-SHS. The dynamic forces between the same silica sphere and a laterally smooth mica surface showed that the fitted Navier slip lengths using the Reynolds lubrication model were an order of magnitude larger than the length scale of the sphere surface roughness. The surface roughness and the lateral heterogeneity of the SNP-SHS hindered attempts to characterize the dynamic response using the Reynolds lubrication model even when augmented with a Navier slip boundary.

1. Introduction

The term "soft matter" has been used to describe a very broad range of materials from the synthetic to the biological with established and emerging applications. An important and unique property of soft matter materials is that their geometric deformations and the magnitude of their interactions with their environment have to be determined self-consistently. This requires knowing how soft matter deforms under external perturbations such as mechanical forces, hydrodynamic flow fields, electric fields, and osmotic pressure gradients from chemical species. Previously,

[a]*Department of Mathematics and Statistics, The Particulate Fluid Processing Centre, The University of Melbourne, Parkville, Victoria, 3010, Australia. E-mail: D.Chan@unimelb.edu.au*
[b]*School of Chemistry, The Particulate Fluid Processing Centre, The University of Melbourne, Parkville, Victoria, 3010, Australia*
[c]*Department of Chemical and Biomolecular Engineering, The Particulate Fluid Processing Centre, The University of Melbourne, Parkville, Victoria, 3010, Australia. E-mail: rrd@unimelb.edu.au*

we have focused on interactions involving drops and bubbles that are the basic building blocks in emulsions and foams to examine the coupling between geometric deformation to both equilibrium[1–8] and dynamic[9–20] forces on the nanometre scale. We also studied how adsorbed molecules at interfaces can modify these interactions by changing equilibrium surface forces and altering hydrodynamic boundary conditions of the flow field at liquid–liquid or liquid–gas interfaces.[9,11–13,19] This paper examines another type of soft material, a super-hydrophobic surface, where previous knowledge of the interactions between drops and bubbles is critically relevant to understanding nanometre scale interactions involving a super-hydrophobic surface in an aqueous environment.

Perhaps the most well-known example of a super-hydrophobic surface (SHS), commonly defined as any surface on which the contact angle of a sessile water drop is larger than $140°–150°$,[21] is the lotus leaf.[22] Originally described by Cassie and Baxter,[21] it is the microstructure of the surface as well as the surface chemistry that impart super-hydrophobicity to the surface. The remarkable self-cleaning property of the lotus leaf has inspired a large number of approaches[23–25] to develop hierarchical or patterned surfaces to create synthetic super-hydrophobic surfaces to be used as anti-fouling coatings[26,27] and in novel applications in microfluidic devices.[28–30] Here we study dynamic interactions involving SHS made from silica nanoparticles (Fig. 1) that is being developed for use in anti-fouling coatings.[26,31,32] These SHS are of interest because their method of preparation is quite simple compared to processes involved in the manufacture of other types of SHS.[23–25] The microstructure of the surface of these materials have a high degree of porosity and roughness which impart to them the propensity to trap air within the porous coating in aqueous environments. Thus they have a unique combination of a rough and porous surface structure mixed with soft, trapped air bubbles in a single surface.

To investigate the dynamic behaviour of this soft matter material, we made use of atomic force microscopy (AFM) measurements and theoretical modelling of the dynamic interactions between these super-hydrophobic surfaces and a well-characterized rigid "probe" particle in aqueous solutions. We attempt to quantify the complex interplay between the hierarchical surface structure and the presence of trapped air within the coating that mediate the dynamic behaviour of this soft matter

Fig. 1 Two views of the same glass plate where the slightly opaque half has been made into a silica nano-particle super-hydrophobic surface; the clear half is uncoated glass. The drops of water, tinted in pink, have a high contact angle on the super-hydrophobic side but spread on the uncoated side.

coating by examining the dependence of the interactions with relative velocity, solution conditions and the types of molecules present. In this context, our previous work in AFM studies through novel force measurement methods and detailed modeling of the dynamic interactions in deformable systems is particularly relevant. Our initial studies focused on how equilibrium interactions affect droplet deformations[1–8,33] that complement the work of other researchers[34–38] and build on earlier studies on bubbles which had less detailed analysis.[39–43] More recent advances extended to the study of dynamic forces involving two drops[9,13–15,18] or a particle and a drop.[16,44] Earlier observations of dynamic deformations of a deformable mercury interface near a mica surface due to mechanical and electrical perturbations has also been modelled with quantitative success with the same theoretical framework.[19,20] The model also had success in predicting dynamic deformations in non-aqueous systems of glycerine in silicone oil[45] and has demonstrated that forces between deformable drops can be measured by simply measuring geometric deformations.[46] For all of the above, the observed hydrodynamic forces are consistent with the no-slip boundary conditions at solids and liquids interfaces.

Of particular relevance to the present work is our earlier studies of interactions involving bubbles[10–12] where, in the presence of surfactant, a no-slip boundary condition is observed whereas for a very clean air–water interface, the hydrodynamic boundary condition is more consistent with a Marangoni boundary condition rather than the no-slip condition. The dynamic interaction forces between a super-hydrophobic surface and a probe particle are expected to exhibit the effects of a combination of factors including the surface roughness, the hydrophobic character of the surface, and the presence of trapped bubbles with larger interfacial areas. This raises the question as to whether this combination of factors can be adequately captured under the single concept of 'slip'. However, what is clear is that the hierarchical nature of the surfaces requires careful characterization of the surfaces and the deployment of large colloid probes in the AFM measurement to maximize the magnitude of the interaction.

2. Materials and methods

2.1. Preparation of silica nano-particle super-hydrophobic surface (SNP-SHS)

The surfaces were prepared by spin-coating a solution of 40 nm silica nano-particles with methyltriacetoxysilane (MTAS) linker and polydimethylsiloxane (PDMS) in a hexane solvent on 1 mm thick, 2.54 cm diameter glass discs. The discs were baked in a furnace at 150 °C for 15 min. A more detailed description of the preparation methods can be found elsewhere.[31,32] The advancing contact angles of sessile drops of various aqueous solutions on these surfaces were measured using a Dataphysics OCA 20 tensiometer and goniometer system. For the 1 mM $NaNO_3$ solution the advancing and receding contact angles were 143° ± 5° and 124° ± 5°, respectively, and for the 5 mM sodium dodecyl sulfate (SDS) solution the advancing and receding contact angles were 117° ± 5° and 95° ± 5°, respectively. As these films were prepared by depositing hydrophobized nano-scale particles, the result is a surface with an extreme degree of roughness. The water contact angle of an equivalent smooth surface of the polysiloxane cross-linker was measured previously to be 75°.[26]

2.2. Method

The super-hydrophobic surfaces were imaged with an Asylum MFP-3D AFM (Asylum Research, Santa Barbara) both in air and in liquid with a closed fluid cell in AC mode. The cantilevers used for the imaging in air were rectangular silicon cantilevers (Budget Sensors, Sofia), and V-shaped non-conductive silicone nitride cantilever (Veeco, Santa Barbara) were used for AC mode in water. The cantilever and the SNP-SHS were cleaned in an ozone atmosphere for at least 20 min just before the imaging experiments. The liquid measurements were first performed in 1 mM NaNO$_3$ aqueous solution. Then the substrate was washed with 30 ml of 20 mM SDS solution and the imaging was performed at approximately the same position on the sample in the same SDS solution three times over several hours.

For force measurements, a silica sphere of 45 ± 2 μm radius was attached to a custom-manufactured rectangular silicon AFM cantilever (dimensions: 450 μm × 50 μm × 2 μm) by using a two part epoxy adhesive.[47] The spring constant of the cantilever, K (0.2 ± 0.02 N m^{-1}) was determined by the thermal method.[48] The radius of the attached sphere was measured by video microscopy using a 50× objective. The surface roughness of the attached sphere was measured by a reverse image on a spiked grating (NT-MDT, Moscow) at the completion of the force measurements.

Force measurements were carried out using an Asylum MFP-3D AFM equipped with a linear variable displacement transducer (LVDT) in a closed fluid cell between a silica sphere and a freshly cleaved mica surface or a super-hydrophobic surface either in aqueous NaNO$_3$ or SDS solutions.

Force curves (an approach and retract force–distance cycle) were taken at a series of approach and retract piezo scan rates between 500 nm s^{-1} to 2 μm s^{-1} over a 70 μm × 70 μm scan area with 9 points on each line (a total of 81 force curves per force map). At higher speeds, up to 50 μm s^{-1}, the force curves were taken at manually selected 5 to 10 separate locations on the substrates. Results for the measured force normalised by the radius of the silica sphere were independent of the sphere size for larger spheres (35–50 μm), although there were irregularities with the smaller spheres (radius ~19 μm).

The deflection of the cantilever was converted from voltage signal to distance based on the constant compliance region (CCR) of force curves at slow (~1 μm s^{-1}) scan rates. The determination of the separation distance between the sphere and the surface using a CCR analysis has a number of difficulties for dynamic measurements. Further analysis of these data used an AFM force balance model that is discussed below.

3. Results

3.1. Surface characterization

3.1.1. Images of the silica nano-particle super-hydrophobic surface (SNP-SHS).
Characterization of geometries involved in direct force measurements is a key requirement for quantitative comparison with theoretical models. In the case of super-hydrophobic surfaces with unusual surface morphology, characterization of the surfaces in air to visualize the intrinsic surface topography, as well as observation in liquid at the same aqueous solution conditions as the force measurement, is vital because of the possibility of trapped air bubbles on the surfaces. The addition of surfactant in the solution may also change the state of any trapped air on the super-hydrophobic surface, so visualization before and after the addition of surfactant in the same region is important.

3.1.2. In air.
The topography of this type of super-hydrophobic surface is known to have significant surface roughness on the micrometre scale in the z-direction.[26] This was observed for the surfaces used in this study as well, where the roughness for different scan size images is given in Table 1. On a 50 × 50 μm scan the roughness scale is several micrometres. On the scale of several micrometres, the surface roughness is still of the order of half a micrometre. The surfaces also exhibit a large degree of lateral heterogeneity as shown in Fig. 2. Due to the large scale roughness of these samples, there may be some tip convolution effects when imaging the deepest valleys in air. As described below, these effects are less of an issue when imaged in liquid.

3.1.3. In aqueous solution.
A series of AFM images for a SHS submerged in a 1 mM NaNO$_3$ electrolyte is given in Fig. 3 at successively higher magnifications.

Table 1 Average peak-to-valley roughness values of SNP-SHS from AFM images at a range of scan sizes

Scan size/μm	In air/μm	In 1 mM NaNO$_3$ electrolyte/μm
50 × 50	2.5	2.0
10 × 10	1.9	1.6
2 × 2	0.7	0.5

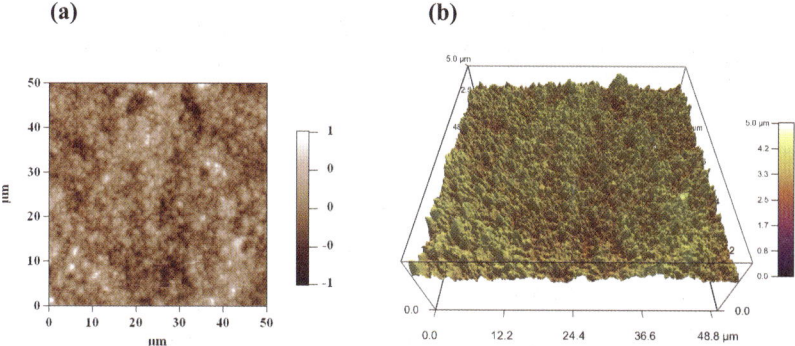

Fig. 2 AFM topography images of a silica nano-particle super-hydrophobic surface in air: (a) top view and (b) 3D view.

There are significant topographical differences compared to images in air, where the troughs and valleys of the images in air have been replaced by smooth interfaces of trapped air pockets that bridge the peaks of the surfaces. These features are visible regardless of magnification and these flat regions, with height variations of less than 2 nm, exhibit a clear and consistent contrast in the phase and amplitude images as well, indicating that these regions are not an artefact from tip convolution effects. When imaged at higher forces, these interfaces become depressed and then restore to their original height with a decrease in force. It is important to note the size of these air pockets is on the scale of hundreds of nanometres or smaller.

To investigate possible differences in the nature of the trapped air on the SNP-SHS in sodium nitrate and in SDS solutions, the same region was imaged before and after the addition of surfactant. These images are shown in Fig. 4 where a large bubble is also observed. The bubble appears to have a textured surface. This is an artefact of the imaging process, where the pressure from the tip deforms the bubble slightly by pressing on the topography underneath the air–water interface. Large bubbles of this size were observed in only about 10% of the 50 × 50 μm scans, whereas bubbles trapped within the peaks and valleys of the surface were observed in all scans. The addition of surfactant is expected to change the geometry of the bubble slightly as expected from adsorption of the SDS to the air–water interface (decreasing the surface tension from 72 mN m^{-1} to 39 mN m^{-1}), but the surfactant does not remove the trapped air from the SNP-SHS.

3.1.4. Silica sphere. This study employed a very rough sphere in the context of high precision AFM force measurements. It is well known that larger silica spheres can have significant roughness compared to smaller spheres, where a recent study also quantified this variability.[49] Reverse imaging of the sphere (radius ∼45 μm) in air reveals a peak to trough roughness of approximately 22 nm and a similar root–mean–squared roughness over a 1 μm square region. This is significantly rougher than smaller silica spheres (radius ∼5 μm) which commonly have

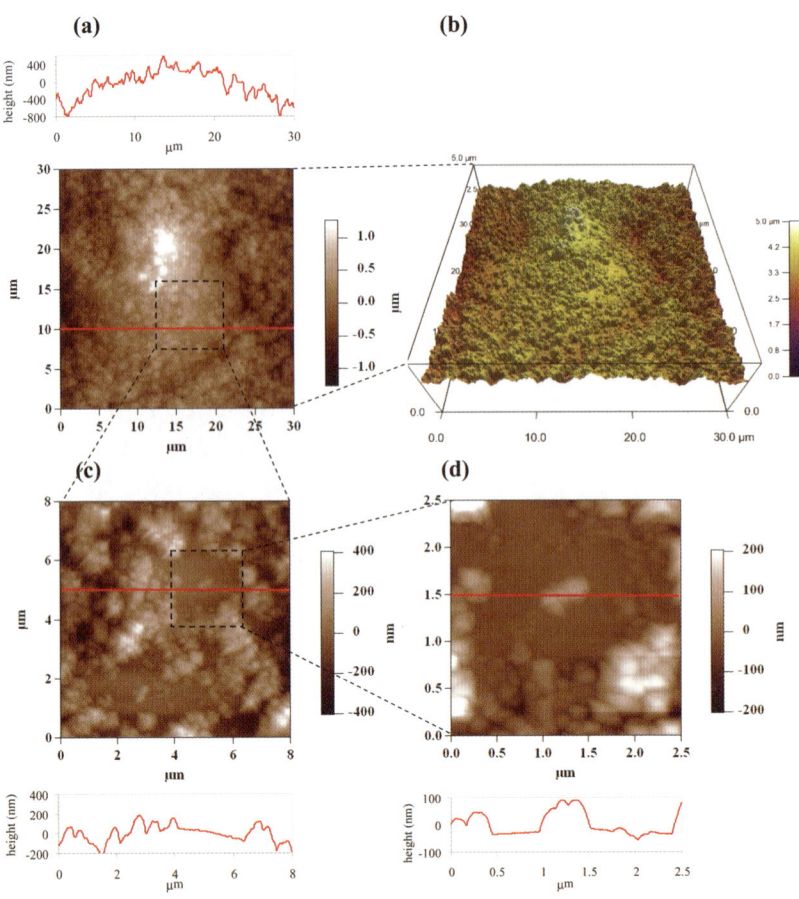

Fig. 3 AFM topography images of a silica nano-particle super-hydrophobic surface in 1 mM NaNO$_3$ aqueous solution: (a) top view and (b) 3D view with a scan size of 30 × 30 μm; (c) a zoomed in image of (a) with a scan size 8 × 8 μm; and (d) a zoomed in image of (c) with a scan size 2.5 × 2.5 μm. The solid lines on figures (a), (c) and (d) correspond to the cross-sections shown in the insets of the respective figures.

a root–mean–squared roughness under 3 nm.[49,50] The local curvature on this lateral scale is negligible due to the larger radius of the sphere.

3.2. Dynamic force measurements

In comparison to previous hydrodynamic drainage measurements on rigid surfaces, a larger sphere was used (radius ∼45 μm) for several reasons. Both the sphere and the SNP-SHS have significant surface roughness. Surface roughness has been observed to decrease the hydrodynamic drainage force between a sphere and plate geometry measured using AFM.[51,52] In addition, previous studies using AFM often used additives such as sucrose to increase the viscosity of the solution and hence the magnitude of the hydrodynamic drainage force and experimentally accessible shear rates,[51,53,54] whereas this work was conducted without a viscosity modifier. In the regime of low Reynolds number lubrication hydrodynamics pertinent to the present experimental system, the hydrodynamic drainage force is expected to be linearly proportional to viscosity and has a quadratic dependence on the radius of the sphere for this geometry (see eqn (1)). In addition, the length scale of the surface roughness

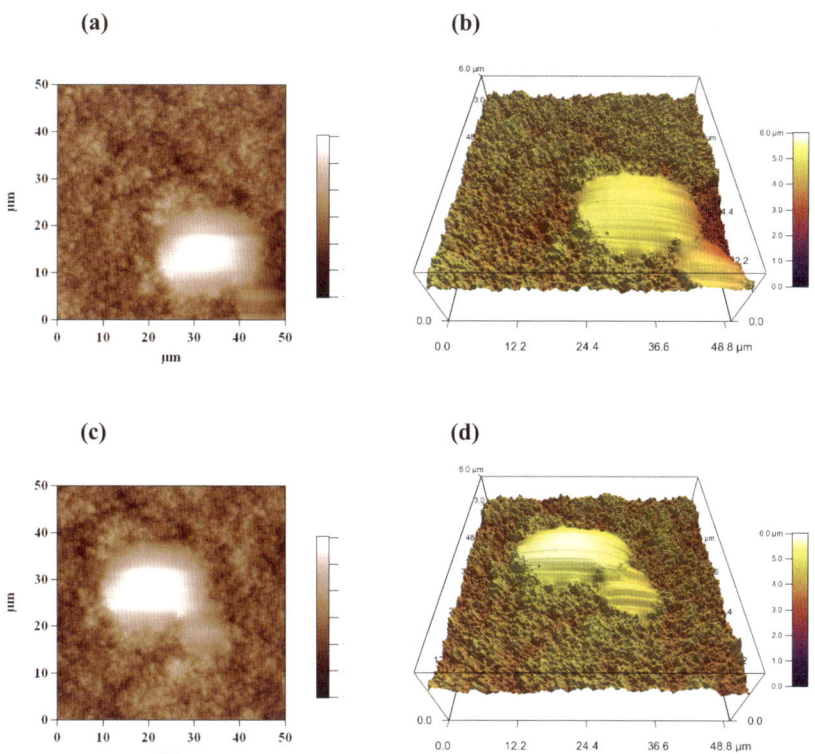

Fig. 4 AFM 50 × 50 μm scan size topography images of silica nano-particle super-hydrophobic surface in aqueous solutions: (a) top view in 1 mM NaNO$_3$ electrolyte; (b) 3D view of (a); (c) the same region of the surface after replacing the 1 mM NaNO$_3$ electrolyte with a 20 mM SDS; (d) 3D image of (c). The large egg-shaped feature in all images is a large air bubble on the surface. The lines on the bubble are scan line artefacts from the imaging process.

of the SNP-SHS and lateral heterogeneity are on the scale of micrometres, therefore the probe particle radius must be significantly larger than these length scales.

3.2.1. Silica sphere—mica surface. The large silica sphere has a non-negligible surface roughness. To isolate the effects of surface roughness of the large silica sphere prior to studying the SNP-SHS, the hydrodynamic drainage forces between the sphere and a model smooth surface, freshly cleaved mica, in aqueous solutions were examined. Typical dynamic force data as a function of time are shown in Fig. 5(a) at a series of increasing scan rates in 1 mM NaNO$_3$. The time axis for each scan rate has been scaled by β, the ratio between the scan rate and the lowest scan rate of the series. It is important to note this is a dynamic measurement and so all quantities such as the force, piezo position, separation and the velocity of the sphere are all parametric in time. The piezo velocity, dX_{LVDT}/dt, is not constant due to the non-linear motion of the piezo, but the LVDT records the actual position of the piezo with time which will be used directly in the analysis and model calculations. In addition, the velocity of the sphere will vary significantly with time in the proximity of the mica. The large hysteresis between the approach and retract curves and dependence of the force on approach and retract scan rate is clear. Both the magnitude of the repulsion on approach and the magnitude of the smoothly varying minimum upon retraction have a strong dependence on the piezo scan rate. The overall functional form of these data is similar to previous AFM measurements

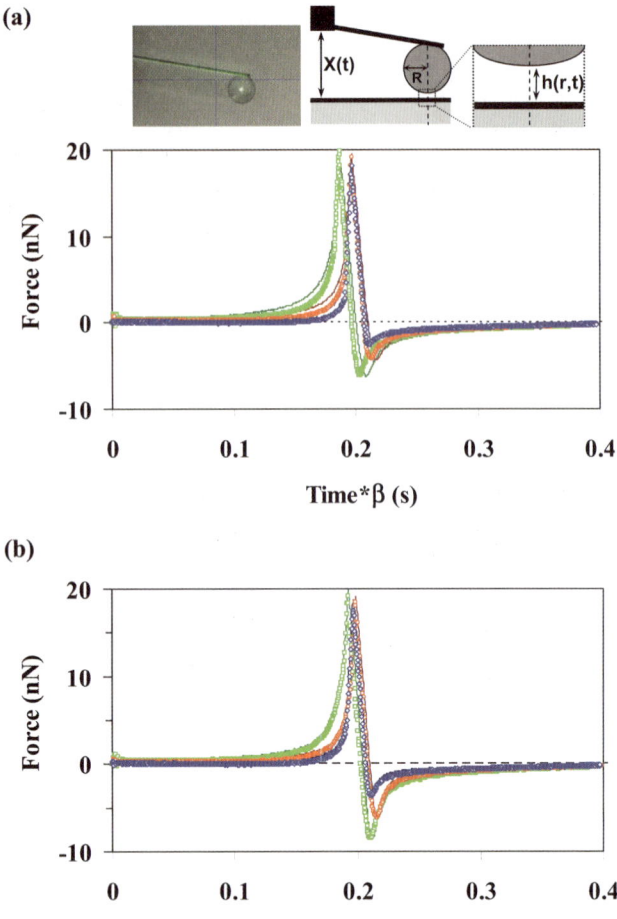

Fig. 5 Measured forces between a silica sphere (45 μm radius) and mica in aqueous solutions of (a) 1 mM NaNO$_3$ at average tip velocities of 11.8 (blue ◇), 23.3 (orange ○) and 34.5 (green □) μm s^{-1}; and (b) 20 mM SDS at average tip velocities of 11.7 (blue ◇), 23.4 (orange ○) and 35.1 (green □) μm s^{-1}. The time is scaled by a velocity ratio β ($\beta = V_h/V_l$, where V_h is 11.8, 23.3 and 34.5 μm s^{-1}, or 11.7, 23.4 and 35.1 μm s^{-1} and V_l is 11.8 or 11.7 μm s^{-1} for (a) and (b), respectively). The solid lines in (a) and (b) correspond to an AFM force balance model fit to these data. The initial separation, $h(0)$, is determined from the fit. The $h(0)$ value from lowest to highest scan rates are 2300, 2300 and 2185 nm for (a) and 2256, 2275 and 2208 nm for (b). The insets in (a) show the photograph of the sphere attached on a rectangular cantilever and schematic diagram of the sphere–surface configuration showing the cantilever displacement function, $X(t)$; sphere radius, R; and separation, $h(r, t)$.

using smooth particles and flat surfaces.[51–54] A rigorous comparison to theory will be discussed below.

In the presence of a 20 mM SDS solution, dynamic force measurements were carried out using the same sphere and on the same sample in the same region as the measurements in Fig. 5(a). Typical results are shown in Fig. 5(b) and show a similar behaviour to Fig. 5(a). The electrical double layer forces are screened to a larger extent as the electrolyte concentration is an order of magnitude higher, the Debye length has decreased by two thirds and the surface potentials are expected to be lower (see Appendix). At this SDS concentration, adsorption of the SDS is not expected to either the mica surface or the silica surface as they are both negatively charged,[55,56] so differences between the two measurements (Fig. 5(a) and (b)) are largely attributed to the differences in electrical double layer repulsion.

3.2.2. Silica sphere–silica nano-particle super-hydrophobic surface (SNP-SHS).

The same silica sphere used in the measurements presented in Fig. 5 was used to measure the hydrodynamic forces between a sphere and SNP-SHS in the presence of 1 mM NaNO$_3$, 5 mM SDS and 20 mM SDS for a series of scan rates. Typical force data of measured forces *versus* time are shown in Fig. 6(a), (b) and (c) for these three solution conditions. In contrast to the mica data in Fig. 5, there is almost a complete absence of any speed dependent repulsion in Fig. 6(a) for the 1 mM NaNO$_3$ solution conditions and the absence of a smoothly varying hydrodynamic minimum in the retraction phase. Force measurements taken in the same region of the surface change significantly upon the exchange of 5 mM SDS for the NaNO$_3$ solution as shown in Fig. 6(b). The exchange of 20 mM SDS results in an additional increase of the dependence of the observed forces on scan rate shown in Fig. 6(c). The oscillations in the retraction curve are from a ringing in the piezo and tip holder that occur at high scan rates around the start of the retraction phase. This type of artefact has been observed previously in dynamic force measurement between rigid surfaces and is not expected to adversely affect the measurements.[53]

4. Discussion

4.1. Silica sphere–mica surface

Model comparison. The dynamic forces observed between the silica sphere and the mica surface have been compared to a force balance model of the AFM measurement that accounts for the electrical double layer forces between the silica and the mica, the motion of the cantilever in the AFM measurement, and the hydrodynamic drainage forces between the surfaces. Key points of the model are discussed below with the detailed description of the theory and relevant model equations provided in the Appendix. To describe the hydrodynamic drainage force between the two surfaces, the choice of the boundary condition of the velocity of the liquid adjacent to each surface is critical. The traditional assumption is that the tangential component of the velocity of the liquid at the surface is the same as that of the surface, commonly referred to as the no-slip boundary condition. Using Reynolds's lubrication theory, the hydrodynamic drainage force can be given by:

$$F_{\text{no-slip}} = -6\pi\mu R^2 \frac{1}{h}\frac{dh}{dt} \qquad (1)$$

where μ is the dynamic viscosity, R is the radius of the sphere, h is the separation and t is time. A comparison between this no-slip model and experimental dynamic force data plotted as a function of time for mica–sphere interaction is shown in Fig. 7 for the 1 mM NaNO$_3$ case. The deviation observed between the calculations is typical for all of the data from Fig. 5(a) and (b) and is evidence that the hydrodynamic drainage force is significantly reduced from what is expected between two smooth surfaces following a no-slip boundary condition. In fitting the model to the force *versus* time data, the model is used to determine the mica–sphere separation.

This approach is significant as the determination of the separation between interacting surfaces is a key step in AFM force measurements. Traditionally, the force–displacement relation in the constant compliance region (CCR) is used to calibrate the instrument and to determine the location of hard contact and zero separation[57–59] for each force curve. However, for rough surfaces, hard contact occurs between the highest asperities of the surfaces and may lead to sliding or twisting of the cantilever upon further loading. In this work we use a static (low scan rate) measurement, to calibrate the instrument to process measurements at higher scan rates under similar

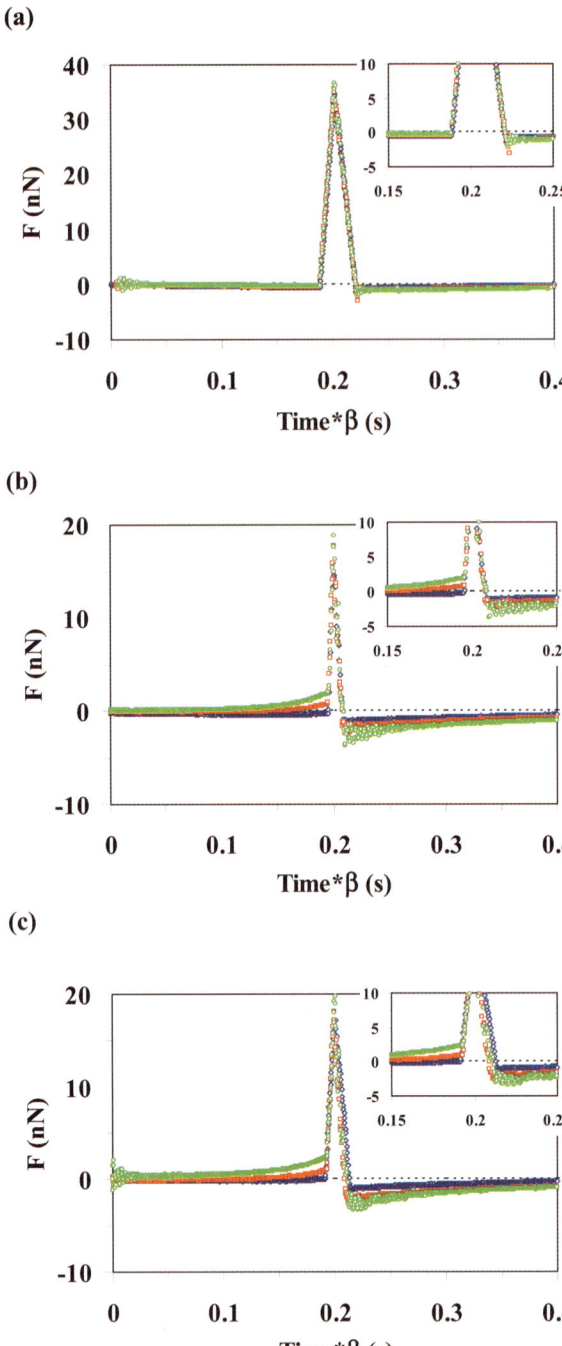

Fig. 6 Measured forces between a silica sphere (45 μm, radius) and a silica nano-particle superhydrophobic surface in aqueous solutions of (a) 1 mM NaNO$_3$ at average tip velocities of 11.9 (blue ◇), 35.0 (orange □) and 58.0 (green ○) μm s^{-1}; (b) 5 mM SDS at average tip velocities of 11.4 (blue ◇), 34.3 (orange □) and 57.3 (green ○) μm s^{-1}; and (c) 20 mM SDS at average tip velocities of 11.3 (blue ◇), 35.3 (orange □) and 55.5 (green ○) μm s^{-1}. The time is scaled by a velocity ratio β ($\beta = V_h/V_l$, where V_h is 11.9, 35.0 and 58.0 μm s^{-1}, or 11.4, 34.3 and 57.3 μm s^{-1}, or 11.3, 35.3 and 55.5 μm s^{-1} and V_l is 11.9 or 11.4 or 11.3 μm s^{-1} for (a), (b) and (c), respectively.

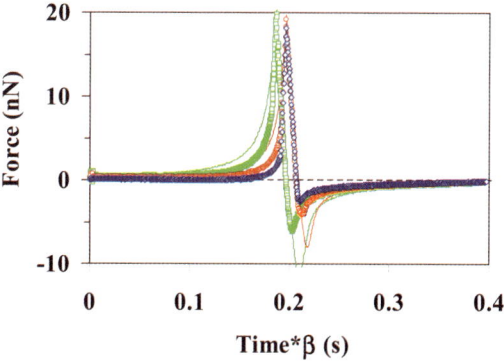

Fig. 7 A comparison of the measured forces between a silica sphere (45 μm radius) and mica in aqueous solutions of 1 mM NaNO$_3$ at average tip velocities 11.8 (blue ◇), 23.3 (orange ○) and 34.5 (green □) μm s^{-1} from Fig. 5(a) to an AFM force balance model using a no-slip boundary condition for the hydrodynamic drainage force. The solid lines are the model calculation. The time is scaled by a velocity ratio β ($\beta = V_h/V_1$, where V_h are 11.8, 23.3 and 34.5 μm s^{-1} and V_1 is 11.8 μm s^{-1}).

conditions and use a hydrodynamic model to determine separation. Furthermore, previous force measurements of hydrodynamic drainage have had difficulties in determining the absolute separation that led to the conclusion that there is slip on a hydrophilic surface.[57,60,61] Recent AFM measurements that incorporated an evanescent wave scattering method to determine hard contact have shown that the observation of slip on hydrophilic surfaces measured using AFM to be an artefact.[54,62] Thus, the approach using force *versus* time may circumvent some of the above difficulties, depending on the accuracy of the hydrodynamic drainage model.

One approach to account for the effects of surface roughness on hydrodynamic drainage has been to use the Navier slip model and invoke the concept of a slip length, where the velocity of the liquid adjacent to a surface is assumed to be proportional to the shear stress at the surface.[51,52,63,64] The concept of a slip length in the context of thin film drainage[52,65–67] was originally proposed to explain deviations in hydrodynamic drainage forces for smooth surfaces but has also been used to characterise rough surfaces.[51,52] We compared our mica–sphere experiment results to a hydrodynamic drainage model that allows slip on one surface, in this case the silica sphere, as we know from independent experiments that the no-slip boundary condition is appropriate for the molecular mica surface.[11,68,69] The hydrodynamic drainage force for this model is given by

$$F_{slip} = -6\pi\mu R^2 \frac{f(h)}{h} \frac{dh}{dt} \quad (2)$$

where $f(h)$ is a function that allows for a slip length on one of the surfaces (see Appendix). The resulting comparisons of the model to the data for the 1 mM NaNO$_3$ and 20 mM SDS are given in Fig. 5(a) and (b), respectively. As the mica–surface separation is parametric in time, one can extract the variations of this separation from the resulting model fit. The fitted initial separation for each experiment, $h(0)$, are reported in Fig. 5. The incorporation of a slip model for the silica surface, has improved the agreement significantly, but there are still deviations as the force approaches the turn-around point, which occurs at small separations. The magnitude of the force data is consistent with the theoretical calculations, but the time position of the minimum in the forces is often predicted after the minimum in the data. Previous studies of hydrodynamic forces focused mainly on analysing the approach curves,[51,54] whereas the comparison in this work would suggest the

Table 2 The fitted Navier slip lengths for the silica sphere surface for silica–mica force data presented in Fig. 5

Speed[a]	11 µm s^{-1}	23 µm s^{-1}	35 µm s^{-1}
1 mM NaNO$_3$	90 nm	120 nm	120 nm
20 mM SDS	240 nm	325 nm	350 nm

[a] These are nominal velocities, exact value given in Fig. 5.

importance of examining the retraction curve as well to evaluate the accuracy of any model. The slip lengths for each system, NaNO$_3$ or SDS, given in Table 2 that fit each force curve varies by around ±10% over the range of scan rates examined. The magnitude of the slip lengths is much larger than the length scale (~20 nm) of the surface roughness of the silica sphere. The degree of agreement between experiments and predictions of the slip model and the deduced slip length can be affected by the extent that surface roughness can be described by the Navier slip model.

There have been recent advances in quantifying the effects of surface roughness on the measurement of static colloidal forces.[50,70,71] The most general consideration is that contact between rough surfaces is determined by the tallest asperities and therefore the magnitude of forces at contact will be reduced compared to those between smooth surfaces. In interpreting measurements of hydrodynamic forces between colloidal particles, the assumption of axisymmetric flow between smooth surfaces becomes problematic, particularly when the surface roughness becomes comparable to the mean separation, as in the case of the SNP-SHS in the present experiments.

Several studies have examined hydrophobic surfaces that were either chemically modified[52] or were polymeric surfaces[63,64] with intrinsic roughness. Only one AFM study has examined hydrophilic surfaces with rough surfaces by chemical modification of a silica surface.[51] The Navier slip model was used to interpret the measurements and the fitted slip lengths were also up to an order of magnitude larger than the length scale of the surface roughness where the previous study used a similar slip model. Vinogradova has proposed modifying the slip model by adding an additional parameter, which acts effectively as an offset in the separation, but this adds an additional fitting parameter to the model.[52] In addition, it is difficult to quantify the effects of random lateral heterogeneities that may give rise to low resistance lateral drainage pathways and hence reduce the magnitude of the hydrodynamic interaction relative to that between ideal smooth axisymmetric surfaces. Furthermore, even between surfaces with similar mean peak-to-trough roughness, the detailed surface topography can be quite different, so the lack of correlation between the deduced Navier slip length and the peak-to-trough roughness suggests that the model is at best empirical and a unified theory is yet to be developed.

4.2. Silica sphere–silica nano-particle super-hydrophobic surface (SNP-SHS)

The SNP-SHS possesses a surface roughness (~1–2 µm) several orders of magnitude greater than the silica sphere (~20 nm), which makes a detailed analysis of the hydrodynamic drainage force much more complicated. Indeed, the measured dynamic forces involving an SNP-SHS are significantly smaller than that involving mica and in some cases are almost undetectable. The approach used by previous researchers to interpret measured forces between a rough silica sphere and a smooth mica surface was to apply a Navier Slip model and treat the surface roughness as a perturbation to a smooth surface. The AFM images of the SNP-SHS indicate that the SNP-SHS roughness is extreme in comparison to the silica sphere and the surface itself is extremely porous. The SNP-SHS images in water do indicate that the topography may appear flatter due to the presence of trapped air within the surface, but the surface roughness is still several hundred nanometres in range, even on a lateral scale of 1–2 μm. The large variations in lateral heterogeneity coupled with the magnitude of the roughness in the normal direction indicate that the axisymmetric lubrication model with smooth surfaces used for the silica sphere–mica system may not be extended to the SNP-SHS system.

The force data with the SNP-SHS exhibit some dynamic force behavior, but not for every solution condition. As described above, the forces with the SNP-SHS in the presence of $NaNO_3$ exhibit almost no dependence on scan rate, whereas the cases for 5 mM and 20 mM SDS show increases in the magnitude of a scan rate dependent force. To demonstrate that the forces in the SDS case arise from hydrodynamic drainage, the data from Fig. 6(a) and (c) were rescaled and plotted in Fig. 8. The cantilever deflection, which is proportional to the force, was divided by the average velocity and plotted against the same scaled time used in the insets of Fig. 6. For both Fig. 8(a) and (b) the approach points of the force curves roughly collapse on one another nearly forming a master curve. In the case of Fig. 8(a), the $NaNO_3$ case, there appears to be very little interaction, dynamic or static. In the case of the SDS in Fig. 8(b), this suggests that the deflection is dominated by hydrodynamic drainage effects. The variation in the overlap is primarily due to the variations in both starting separation and the variable nature of the rough SNP-SHS.

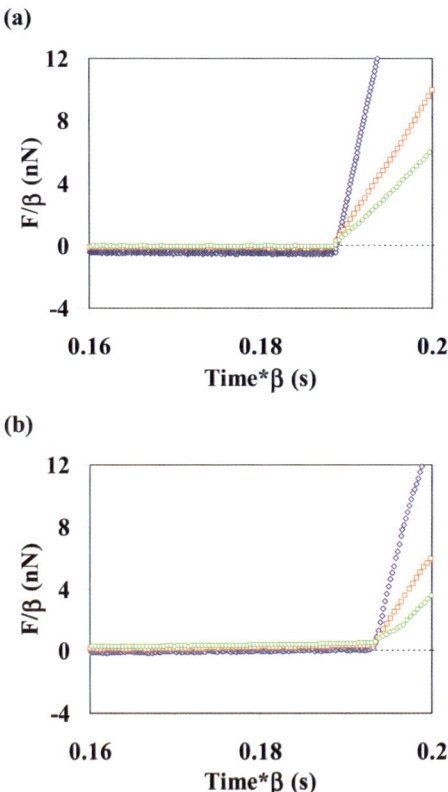

Fig. 8 Force *vs.* time curves for a silica sphere (45 μm, radius) and a silica nano-particle super-hydrophobic surface in aqueous solutions of (a) 1 mM $NaNO_3$ at average tip velocities of 11.9 (blue ◇), 35.0 (orange □) and 58.0 (green ○) μm s^{-1}; and (b) 20 mM SDS at average tip velocities of 11.3 (blue ◇), 35.3 (orange □) and 55.5 (green ○) μm s^{-1}. Both the force and time are scaled by a velocity ratio β ($\beta = V_h/V_l$, where V_h are 11.9, 35.0 and 58.0 μm s^{-1}, or 11.3, 35.3 and 55.5 μm s^{-1} and V_l is 11.9 or 11.3 μm s^{-1} for (a) and (b), respectively.

The SNP-SHS makes it difficult to develop further quantitative analysis using lubrication theory, but the rescaling of the SDS data and the lack of any dynamic effects in the absence of surfactant suggests that the trapped air in the SNP-SHS has an important role in affecting hydrodynamic drainage behavior. There are several possible reasons that the addition of surfactant can cause an increase in the drainage force. The first possible effect is from the change in shape and position of the air–water menisci of the many trapped air pockets on the SNP-SHS. As mentioned above, the contact angle of water on the SNP-SHS changes with surfactant present (from approximately 140° to 120°) as well as the change in surface tension with the addition of SDS. The AFM images before and after the addition of surfactant show that the trapped air is not removed, but the shape of the air–water menisci may change reducing some of the rough character of the surface. Both previous work and the measurements in this work on mica indicate that as surface roughness is reduced the hydrodynamic drainage forces will increase. Even with SDS in the system, the magnitudes of the dynamic forces are still considerably smaller than that observed for smooth surfaces such as mica.

The second possible reason for the changes in drainage behavior is from the boundary condition at the multitude of air–water menisci on the SNP-SHS. In the absence of any surface active materials the air–water interface is expected to match the velocity of the fluid adjacent to it, referred to as a completely mobile interface, which has the effect of significantly reducing the resistance to flow in the thin film between the sphere and the SNP-SHS. The addition of surface active materials at sufficiently high concentrations will saturate these interfaces, causing them to become immobile and result in an increase in the hydrodynamic drainage forces between the sphere and the SNP-SHS.

In our previous studies, we have been able to produce bubbles attached to an AFM cantilever with similar radii to the silica sphere in this work and examine the hydrodynamic drainage behavior during the velocity-dependent approach and retract above a mica plate.[10–12] In the presence of sufficient surfactant to saturate the air–water interface, the measured hydrodynamic forces change from the air–water interface being consistent with a partially mobile Marangoni surface to a no-slip surface. An example of the agreement between theory and measurement for the interaction forces between a bubble and flat mica plate is shown in Fig. 9 for a series of approach and retract scan rates. This result is consistent with what is expected for the case of a system with surfactant present and is consistent with our previous work studying the hydrodynamic interactions between oil droplets.[9,13–18]

Force measurements in the absence of any surfactants exhibited behavior that was intermediate between that of an air–water interface with a completely mobile

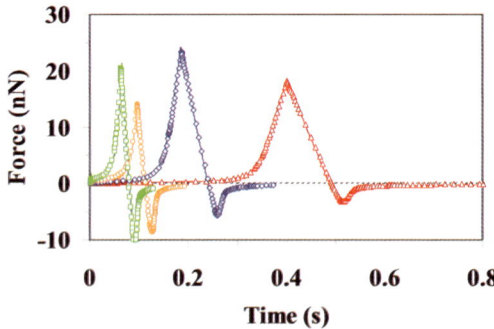

Fig. 9 Dynamic forces between a bubble and a mica plate in 10 mM SDS at tip velocities 5 (red △), 10 (blue ◇), 20 (orange ○) and 30 (□) μm s^{-1} and the model with no-slip boundary condition (solid lines).[11]

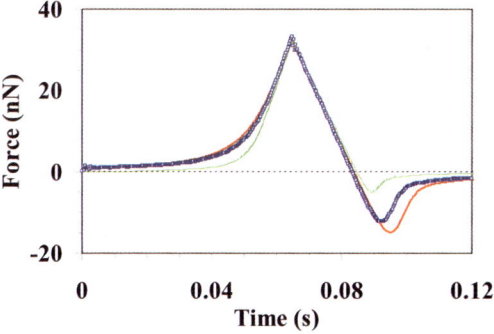

Fig. 10 Dynamic forces between a bubble and a mica plate in 1 mM NaNO$_3$ (no SDS) at tip velocity 30 μm s^{-1} (blue □) and the model with no-slip (solid line) and full-slip (dotted line) boundary conditions.[11]

interface and a completely immobile interface. An example of this is shown in Fig. 10. The observed force behavior was consistent with a partially mobile interface where both surface diffusion and convection were present due to the transport of ultra low levels of contaminants at the air–water interface. The authors proposed a method to model this quantitatively using a simple adsorption isotherm. These data and calculations demonstrate that on smooth surfaces the effects of surfactant can affect the hydrodynamic drainage forces in geometries of comparable sizes to this study by as much as fifty percent through altering the air–water boundary condition from a partially mobile interface to an immobile interface. The conclusion is that it is difficult to obtain air–water interfaces that are clean enough with such small surface areas to observe a fully mobile interface and this is consistent with difficulties in obtaining completely clean bubbles in bubble rise and bubble coalescence studies.[72–74]

The pronounced lack of drainage forces in the NaNO$_3$ solution and then increases in the magnitude of drainage forces with SDS solutions would suggest that the mobility of the air–water interfaces of the surface air pockets may play a significant role in the drainage behavior. It is worth noting that in the case of a SNP-SHS there is a large interfacial area of trapped air pockets such that contamination levels at the air–water interface may be quite low. Previous drainage studies by Charlaix and co-workers[75,76] between a millimetre size glass sphere and an order array of bubbles referred to as a "mattress of bubbles" have demonstrated the importance of the effects of a large number of bubbles at an air–liquid interface, but this study did not examine the effects of added surfactant and it employed viscosity modifiers with mixtures of water and glycerol. The added complexity of the SNP-SHS makes it difficult to identify if the geometric changes in the air–water interface are more or less important than an alteration of the flow boundary conditions at the air–water interfaces, but in either case the dynamic forces are clearly mediated by the presence and effect of surfactant on the air–water interface.

5. Conclusion

Surfaces that are used in practice that possess large scale surface roughness and heterogeneity can significantly reduce hydrodynamic effects that may dominate the dynamic interaction between relatively smooth surfaces. Previous work for hydrodynamic interactions between bubbles and surfaces has shown that the role of molecules at interfaces in changing hydrodynamic boundary conditions may be relevant to this SNP-SHS as well because of the presence of air pockets on such a surface. In addition the high degree of surface roughness of the SNP-SHS also

Table 3 System parameters for 1 mM NaNO$_3$ and 20 mM SDS

Parameter	Experimental	Model
Viscosity, μ/Pa s	10^{-3}	10^{-3}
Spring constant/N m^{-1}	0.20 ± 0.02	0.22
Mica surface potential (NaNO$_3$)[77]/mV	-80 ± 10	-80
Mica surface potential (SDS)[77]/mV	-70 ± 10	-70
Silica sphere surface potential (NaNO$_3$)[78]/mV	-50 ± 10	-50
Silica sphere surface potential (SDS)[78]/mV	-20 ± 10	-20
Radius of silica sphere/μm	45 ± 2	43

has an important role. One must use caution when attempting to extrapolate ideas gained about hydrodynamic drainage models developed for smooth surfaces to practical situations in either super-hydrophobic surfaces or even in other applications such as micro-fluidics, where surface roughness is often a key feature that is not well modeled by the concept of a simple Navier slip.

Appendix

The equation that governs the evolution of the separation, $h(t)$, between the silica sphere and the mica or super-hydrophobic surface is obtained by balancing forces on the sphere due to cantilever deflection, colloidal forces and hydrodynamic interaction. It has the form (see Fig. 5(a) in the inset):

$$\frac{dh}{dt} = \frac{h}{6\pi\mu R^2 f(h)} \left\{ K[\Delta X(t) - h(t) + h(0)] + 2\pi R\, E(h) \right\} \quad (A1)$$

where

$$f(h) = \frac{1}{4}\left\{ 1 + \frac{3h}{2b}\left[\left(1 + \frac{h}{4b}\right)\log\left(1 + \frac{4b}{h}\right) - 1\right]\right\}$$

is the function that characterizes the hydrodynamic boundary condition of no-slip on the mica surface and Navier slip on the silica sphere[52,65] with slip length b. Interactions due to surface forces such as electrical double layer or van der Waals interactions are represented by the Deryaguin method in terms of the interaction energy per unit area, $E(h)$. The piezo displacement function $\Delta X(t)$ is taken from the LVDT output of the AFM. The cantilever deflection $[\Delta X(t) - h(t) + h(0)]$ can be obtained by solving eqn (A1) with a suitable choice of the initial separation, $h(0)$.

In our modeling we use the Poisson–Boltzmann theory to estimate the electrostatic repulsion for a given surface potential on each surface. van der Waals interactions are negligible for the range of separation encountered in our modeling.

The specific parameters used in modeling the fits in Fig. 4(a) and (b) are given below in Table 3.

References

1 R. R. Dagastine, D. C. Prieve and L. R. White, *J. Colloid Interface Sci.*, 2004, **269**, 84–96.
2 R. R. Dagastine and L. R. White, *J. Colloid Interface Sci.*, 2002, **247**, 310–320.
3 D. Y. C. Chan, R. R. Dagastine and L. R. White, *J. Colloid Interface Sci.*, 2001, **236**, 141–154.
4 L. Y. Clasohm, I. U. Vakarelski, R. R. Dagastine, D. Y. C. Chan, G. W. Stevens and F. Grieser, *Langmuir*, 2007, **23**, 9335–9340.
5 S. A. Nespolo, D. Y. C. Chan, F. Grieser, P. G. Hartley and G. W. Stevens, *Langmuir*, 2003, **19**, 2124–2133.

6 P. G. Hartley, F. Grieser, P. Mulvaney and G. W. Stevens, *Langmuir*, 1999, **15**, 7282–7289.
7 P. Mulvaney, J. M. Perera, S. Biggs, F. Grieser and G. W. Stevens, *J. Colloid Interface Sci.*, 1996, **183**, 614–616.
8 R. R. Dagastine, T. T. Chau, D. Y. C. Chan, G. W. Stevens and F. Grieser, *Faraday Discuss.*, 2005, **129**, 111–124.
9 G. B. Webber, S. A. Edwards, G. W. Stevens, F. Grieser, R. R. Dagastine and D. Y. C. Chan, *Soft Matter*, 2008, **4**, 1270–1278.
10 I. U. Vakarelski, J. Lee, R. R. Dagastine, D. Y. C. Chan, G. W. Stevens and F. Grieser, *Langmuir*, 2008, **24**, 603–605.
11 O. Manor, I. U. Vakarelski, X. Tang, S. J. O'Shea, G. W. Stevens, F. Grieser, R. R. Dagastine and D. Y. C. Chan, *Phys. Rev. Lett.*, 2008, **101**, 024501.
12 O. Manor, I. U. Vakarelski, G. W. Stevens, F. Grieser, R. R. Dagastine and D. Y. C. Chan, *Langmuir*, 2008, **24**, 11533–11543.
13 R. R. Dagastine, R. Manica, S. L. Carnie, D. Y. C. Chan, G. W. Stevens and F. Grieser, *Science*, 2006, **313**, 210–213.
14 S. L. Carnie, D. Y. C. Chan, C. Lewis, R. Manica and R. R. Dagastine, *Langmuir*, 2005, **21**, 2912–2922.
15 R. R. Dagastine, G. W. Stevens, D. Y. C. Chan and F. Grieser, *J. Colloid Interface Sci.*, 2004, **273**, 339–342.
16 G. B. Webber, R. Manica, S. A. Edwards, S. L. Carnie, G. W. Stevens, F. Grieser, R. R. Dagastine and D. Y. C. Chan, *J. Phys. Chem. C*, 2008, **112**, 567–574.
17 R. Manica, J. N. Connor, R. R. Dagastine, S. L. Carnie, R. G. Horn and D. Y. C. Chan, *Phys. Fluids*, 2008, **20**, 032101.
18 S. L. Carnie, D. Y. C. Chan and R. Manica, *ANZIAM J.*, 2005, **46**(E), C805–C819.
19 R. Manica, J. N. Connor, L. Y. Clasohm, S. L. Carnie, R. G. Horn and D. Y. C. Chan, *Langmuir*, 2008, **24**, 1381–1390.
20 R. Manica, J. N. Connor, S. L. Carnie, R. G. Horn and D. Y. C. Chan, *Langmuir*, 2007, **23**, 626–637.
21 A. B. D. Cassie and S. Baxter, *Trans. Faraday Soc.*, 1944, **40**, 546–551.
22 W. Barthlott and C. Neinhuis, *Planta*, 1997, **202**, 1–8.
23 R. Fuerstner, W. Barthlott, C. Neinhuis and P. Walzel, *Langmuir*, 2005, **21**, 956–961.
24 X. Zhang, F. Shi, J. Niu, Y. Jiang and Z. Wang, *J. Mater. Chem.*, 2008, **18**, 621–633.
25 P. Roach, N. J. Shirtcliffe and M. I. Newton, *Soft Matter*, 2008, **4**, 224–240.
26 H. Zhang, R. Lamb and J. Lewis, *Sci. Technol. Adv. Mater.*, 2005, **6**, 236–239.
27 J. Genzer and K. Efimenko, *Biofouling*, 2006, **22**, 339–360.
28 A. Chunder, K. Etcheverry, G. Londe, H. J. Cho and L. Zhai, *Colloids Surf., A*, 2009, **333**, 187–193.
29 F.-M. Chang, Y.-J. Sheng, S.-L. Cheng and H.-K. Tsao, *Appl. Phys. Lett.*, 2008, **92**, 264102.
30 J. Hyvaluoma and J. Harting, *Phys. Rev. Lett.*, 2008, **100**, 246001.
31 H. Zhang and R. N. Lamb, *PCT Int. Appl.*, 2005, p. 36.
32 R. N. Lamb, H. Zhang and C. L. Raston, *PCT Int. Appl.*, 1998, p. 20.
33 R. R. Dagastine and G. W. Stevens, in *Interfacial Nanochemistry: Molecular Science and Engineering at Liquid–Liquid Interfaces*, ed. H. Watarai, N. Teramae and T. Sawada, Kluwer Academic, New York, 2005, pp. 77–95.
34 G. Gillies, C. A. Prestidge and P. Attard, *Langmuir*, 2002, **18**, 1674–1679.
35 D. E. Aston and J. C. Berg, *J. Colloid Interface Sci.*, 2001, **235**, 162–169.
36 B. A. Snyder, D. E. Aston and J. C. Berg, *Langmuir*, 1997, **13**, 590–593.
37 A. P. Gunning, A. R. Mackie, P. J. Wilde and V. J. Morris, *Langmuir*, 2004, **20**, 116–122.
38 D. Bhatt, J. Newman and C. J. Radke, *Langmuir*, 2001, **17**, 116–130.
39 W. A. Ducker, Z. Xu and J. N. Israelachvili, *Langmuir*, 1994, **10**, 3279–3289.
40 M. Preuss and H.-J. Butt, *Langmuir*, 1998, **14**, 3164–3174.
41 G. Gillies, M. Kappl and H.-J. Butt, *Adv. Colloid Interface Sci.*, 2005, **114–115**, 165–172.
42 M. L. Fielden, R. A. Hayes and J. Ralston, *Langmuir*, 1996, **12**, 3721–3727.
43 N. Yap, D. Feng, R. R. Dagastine, G. C. Lukey and J. S. J. van Deventer, *Publ. Australas. Inst. Min. Metall.*, 2005, **5/2005**, 643–649.
44 D. E. Aston and J. C. Berg, *Ind. Eng. Chem. Res.*, 2002, **41**, 389–396.
45 R. Manica, E. Klaseboer and D. Y. C. Chan, *Soft Matter*, 2008, **4**, 1613–1616.
46 D. Y. C. Chan, O. Manor, J. N. Connor and R. G. Horn, *Soft Matter*, 2008, **4**, 471–474.
47 W. A. Ducker, T. J. Senden and R. M. Pashley, *Nature*, 1991, **353**, 239–241.
48 J. L. Hutter and J. Bechhoefer, *Rev. Sci. Instrum.*, 1993, **64**, 1868–1873.
49 P. J. van Zwol, G. Palasantzas, M. van de Schootbrugge, J. T. M. de Hosson and V. S. J. Craig, *Langmuir*, 2008, **24**, 7528–7531.
50 R. R. Dagastine, M. Bevan, L. R. White and D. C. Prieve, *J. Adhes.*, 2004, **80**, 365–394.
51 E. Bonaccurso, H.-J. Butt and V. S. J. Craig, *Phys. Rev. Lett.*, 2003, **90**, 144501.

52 O. I. Vinogradova and G. E. Yakubov, *Phys. Rev. E: Stat., Nonlinear, Soft Matter Phys.*, 2006, **73**, 045302.
53 C. D. F. Honig and W. A. Ducker, *Phys. Rev. Lett.*, 2007, **98**, 028305.
54 C. Neto, V. S. J. Craig and D. R. M. Williams, *Eur. Phys. J. E*, 2003, **12**, 71–74.
55 J. Penfold, E. Staples, I. Tucker and R. K. Thomas, *Langmuir*, 2002, **18**, 5755–5760.
56 J. C. Schulz and G. G. Warr, *Langmuir*, 2000, **16**, 2995–2996.
57 E. Bonaccurso, M. Kappl and H.-J. Butt, *Curr. Opin. Colloid Interface Sci.*, 2008, **13**, 107–119.
58 T. J. Senden, *Curr. Opin. Colloid Interface Sci.*, 2001, **6**, 95–101.
59 I. M. Nnebe, R. D. Tilton and J. W. Schneider, *J. Colloid Interface Sci.*, 2004, **276**, 306–316.
60 C. Neto, D. R. Evans, E. Bonaccurso, H.-J. Butt and V. S. J. Craig, *Rep. Prog. Phys.*, 2005, **68**, 2859–2897.
61 E. Lauga, M. P. Brenner and H. A. Stone, in *Handbook of Experimental Fluid Dynamics*, ed. J. Foss, C. Tropea and A. Yarin, Springer, New York, 2005.
62 E. Bonaccurso, M. Kappl and H.-J. Butt, *Phys. Rev. Lett.*, 2002, **88**, 076103.
63 R. Pit, H. Hervet and L. Léger, *Phys. Rev. Lett.*, 2000, **85**, 980–983.
64 Y. Zhu and S. Granick, *Phys. Rev. Lett.*, 2002, **88**, 106102.
65 O. I. Vinogradova, *Langmuir*, 1995, **11**, 2213–2220.
66 O. I. Vinogradova, *Langmuir*, 1996, **12**, 5963–5968.
67 C. Cottin-Bizonne, A. Steinberger, B. Cross, O. Raccurt and E. Charlaix, *Langmuir*, 2008, **24**, 1165–1172.
68 D. Y. C. Chan and R. G. Horn, *J. Chem. Phys.*, 1985, **83**, 5311–5324.
69 J. N. Israelachvili, *J. Colloid Interface Sci.*, 1986, **110**, 263–271.
70 L. Suresh and J. Y. Walz, *J. Colloid Interface Sci.*, 1997, **196**, 177–190.
71 J. Y. Walz, L. Suresh and M. Piech, *J. Nanopart. Res.*, 1999, **1**, 99–113.
72 C. L. Henry, C. N. Dalton, L. Scruton and V. S. J. Craig, *J. Phys. Chem. C*, 2007, **111**, 1015–1023.
73 C. L. Henry, L. Parkinson, J. R. Ralston and V. S. J. Craig, *J. Phys. Chem. C*, 2008, **112**, 15094–15097.
74 R. S. Allan, G. E. Charles and S. G. Mason, *J. Colloid Sci.*, 1961, **16**, 150–165.
75 A. Steinberger, C. Cottin-Bizonne, P. Kleimann and E. Charlaix, *Phys. Rev. Lett.*, 2008, **100**, 134501.
76 A. Steinberger, C. Cottin-Bizonne, P. Kleimann and E. Charlaix, *Nat. Mater.*, 2007, **6**, 665–668.
77 P. J. Scales, F. Grieser and T. W. Healy, *Langmuir*, 1990, **6**, 582–589.
78 P. G. Hartley, I. Larson and P. J. Scales, *Langmuir*, 1997, **13**, 2207–2214.

General discussion

Professor Fery opened the discussion of the paper by Professor Kornyshev: Firstly, for rod-like particles, transitions from rods being aligned parallel to the interface to rods being aligned perpendicular to the interface are observed at high densities. Can you access densities in this range in order to provoke such transitions? What is the limit of the densities?

Secondly, have you investigated the effect of interfacial position for *e.g.* gold nanoparticles on plasmon resonance? This should be strongly influenced by interfacial position (the dielectric environment changes dramatically) and particle density (particle–particle coupling).

Professor Kornyshev answered: To your first question: yes, it will be possible to realize such a situation. When particles are adsorbed at the interface, the depth of the adsorption wells is typically much greater than the thermal energy. Thus the adsorption isotherm in practice does not depend on the concentration of particles in the bulk—entropic effects become unimportant here. The density of the particles in the adsorbed layer will then depend on the depth of adsorption well and the interaction between the particles at the interface. We have not studied the adsorption of non-spherical particles so far, but this may be an interesting project for the future.

The answer to question two is 'yes', but I am not in a position to disclose any details about it. Please wait for the next paper by Flatté, Kornyshev and Urbakh, which we are currently preparing for publication.

Dr Titmuss asked: You present a nice calculated isotherm for nanoparticle adsorption at interfaces; do you have any comparisons with experiment, as we are currently conducting such experiments?

You mentioned Raman enhancement by gold nanoparticles at the interface—do you have an order of magnitude for this enhancement? Could you use this effect to make experimental measurements of the isotherm?

Professor Kornyshev replied: Voltage controlled adsorption isotherms have been reported in the paper by Hubert Girault's team[1-3] for small nanoparticles. They detected their presence through SHG signals. The comparison of the theory with those experiments was discussed in another paper of ours[4] and it is not bad. Such experiments, however, have not been performed for large nanoparticles (20–50 nm in radius). As for the enhanced Raman signals, we just started to work on them and I cannot give you any definite figures at the moment. Of course, when properly calibrated this could be a good method for the study of adsorption, but it will benefit from a pertinent theory.

1 B. Su, J.-P. Abid, D. J. Fermin, H. H. Girault, H. Hoffmannova, P. Krtil and Z. Samec, *J. Am. Chem. Soc.*, 2004, **126**, 915.
2 B. Su, D. J. Fermin, J.-P. Abid, N. Eugster and H. H. Girault, *J. Electroanal. Chem.*, 2005, **583**, 241.
3 J.-P. Abid, M. Abid, C. Bauer, H. H. Girault and P.-F. Brevet, *J. Phys. Chem. C*, 2007, **111**, 8849.
4 M. E. Flatté, A. A. Kornyshev and M. Urbakh, *J. Phys.: Condens. Matter*, 2008, **20**, 073102.

Professor Ikkala asked: Professor Kornyshev, you describe an interesting concept to localize nanoparticles at the interface of two immiscible liquids for functions. My question is mostly technical: in order to exploit your effects, would it be necessary to control also the dispersed structure within the heterogeneous liquid mixture (since *e.g.* emulsions can be process- and history-dependent)?

Professor Kornyshev responded: As long as you do not emulsify water and oil (by, say, applying very large electric fields), and have a flat interface between them, I don't see where the term 'emulsion' becomes pertinent. I would not call nanoparticles in solution an emulsion. If you use the term 'emulsions' in the meaning of Pickering emulsions, speaking essentially about the adsorbed layers of nanoparticles at ITIES, the answer to your question is, generally, yes. The barriers confining the adsorption well can give rise to hysteresis and delay of response—some kind of 'memory'. In order to avoid this we must avoid situations with barriers. There will be more in our next paper.

Mr Carew enquired: What is the benefit of assembling nanoparticles at electrolyte interfaces instead of between solid electrodes?

Professor Kornyshev replied: Electric-field controlled assembly of nanoparticles at liquid/liquid interfaces (ITIES) has the benefit of ideal reversibility, because the environment of the nanoparticle in the 'adsorbed' state is liquid from all sides. However, one must be careful here, as this may not always be true. For instance, let us first polarize the aqueous phase so negatively relative to the oil phase that the adsorption well at the interface is very deep and there is no barrier separating the well from the bulk water. Let us then reduce the absolute value of the voltage drop, so that the potential well for adsorption becomes metastable (*i.e.* in its minimum the energy of the nanoparticle is higher than in the water bulk and the well is separated from the water bulk by a barrier). Within this region of voltage variation, if the barrier is much higher than the thermal energy, the nanoparticle field-induced adsorption–desorption may exhibit hysteresis. Such situations should be and can be avoided, simply by using a larger voltage variation.

The second benefit is that ITIES is accessible to light from both sides. Going back from the ITIES to a classical solid electrode/electrolyte system, using transparent electrodes, is actually one of the targets in our research. The challenge here is to avoid the irreversibility of adsorption. This is not impossible, and our team is working on it. Unfortunately, again I cannot disclose at this stage any further details; they will be discussed in our next paper.

Professor Steiner opened the discussion of the paper by Professor Ikkala: In your paper you demonstrate that cellulose can be disassembled into individual fibrils and then reassembled to form hydrogels or aerogels. Can the solution of disassembled nanofibres be spun so that the nanofibres form a macroscopic continuous fibre?

Professor Ikkala replied: First a comment—there exists already extensive literature on cellulosics that have been molecularly dissolved in solvents and thereupon form gels. By contrast, the present method aims to preserve the smallest nanometer scale constituent crystalline nanofibers. On dissolution the very feasible mechanical properties of the native crystal structure of cellulose I would be lost. Therefore, the challenge is to cleave them from the macroscopic fibers without leading to complete dissolution. Feasible routes have emerged recently to allow long cellulose I-containing nanofibers (see ref. 35 and 40 in our paper). Such mechanically strong fibers have been "reassembled" on surfaces to allow sheet-like "nanopaper" with extraordinary mechanical properties (see ref. 36 in our paper). Therefore your question is proper. It is a natural question to try to reassemble them also to form new more macroscopic composite fibers, aiming at very good mechanical properties. Such work, however, has not yet been reported to my knowledge.

Professor Reinhoudt continued the discussion of Professor Kornyshev's paper: First: your system reminds me of CHEMFETs or ion selective electrodes. Have you studied your system with impedance spectroscopy?

Second: have you varied the concentration of the hydrophilic salts in the aqueous phase and do you observe a variation in the overall potential?

Professor Kornyshev replied: Firstly, of course they have been studied by the groups of Schiffrin,[1] Samec[2] and Girault.[3]

Your second question is an important one. For the ITIES system at equilibrium, there is no electric field in the bulk. It is totally screened by the double layers at the corresponding interfaces. The electric field localized in the double layer at the interface for a given voltage across the ITIES is, of course, larger the more compact the double layer is; it is thus increased with the increase of electrolyte concentration. However, to amplify the effect of the electric field we need to have as many ligands ionized as possible, distributed over the nanoparticle surface. For this reason, in order to create a deep well for a large nanoparticle, it may be beneficial to have a weaker but less localized field. Calculation gives an estimate as to where the optimum lies. For instance, 2 nm particles can be localized with 0.1M LiCl ITIES, whereas if we want the same effect for 20 nm particles we must use much lower electrolyte concentrations. There will be more about this in our next paper.

1 H. H. Girault and D. H. Schiffrin, in *Electroanalytical Chemistry*, ed. A. J. Bard, Marcel Dekker, New York, 1989, vol. 15, p. 1.
2 Z. Samec and T. Kakiuchi, in *Advances in Electrochemistry and Electrochemical Science*, ed. E. Gersicher and C. W. Tobias, VCH, Weinheim, 1995, p. 297.
3 H. H. Girault, in *Modern Aspects of Electrochemistry*, ed. J. O'M. Bockris, B. E. Conway and R. E. White, Plenum Press, New York, 1993, vol. 25, p. 1.

Professor Woolfson continued the discussion of the paper by Professor Ikkala: Your systems are essentially bioinspired or *ex vivo* biomaterials. This is a rapidly growing field with many different materials emerging. Do you see any consensus (or champion) materials emerging with particularly high potentials for applications? Also, how do you propose to functionalise your materials leading to applications?

Professor Ikkala replied: Bioinspiration is really the main underlying theme, in an effort to combine some aspects of biological materials with rational bottom-up construction. First, as described in the article, indirect evidence suggests that the native cellulose nanofibers are expected to have good mechanical properties due to the parallel polymer chains that are mutually interlocked by hydrogen bonds. The hydroxyl groups on the nanofiber surface also allow facile chemical modifications for various functionalities. Incidentally, a high elastic modulus of *ca.* 150 GPa was recently indeed confirmed using a specific form of nanocellulose fibers (tunicates) by Iwamoto *et al.*[1] For widespread applications it is essential to develop a facile and scalable preparation and processing of nanocellulose and at present there are several routes being explored in a number of academic and industrial laboratories. Also, regarding the sustainability, I see native nanocellulose as a potential "champion" material for applications. On the other hand, regarding the multilevel hierarchical constructs that were also described in the article based on a combination of block copolypeptides, guanine monophosphates, and metal cations, the aim was mostly to explore novel systematic routes for hierarchies. The approach seems to allow generalizations to allow rational routes for several type of hierarchies.[2] In general, one could foresee applications for electroactive material due to control of self-assembly of guanine quartets and for scaffolds for bioapplications. At the present early stage, however, we cannot yet provide comparisons with other routes.

1 S. Iwamoto, W. Kai, A. Isogai and T. Iwata, *Biomacromolecules*, 2009, DOI: 10.1021/bm900520n.
2 Houbenov *et al.*, in preparation.

Dr Titmuss remarked: The Young's modulus you present seems very high, when considered in terms of a stored energy density. What is the origin of the high Young's modulus? Does the modulus remain so high over long time-scales? Does the material creep?

Professor Ikkala answered: The predicted Young's modulus refers to the fully crystalline native cellulose I nanocrystallites, where the parallel, relatively rigid cellulose chains are interlocked with each other by a dense set of hydroxylic hydrogen bonds, thus leading to very compact and strong crystalline whiskers with nanometre-scale diameters and high aspect ratios. Their mechanical properties are not yet known in detail, and the values quoted so far are predicted values. In the present long and entangled native cellulose I nanofibers, as dealt with in the current work, such whiskers remain linked end-to-end, as the less-ordered domains between them have not been purged in the selected process to cleave the nanofibers from the macroscopic fibers. Also the mechanical and creep properties of such long nanofibers are at present still not known. That the native nanocellulose nanofibers can have very feasible mechanical properties must be deduced at present only indirectly: Nakagaito and Yano[1] showed in phenolic resin composites a modulus of 18–20 GPa, and strengths of *ca.* 350 MPa which approach that of steel. See also our ref. 36.

1 A. N. Nakagaito and H. Yano, *Appl. Phys. A*, 2004, **78**, 547.

Dr Bittner asked: Concerning atomic layer deposition (ALD), what is the state of the art for the range of materials (perhaps including oxides and metals) and temperatures (perhaps below 50 °C)? See Knez *et al.*[1] and Lee *et al.*[2] for ALD on biological matter.

1 M. Knez, A. Kadri, C. Wege, U. Gösele, H. Jeske and K. Nielsch, *Nano Lett.*, 2006, **6**, 1172.
2 S.-M. Lee, E. Pippel, U. Gösele, C. Dresbach, Y. Qin, C. Vinod Chandran, T. Bräuniger, G. Hause and M. Knez, *Science*, 2009, **324**, 488.

Professor Ikkala answered: As far as I understand, your question is concerned with the fact that in connection of ALD-coating of polymeric and biological materials one has limit on sufficiently low temperatures. This is naturally a limitation but there are several low temperature materials and processes dealing oxides that are proper in this respect, such as Al^2O^3 and TiO^2. After such treatments, the organic template can be removed *e.g.* by heating, and subsequently materials requiring higher temperatures are available. You pointed out two interesting developments by Knez *et al.* and Lee *et al.* for ALD on biological matter. We fully agree that they are very interesting examples which show the power of ALD in connection with soft matter, encouraging further work.

Professor Matile opened the discussion of the paper by Professor Huskens: I find this layer-by-layer assembly with molecular recognition absolutely spectacular! My questions are: is spherical shape required to prevent saturation of all sites and keep the process going? How many layers can be added? Are the kinetics with molecular recognition as fast and simple as with polyelectrolyte layer-by-layer? Do single dendrimers placed on an acceptor layer roll around or are they irreversibly trapped?

Professor Huskens replied: The spherical shape is indeed employed here to prevent site saturation. An initial study using guest-functionalized polymers[1] showed that such polymers flattened completely, resulting in efficient use of all sites and absence of interaction with cyclodextrin-functionalized gold nanoparticles. However, other options exist which leave sticky ends. Another steric solution may be to make an intentional mismatch between the repeat distances in polyvalent guest- and host-functionalized building blocks. A more delicate solution lies in weakening(!) the individual interaction strength. In another study,[2] employing orthogonal host–guest and metal–ligand interactions, we have shown that a weaker binding metal ion leads to more "sloppy" binding and thus to sticky ends, here shown in the aggregation of host-functionalized vesicles. In the layer-by-layer scheme under discussion here, we have built up to 30 bilayers without noticing a change in the amount of material added per bilayer (linear growth), although the roughness increased somewhat upon increase of the number of layers. Kinetics has not been carefully investigated, but the choice of the concentrations of the components plays a major role both in adsorption times and in the degree of nonspecific adsorption.[3] The suggestion about rolling dendrimers is a striking thought! We are currently investigating the option of multivalent molecules "walking" along a multivalent interface. From those initial results, it is probably fair to say that the interaction strength of the dendrimers at hand is too high to allow rolling, but tuning this interaction to make it possible is definitely an option!

1 O. Crespo-Biel, M. Péter, C. M. Bruinink, B. J. Ravoo, D. N. Reinhoudt and J. Huskens, *Chem. Eur. J.*, 2005, **11**, 2426.
2 C. W. Lim, O. Crespo-Biel, M. C. A. Stuart, D. N. Reinhoudt, J. Huskens and B. J. Ravoo, *Proc. Natl. Acad. Sci.*, 2007, **104**, 6986.
3 O. Crespo-Biel, B. Dordi, D. N. Reinhoudt and J. Huskens, *J. Am. Chem. Soc.*, 2005, **127**, 7594.

Mr Carew asked: How do the mechanical properties of templated assemblies gold nanoparticles compare to the mechanical properties of untemplated assemblies of gold nanoparticles?

Professor Huskens replied: Mechanical properties have so far only been assessed for composite structures containing polystyrene particle structures which have been filled with the gold nanoparticle glue materials,[1] not for the capsule structures discussed in the current study. Also the mechanical properties of untemplated gold nanoparticle structures have not been measured to date, to my knowledge.

1 X. Y. Ling, I. Y. Phang, H. Schönherr, D. N. Reinhoudt, G. J. Vancso and J. Huskens, *Small*, 2009, **5**, 1428.

Mr Ahangar continued the discussion of the paper by Professor Kornyshev: Do you think that your concept can be applied to the treatment of oil slicks as this also involves interaction at surfaces?

Professor Kornyshev answered: I have no idea how.

Dr Titmuss returned to the discussion of the paper by Professor Huskens: In the paper you mention using layer-by-layer assembly to form layers comprising 30 bilayers—what would be the minimum number of bilayers that would lead to a mechanically stable film? Is there any correlation between the size of the particles and the mechanical properties of the resulting films? Do smaller particles give stronger films by creating more internal surface area for volume of material?

Professor Huskens answered: Real mechanical measurements were only done on the free-standing bridges,[1] so before polystyrene core removal. These were done with different numbers of bilayers: 1 (only dendrimer), 10, 20 and 30. Using only the dendrimer was insufficient for mechanical stability. For 10, 20 and 30 bilayers, stable structures were obtained but mechanical AFM measurements showed that the strength did increase upon increase of the number of bilayers. When making the hollow capsule ribbons in our paper, only 30 bilayers was attempted and the only indications about strength come from visual inspection. The ribbons are relatively stiff and they fracture into pieces, typically of several tens of microns. It is probably true that smaller particles give stronger films as the internal surface area increases, but the number of interactions between a single particle and a dendrimer (multivalency effect) is also important. This has been illustrated before,[2] showing that a smaller generation of dendrimer resulted in a lower transfer printing yield indicating a smaller strength of the printed structures.

1 X. Y. Ling, I. Y. Phang, H. Schönherr, D. N. Reinhoudt, G. J. Vancso and J. Huskens, *Small*, 2009, **5**, 1428.
2 X. Y. Ling, I. Y. Phang, D. N. Reinhoudt, G. J. Vancso and J. Huskens, *ACS Appl. Mater. Interface*, 2009, **1**, 960.

Dr Faul asked: What would the influence be on changing (i) the type of materials used and, maybe more importantly, (ii) the shape (specifically use of anisotropic shapes) of particles used in this double-templating strategy?

Professor Huskens responded: The main concept in our work on supramolecular assembly is that all kinds of building blocks can be used, with varying core material, size, shape, *etc.*, as long as they can be functionalized with host and/or guest moieties. To stress this versatility, we have so far employed a wide range of building blocks in supramolecular assembly (although only a limited set in the discussed free-standing or free-floating assemblies) such as: gold nanoparticles (5 nm), silica nanoparticles (50–500 nm), polystyrene (300–500 nm), dendrimers (2–6 nm), polymers, synthetic "small" molecules, *etc*. So, the idea is that the properties of the final structure can be tuned by changing the core materials. One of our future goals is, for example, to use the bridge structures[1,2] as photonic bandgap or plasmonic materials, by varying the infiltrating and/or particle core materials. Using anisotropic building blocks may definitely be interesting, but we have no experience in this direction yet.

1 X. Y. Ling, I. Y. Phang, W. Maijenburg, H. Schönherr, D. N. Reinhoudt, G. J. Vancso and J. Huskens, *Angew. Chem., Int. Ed.*, 2009, **48**, 983.
2 X. Y. Ling, I. Y. Phang, H. Schönherr, D. N. Reinhoudt, G. J. Vancso and J. Huskens, *Small*, 2009, **12**, 1428.

Professor Sen continued the discussion of the paper by Professor Kornyshev: What are the differences between your system and the typical Pickering emulsions? Can you get the particles to actively diffuse towards the oil/water interface?

Professor Kornyshev replied: There are two questions here. As explained in our paper, one of the components that drives nanoparticle localization at the interface is the surface tension. This is the main driving force in the Pickering systems as well. However, here we also have highly charged nanoparticles and thus we have the effect of an applied electric field which can bring nanoparticles to or move them away from the surface. You may call this process 'active'. In reality, since the electric field is localized near the surface and there is no Ohmic drop—as there is no dc current across the interface for the voltages of interest—the particles can reach the 'trap' at the interface mainly by bulk diffusion. We discussed this issue in the current paper, but there will be more on what it will mean for a practical device in the paper to follow.

Professor Dr van Esch continued the discussion of the paper by Professor Huskens: It was mentioned that in layer-by-layer (LBL) constructs by using host–guest interactions it remains a challenge to achieve full reversal of interactions at the interface by binding of multivalent host or guest compounds, especially with flexible polymers. However, this has never been an issue in LBL deposition using polyelectrolytes. What are the major reasons for this discrepancy, especially in terms of the differences in adsorption thermodynamics and kinetics? Is it really necessary to use steric constraints like the multivalent spheres used in your work? Or can such an affinity reversal also be achieved by increasing the concentration of the host (guest) multivalent compound resulting in a kinetically trapped state with excess of host (guest) groups exposed at the new interface?

Professor Huskens replied: There is only one study on a flexible polymer using host–guest interactions[1] which shows that a multivalent polymer very efficiently uses all interactions. This led to a strongly flattened polymer layer of less than 1 nm which did not interact any more with host-functionalized particles. Thermodynamically, there is a very high driving force for intramolecular binding at a multivalent interface compared to intermolecular binding. It is doubtful whether the concentration can practically be chosen high enough to promote intermolecular binding and kinetic trapping. The comparison between electrostatic and host–guest LBL is an interesting one. Individual electrostatic interactions (between a monovalent cation and anion) are weaker than the cyclodextrin–adamantyl host–guest pair but work more long-range (although tunable by *e.g.* salt concentration). Moreover, the conformational flexibility of electrostatic polymers is in large part determined by repulsion between the subunits within the polymer and potential charge screening by counterions and additional salt, whereas in host- or guest-functionalized polymers only steric interactions play a significant role. Therefore, I think sterics is the main game to play with host–guest polymer systems, but there are definitely other options than the spherical solution chosen by us. For example, choosing a higher or mismatching fraction of functionalized monomer units in the polymer may lead to unused sites after adsorption allowing subsequent assembly steps. Another option is to use different types of host–guest interactions in the same polymer.

1 O. Crespo-Biel, M. Péter, C. M. Bruinink, B. J. Ravoo, D. N. Reinhoudt and J. Huskens, *Chem. Eur. J.*, 2005, **11**, 2426.

Professor Fery asked: Do the free-standing assemblies deform elastically or plastically? What is the plasticity limit and how do the structures deform when the elastic regime is left?

Professor Huskens replied: So far we have only done mechanical measurements on the free-standing particle bridges, not on the free-floating hollow ribbons described in the current *Faraday Discussions* paper. These bridges were first described in ref. 1, but the mechanical measurements were described in detail recently in ref. 2. The particle bridges deform plastically when forces are applied up to about 5000 nN. Deflection is maximal around 150 nm. At higher forces, irreversible processes occur such as the slippage of a particle layer.

1 X. Y. Ling, I. Y. Phang, W. Maijenburg, H. Schönherr, D. N. Reinhoudt, G. J. Vancso and J. Huskens, *Angew. Chem., Int. Ed.*, 2009, **48**, 983.
2 X. Y. Ling, I. Y. Phang, H. Schönherr, D. N. Reinhoudt, G. J. Vancso and J. Huskens, *Small*, 2009, **12**, 1428.

Professor Colquhoun continued the discussion of the paper by Professor Ikkala: Many important applications of high-modulus fibres are in polymer-composite materials. Can you foresee your cellulose nanofibres finding use in such materials, and if so how? Also, what is the maximum length of cellulose nanofibres achieved in your work?

Professor Ikkala replied: As I already pointed out in relation to the question by Professor Woolfson, the mechanical properties of long native cellulose nanofibers are expected to be very favourable based on indirect evidence, and even direct evidence thereof is starting to accumulate.[1] Also, as the fibers have hydroxyl groups on the surface, allowing chemical reactions for functionalization, the polymer composites are presently extensively pursued. As referenced in the article, promising results have already been obtained *e.g.* in phenolic resins. Our fibers form network structures, where the junction points can be formed due to domains within larger fibers that have not been cleaved into nanofibers, hydrogen bonds, or physical entanglements. Therefore, it has still turned out to be nontrivial to characterize the length of the nanofibers and at the moment we have only qualitatively information that they are drastically longer than the previously extensively used whisker-like cellulose nanocrystals. This should provide considerable benefits in composites.

1 S. Iwamoto, W. Kai, A. Isogai and T. Iwata, *Biomacromolecules*, 2009, DOI: 10.1021/bm900520n.

Professor Steiner continued the discussion of the paper by Professor Kornyshev: The application of an electric field across a dielectric interface triggers a capillary instability (a so-called electrohydrodynamic instability). Is the planar conformation of the interfaces you describe in your paper stable once an electric field is applied?

Professor Kornyshev responded: It is a just question. In our earlier paper,[1] we have studied the effect of electric field on the full spectrum of dynamic surface corrugations. The mean squared amplitude of the capillary waves first grows as a square of the voltage across the interface, but then it starts to rise more steeply up to a critical point of electric-field-induced emulsification: this has been experimentally observed by others. But in our systems with nanoparticles we should never go to such large fields. Fluctuations with modest amplitudes (smaller than the size of nanoparticles) and short wavelengths will, presumably, be unnoticeable in the optical effects with nanoparticles. Capillary wave can also be excited by ion transport across the interface[2] but, again, we must stay away from the ITIES ion-breakdown voltages.

1 L. I. Daikhin, A. A. Kornyshev and M. I. Urbakh, *J. Electroanal. Chem.*, 2000, **483**, 68–80.
2 L. I. Daikhin, A. A. Kornyshev, A. M. Kuznetsov and M. Urbakh, *Chem. Phys.*, 2005, **319**, 253–260.

Dr Shaffer continued the discussion of the paper by Professor Ikkala: If you continue to dilute gels based on your long cellulose nanofibres, can individual structures be eventually identified? If so, what are their lengths, if not, do the nanofibres remain bound by undigested material, interfibre interactions (H-bonds), or physical entanglements? Related problems have been observed in high-aspect-ratio carbon nanotube samples, for which dissolution or suspension generally requires cutting into short lengths, detrimental to mechanical properties. One solution emerging is the alignment of long nanotubes, prepared in the gas phase, as a dilute 'aerogel', with a low density of physical cross-links. This strategy might be considered analogous to gel-drawing of ultra-high molecular weight polymers (particularly polyethylene) to produce high performance fibres. Are these approaches relevant to your cellulose nanofibre gels? Is it possible to draw a highly aligned macroscopic fibre from your materials?

Professor Ikkala replied: The native cellulose nanofibers described in the text form gels in aqueous medium even at the lowest so far investigated concentration of 0.125 wt%. The nanofibers form networks, and so far individual fibers have not been clearly observed. The junction points of the gel network can result based on non-ideal cleavage of the nanofibers from the macroscopic fibers due to so far non-ideal processing, hydrogen bonds, and even entanglements. More understanding is being gained by varying the processing conditions and concentrations. The length of the fibers cannot thus be quantitatively answered as yet, even if qualitatively it is known that they are much longer than the cellulose nanocrystals. On the other hand, recently it was shown that corresponding cellulose nanofibers incorporating negative charges allowed to have a clear dispersion by centrifugation and allowed multilayer assemblies.[1] Also, our recent rheological evidence shows that negatively charged nanocellulose hydrogels cease to form gels at the low concentration *e.g.* 0.125 wt%, unlike the charge-neutral nanocellulose hydrogels.[2] These observations point towards Coulombic colloidal stabilization effects within charged nanocellulose, thus facilitating individualization of fibers. In conclusion, the individualization of the fibers from the network, the reversibility of the gel, and the fiber length depend in a complicated way on processing, starting materials, and their functionalization, and a satisfactory answer cannot be given at the moment. Your final comment on the gel drawing is most proper and is, in fact, being pursued at present.

1 L. Wågberg, G. Decher, M. Norgren, T. Lindström, M. Ankerfors and K. Axnäs, *Langmuir*, 2008, **24**, 784.
2 Wang *et al.*, in preparation.

Dr Munz continued the discussion of the paper by Professor Huskens: Firstly, could the free-standing hybrid particle composites possibly be utilised for chemical sensing applications, similar to nanowires or carbon nanotubes? Upon adsorption of molecules, such linear nanostructures tend to show strong changes in electrical conductivity.

Secondly, if there is no direct electrical contact between adjacent particles, is there a chance for a significant tunnelling current?

Professor Huskens replied: At least three different sensing/detection schemes seem plausible: (i) optical (as photonic bandgap or plasmonic material), (ii) electrical, or (iii) mechanical. So far, only a few mechanical measurements have been done,[1] and optical detection is our first plan on the line in the future. The current structures are most likely not conducting. The cyclodextrin–adamantyl couple is about 1 nm thick, and the dendrimer core another 2–3 nm, making the insulating layer between two adjacent gold nanoparticles at least several nm thick. It also clear from optical (UV/Vis absorption) measurements that coupling between the gold nanoparticles is limited.[2] So, even tunnelling seems unlikely, also because the structures represent many tunnelling barriers in series. Most likely, the structures can be made conductive, *e.g.* by some annealing procedure or by using them as a template in electroless deposition. This should be tuned to get structures with some, but not fully metallic, conductive properties to achieve sensitivity. The openness of the 3D structures made here may assist in creating a large sensing surface.

1 X. Y. Ling, I. Y. Phang, H. Schönherr, D. N. Reinhoudt, G. J. Vancso and J. Huskens, *Small*, 2009, **5**, 1428.
2 O. Crespo-Biel, B. Dordi, D. N. Reinhoudt and J. Huskens, *J. Am. Chem. Soc.*, 2005, **127**, 7594.

Mr Ahangar opened the discussion of the paper by Professor Chan: Do you think that the phenomenon of the air pocket slip can be changed if we use a different solvent other than water which means that we have to focus on solvent interaction chemistry?

Professor Chan responded: The answer depends on the property of the solvent and how that wets or does not wet the rough silica nanoparticle surface. If the solvent/air surface tension is high, then one might expect the behaviour would be similar to that with water in the absence of surfactant. The mobility of the air/liquid interface is controlled by the surface active material in the solution. The addition of a surface active agent to the solvent would be likely to give an immobile air/liquid interface again. However, one could choose solvents that have very low surface tensions (*e.g.* fluorinated aromatic hydrocarbons) that are good at wetting almost any surface. This may then remove the trapped air pockets which perhaps exhibit different hydrodynamic boundary conditions.

Professor Steiner remarked: Two questions—first, what is the surface chemistry of the silica nanoparticles? Silicon oxide has a high surface energy and is wetted by water. According to the Cassie–Baxter theory, trapped air bubbles are not supported by a rough silica surface. Instead of superhydrophobicity, enhanced wetting should be observed.

Second, when adding SDS, are the air bubbles still in place? Is, maybe, the transition from slip to stick a transition from a surface state with trapped air bubbles to one that is fully wetted by water?

Professor Chan replied: In response to question 1: after the silica nanoparticles are deposited on to the substrate, the silane linker is still present on the silica nanoparticles. This will make the coated surface hydrophobic. Furthermore, the silica nanoparticle layer is rough and also forms a hierarchical structure of varying degrees of roughness and porosity; this is commonly accepted as one of the key criteria for creating a superhydrophobic surface in the manner of a lotus leaf. On the other hand, the silica colloid probe, though rough, is known to behave like a solid surface in many force measurement studies and therefore is not expected to trap any air bubbles on its surface.

In response to question 2: as shown in the AFM images in liquid in Fig. 3 and 4 of our paper, the same trapped air bubbles are observed both before and after the addition of SDS. As discussed in the paper, it is expected that the increased hydrodynamic repulsion seen upon the addition of SDS is most likely to be the result of a change in the hydrodynamic boundary condition at the air-water interface of the trapped air bubbles. The presence of the SDS is known to cause the air/water interface to behave as an immobile (no-slip) interface as we have demonstrated using dynamic AFM force measurements using a bubble as a colloid probe in the absence and presence of added SDS. Please see ref. 1 for more details.

1 O. Manor, I. U. Vakarelski, X. Tang, S. J. O'Shea, G. W. Stevens, F. Grieser, R. R. Dagastine and D. Y. C. Chan, *Phys. Rev. Lett.*, 2008, **101**, 024501.

Mr Aufderhorst-Roberts opened the discussion of the paper by Professor Fery: Firstly, you mention potential applications of this approach as including nanocircuitry and magnetic data storage. Both of these demand significantly complex patterning. Can this be achieved to the level required, using this method?

Secondly, the presence of junctions in the patterns is attributed to surface defects. Can you control the positions in which they occur and quantities in which they occur?

Professor Fery answered: Firstly, in the paper, the relatively simple deformation situation of uni-axial strain of a homogeneous elastic substrate was studied, since it was not the pattern formation but the templating of nanoparticle deposition which was the center of our interest. However, more complex patterns can be achieved by using other strain situations (like biaxial strain or embossing as described in ref. 1).

Secondly, defect positions should be controllable by using substrates with patterned elastic properties.

1 C. Lu, H. Möhwald and A. Fery, *Soft Matter*, 2007, **3**, 1530–1536.

Professor Ikkala asked: Professor Fery, you describe wrinkled surfaces to allow overall alignment of rod-like moieties. The prior art extensively describes simple concepts based on rubbing of surfaces which is used for example in liquid crystal technology. Could you comment on the benefits of your method relative to rubbing?

Professor Fery replied: Rubbing allows one to achieve orientation, but our method has the additional advantage that we can very well control lateral spacing between the virus lines as well as the width of the virus lines. This is not possible by rubbing.

Dr Bittner commented: The alignment of nanoparticles by shear forces, *e.g.* by rubbing a solution on a flat surface, is well known (see also DNA stretching). For rod-like particles it can occur automatically when preparing samples for TEM and AFM. For the tobacco mosaic virus used by Professor Fery, my group reported alignment in 2005,[1] but the TEM pioneers already published similar images.[2]

1 A. M. Bittner, X. C. Wu, S. Balci, M. Knez, A. Kadri and K. Kern, *Eur. J. Inorg. Chem.*, 2005, 3717.
2 H. Ruska, *Arch. Gesamte Virusforsch.*, 1943, **2**, 480.

Professor Fery responded: Alignment using our method allows for precise control of lateral separation of virus lines (by choosing the periodicity of the wrinkles) and a large coverage as well as the possibility of transferring patterns by means of printing to other substrates. This combination is unique and cannot be achieved by the methods mentioned.

Dr Bittner then asked: What has to be controlled to avoid virus adsorption on top of the wavy pattern? What is the mechanism? Can one exclude van der Waals interaction with the surface? My group interpreted a similar behaviour of rod-like viruses on structured and hydrophilised PDMS surfaces to discontinuous dewetting; this phenomenon should be based on capillary forces in confined pockets. Note that this works equally well for tobacco mosaic virus (see Balci *et al.*[1]) as for spherical particles (see the supplementary information of ref. 1).

1 S. Balci, D. M. Leinberger, M. Knez, A. M. Bittner, F. Boes, A. Kadri, C. Wege, H. Jeske and K. Kern, *Adv. Mater.*, 2008, **20**, 2195.

Professor Fery responded: Firstly the virus must be non-adhesive on the substrate. If there are adhesive interactions, the topographical pattern will be ignored. We believe (which is currently under investigation) that the main effect for the assembly process is not direct interaction between the particles and the surface, but that rather the dewetting of the liquid layer determines the quality of the assembly. This is in agreement with the work mentioned.

Professor Huck opened the discussion of the paper by Professor Hu: Your Monte Carlo simulation probes the relation between volume size and crystallization and clearly demonstrates the influence of the surface on the orientation of crystals. How small does the polymer droplet need to be to guarantee a single, stochastic nucleation event at room temperature, and hence force the generation of well-defined single crystals?

Professor Hu replied: The single nucleation event for single crystal in each droplet is very much related to the chosen temperatures, which should be high enough to make crystal nucleation a rare event in small polymer droplets. At low temperatures, even only a single polymer in solution can still perform multiple nucleation events, as demonstrated by the report of Muthukumar's group for molecular simulations of polymer crystallization.[1] Our present simulations of 5-chain droplets also show multiple crystallites formed at low temperatures.

1 M. Muthukumar and P. Welch, *Polymer*, 2000, **41**, 8833–8837.

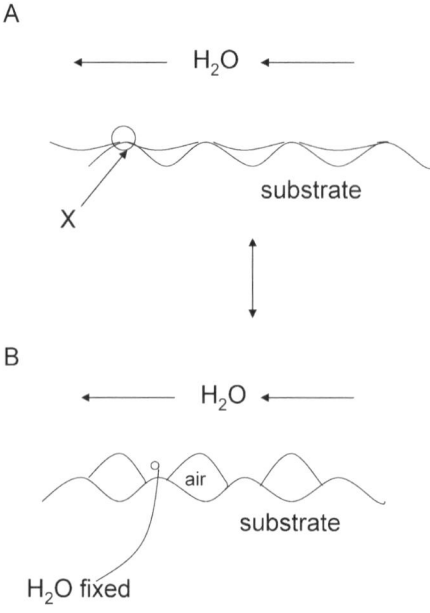

Fig. 1

Professor Whitesides continued the discussion of Professor Chan's paper: What happens in regions such as X in Fig. 1? Also, how does either case A or case B lead to a slip or non-stick surface?

Professor Chan answered: The no-slip hydrodynamic boundary condition is expected to occur from the immobile air/water interfaces of the trapped air pockets when SDS is added. Prior to the addition of SDS, the air/water interface is expected to be mobile and significantly reduces the drainage forces. One would expect that if the bubbles had a high curvature, the roughness would decrease the drainage forces. AFM imaging suggests that aside from an occasional isolated large bubble, the air/water interfaces are flat and have low curvature and bridge the cracks and crevices on the SNP-SHS surface. Aside from the mobility of the air/water interface our measurement cannot infer any detailed information about the possibility of origin of a slip/non-stick surface being the result of small scale protrusions as suggested in Fig. 1. The reason is that our measurements use a large colloid probe (to maximise sensitivity or magnitude of the measured forces) which therefore only provides an average result over the effective interaction area. Any explanation or mechanism for the slip/non-stick condition should also explain the increase in drainage resistance upon the addition of SDS. One fact that we do know is that the adsorption of SDS at the air/water interface will change it from the slip to a no-slip surface, as I said to Professor Steiner earlier.

Dr Titmuss asked: The images in the paper show that the surfactant appears to exert no effect on the height of the bubbles or their cross-sectional area. I would have expected the change in the surface tension to have caused a change in the contact angle and a concomitant change in the spreading of the bubble, and hence its height/cross-section—why is this not the case?

In trying to understand the reasons why the model calculations shown in your Fig. 7 fit the data so poorly around the minima in the force $vs.$ $t\beta$, I am wondering how justified the assumption of no slip really is? Your justification seems to be based on drainage experiments performed using the SFA, yet there are more recent experiments by Klein and co-workers[1] that probe the friction across thin films of surfactants between molecularly smooth mica interfaces. Their work suggests that, for the interface between a surfactant film and mica, a slip boundary condition might be more appropriate? How do you reconcile these observations?

1 W. H. Briscoe, S. Titmuss, F. Tiberg, R. K. Thomas, D. J. McGillivray and J. Klein, *Nature*, 2006, **444**, 191–194.

Professor Chan responded: First to address the addition of SDS: certainly the interfacial tension would change and it is conceivable that the contact angle may as well. In most instances the trapped air pockets are bridging cracks and crevices where the curvature of the air/water interface is quite low and the three-phase contact appears to be pinned by the rough surface. This is not to say one should not consider contact angle effects. In fact, in the paper we discussed the fact that changes in the contact angle upon the addition of the SDS could cause rearrangement of the trapped air pocket interfaces which in turn cause the effective surface roughness to decrease and thereby increase the hydrodynamic drainage forces. This possibility cannot be ruled out completely, but since the trapped air interfaces seem relatively flat (as the large bubble in our Fig. 4 is only observed occasionally) one would expect the surfactant modifying the mobility of the air/water interface to be a larger effect on the hydrodynamic drainage forces.

Question of slip for the silica sphere–mica surfaces: First, the basis for assuming no-slip on a flat mica or silica surface is not based solely on SFA measurements,[1] but a series of recent AFM measurements with an independent measurement of the inter-surface separation distance[2-4] which are consistent with the SFA measurements. There has been much debate in the literature in this area, but it seems that the most recent results now confirm what seemed to be the consensus twenty years ago.

The question does raise some interesting issues about what is referred to as slip. As pointed out in the question, the turn-around point in Fig. 7 is poorly fitted by a - no-slip model. As discussed in the paper, we do employ a Navier slip model and while the Figures are not given in the paper, the fits are much better with a Navier slip model as there is an additional adjustable parameter. In this instance, the slip lengths turned out to be an order of magnitude larger than the roughness on the silica sphere, so while one does achieve a "good fit" to the data, one would question whether the Navier slip model is accurately representing the physics at hand. More to the point, when we discuss rough surfaces and invoke the term slip we are really applying models developed for flat interfaces and applying them to rough interfaces because we know that there has to be some deviation. Simply applying a Navier slip model does not confirm that there is motion of the liquid molecules at the solid/liquid interface. We are simply replacing a rough surface by a mathematically flat surface that exhibits slip. If we examine measurements on smooth surfaces and extrapolate the molecular level description to a rough surface, there does not seem to be an obvious molecular argument for slip on a silica or mica surface. Instead the surface roughness can cause surfaces to come into hard contact at the peaks of the roughness while the separations between the valleys of the surfaces can still be large. Such gaps will provide a channel for the intervening liquid film to drain so the hydrodynamic resistance to approach would appear to be less than that for smooth surface at the same separation. Furthermore, the Navier model assumes axial symmetry whereas for rough surfaces, where there are large random variations of the separation between the surfaces in the lateral direction, the assumption of axial symmetry may not be sustainable.

In response to the work by Kelin and co-workers,[5,6] friction does raise an interesting and different view of the situation. Most of the work by Klein, as well as the extensive AFM studies by Higashitani and co-workers,[7-10] show the hydrated ions are important in sub-nanometre films and behave as ball-bearings and provide an additional lubricity to the film. This is an interesting point at contact, but does not really address most of the measured force range, where the film thicknesses are well over a nanometre.

1 C. D. F. Honig and W. A. Ducker, *J. Phys. Chem. C*, 2007, **111**, 16300–16312.
2 C. D. F. Honig and W. A. Ducker, *Phys. Rev. Lett.*, 2007, **98**, 028305(1–4).
3 C. D. F. Honig and W. A. Ducker, *J. Phys. Chem. C*, 2008, **112**, 17324–17330.
4 C. Neto, D. R. Evans, E. Bonaccurso, H.-J. Butt and V. S. J. Craig, *Rep. Prog. Phys.*, 2005, **68**, 2859–2897.
5 J. Klein, U. Raviv, S. Perkin, N. Kampf and S. Giasson, *Mater. Res. Soc. Symp. Proc.*, 2004, **790**, 325–332.
6 U. Raviv and J. Klein, *Science*, 2002, **297**, 1540–1543.
7 B. C. Donose, I. U. Vakarelski and K. Higashitani, *Langmuir*, 2005, **21**, 1834–1839.
8 B. C. Donose, I. U. Vakarelski, E. Taran, H. Shinto and K. Higashitani, *Ind. Eng. Chem. Res.*, 2006, **45**, 7035–7041.
9 E. Taran, B. C. Donose, I. U. Vakarelski and K. Higashitani, *J. Colloid Interface Sci.*, 2006, **297**, 199–203.
10 E. Taran, Y. Kanda, U. Vakarelski Ivan and K. Higashitani, *J. Colloid Interface Sci.*, 2007, **307**, 425–432.

Dr Neto said: I have two questions: Firstly, what is your view of the reason why the aqueous solution should slip at the rough silica sphere, but not at the smooth mica surface?

Secondly, can you extract a slip length from the measurements made on the super-hydrophobic surfaces, both before and after the introduction of the SDS in solution?

Professor Chan replied: In response to question 1: the use of the Navier slip model in this instance does not mean one would expect actual "slip" at a silica interface in an aqueous solution. We only wanted to see what slip length we would get if we were to fit our measurements with the Navier slip model, and whether that slip length would be related to some characteristic length of the rough surfaces. As pointed out in the paper, the fitted slip lengths in Table 2 for the silica sphere mica interaction are an order of magnitude larger than the surface roughness of the silica sphere. This would suggest that the Navier slip length model is somewhat ineffective in capturing the relevant physical situation that caused the reduction in drainage forces.

It has been shown recently,[1–3] in contrast to the previous results of slip measurements,[4] that one would not expect "slip" at a smooth silica/water interface or at a smooth mica/water interface when measured using the AFM. Studies using the surface forces apparatus over 20 years ago also showed that the flow of water between two mica surfaces at sub-nanometre separation is consistent with the no-slip boundary condition at the mica/water interface.[5] Historically, the earliest application of the Navier slip model in interpreting direct force measurements was an attempt to provide an explanation for the so called long-range hydrophobic force in terms of "hydrodynamic slippage".

In response to question 2: given that application of the Navier slip model on a rough surface produced slip lengths that cannot be related to measured characteristic roughness scales of the surfaces, it may not be a question of whether one can, but more whether one should use the Navier slip model. Our SNP-SHS surface roughness is two orders of magnitude higher than the silica sphere roughness, so one would question the physical meaning of using the Navier model to attempt a fit to the experimental data. In addition, the drainage forces prior to the addition of SDS are negligible and the velocity dependent forces observed after the addition of SDS are larger but quite small in comparison to even the silica sphere–mica measurement. I also refer you to my earlier response to Professor Steiner.

1 C. D. F. Honig and W. A. Ducker, *J. Phys. Chem. C*, 2007, **111**, 16300–16312.

2 C. D. F. Honig and W. A. Ducker, *Phys. Rev. Lett.*, 2007, **98**, 028305.
3 C. D. F. Honig and W. A. Ducker, *J. Phys. Chem. C*, 2008, **112**, 17324–17330.
4 C. Neto, D. R. Evans, E. Bonaccurso, H.-J. Butt and V. S. J. Craig, *Rep. Prog. Phys.*, 2005, **68**, 2859–2897.
5 D. Y. C. Chan and R. G. Horn, *J. Chem. Phys.*, 1985, **83**, 5311–5324.

Professor Woolfson continued the discussion of the paper by Professor Fery: What are the prospects for making gene fusions to express bioactive peptides/proteins (*e.g.* cell-adhesion peptides or growth factors) on the outside of the virus? In this way it might be possible to direct cell growth and movement on surfaces.

Professor Fery replied: Bioactive peptides (like RGD peptide) can be expressed on the surface of the viruses for promoting the cell adhesion and directing the cell migration. For a recent related paper see ref. 1. Therefore we believe that a perspective in this direction is possible.

1 J. Rong, L. A. Lee, K. Li, B. Harp, C. M. Mello, Z. Niu and Q. Wang, *Chem. Commun.*, 2008, 5185–5187.

Professor Huskens continued the discussion of the paper by Professor Hu: You conclude that (i) nucleation at the edges of a droplet occurs with alignment while it occurs random in the center of the droplet, and (ii) probabilities for nucleation of crystallization are similar at the edges and in the bulk. You assume a perfectly flat surface in your simulations. Please comment on the potential role of surface defects (step edges, vacancies, *etc.*) on alignment and probabilities for nucleation of crystallization.

Professor Hu responded: This is a very important issue! Although the droplets are so small as to be in the nano-scale, they are viable to capture surface defects upon their practical formation process such as dewetting from the solid substrate. Such defects may turn the expected stochastic event of homogeneous crystal nucleation near interfaces or in the bulk phase into a defect-induced event of heterogeneous crystal nucleation. If we initiate crystal nucleation from the well-designed defects, we may be able to control the uniform crystal orientations among nano-droplets.

Professor Huskens returned to the discussion of the paper by Professor Fery: Oxidation of micropatterned (*e.g.* line-patterned) PDMS is known to lead to wavy structures orthogonal to the micro-lines. Have you tried oxidation in the stretched state with micropatterned PDMS instead of flat PDMS to see whether the defect density is less?

Professor Fery replied: No, this was not tested.

Mr Barbero asked: The process by which the virus patterns are created is a complex process. What determines the formation of single or double lines of the virus on the structured substrates? The spinning rate and time must be important variables for the organization and adhesion of the strands. How long do you spin, and how do you measure the presence of water on the substrate after spinning?

In Fig. 2, you show AFM scans of single and double strands in the grooves of the substrate. Why are the double strands more shallow than the single ones?

Professor Fery answered: The spinning speed (as reported in the paper) and the periodicity of the wrinkle patterns are governing whether single or double lines develop. The presence of water after spinning cannot be measured, but we expect that water is still present and that the assembly process is influenced by dewetting. The double strands appear more shallow because of tip–sample convolution effects.

Mr Carew remarked: Two questions: First, what is the range of line spacings and defect densities that can be achieved using this technique? Second, what is the range of line spacings and defect densites that you have observed for your system?

Professor Fery answered: Theoretically, line spacing should be possible down to nanoscale dimensions. The concept behind wrinkle formation is based on continuum mechanics, which applies down to nanometre length scales. To my knowledge there are no theoretical considerations on the defect densities.

Secondly, we have covered a range on line spacings between 150 nm and tens of microns. The range between 150 nm and 1 micron can be achieved with plasma oxidation of PDMS, for larger wavelengths, polymer coatings or UV–ozone treatment can be used. We find that defect density depends on the nature of the coating used in the wrinkle formation process and on the wavelengths. For wrinkles of small wavelengths on the order of 300 nm, a typical defect density is 1 defect per 20 square microns. For larger wavelengths defect density is greatly reduced.

Dr Shaffer commented: You report a very specific spin speed for achieving tobacco mosaic virus (TMV) deposits with a high degree of order, yet the forces on the dispersion are highly dependent on both the distance from the axis, and the local orientation of the 'wrinkles' which must vary from radial to tangential. In other high aspect ratio systems, complex orientation distributions have been identified, after spin coating. Have you observed any systematic variations in TMV orientation as a function of position or wrinkle orientation during the spin-coating process? The biggest potential advantage of your approach is the applicability to large areas; how do you envision depositing the TMV suspension uniformly at the square metre scale?

Professor Fery replied: We find that TMV orientation does not strongly depend on lateral position but rather is dominated by wrinkle orientation. This is understandable if one assumes that the function of spin coating is mainly to set a film thickness (of the TMV-containing solution) and that the orientation of the viruses is governed by a dewetting process. Film thickness is not uniform in the spin coating process, but varies over the sample. Still it is of the same order of magnitude for the samples investigated here and we find similar structures over the sample area. We believe that the key for further upscaling is mainly developing an understanding of the underlying dewetting mechanism. An *in situ* control of the deposition process has so far not performed, but would be helpful in this respect.

Professor Chan said: In evaporative lithography using solutions of nanoparticles, the drying rate plays a key role in determining the morphology of the assembled structure, see for example ref. 1. Do you seem similar effects in assembly on crinkled substrates?

1 I. U. Vakarelski, D. Y. C. Chan, T. Nonoguchi, H. Shinto and K. Higashitani, *Phys. Rev. Lett.*, 2009, **102**, 058303.

Professor Fery replied: Yes, we believe that the morphology of the networks is mainly determined by the dewetting process.

Mr Hopkinson asked: Presumably it is possible to apply the strain in two dimensions, thus forming an inverted grid pattern on the surface. If this is the case, then the attachment behaviour of particles at the junctions would be of interest; so do you see preferential attachment at groove junctions with TMV, or in the studies you described using spheres? I imagine this could be exploited as a simple non-lithographic method for the preparation of nanoscale devices.

Professor Fery responded: Firstly, yes, strain can be applied biaxially, which causes vermiculate wrinkle patterns. As to your question on attachment, we have no experimental evidence so far for this effect, but I agree that this is worth a closer look. In general, the junctions are of interest, since they break the symmetry of the system.

Dr Browne said: In the procedure used to form the wrinkles the PDMS is stretched along the long axis. The strain would be expected to lead to a wasp-waisting of the PDMS piece. Does this lead to variation/uniformity in the wavelength of the ridges formed, and if not, then why not?

Professor Fery answered: In the studies, we used pieces from the center part of the substrate, where strain is uniform. The wavelength is not very sensitive to the absolute strain, but rather depends mainly on the thickness of the top-layer and elastic properties of layer and substrate. Thus wavelength is found to be uniform even outside this area. The orientation becomes less well defined.

Dr Steinke enquired: Is it possible to use gradients of elasticity to produce more complex "wrinkle" patterns? What would be the most important parameter limiting the level of complexity of patterns that can be generated in this way?

Professor Fery answered: The wrinkle wavelength depends on the elastic constant of the substrate. Thus, one would expect that patterns of higher complexity can be achieved in this way. This is currently studied in our group.

Mr Barbero asked: The wavelength of the wrinkles is determined by the thickness of the film and the modulus of elasticity of the film and the substrate. What is the influence of the surface energy (if any) on the wavelength of the wrinkles?

Professor Fery responded: This is an interesting effect, since indeed for very thin films surface energy contributions should become important. In the regime which is investigated in this paper, mechanical deformation energies are large when compared to surface energies. Therefore, we expect no significant influence.

Mr Ahangar remarked: I think that during the filling of grooves with tobacco mosaic virus (TMV), it will always have a certain extent of disorder as its attachment is highly dependent on the physical forces present during the spin coating process and whether the alignment stays for a longer time. I think that there will always be a surface disruptive force of entropy which will not allow one to maintain the functionalisation for high aspect ratios for longer times; if it is true, will that affect the reliability of your results?

Professor Fery responded: The alignment appears to be permanent. We observe no particle mobility after the structures have been produced. The structures are "quenched" and therefore no disorder occurs on relevant timescales. Naturally physical forces (like dewetting) are governing the assembly process and defects can be further reduced by understanding and consequently optimizing the approach.

Dr Channon addressed the conference: In most of the examples of self-assembly presented in these discussions (and probably the literature in general) one is presented with a designed monomer unit that subsequently assembles under some given condition to form a "nano-object" of one description or another. If some further organisation is required it is most often imposed in a "top-down" fashion, through lithography or micro-contact printing, for example. This is a fundamentally different mechanism to that observed in nature, where layers of hierarchical, "bottom-up" self-assembly are employed to form macroscopic structures. The problem seems to be one of anticipating and designing interactions on a range of length-scales (making the things interact where they are designed to, and not interact where they are not supposed to), as required for hierarchical assembly. This problem seems to be fundamental to the realisation of many of the promised advantages of nano-scale self-assembly as a technologically useful enterprise, so my question is: does anyone see a route forward in the design of truly hierarchical self-assembled systems? Will it require a fundamental breakthrough in thinking, or just faster computers to simulate more possibilities in more detail?

Professor Huskens answered: This question is at the heart of the analogy between supramolecular and biological assembly. There are three main differences. First, nature works with compartments: practically all assembly occurs at interfaces, steered by conglomerates of proteins/enzymes, thus allowing spatial control. Second, nature employs a range of different specific interaction types simultaneously so that orthogonal and sequential assembly steps become possible. Third, nature employs so-called dynamic self-assembly, *i.e.* assembly is steered by energy-consuming processes (*e.g.* using ATP conversion) thus allowing temporal control. So, I think the answer to your question, whether we can mimic this in artificial systems, is "yes", but only when we become capable of designing and using truly complex systems in which all these ingredients are encompassed in a collective and cooperative manner.

Professor Mao replied: Our understanding and exploration of molecular self-assembly is far from complete. Most of what we have explored relies on single (or a small set of) strong interactions, for example, DNA base-pairing. For hierarchical assembly, we need, I believe, to explore even weaker interactions and multivalent interactions. Lower-level assembly brings multiple weak interaction partners together and leads to overall strong interactions.

Professor Hu answered: Complementary to other panel members' comments, I think that Professor Mao's work on DNA self-assembly may be a good example in answering your first question. For your second question, I think that current investigations of self-assembly focus too much on the hierarchical structures, and a fundamental breakthrough of our thinking may be required to address the hierarchical functioning. The latter may hold more practical importance for self-assembly in the field of, for example, biomimetics. Computer simulations are useful tools of our investigations. It becomes our faith that computer simulations are not just exploring more possibilities in more details but provide us with insights into a better understanding of parallel reality.

Dr Bittner continued the discussion of the paper by Professor Hu: Some of your images suggest a rather flat and at least locally crystalline surface of the polymer droplets. Was AFM with molecular resolution demonstrated on this or similar systems?

Professor Hu answered: Yes. One of my collaborators, Dr Hobbs' group, did observe the rather flat and locally crystalline surface of polymer droplets with AFM as reported in ref. 9 of our paper, as well as Professor Dalnoki-Veress' group in ref. 6, although the real droplet sizes are slightly larger than our simulation systems and the resolution of AFM may not be at the molecular level.

Counterion-activated polyions as soft sensing systems in lipid bilayer membranes: from cell-penetrating peptides to DNA

Toshihide Takeuchi, Naomi Sakai and Stefan Matile*

Received 6th January 2009, Accepted 31st March 2009
First published as an Advance Article on the web 21st July 2009
DOI: 10.1039/b900133f

Activated by amphiphilic anions, cationic oligo- and polymers can phase transfer into and across bulk and lipid bilayer membranes. In biology, this system accounts for the voltage gating of ion channels and the ability of cell-penetrating peptides (CPPs) to cross membrane barriers. In soft nanotechnology, the same system has been used for multianalyte sensing in complex matrices (cholesterol sensing in blood and eggs, for example). Would inversion of all involved charges provide access to similarly versatile functional supramolecular systems? Activation of polyanions by amphiphilic cations is well known as the basis of non-viral gene transfection. Could DNA/RNA nanotechnology thus be applied to produce smart, stimuli-responsive cation transporters that can function as multianalyte sensing in fluorogenic vesicles? Preliminary results suggest that the answer to these questions will be yes.

Introduction

We have used the term "dynamic biphilicity" to name molecules that can be both hydrophilic and lipophilic, depending on the circumstances.[1,2] Biphiles are not amphiphiles. Amphiphiles, such as phospholipids or fatty acids, are molecules with both hydrophilic and lipophilic parts. To dissolve in water, amphiphiles have to hide their hydrophobic parts. Self-assembly into micelles and bilayers with hydrophilic surfaces and hydrophobic cores is the consequence. Self-assembly into inversed architectures with lipophilic surfaces and lipophobic cores makes amphiphiles soluble in nonpolar media. In sharp contrast, biphiles, like chameleons, can adapt as monomers to any environment. They are neither hydrophobic nor lipophobic, they like both. But how is this possible? And why should we care?

The multiple activities of cationic and anionic polymers in lipid bilayer membranes have created much confusion in the past, particularly in the life sciences, and several probing questions remain disputed until today.[3-28] In the following, we elaborate on counterion-mediated biphilicity as a unifying model to first rationalize and then actually use these complex processes to create new functional systems. Generality of this model is further supported by preliminary proof-of-principle experiments, which show the ability of not only polycations such as cell-penetrating peptides[1-26] but also polyanions such as plasmids or aptamers to function as multianalyte sensors in fluorogenic vesicles.

Department of Organic Chemistry, University of Geneva, Geneva, Switzerland. E-mail: stefan. matile@unige.ch; Web: www.unige.ch/sciences/chiorg/matile/; Fax: +41 22 379 5123; Tel: +41 22 379 6523

Polyion–counterion complexes

Biphilicity should originate from a proximity effect that takes place with weakly acidic polycations (Fig. 1).[1,2] In neutral water, more acidic polycations such as polylysine can liberate protons to minimize the repulsion between proximal cations. The involved ammonium cations have an intrinsic $pK_a = 10.5$, charge repulsion from proximal cations can result in a reduction to $pK_a < 7$.[29] However, in weakly acidic polyions such as polyarginine, the reduction of pK_a is probably not enough to reduce the charge repulsion, a change in pK_a of guanidinium cation from the intrinsic 12.5 to <7 is unlikely. The only alternative for proximal guanidinium cations to neutralize or even invert their charge is to capture counteranions. As a result, polyion–counterion complexes with low-acidity polycations are exceptionally stable.[1,2,30–34]

Because of their need to minimize intramolecular charge repulsion, low-acidity polycations such as polyarginine never exist alone, tightly bound counteranions will always be there and matter.[1,2] To dissolve as monomers in water, low-acidity polycations will always coexist with hydrophilic counteranions. They can also dissolve in bulk or lipid bilayer membranes by capturing amphiphilic counteranions.[2] Therefore, low-acidity polycations could exchange some of their hydrophilic counteranions with amphiphilic ones to phase transfer from water into bulk or lipid bilayer membranes, and then amphiphilic ones with hydrophilic ones to back-transfer into water.

To successfully adapt to changing environments while moving, for example, across a lipid bilayer, low-acidity polycations have to exchange their counterions fast. This is possible despite exceptional affinity because the velocity of counteranion exchange is not much affected by the powerful proximity effects described above. Polycation–counteranion complexes have exceptional thermodynamic but poor kinetic stability. Dynamic biphilicity originates from polyion–counterion complexes that are stable and labile. This rapid anion hopping along their sticky scaffold is the reason why low-acidity polycations can adapt to every environment provided the right counteranions are within reach. It is also the reason why the functional consequences of dynamic biphilicity are so very difficult to study.

Biphilic activity

The creation of most of the guanidinium-rich cell-penetrating peptides (CPPs) known today has been inspired by the transduction domain RKKRRQRRR of HIV-1 Tat, the human immunodeficiency virus type 1 transactivator of

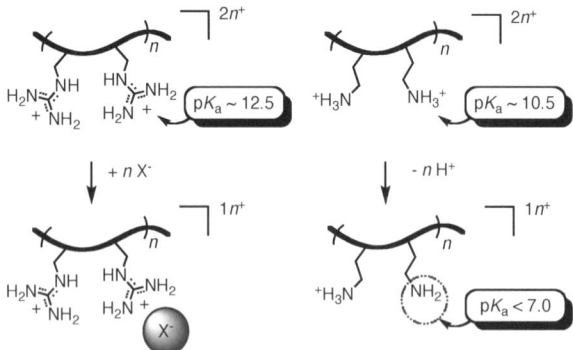

Fig. 1 Proximal charge repulsion as the origin of the stability of polyion–counterion complexes. Acidic polycations such as polylysine minimize intramolecular charge repulsion by reducing the pK_a's of the individual ammonium cations for partial proton release in neutral water (*right*). Less acidic polycations such as polyarginine do not have this possibility and capture counteranions X$^-$ instead (*left*).

transcription.[1-26] Their ability to apparently transport almost everything linked across cell membranes, including objects as large as GFP or nanoparticles, has attracted intense attention and caused much scientific debate. The only structural requirement of CPPs seems to be the presence of several guanidinium cations. Biological R-rich protein transduction domains (*Drosophila* Antennapedia transcription factor, herpes simplex virus type-1 VP22 transcription factor), linear, branched, helical, stereoisomeric or hydrophobized oligoarginines, guanidinium-rich peptoids, oligocarbamates, peptide nucleic acids, synthetic polymers, and so on have all been validated as functional CPPs.

The probing question how cell-penetrating peptides (CPPs) can cross biomembranes has recently, after a decade of controversy, been answered with dynamic biphilicity.[1,2,35-41] Polyarginine (pR), octaarginine (R_8) and other minimalist CPP models were found to readily dissolve as monomers in organic solvents such as chloroform in the presence of amphiphilic counterions.[2] Counterion-activated CPPs could carry hydrophilic anions such as the fluorescent 5(6)-carboxyfluorescein (CF) during their transfer from water into chloroform. In the U-tube, counterion-activated CPPs could carry hydrophilic CF anions across bulk chloroform layers. Moving on from bulk to lipid bilayer membranes, counterion-activated CPPs were found to mediate the fluorogenic CF export from CF-loaded vesicles.

This exhaustive evidence for the functional relevance of counterions naturally encouraged the introduction of new counterions[36,37] to improve existing[38] and create new functions.[39,40] We found, for example, that CPP uptake by HeLa cells increased up to 5-times at room temperature in the presence of pyrenebutyrate **1** (Fig. 2).[38] Up to 25-times increased uptake at 4 °C and little uptake in depolarized HeLa cells demonstrated that counterion-mediated CPP delivery occurs by membrane-potential driven, passive diffusion across the plasma membrane and not by endocytosis. Confocal fluorescence microscopy images confirmed the validity of this interpretation. Rapid cytosolic delivery of proteins as large as enhanced green fluorescent proteins (EGFPs) attached to CPPs implied that pyrenebutyrate–CPP complexes form giant but transient pores during action (*vide infra*). Delivered with pyrenebutyrate–CPP complexes, apoptosis-inducing peptides exhibited normal intracellular function (*i.e.*, initiation of programmed cell death by mitochondrial membrane disruption). So far, it is unclear why pyrenebutyrate **1** works best as a counterion activator of arginine-rich CPPs **2**. Possible explanations include arene-templated ion pairing plus π,π-interactions to stabilize the counterion–polyion complexes **3**,[1,42-44] and the preferential partitioning of arenes at the membrane interface to increase membrane affinity and facilitate translocation (Fig. 2).[1,45-47]

The unusual behavior of oligo- and polyarginines in bilayer membranes is, however, not limited to cytosolic delivery. Since the crystal structures of biological potassium channels appeared, there has been much debate on how membrane-facing

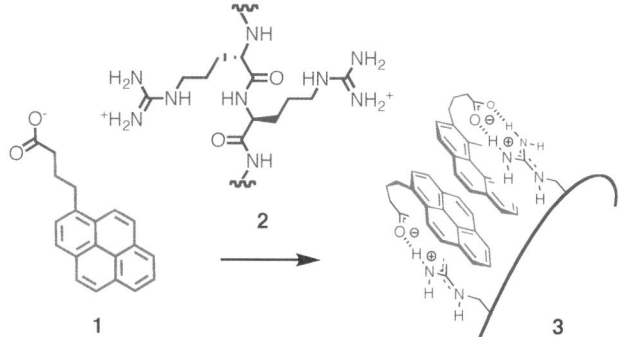

Fig. 2 Pyrenebutyrate as amphiphilic counterion activator of arginine-rich polyions **2** produces polyion–counterion complexes **3** that are particularly active.

guanidinium cations of voltage gating paddles could cross the membrane.[27,28] Translocation of large arginine-rich paddles has been criticized as energetically unrealistic because "lipid membranes are a forbidden zone for charged molecules."[48] As we have predicted in the same year,[2,35,37] anionic lipids were found later on to be essential for the voltage gating paddle to cross the membrane and thus, for its function.[28]

Many more examples for counterion-mediated function of arginine-rich oligo-/polymers in lipid bilayer membranes exist. Partitioning of guanidinium-containing pores, for example, does not saturate, whereas that with ammonium-containing pores does.[49] Pores formed by amphiphilic guanidinium-rich polycations are cation rather than anion selective.[50,51] Polycationic pores attracting cations rather than anions are quite remarkable and can only be explained by charge inversion of the interior of the pore by firmly immobilized hydrophilic counteranions. In agreement with this interpretation, arginine-rich pores have unusually low conductance and undergo inversion of anion/cation selectivity under mildly acidic conditions.[51]

Biphilic action in biomembranes: a unifying model

The multifunctionality of polyion–counterion complexes in bulk and lipid bilayer membranes can be readily summarized in a general model (Fig. 3).[1] Maximal four types of players are involved: Biphilic polyions, amphiphilic counterions, hydrophilic low-affinity ("weak") counterions and hydrophilic high-affinity ("strong") counterions. Hydrophilic counterions prefer the aqueous phase, amphiphilic counterions the membrane–water interface. Depending on the available counterions and the applied concentration gradients, counterion hopping of (and on) biphilic polyions added to triphasic vesicular systems will follow the principle of Le Chatelier.[52] With weak external counterions and stronger, supramolecularly polyvalent amphiphilic counterions, for example, polyions can exchange some counterions and enter into the membrane (Fig. 3a).[1,2,20,35] Strong external counterions, in contrast, will keep polyions away from the membrane and cause external release of membrane-bound polyions (Fig. 3b).[35]

In the absence of strong internal or external counterions, polyions will stay bound to the amphiphilic counterions in the membrane.[2,35] The polyion–counterion

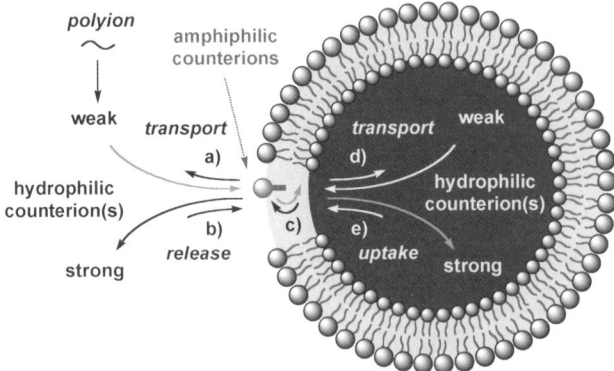

Fig. 3 A unifying model for counterion hopping of (and on) biphilic polyions in triphasic vesicular systems. Polyions can enter membranes by exchanging low-affinity ("weak") external (equilibrium a) or internal (d) hydrophilic counterions with amphiphilic counterions. Polyions can leave membranes by exchanging the latter with strong external (b) or internal (e) counterions. With amphiphilic counterions, polyions can (c) mediate transmembrane transport of weak (a, d) but not strong (b, e) hydrophilic counterions. Polyions can penetrate membranes by exchanging weak (a) but not strong (b) external counterions with amphiphilic counterions and exchanging the latter then with (e) strong but not (d) weak internal counterions.

complexes can shuttle from one side to the other and transport hydrophilic counterions across the membrane (Fig. 3c, 4 and 5).[1,2,35–40] In this situation, polyions function as ion transporters that can be inactivated by strong, internal or external hydrophilic counterions. Sensing applications of this activity will be described later on.[39–41]

With weak internal counterions, polyions will stay with the amphiphilic counterions in the membrane and continue functioning as transporters of weak counterions (Fig. 3d).[9,35] Only strong internal counterions will cause uptake and keep the internalized polyions away from the membrane (Fig. 3e). To really penetrate cells and avoid kinetic competition from endocytosis, the presence of intrinsic (anionic lipids) or added amphiphilic counterion activators (**1**) is as essential as that of internal high-affinity counterion acceptors (RNA, ATP, *etc*).[38] From this biphilicity point of view, CPPs enter cells by multi-ion hopping from weak external counterions to amphiphilic counterion activators in the membranes (Fig. 3a) and from there to strong counterion acceptors in the cytosol (Fig. 3e).

The actions of polyion–counterion complexes in lipid bilayers will depend on their concentration.[53–55] At low concentration, the complexes might simply sit at the membrane surface without exhibiting any activity (Fig. 4a). With increasing concentration or with better counterion activators, the complexes might cause local micellar defects in the bilayer (Fig. 4b). These micellar defects would allow translocation across the membrane without entering the hydrophobic core and without forming pores.

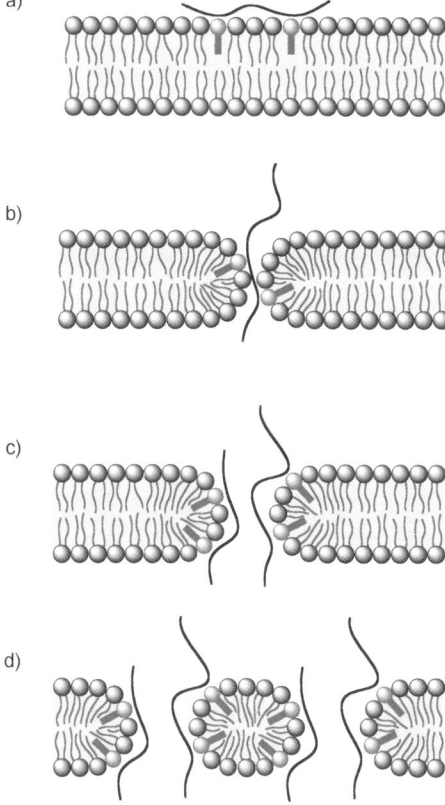

Fig. 4 The actions of polyion–counterion complexes in lipid bilayer membranes include, with increasing concentration, (a) binding at the interface, (b) the formation of micellar ion transporters, (c) formation of micellar pores, and (d) vesicle destruction into disc micelles.

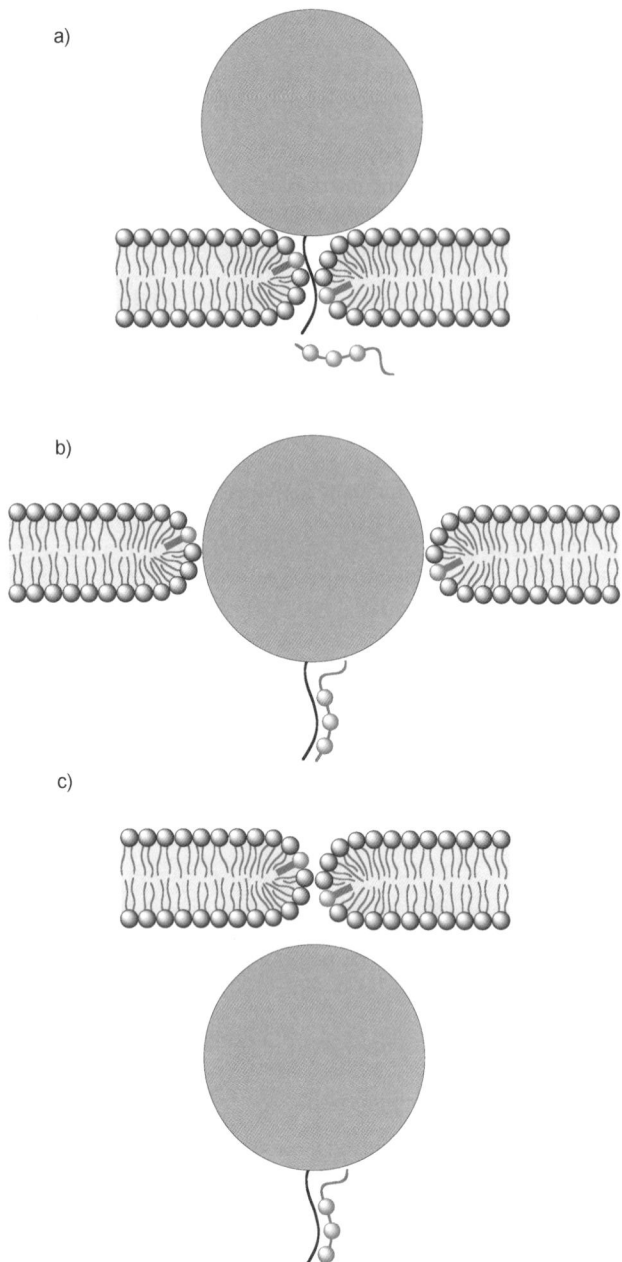

Fig. 5 Moving big objects (faded big circles) with small transporters (solid dark lines) through small holes, assisted by amphiphilic activators (shaded circles) and hydrophilic acceptors (three shaded circles): Polyion–counterion complexes produce micellar defects (a) that can swell (b), shrink (c) and heal; see Fig. 3 and 4 and text for details.

The formation of micellar ion transporters by polyion–counterion complexes could be easily confirmed (from their coinciding flippase activity),[53–57] improved (with cone-shaped amphiphilic counterion activators) and inhibited (with inversed-cone-shaped activators).[58,59] The same micellar polyion–counterion translocation is expected to account for the voltage gating of biological potassium channels.[1,2,27,28,35-37] The unifying model for biphilic action in biomembranes identifies amphiphilic counterion activators and inactivators as a new attractive target for drug discovery in neurodegenerative diseases and beyond.

At increasing concentration, polyion–counterion complexes may transform from micellar transporters to transient micellar pores (Fig. 4c).[53-56,10] Pore formation requires charge repulsion between neighboring polyion–counterion complexes (of maybe intramolecular variations thereof) to generate water-filled transmembrane space.[53-56,60] The formation of micellar (or "toroidal")[56] pores has been proposed to occur for several systems reaching from simply stressed bilayers to amyloid pores and CPP-complexes.[10,53-56] The characteristics of micellar pores include ultrashort lifetimes and irregular, heterogeneous conductances of single pores, interfacial pore location as well as flippase activity. Further increasing concentration will result in lysis, that is the destruction of the membrane into disk micelles (Fig. 4d).[58,59]

The puzzling observation that small CPPs can carry giant objects into cells is understandable considering polyion–counterion complexes acting as micellar transporters. Once produced by polyion–counterion complexes (Fig. 5a), micellar defects can first swell to let large objects pass (Fig. 5b) and then shrink (Fig. 5c) and finally heal. The same dynamics is applicable to micellar pores.

Screening and sensing

According to the unifying model of biphilic action in lipid bilayer membranes, polyions should be able to serve as adaptable, non-invasive, cost-efficient, commercially available optical (a) detectors of reactions, (b) detectors of enzyme activity (including applications to inhibitor screening for drug discovery), and (c) multianalyte sensors in complex matrices.

The possibility to detect enzyme activity and its inhibitors with polyion–counterion complexes has been confirmed using hyaluronidase as an example.[39] The substrate of hyaluronidase is hyaluronan. This high-molecular weight anionic polysaccharide is a major component of the extracellular matrix. It is used in cosmetics and plastic surgery and plays a key role in several (patho)physiological processes including fertilization, tumor growth and metastasis. For the detection of the activity of hyaluronidase, pR was used as polyion and dodecylphosphate (DP) as counterion activator in fluid-phase CF-loaded DPPC vesicles (shelf-life of 3.5 years in the gel-phase). Ion exchange of the amphiphilic DP activators by the high-affinity hydrophilic hyaluronan substrate inactivated pR polyion transporters. The low-affinity tetrasaccharide products of hyaluronan hydrolysis with testicular hyaluronidase, however, failed to inactivate pR-DP complexes. The action of hyaluronidase thus resulted in formal activation of pR-DP complexes and could be visualized for the naked eye with fluorogenic or chromogenic vesicles. Compatibility of CPP-counterion enzyme assays with inhibitor screening was illustrated with cromolyn and heparin as representative hyaluronidase inhibitors.

The use of CPP-counterion complexes as multianalyte sensors[61-72] appeared less promising for a long time. Straightforward with unproblematic analytes,[39] CPPs failed to compete with related methods when more demanding questions were asked. Examples include less successful discrimination between ATP and ADP[39] or IP$_7$ and phytate.[72] However, recently we found that CPPs are perfect sensors of an important class of analyte that is problematic for alternative sensing systems in lipid bilayer membranes, *i.e.*, hydrophobic analytes.[40] A cholesterol sensor was made to illustrate the concept.

To sense cholesterol **4** with CPPs in complex matrices such as eggs, caviar or blood, the analyte was first converted with cholesterol oxidase (Fig. 6). This step is referred to as signal generation because the selectivity of the enzyme assures that differential sensing before and after enzymatic treatment will report exclusively on cholesterol levels. Covalent capture of cholestenone product **5** with hydrazide **6** gives hydrazone **7**. This step is referred to as signal amplification because it converts undetectable hydrophobic analytes **5** and inactive hydrophilic anions **6** into amphiphilic

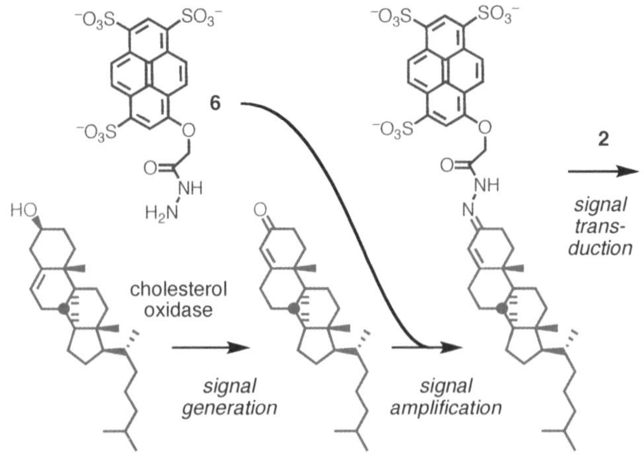

Fig. 6 Cholesterol sensing with CPPs in fluorogenic vesicles. Selectivity is assured by enzymatic signal generation with cholesterol oxidase, sensitivity by signal amplification with hydrazide **6**, and signal transduction by activation of CPP transporters **2** with the resulting amphiphilic counterion activators **7**.

Table 1 Characteristics of counterion activators of polyion transporters in EYPC LUVs at constant lipid and polyion concentration.[a,b,c]

Entry	Counterion	Polyion[d]		Y_{MAX} (%)[e]	EC_{50} (μM)[f]
1	1	polyarginine	2	78 ± 2[b]	44 ± 2[b]
2	7	polyarginine	2	60 ± 6[c]	0.6 ± 0.03[c]
3	12	polyarginine	2	NA[b,g]	NA[b,g]
4	13	polyarginine	2	27 ± 1[b]	16 ± 1[b]
5	14	polyarginine	2	61 ± 3[b]	19 ± 1[b]
6	15	calf thymus DNA	8	63 ± 2	147 ± 3
7	15	polyG		54 ± 1	68 ± 2
8	16	calf thymus DNA	8	65 ± 2	53 ± 3
9	16	polyG		64 ± 2	34 ± 3
10	16	polyphosphate	9	66 ± 4	140 ± 15
11	16	polyglutamate	10	NA[g]	NA[g]
12	16	hyaluronan	11	NA[g]	NA[g]

[a] From DPX/HPTS export from LUVs⊃DPX/HPTS, see Fig. 9 and 10. [b] From CF export from LUVs⊃CF, from ref. 37. [c] Similar to (b), optimized conditions with submicellar Triton X-100 solubilizers, from ref. 40. [d] Concentrations: 12.5 μg ml^{-1} **8** (~41 μM phosphate), 15.0 μg ml^{-1} polyG (~44 μM phosphate), 36.0 μg ml^{-1} **9** (~383 μM phosphate), 15.0 μg ml^{-1} **10** (~116 μM carboxylate), 17.5 μg ml^{-1} **11** (~47 μM carboxylate). [e] Maximal activity accessible with activator. [f] Effective activator concentration needed to reach 50% activity. [g] Inactive (Y_{MAX} < 15%).

counterion activators **7**. Amphiphilic anions **7** were found to activate biphilic polyions **2** for signal transduction with submicromolar sensitivity (EC_{50} = 0.6 μM, Table 1).

Charge inversion

According to the unifying model of biphilic action in lipid bilayer membranes, the multifunctional polyion does not have to be a polycation and the counterion activator does not have to be an amphiphilic anion (Fig. 3). In biology, polycations are used to exploit dynamic biphilicity because biomembranes are anionic and contain intrinsic amphiphilic counterion activators. These biological anionic lipids are used by HIV-1 Tat to penetrate cells[3–26] or by potassium channels to sense membrane polarization.[27,28] On the other hand, non-viral gene transfection, *i.e.*, the delivery of RNA as representative polyanions is more difficult because the complementary cationic counterion activators do not exist in biomembranes.[73–77] Many amphiphilic cations have been proposed since the introduction of DOTMA to overcome this charge imbalance of biomembranes and enable the non-viral cellular uptake of RNA and other oligonucleotides.[73–77] Here, we propose that the inversion of all involved charges in multifunctional polyion–counterion complexes should provide access to a counterion-mediated functional plasticity for polyanions that is comparable to that of arginine-rich peptides. This concept is attractive because it suggests that DNA/RNA nanotechnology (*e.g.*, aptamers and ribozymes)[64–66] could be used to produce smart, stimuli-responsive cation transporters that can function as multianalyte sensors. Moreover, the complementarity with CPPs offers a stimulating new approach to non-viral gene transfection.

Application of the lessons learned for polycations suggests that polyanions that are compatible with biphilic action must be very weak bases (Fig. 7). This applies for oligonucleotides **8** or polyphosphate **9** but not for polyglutamate **10** or hyaluronan **11**. To minimize intramolecular charge repulsion in neutral water, the latter, more basic polyanions could capture protons rather than countercations. The conjugate acids of the carboxylate anion, *i.e.*, carboxylic acids, have an intrinsic $pK_a \sim 4.5$. Their acidity can decrease up to $pK_a > 7$ to minimize charge repulsion from proximal anions.[29,60] Less basic polyanions such as oligonucleotides are incapable of reducing intramolecular charge repulsion by protonation. With an intrinsic $pK_a \sim 1$, the acidity of the conjugate acid of phosphodiesters is clearly too high for this purpose. Complementary to the situation with low-acidity polycations, the only solution with low-basicity polyanions appears to be the tight binding of countercations.

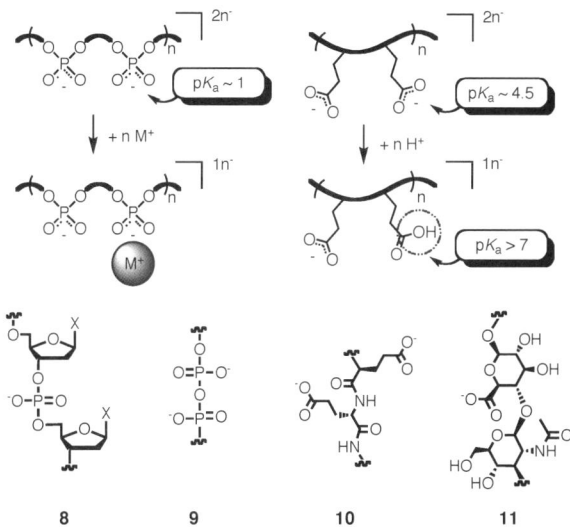

Fig. 7 Application of the lessons learned with biphilic polycations to polyanions. Basic polyanions such as polyglutamate **10** or hyaluronan **11** could minimize intramolecular charge repulsion by increasing the pK_a's of their individual conjugate acids for partial protonation in neutral water (*top right*). Less basic polyanions such as oligonucleotides **8** (X = nucleobases) or polyphosphate **9** should not have this possibility and capture countercation M⁺ instead (*top left*).

As with low-acidity polycations, the conclusion with low-basicity polyanions is that countercations will always be there and matter. The functional consequences of cation hopping of and on biphilic polyanions will be the same as for polycations. This means that amphiphilic countercations will mediate cellular uptake of polyanions. This process is extensively studied for non-viral gene transfection.[73–77] Consideration of this important process from a new point of view, that is counterion-mediated multifunctionality of biphilic polyions (Fig. 3), may inspire new approaches. Moreover, this also means that complexes formed between biphilic polyanions and amphiphilic countercations could act as transporters of low-affinity hydrophilic cations in lipid bilayers as outlined in Fig. 4. Moreover, these polyanion–countercation transporters could be of use for multianalyte sensing in complex matrices. This perspective is particularly attractive considering the possibility of unifying signal generation and transduction with aptamers[64–66] or ribozymes as countercation-activated transmembrane cation transporters. This process is essentially unexplored, a preliminary feasibility assessment is reported in the next paragraph.

Activators

The efficiency of amphiphilic counteranions was determined from their ability to activate polycations such as polyarginine to export CF from egg yolk phosphatidylcholine large unilamellar vesicles (EYPC-LUVs⊃CF).[2,35–37,39–41,55] Loaded at concentrations sufficient for self quenching, the local CF dilution during CF export is monitored as an increase in CF fluorescence. Activators are characterized by their Y_{MAX} and EC_{50}. Simply speaking, the former describes activity, the latter sensitivity.

The Y_{MAX} is the maximal relative fluorescence intensity accessible with the polyion–activator complex compared to that achieved by complete destruction of the CF-loaded vesicles with a detergent. The Y_{MAX} relates to the fraction of vesicles that is affected by the active transporter and depends on a complex mix of parameters including delivery, partitioning, intervesicular transfer, precipitation from the water phase, and so on.[37,78] $Y_{MAX} > 30\%$ is required for meaningful applications, $Y_{MAX} > 60\%$ is ideal.

The EC_{50} is the effective concentration needed to reach 50% of Y_{MAX}. The EC_{50} is related to the dissociation constant of the polyion–counterion complex, although stoichiometirc binding[78–80] often dominates because of the exceptionally high affinities involved. Both Y_{MAX} and EC_{50} depend on the nature and concentration of the vesicles, the nature and concentration of the polyion, temperature, pH, ionic strength, and so on.[37,78]

An outstanding $Y_{MAX} = 78\%$ was found for pyrenebutyrate **1** as a benchmark activator of polyarginine (Table 1, entry 1).[37] Replacement of the aryl tail with an alkyl tail in laurate **12** caused a dramatic loss in activity to negligible $Y_{MAX} = 10\%$ (Table 1, entry 3, Fig. 8). This loss in activity occurred although the sensitivity even increased slightly from $EC_{50} = 44$ µM for **1** to $EC_{50} = 34$ µM for **12**. The properties of the resulting counterion–polyion complexes rather than their formation thus account for the ultimately observed activity.

Replacement of the carboxylate in laurate **12** with a sulfonate or a phosphate partially ($Y_{MAX} = 27\%$) or fully restored activity ($Y_{MAX} = 61\%$, Table 1, entries 4 and 5), respectively.[37] Both dodecylsulfonate **13** and dodecylphosphate **14** recognized polyarginine at EC_{50}s below those of carboxylates. The high activity of activator **7** at nanomolar sensitivity could originate from the multivalency effects conceivable with the triply charged headgroup and hydrophobic matching of the steroidal tail in lipid bilayers (Table 1, entry 2).[40] However, these values are not fully comparable because experimental conditions were optimized for sensing applications (*e.g.*, submicellar Triton X-100 was used for delivery). Taken together, the selected data suggest that, under given conditions, aryl tails are better than alkyl tails, and phosphates possibly better than sulfates and carboxylates. However, these

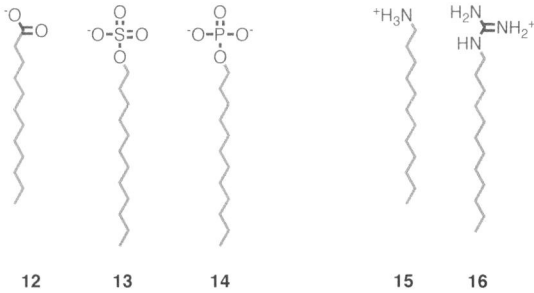

Fig. 8 Selected counterion activators for polycations such as CPPs (**12–14**) and for polyanions such as DNA (**15, 16**).

conclusions are far from general. Quite naturally, complex and dynamic functional systems often produce conflicting trends. Many different counterion activators have been characterized for polyarginine transporters under many different conditions.[37] Consistent with the limitation of biphilic multifunctionality as defined in Fig. 3 to weakly acidic polycations, polylysine showed no activity under conditions that worked well for polyarginine (Fig. 1).[2]

DNA as cation transporter

To elaborate on the concept of charge inversion in the context of biphilic multifunctionality, dodecylammonium cations **15** (DA) and dodecylguanidinium cations **16** (DG) were selected as model activators (Fig. 8). DNA **8**, RNA (polyG) and polyphosphate **9** were considered as low-basicity polyanions, polyglutamate **10** and hyaluronan **11** as more basic examples (Fig. 7).

To assure detectability of the export of both cations and anions, EYPC LUVs were loaded with the anionic fluorophore HPTS (8-hydroxy-1,3,6-pyrenetrisulfonate) together with the cationic quencher DPX (*p*-xylene-bis-pyridinium bromide). In general, the availability of assays with little intrinsic selectivity is vital to successfully detect unknown functions.[55] HPTS was ideal as a bifunctional fluorophore that also reports on pH. EYPC-LUVs⊃DPX/HPTS were prepared by routine freeze–thaw–extrusion methods. External probes were removed by gel filtration under iso-osmotic conditions. Addition of either DG or DNA to EYPC-LUVs⊃DPX/HPTS did not cause an increase in HPTS emission (Fig. 9a, $t < 0$). This demonstrated that neither polyanion nor countercation alone are capable to mediate the export of either DPX or HPTS from EYPC-LUVs⊃DPX/HPTS. Addition of both DG and DNA together caused rapid export of DPX or HPTS from EYPC-LUVs⊃DPX/HPTS (Fig. 9a, $t > 0$).

Dose–response curves at constant DNA concentration revealed with $Y_{MAX} = 65\%$ excellent activity with DG (Fig. 10, ●; Table 1, entry 8). The $EC_{50} = 53$ μM revealed appreciable sensitivity for DG activators under these conditions. Replacement of the guanidinium with an ammonium cation in activator **15** gave with $Y_{MAX} = 63\%$ similarly satisfactory activity for the polyion–counterion complex obtained with DNA (Table 1, entry 6). However, the sensitivity of ammonium activator **15** was with $EC_{50} = 147$ μM almost 3-times below that of the guanidinium activator **16** (Fig. 10, X; Table 1, entries 6 and 8). This result was consistent with the preference of phosphates for guanidiniums over ammoniums for hydrogen-bond assisted ion pairing.

The activation of polyG as representative RNA transporter was with $Y_{MAX} \sim 60\%$ similarly successful (Table 1, entries 7 and 9). The EC_{50}s with RNA were up to 2-times better than with DNA transporters. Similar to DNA, the guanidinium amphiphiles **16** activated RNA transporters better than ammonium

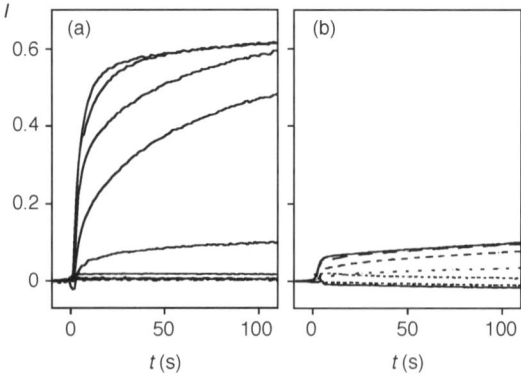

Fig. 9 Activation of DNA (a) and polyglutamate (b) transporters with DG activators. Changes in fractional HPTS fluorescence emission I (λ_{ex} 413 nm, λ_{em} 511 nm) as a function of time during addition of DG (at $t < 0$ s; 1.5, 5, 15, 25, 75, 100, 150 and 250 µM) and transporter (at $t = 0$ s; (a) 12.5 µg ml^{-1} calf thymus DNA, (b) 15 µg ml^{-1} polyglutamate) to EYPC-LUVs⊃HPTS/DPX (~125 µM lipid), calibrated by final lysis (at $t > 150$ s, $I = 1$ with excess Triton X-100). Conditions: 1 mM HPTS, 3.3 mM DPX, 10 mM Tris, 100 mM NaCl, pH 7.4 (inside); 10 mM Tris, 107 mM NaCl, pH 7.4 (outside).

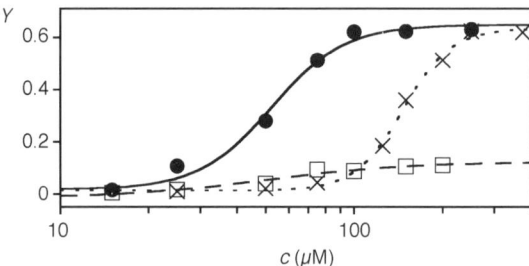

Fig. 10 Dose–response curves for DG (●, □) and DA (×) at constant concentration of DNA (●, from Fig. 9a; ×) and polyglutamate (□, from Fig. 9b) transporters in EYPC-LUVs⊃HPTS/DPX, with curve fit to the Hill equation.

amphiphiles **15**. Activation of polyvinylsulfate (not shown) and polyphosphate **9** with DG gave excellent results (Table 1, entry 10). The relatively high EC$_{50}$ = 140 µM for polyphosphate originated presumably[79] from the high polyanion concentration used and should not be interpreted as poor sensitivity.

DG did not activate poly-D-glutamate as transporters in EYPC-LUVs⊃DPX/HPTS (Fig. 10, □; Table 1, entry 11). The same insensitivity toward DG activators was found for hyaluronan (Table 1, entry 12) and polyaspartate (not shown). Overall, polyanions with low basicity such as DNA, RNA, polyphosphate and polyvinylsulfate could reproducibly be activated, whereas more basic polyanions such as polyglutamate, polyaspartate and hyaluronan did not respond (Table 1). These results provided excellent experimental support for the here proposed generalized concept of polyion biphilicity with reduction of intramolecular charge repulsion as the general origin of the counterion-mediated function (Fig. 3 and 7).

With increasing concentration, polyion–counterion complexes are expected to act in lipid bilayer membranes as micellar ion transporters, transient micellar pores, and membrane destructors (Fig. 4). The frequent, successful and non-cytotoxic use of amphiphilic countercations as non-viral gene vectors[73–77] implies that polyanion–countercation complexes can cross bilayer membranes without causing irreversible damage. To secure more direct evidence for ion transport activity, the activity of

DNA–DG complexes in EYPC-LUVs⊃HPTS/DPX and EYPC-LUVs⊃CF was compared (Fig. 11). Activity in the EYPC-LUVs⊃HPTS/DPX assay is compatible with cation transport (*i.e.*, DPX export), anion transport (*i.e.*, HPTS export) and membrane destruction. Activity in the EYPC-LUVs⊃CF assay is compatible with anion transport and membrane destruction only. Identical activity in the two assays would thus suggest that polyanion–countercation complexes do not act as cation transporters, different activity in the two assays would support that polyanion–countercation complexes can indeed act as cation transporters.

At high vesicle concentrations, the two assays gave quite similar results (Fig. 11a, $EC_{50} = 52.6 \pm 3.0$ μM *vs* $EC_{50} = 67.8 \pm 1.4$ μM). However, the EC_{50}s for DNA activation by DG decreased in EYPC-LUVs⊃HPTS/DPX with decreasing vesicle concentration, whereas the EC_{50}s increased in EYPC-LUVs⊃CF under the same conditions (Fig. 11, a–d). At highest dilution, the found $EC_{50} = 20.3 \pm 1.7$ μM in EYPC-LUVs⊃HPTS/DPX was more than 4-times better than the $EC_{50} = 91.8 \pm 6.1$ μM in EYPC-LUVs⊃CF (Fig. 11d). Under these conditions, DNA–DG complexes are likely to act as cation transporters at DG concentrations around 20 μM but to destroy vesicles at concentrations >90 μM. The different response obtained with different assays provides compelling experimental support

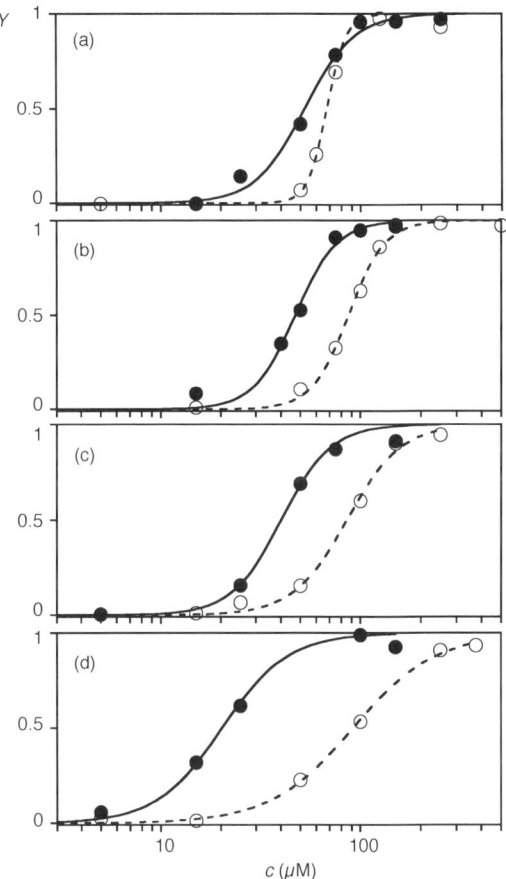

Fig. 11 Dose–response curve for DG activation of calf thymus DNA (12.5 μg ml^{-1}) to EYPC-LUV⊃HPTS/DPX (●) and EYPC-LUV⊃CF (○) at different LUV concentration obtained by addition of (a) 50 μl (∼125 μM EYPC final), (b) 30 μl (∼75 μM EYPC), (c) 10 μl (∼25 μM EYPC) and (d) 5 μl (∼12 μM EYPC) LUV stock solution to buffer (2 ml final).

that counterion-bound polyanions such as DNA can function as cation transporters in lipid bilayer membranes.

Conclusions

The specific objective of this article was to expand and generalize the concept of counterion-mediated multifunctionality of biphilic polyions from polycations to polyanions. As expected, it was found unproblematic to invert charges completely from polycationic transporters such as cell-penetrating peptides and counteranion activators such as pyrenebutyrate to polyanion transporters (DNA) and counterca-tion activators (dodecylguanidinium cations). Consistent with intramolecular charge repulsion as the origin of biphilicity, other low-basicity polyanions such as RNA, polyphosphate or polyvinylsulfonate were also active, whereas the more basic polyglutamate, polyaspartate or hyaluronan were inactive. Higher activity found in assays that report on cation transport than in assays that report on anion transport or lysis support that polyion–counterion complexes change their mode of action with increasing concentration from "micellar" ion transporters to "micellar" pores and membrane destructors. These preliminary proof-of-principle data suggest that DNA/RNA nanotechnology (*e.g.*, aptamers and ribozymes)[64–66] can be used to produce smart, stimuli-responsive cation transporters that can function as multianalyte sensing in fluorogenic vesicles. These conclusions open several attractive perspectives, not only with regard to sensing applications but also with regard to "biphilic" non-viral gene transfection and slow release.

Experimental

Materials and methods

As in previous reports on the topic,[37,39–41,55] all compounds used were commercially available.

Vesicle preparation

A thin lipid film was prepared by evaporating a solution of 25 mg EYPC in 1 ml MeOH/CHCl$_3$ (1:1) on a rotary evaporator (r.t.) and then *in vacuo* overnight. The resulting film was hydrated with 1.0 ml buffer [1 mM HPTS, 3.3 mM DPX, 10 mM Tris, 100 mM NaCl, pH 7.4 (for EYPC-LUVs⊃HPTS/DPX) or 50 mM CF, 10 mM Tris, 10 mM NaCl, pH 7.4 (for EYPC-LUVs⊃CF)] for more than 30 min, subjected to freeze–thaw cycles (5×) and extrusions (15×) through a polycarbonate membrane (pore size, 100 nm). Extravesicular components were removed by gel filtration (Sephadex G-50) with 10 mM Tris, 107 mM NaCl, pH 7.4. Final conditions: ~5 mM EYPC; inside: 1 mM HPTS, 3.3 mM DPX, 10 mM Tris, 100 mM NaCl, pH 7.4 (for EYPC-LUVs⊃HPTS/DPX) or 50 mM CF, 10 mM Tris, 10 mM NaCl, pH 7.4 (for EYPC-LUVs⊃CF); outside: 10 mM Tris, 107 mM NaCl, pH 7.4.

Vesicle experiments

EYPC-LUV stock solutions were diluted with buffer (10 mM Tris, 107 mM NaCl, pH 7.4, Fig. 11), placed in a thermostatted fluorescence cuvette (25 °C) and gently stirred. Changes in emission were followed at λ_{em} 511 nm (λ_{ex} 413 nm for HPTS and 492 nm for CF) as a function of time during addition of cationic activators at varied concentration, polyanions and aqueous Triton X-100 (0.024%). Data were normalized to fractional emission intensity I using eqn (1)

$$I = (I_t - I_0)/(I_\infty - I_0) \qquad (1)$$

where $I_0 = I_t$ at polyanion addition, $I_\infty = I_t$ at saturation after lysis. Effective concentration EC_{50} and Hill coefficient n were determined by plotting the fractional activity $Y (= I$ just before lysis) as a function of the cation concentration c_{CATION} and fitting them to the Hill eqn (2)

$$Y = Y_0 + (Y_{MAX} - Y_0)/\{1 + (EC_{50}/c_{CATION})^n\} \quad (2)$$

where Y_0 is Y without cations, Y_{MAX} is a value with an excess cation at saturation.

Acknowledgements

We thank the University of Geneva and the Swiss NSF for financial support.

References

1 N. Sakai, S. Futaki and S. Matile, *Soft Matter*, 2006, **2**, 636–641.
2 N. Sakai and S. Matile, *J. Am. Chem. Soc.*, 2003, **125**, 14348–14356.
3 I. Nakase, T. Takeuchi, G. Tanaka and S. Futaki, *Adv. Drug Delivery Rev.*, 2008, **60**, 598–607.
4 K. M. Stewart, K. L. Horton and S. O. Kelley, *Org. Biomol. Chem.*, 2008, **6**, 2242–2255.
5 A. Verma, O. Uzun, Y. H. Hu, Y. Hu, H. S. Han, N. Watson, S. L. Chen, D. J. Irvine and F. Stellacci, *Nat. Mater.*, 2008, **7**, 588–595.
6 A. Ziegler, *Adv. Drug Delivery Rev.*, 2008, **60**, 580–597.
7 E. K. Esbjorner, P. Lincoln and B. Norden, *Biochim. Biophys. Acta, Biomembr.*, 2007, **1768**, 1550–1558.
8 S. M. Fuchs and R. T. Raines, *ACS Chem. Biol.*, 2007, **2**, 167–170.
9 T. Shimanouchi, P. Walde, J. Gardiner, Y. R. Mahajan, D. Seebach, A. Thomae, S. D. Kraemer, M. Voser and R. Kuboi, *Biochim. Biophys. Acta, Biomembr.*, 2007, **1768**, 2726–2736.
10 M. Tang, A. J. Waring and M. Hong, *J. Am. Chem. Soc.*, 2007, **129**, 11438–11446.
11 J. M. Gump and S. F. Dowdy, *Trends Mol. Med.*, 2007, **13**, 443–448.
12 E. P. Holowka, V. Z. Sun, D. T. Kamei and T. J. Deming, *Nat. Mater.*, 2007, **6**, 52–57.
13 S. Pujals, J. Fernandez-Carneado, C. Lopez-Iglesias, M. J. Kogan and E. Giralt, *Biochim. Biophys. Acta, Biomembr.*, 2006, **1758**, 264–279.
14 I. Tsogas, D. Tsiourvas, G. Nounesis and C. M. Paleos, *Langmuir*, 2006, **22**, 11322–11328.
15 Y. Wolf, S. Pritz, S. Abes, M. Bienert, B. Lebleu and J. Oehlke, *Biochemistry*, 2006, **45**, 14944–14954.
16 S. Afonin, A. Frey, S. Bayerl, D. Fischer, P. Wadhwani, S. Weinkauf and A. S. Ulrich, *ChemPhysChem*, 2006, **7**, 2134–2142.
17 E. Barnay-Wallje, A. Anderson, A. Graslund and L. Maler, *J. Biomol. NMR*, 2006, **35**, 137–147.
18 E. Goncalves, E. Kitas and J. Seelig, *Biochemistry*, 2005, **44**, 2692–2702.
19 R. Fischer, M. Fotin-Mleczek, H. Hufnagel and R. Brock, *ChemBioChem*, 2005, **6**, 2126–2142.
20 J. B. Rothbard, T. C. Jessop, R. S. Lewis, B. A. Murray and P. A. Wender, *J. Am. Chem. Soc.*, 2004, **126**, 9506–9507.
21 S. Futaki, *Curr. Protein Pept. Sci.*, 2003, **4**, 87–157.
22 M. Lindgren, M. Hällbrink, A. Prochiantz and U. Langel, *Trends Pharmacol. Sci.*, 2000, **21**, 99–103.
23 H. J. Ryser and R. Hancock, *Science*, 1965, **150**, 501–503.
24 E. M. Kolonko and L. L. Kiessling, *J. Am. Chem. Soc.*, 2008, **130**, 5626–5627.
25 B. A. Smith, D. S. Daniels, A. E. Coplin, G. E. Jordan, L. M. McGregor and A. Schepartz, *J. Am. Chem. Soc.*, 2008, **130**, 2948–2949.
26 T. Jiang, E. S. Olson, Q. T. Nguyen, M. Roy, P. A. Jennings and R. Y. Tsien, *Proc. Natl. Acad. Sci. U. S. A.*, 2004, **101**, 17867–17872.
27 J. Jiang, V. Ruta, J. Chen, A. Lee and R. MacKinnon, *Nature*, 2003, **423**, 42–48.
28 D. Schmidt, Q. X. Jiang and R. MacKinnon, *Nature*, 2006, **444**, 775–779.
29 T. Bugg, *An Introduction to Enzyme and Coenzyme Chemistry*, Blackwell, Oxford, 1997, 35ff.
30 K. Ariga and T. Kunitake, *Acc. Chem. Res.*, 1998, **31**, 371–378.
31 K. A. Schug and W. Lindner, *Chem. Rev.*, 2005, **105**, 67–114.
32 M. D. Best, S. L. Tobey and E. V. Anslyn, *Coord. Chem. Rev.*, 2003, **240**, 3–15.

33 T. Iwata, S. Lee, O. Oishi, H. Aoyagi, M. Ohno, K. Anzai, Y. Kirino, R. J. Austin, T. Xia, J. Ren, T. T. Takahashi and R. W. Roberts, *J. Am. Chem. Soc.*, 2002, **124**, 10966–10967.
34 B. P. Orner, X. Salvatella, J. Sánchez-Quesada, J. de Mendoza, E. Giralt and A. D. Hamilton, *Angew. Chem., Int. Ed.*, 2002, **41**, 117–119.
35 N. Sakai, T. Takeuchi, S. Futaki and S. Matile, *ChemBioChem*, 2005, **6**, 114–122.
36 F. Perret, M. Nishihara, T. Takeuchi, S. Futaki, A. N. Lazar, A. W. Coleman, N. Sakai and S. Matile, *J. Am. Chem. Soc.*, 2005, **127**, 1114–1115.
37 M. Nishihara, F. Perret, T. Takeuchi, S. Futaki, A. N. Lazar, A. W. Coleman, N. Sakai and S. Matile, *Org. Biomol. Chem.*, 2005, **3**, 1659–1669.
38 T. Takeuchi, M. Kosuge, A. Tadokoro, Y. Sugiura, M. Nishi, M. Kawata, N. Sakai, S. Matile and S. Futaki, *ACS Chem. Biol.*, 2006, **1**, 299–303.
39 T. Miyatake, M. Nishihara and S. Matile, *J. Am. Chem. Soc.*, 2006, **128**, 12420–12421.
40 S. M. Butterfield, T. Miyatake and S. Matile, *Angew. Chem., Int. Ed.*, 2009, **48**, 325–328.
41 A. Hennig, G. J. Gabriel, G. N. Tew and S. Matile, *J. Am. Chem. Soc.*, 2008, **130**, 10338–10344.
42 S. E. Thompson and D. B. Smithrud, *J. Am. Chem. Soc.*, 2002, **124**, 442–449.
43 X. Wang, O. V. Sarycheva, B. D. Koivisto, A. H. McKie and F. Hof, *Org. Lett.*, 2008, **10**, 297–300.
44 J. M. Sanderson and E. J. Whelan, *Phys. Chem. Chem. Phys.*, 2004, **6**, 1012–1017.
45 A. N. J. A. Ridder, S. Morein, J. G. Stam, A. Kuhn, B. de Kruijff and J. A. Killian, *Biochemistry*, 2000, **39**, 6521–6528.
46 W. M. Yau, W. C. Wimley, K. Gawrisch and S. H. White, *Biochemistry*, 1998, **37**, 14713–14718.
47 M. Schiffer, C. H. Chang and F. J. Stevens, *Protein Eng., Des. Sel.*, 1992, **5**, 213–214.
48 F. J. Sigworth, *Nature*, 2003, **423**, 21–22.
49 S. Stankowski, M. Pawlak, E. Kaisheva, C. H. Robert and G. Schwarz, *Biochim. Biophys. Acta, Biomembr.*, 1991, **1069**, 77–86.
50 T. Iwata, S. Lee, O. Oishi, H. Aoyagi, M. Ohno, K. Anzai, Y. Kirino and G. Sugihara, *J. Biol. Chem.*, 1994, **269**, 4928–4933.
51 N. Sakai, N. Sordé, G. Das, P. Perrottet, D. Gerard and S. Matile, *Org. Biomol. Chem.*, 2003, **1**, 1226–1231.
52 "If there is a change in the condition of a system in equilibrium, the system will adjust itself in such a way as to counteract, as far as possible, the effect of that change." K. J. Laidler and J. H. Meiser, *Physical Chemistry*, 3rd edn,Houghton Mifflin, Boston, MA, 1999, p. 169. After the French chemist Henri Louis Le Chatelier (1850–1936).
53 S. Matile, A. Som and N. Sordé, *Tetrahedron*, 2004, **60**, 6405–6435.
54 A. L. Sisson, M. R. Shah, S. Bhosale and S. Matile, *Chem. Soc. Rev.*, 2006, **35**, 1269–1286.
55 S. Matile and N. Sakai, The Characterization of Synthetic Ion Channels and Pores, in *Analytical Methods in Supramolecular Chemistry*, ed. C. A. Schalley, Wiley, Weinheim, 2007, pp. 391–418.
56 K. Matsuzaki, O. Murase, N. Fujii and K. Miyajima, *Biochemistry*, 1996, **35**, 11361–11368.
57 G. Das, H. Onouchi, E. Yashima, N. Sakai and S. Matile, *ChemBioChem*, 2002, **3**, 1089–1096.
58 J. N. Israelachvili, *Intramolecular and Surface Forces*, Academic Press, New York, 1991.
59 D. D. Lasic, *Trends Biotechnol.*, 1998, **16**, 307–321.
60 B. Baumeister, A. Som, G. Das, N. Sakai, F. Vilbois, D. Gerard, S. P. Shahi and S. Matile, *Helv. Chim. Acta*, 2002, **85**, 2740–2753.
61 *Creative Chemical Sensing Systems. Topics in Current Chemistry*, ed. T. Schrader, Springer, Heidelberg, 2007.
62 J. J. Lavigne and E. V. Anslyn, *Angew. Chem., Int. Ed.*, 2001, **40**, 3118–3130.
63 A. Hennig, H. Bakirci and W. M. Nau, *Nat. Methods*, 2007, **4**, 629–632.
64 J. Liu and Y. Lu, *Angew. Chem., Int. Ed.*, 2006, **45**, 90–94.
65 R. Nutiu, J. M. Y. Yu and Y. Li, *ChemBioChem*, 2004, **5**, 1139–1144.
66 I. Willner and M. Zayats, *Angew. Chem., Int. Ed.*, 2007, **46**, 6408–6418.
67 J.-P. Goddard and J.-L. Reymond, *Curr. Opin. Biotechnol.*, 2004, **15**, 314–322.
68 N. A. Rakow and K. S. Suslick, *Nature*, 2000, **406**, 710–703.
69 S. Matile, H. Tanaka and S. Litvinchuk, *Top. Curr. Chem.*, 2007, **277**, 219–250.
70 S. Litvinchuk, H. Tanaka, T. Miyatake, D. Pasini, T. Tanaka, G. Bollot, J. Mareda and S. Matile, *Nat. Mater.*, 2007, **6**, 576–580.
71 S. Hagihara, H. Tanaka and S. Matile, *J. Am. Chem. Soc.*, 2008, **130**, 5656–5657.
72 S. M. Butterfield, D.-H. Tran, H. Zhang, G. D. Prestwich and S. Matile, *J. Am. Chem. Soc.*, 2008, **130**, 3270–3271.
73 P. L. Felgner, T. R. Gadek, M. Holm, R. Roman, H. W. Chan, M. Wenz, J. P. Northrop, G. M. Ringold and M. Danielsen, *Proc. Natl. Acad. Sci. U. S. A.*, 1987, **84**, 7413–7417.
74 K. Kostarelos and A. D. Miller, *Chem. Soc. Rev.*, 2005, **34**, 970–994.

75 V. Bagnacani, F. Sansone, G. Donofrio, L. Baldini, A. Casnati and R. Ungaro, *Org. Lett.*, 2008, **10**, 3953–3956.
76 R. Haag, *Angew. Chem., Int. Ed.*, 2004, **43**, 278–282.
77 A. J. Kirby, P. Camilleri, J. B. F. N. Engberts, M. C. Feiters, R. J. M. Nolte, O. Söderman, M. Bergsma, P. C. Bell, M. L. Fielden, C. L. García Rodríguez, P. Guédat, A. Kremer, C. McGregor, C. Perrin, G. Ronsin and M. C. P. van Eijk, *Angew. Chem., Int. Ed.*, 2003, **42**, 1448–1457.
78 F. Mora, D.-H. Tran, N. Oudry, G. Hopfgartner, D. Jeannerat, N. Sakai and S. Matile, *Chem.–Eur. J.*, 2008, **14**, 1947–1953.
79 O. H. Straus and A. Goldstein, *J. Gen. Physiol.*, 1943, **26**, 559–585.
80 B. K. Soichet, *J. Med. Chem.*, 2006, **49**, 7274–7277.

PAPER

Recognition of sequence-information in synthetic copolymer chains by a conformationally-constrained tweezer molecule†

Howard M. Colquhoun,* Zhixue Zhu,* Christine J. Cardin, Michael G. B. Drew and Yu Gan

Received 13th January 2009, Accepted 31st March 2009
First published as an Advance Article on the web 23rd July 2009
DOI: 10.1039/b900684b

A novel type of tweezer molecule containing electron-rich 2-pyrenyloxy arms has been designed to exploit intramolecular hydrogen bonding in stabilising a preferred conformation for supramolecular complexation to complementary sequences in aromatic copolyimides. This tweezer-conformation is demonstrated by single-crystal X-ray analyses of the tweezer molecule itself and of its complex with an aromatic diimide model-compound. In terms of its ability to bind selectively to polyimide chains, the new tweezer molecule shows very high sensitivity to sequence effects. Thus, even low concentrations of tweezer relative to diimide units (< 2.5 mol%) are sufficient to produce dramatic, sequence-related splittings of the pyromellitimide proton NMR resonances. These induced resonance-shifts arise from ring-current shielding of pyromellitimide protons by the pyrenyloxy arms of the tweezer-molecule, and the magnitude of such shielding is a function of the tweezer-binding constant for any particular monomer sequence. Recognition of both short-range and long-range sequences is observed, the latter arising from cumulative ring-current shielding of diimide protons by tweezer molecules binding at multiple adjacent sites on the copolymer chain.

Introduction

The information needed for all living things to function and reproduce is encoded in the linear sequences of monomer residues which make up the polymer chains of DNA and/or RNA.[1] This information is processed by other molecules which have the ability to interact selectively with, *i.e.* to *recognise*, specific sequences of nucleotides.[2] Over the past few years, a number of reports have appeared describing ways in which the information-processing capabilities of DNA might be exploited in a non-biological context,[3] but in fact there is no reason why the idea of using a copolymer chain as an information storage medium should be restricted to nucleic acids: in principle *any* copolymer would do, since even the simplest AB copolymer chain is the logical equivalent of a string of binary numbers.[4] The real problem is how to read, write and copy sequence-information at a molecular level.

In biology, transcription and copying of nucleic acids, and translating their nucleotide sequences into the amino acid sequences of proteins, are achieved by

Department of Chemistry, University of Reading, Whiteknights, Reading, UK RG6 6AD. E-mail: h.m.colquhoun@rdg.ac.uk; z.x.zhu@rdg.ac.uk; Tel: +44 (0)118 378 6717

† CCDC reference numbers 648369, 717467 and 717568. For crystallographic data in CIF or other electronic format see DOI: 10.1039/b900684b

a quite staggeringly complex set of molecular machinery,[5] although interestingly the only mechanism for *writing* entirely new information to DNA is that of random error, *i.e.* mutation. It is also noteworthy that, in eukaryotic cells, the genetic sequence corresponding to a single polypeptide chain rarely comprises a simple consecutive run of nucleotides—rather the information is scattered in short sections along the DNA strand, and is only brought together in a useable form by dint of enzymes cutting and splicing the initial *m*-RNA transcript from a very much longer DNA sequence.[5] However, even today, when our understanding of biology's fabulously complex gene-translating machinery is reasonably well-developed at the levels of organic and supramolecular chemistry, it is fair to say that no-one has the *faintest idea* how these molecular mechanisms actually originated.[6] Hypothesis and speculation abound, but even in 2006, after pioneering research into the chemistry of life's origins for nearly fifty years and developing the widely-respected concept of a primeval "RNA World",[7] Orgel could still write that, beyond a rudimentary timeline, "almost everything else about the origin of life remains obscure".[6]

To a very large extent of course, life depends on extremely complex but highly organised *polymer* chemistry. As noted above, the idea of using linear copolymers as information-storage media includes, but is conceptually independent of, molecular biology. We have thus recently initiated an experimental programme aimed at exploring the *general* principles on which the processing of polymer sequence information at the molecular level might be based.[8–10] In particular we have designed a novel series of chain-folding synthetic macromolecules and a complementary set of small, tweezer-type molecules[11] which can recognise and bind to specific monomer-sequences in these polymers.[8] The macromolecules used in our work (for example polyimide **1** and copolyimides **2** and **3**, Chart 1) are entirely unrelated to biological polymers and are vastly more stable, with analogues such as **4** and **5** (Chart 1) being used extensively in the electronics and aviation industries as high-temperature resistant films and coatings.[12]

We have shown that electron-rich tweezer-type molecules interact strongly with polymers **1–3**, through formation of donor–acceptor, π–π stacked complexes with the electron-deficient diimide units of the polymer chain.[8–10] In particular, tweezer-molecules **6** and **7** (Chart 1) produce very large complexation shifts of the pyromellitimide protons, as a result of aromatic ring-current shielding by the complexing pyrenyl moieties. Moreover, the three diimide resonances in the spectrum of copolymer **3** (representing three different diimide-centred "triplet" sequences) show very different sensitivities to the presence of tweezer-molecule **1**. These differences were found to arise from minor variations in the steric environment around the pyromellitimide binding site.[10] Quantitative binding-constant analyses showed that steric hindrance resulting from the presence of an *o*-methyl group on one or both of the aromatic rings adjacent to the diimide residue progressively reduces the tweezer-binding constant by almost an order of magnitude for each methyl group, with correspondingly drastic reductions in the complexation shifts $\Delta\delta$ for the diimide resonances.[10] Even more remarkably, at high tweezer-loadings (> 10 mol% of tweezer relative to diimide residues), the singlet resonance associated with unhindered pyromellitimide residues begins to broaden and is eventually resolved into three signals. These were shown to represent higher-order sequence information (up to 27 aromatic rings), and the ability of the tweezer-molecule to differentiate between such extended monomer sequences was found to result from the cumulative ring-current shielding effects of tweezer-molecules bound at diimide sites adjacent to the site being "observed" in the ^{1}H NMR spectrum.[10]

In the present paper, we describe the design and synthesis of a new type of tweezer-molecule in which a strongly preferred "binding" conformation is generated by convergent, intramolecular hydrogen bonding. Experiments with different types of chain-folding copolyimides show that the ^{1}H NMR spectra of these polymers are exceptionally sensitive to the binding of this new type of tweezer-molecule, in that

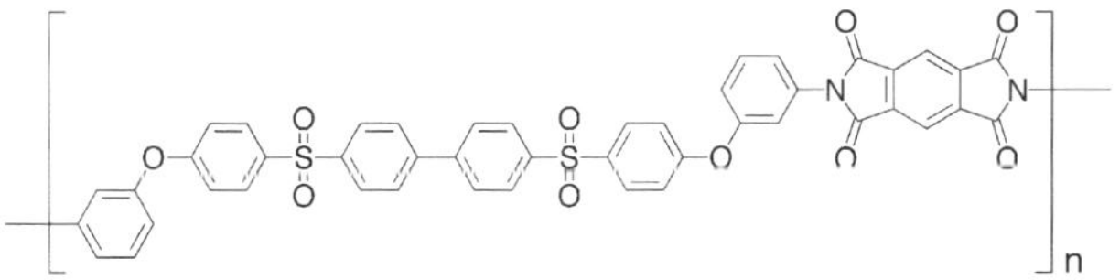

Chart 1 Polyamides **1–5** and tweezer-molecules **6** and **7**.

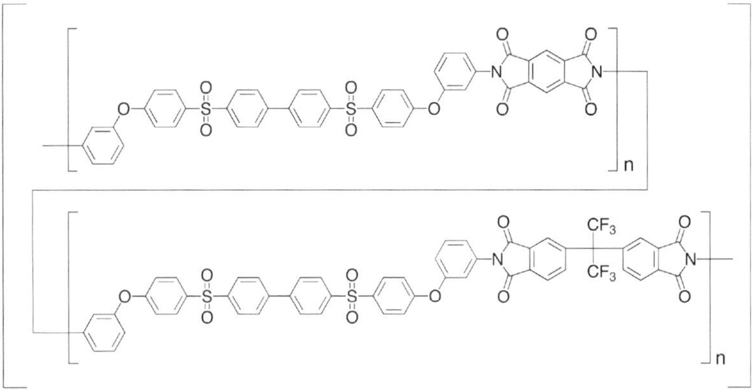

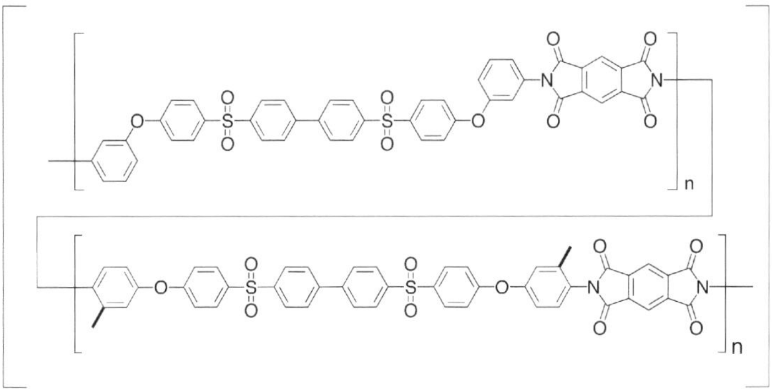

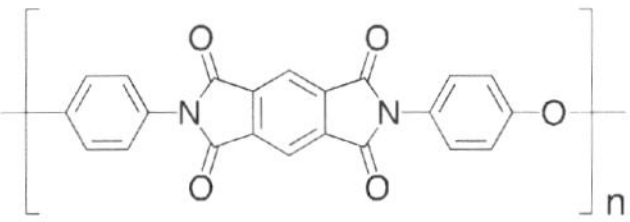

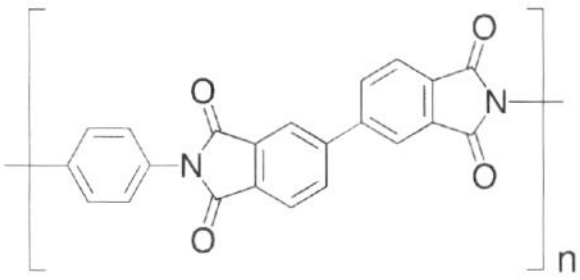

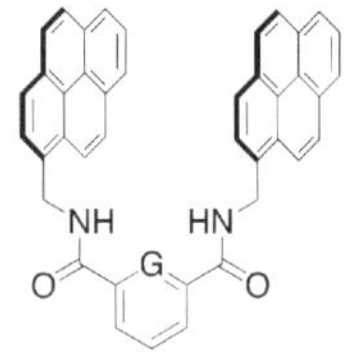

resonances associated with sequences differing only over long range are readily resolved, even in the presence of very low concentrations of the tweezer.

Experimental

Materials

Pyromellitic dianhydride (97%), 4,4-(hexafluoroisopropylidene)diphthalic anhydride (99%), 4,4′-bis[(4-chlorophenyl)sulfonyl]-1,1′-biphenyl (98%), 3-aminophenol (98%), 4-amino-3-methylphenol (97%), 2,2′-iminodibenzoic acid (95%), 1-hydroxypyrene (98%), 1-pyrenemethylamine hydrochloride (95%), 2,6-pyridinedicarbonyl dichloride (97%), potassium carbonate (99%), thionyl chloride (99%), triethylamine (99.5%), *N,N*-dimethylacetamide (DMAC) (99%), *N,N*-dimethylformamide (DMF) (99%) and toluene (99%) were purchased from Aldrich. Dichloromethane, methanol and silica gel (60 Å, 35–70 μm) were purchased from Fisher Scientific. TLC silica gel 60 F_{254} aluminium sheets were purchased from VWR. Deuterated solvents were purchased from Cambridge Isotope Laboratories. 1,1,1,3,3,3-Hexafluoropropan-2-ol (99%) was purchased from Apollo Scientific Ltd. Commercial solvents and reagents were used without purification unless otherwise stated. *N,N*-dimethylacetamide (DMAc) was distilled over calcium hydride before use. Diamines **8** and **9**,[13] diimide model compounds **10** and **11**,[10] tweezer-molecule **6**,[14] and polyimides **1**, **2** and **3**,[10] were prepared (see Chart 2) according to literature procedures.

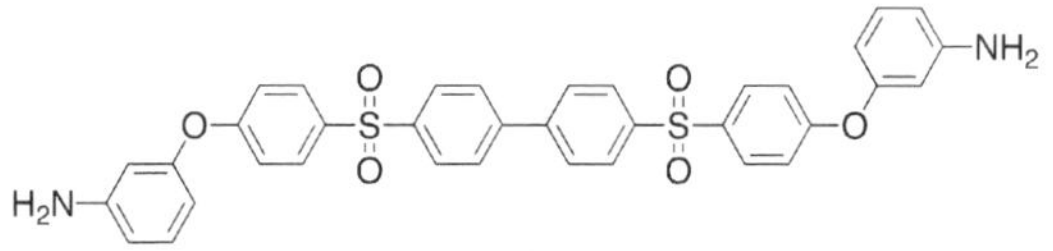

8

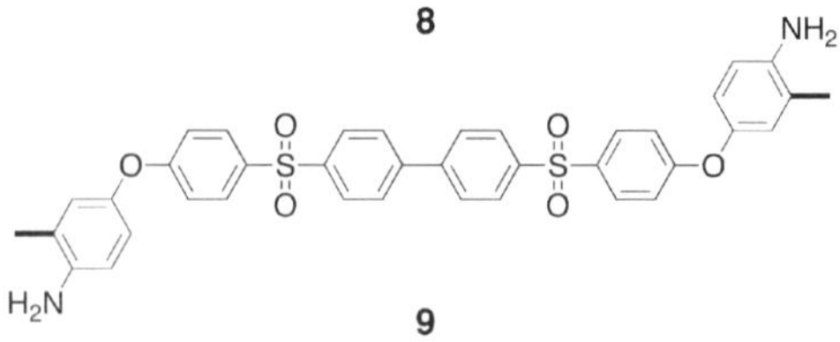

9

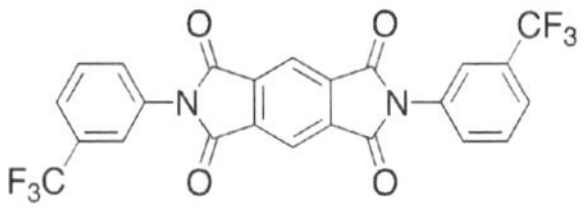

10

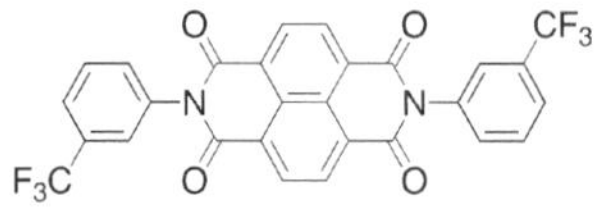

11

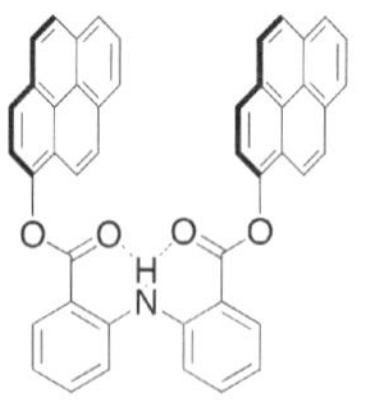

12

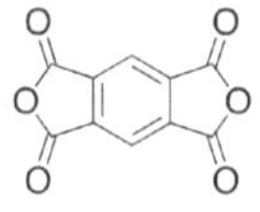

13

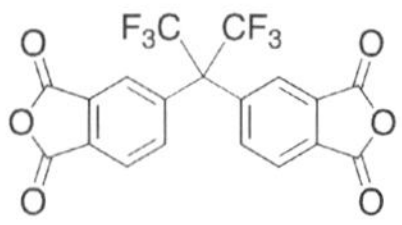

14

Chart 2 Diamines, dianhydrides, model compounds, and the novel tweezer-molecule **12**. Monomer **9** contains sterically-hindering *o*-methyl groups, shown in bold.

General instrumentation

Proton and ^{13}C NMR spectra were recorded on Bruker DPX 250 MHz, and Bruker AV700 spectrometers. Chemical shifts are reported in ppm relative to TMS, and multiplicities as singlet (s), doublet (d), triplet (t), quartet (q), and multiplet (m). Matrix-assisted, laser desorption/ionisation, time-of-flight (MALDI-TOF) mass spectra were obtained using a SAI LT3 LaserTof spectrometer. Samples for MALDI-TOF MS were prepared by mixing and evaporating equal volumes of 1,8,9-trihydroxyanthracene matrix (10 mg mL^{-1} in THF), sodium trifluoroacetate (5 mg mL^{-1} in THF), and analyte (1 mg mL^{-1} in dichloromethane). Infra-red spectra were recorded on a Perkin Elmer 1700 FTIR spectrometer. UV-visible absorption spectra were recorded on a Perkin Elmer Lambda 25 spectrometer. Elemental analyses were provided by Medac Ltd (UK). Melting points and polymer glass transition temperatures were measured under nitrogen by differential scanning calorimetry (DSC) using a Mettler DSC 20 system, at a heating rate of 10 °C min^{-1}. Inherent viscosities (η_{inh}) of polyimides were measured at 25 °C in a thermostatted water bath on 0.1% polymer solutions in 1-methylpyrrolidinone (NMP) as solvent using a Schott-Geräte CT-52 semi-automated viscometer and AVS 470 measurement system. Molecular weights of polyimides relative to polystyrene standards were determined by gel permeation chromatography (GPC) on a Polymer Laboratories PL-220 instrument equipped with a differential refractive index detector and 2 × PL-gel 10 mm mixed-B columns. Analyses were carried out in DMF/LiBr solution (0.05 M in LiBr) at 60 °C, at a flow rate of 1.0 mL min^{-1} and with an injection volume of 100 µL. Samples were run in DMF–LiBr (0.05 M) at a concentration of 1 mg mL^{-1}, and both solvent and sample solutions were filtered through a 0.02 mm PTFE membrane (Whatman) prior to injection. Computational modelling using charge-equilibrated molecular mechanics, (*Cerius*2, v. 3.5, Accelrys Inc., San Diego) was carried out on an SGI-O2 workstation using a custom-modified Dreiding-II force field. Single crystal X-ray data were measured on a MarResearch Image Plate diffractometer (compound **7**), and on an Oxford Diffraction X-Calibur CCD diffractometer (compound **12** and complex [**11** + **12**]).

Synthesis

N,N'-bis(1-pyrenylmethyl)pyridine-2,6-dicarboxamide (**7**). To a mixture of 1-pyrenemethylamine hydrochloride (0.268 g, 1.00 mmol) and triethylamine (0.5 mL) in dichloromethane (10 mL) was added 2,6-pyridinedicarbonyl dichloride (0.096 g, 0.47 mmol) in dichloromethane (10 mL). The mixture was stirred at room temperature overnight, and then extracted with water. The organic extracts were dried over MgSO$_4$ and then the solvent was evaporated. The residue was purified by column chromatography (SiO$_2$: dichloromethane–methanol, 99/1, v/v) to yield **7** as a pale yellow solid (0.230 g, 83%). m.p. 418 °C; ^{1}H NMR (CDCl$_3$/hexafluoropropan-2-ol, 250 MHz) δ (ppm) = 8.19 (d, J = 7.5 Hz, 4H), 8.02 (m, 11H), 7.90 (m, 4H), 7.79 (d, J = 7.9 Hz, 2H), 7.66 (d, J = 7.8 Hz, 2H), 5.15 (d, J = 5.6 Hz, 4H); ^{13}C NMR (CDCl$_3$/hexafluoropropan-2-ol, 62.5 MHz) δ (ppm) = 164.2, 147.6, 139.2, 131.5, 130.7, 129.7, 128.9, 128.8, 128.6, 128.0, 127.5, 126.8, 126.6, 126.1, 125.8, 125.2, 124.9, 124.1, 124.0, 122.2, 119.6, 119.5, 115.0, 42.1 ppm; MS (MALDI-TOF): calc. for [C$_{41}$H$_{27}$N$_3$O$_2$ + Na]$^+$, *m/z* = 616, found 616; Anal. calc. for C$_{41}$H$_{27}$N$_3$O$_2$: C 82.95, H 4.58, N 7.07; found C 82.77, H 4.53, N 6.99%. X-Ray quality single crystals of **7** were grown from a solution in chloroform/hexafluoropropan-2-ol by vapour diffusion with methanol.

Bis-(1-pyrenyl)-2,2'-iminodibenzoate (**12**). A suspension of 2,2'-iminodibenzoic acid (0.129 g, 0.500 mmol) in thionyl chloride (5 mL) was stirred under reflux for 2 h. Evaporation of excess thionyl chloride under reduced pressure gave a pale yellow solid. This was dissolved in dichloromethane (30 mL), and 1-hydroxypyrene

(0.234 g, 1.00 mmol) and triethylamine (0.5 mL) were added to the solution. The mixture was stirred at room temperature overnight, and then extracted with water. The organic layer was dried over magnesium sulfate and the solvent was evaporated. The residue was purified by column chromatography (SiO_2: dichloromethane) to yield **12** as a pale yellow solid (0.248 g, 72%), m.p. 262 °C; ^{1}H NMR (250 MHz, $CDCl_3$) δ (ppm) = 11.09 (s, 1H), 8.55 (dd, J = 8.0 and 1.6 Hz, 2H), 8.21 (dd, J = 7.5 and 1.1 Hz, 2H), 8.13–8.10 (m, 4H), 8.08 (s, 2H), 8.04–7.98 (m, 6H), 7.94 (d, J = 8.0 Hz, 2H), 7.84 (d, J = 8.0 Hz, 2H), 7.77 (d, J = 8.4 Hz, 2H), 7.66 (dt, J = 8.0 and 1.6 Hz, 2H), 7.18 (dt, J = 7.5 and 1.1 Hz, 2H); ^{13}C NMR (62.5 MHz, $CDCl_3$/TFA 6:1, v/v) δ (ppm) = 168.3, 143.3, 142.5, 135.6, 132.9, 131.3, 131.0, 130.2, 128.9, 128.0, 127.1, 127.0, 126.4, 126.0, 125.7, 125.1, 124.5, 124.2, 123.1, 120.4, 119.5, 117.4; IR (Nujol): ν(C=O) 1713, ν(C–O) 1205 cm^{-1}; MS (MALDI-TOF) calc. for $[C_{46}H_{27}NO_4 + Na]^+$, m/z = 690, found 690; Anal. calc. for $C_{46}H_{27}NO_4$: C 84.00, H 4.14, N 2.13; found C 83.46, H 4.22, N 2.17%. X-Ray quality single crystals of **12** were grown from a solution in chloroform/hexafluoropropan-2-ol (6:1 v/v) by vapour diffusion with hexane.

Binding constant by the UV-Vis dilution method[15]. A standard solution containing equimolar amounts of model di-imide **11** (5 mM) and tweezer-molecule **12** in $CHCl_3$–hexafluoropropan-2-ol (6:1, v/v) was made up in a 10 mL volumetric flask, and the absorbance of this solution at maximum absorption (547 nm) was recorded. The solution was then diluted accurately and the absorbance remeasured. This process of accurate dilution and remeasurement of absorbance was repeated six times to obtain seven sets of data. The experiment was repeated with a second standard solution (4.35 mM), and seven-fold dilution yielded eight sets of data.

^{1}H NMR studies of the binding of tweezer-molecule 12 to model di-imide 11 and to polyimides 1–3. A stock solution of the model di-imide (4 mM) or polyimide (4 mM with respect to total diimide residues) in $CHCl_3$–hexafluoropropan-2-ol (6:1 v/v) was prepared in a 5 mL volumetric flask, and 0.8 mL of this solution was added to an NMR tube using a micropipette. The solution was slowly evaporated under nitrogen and the residue dried at 80 °C under vacuum for 4 h. A solution of tweezer molecule **12** in $CDCl_3$–hexafluoropropan-2-ol (6:1 v/v), having the required concentration of tweezer, was similarly prepared in a 5 mL volumetric flask. An aliquot of this solution (0.8 mL) was added to the NMR tube by micropipette and mixed well to re-dissolve the model di-imide or polyimide before carrying out NMR analysis. Peak assignments were made by reference to chemical shift data for known cyclic and linear oligo-imides, by 2D (COSY) analyses, by evaluation of integrals, and by tracking incremental changes in peak positions on progressive addition of tweezer-molecule **12**.

Crystal data. 7: $C_{41}H_{27}N_3O_2 \cdot 0.5(CH_3OH) \cdot 0.25(H_2O)$, M_r = 613.68, tetragonal, space group $I4_1/a$, a = 28.41(3), b = 28.41(3), c = 16.57(2) Å, V = 13374(26) Å^3, Z = 16, ρ_{calcd} = 1.219 g cm^{-3}, μ(Mo-Kα) = 0.77 cm^{-1}, T = 293(2) K; 5541 independent measured reflections, $F(000)$ = 5136, F^2 refinement, R_1 = 0.101, wR_2 = 0.247, 2412 independent observed reflections ($I > 2\sigma(I)$, $2\theta_{max}$ = 26°), 437 parameters. **12**: $C_{46}H_{27}NO_4$, M_r = 657.69, orthorhombic, space group $Iba2$, a = 12.7723(3), b = 16.8766(3), c = 14.4195(3) Å, V = 3108.2(1) Å^3, Z = 4, ρ_{calcd} = 1.405 g cm^{-3}, μ(Mo-Kα) = 0.89 cm^{-1}, T = 150(2) K; 3820 independent measured reflections, $F(000)$ = 1368, F^2 refinement, R_1 = 0.045, wR_2 = 0.102, 3250 independent observed reflections ($I > 2\sigma(I)$, $2\theta_{max}$ = 30°), 285 parameters. **[11 + 12]**: $C_{74}H_{39}F_6N_3O_8 \cdot 3CHCl_3 \cdot 3(CF_3)_2CHOH$. M_r = 2025.32, triclinic, space group P-1, a = 10.744(1), b = 15.004(1), c = 25.720(1) Å, α = 81.728(4), β = 89.451(4), γ = 73.073(5)°, V = 3922.9(3) Å^3, Z = 2, ρ_{calcd} = 1.715 g cm^{-3}, μ(Cu-Kα) = 6.4 cm^{-1}, T = 150(2) K; 8289 independent measured reflections, $F(000)$ = 1242, F^2 refinement,

$R_1 = 0.120$, $wR_2 = 0.331$, 5028 independent observed reflections ($I > 2\sigma(I)$, $2\theta_{max} = 52°$), 808 parameters. A "squeezed" solvent content of 3 chloroform and 3 hexafluoroisopropanol molecules in the asymmetric unit provides the closest match to a squeezed volume of 1477 Å^3 in the unit cell.

Results and discussion

Although previous single crystal X-ray analyses of tweezer-complexes with model compounds showed well-organised, multiple π-stacking interactions, the tweezer-molecules **6** and **7** are not themselves pre-organised in an optimum binding conformation. Rotation about the C–N (amide) bonds is obviously restricted, but there are several other degrees of conformational freedom including rotation about (a) the arene–carbonyl bond, (b) the methylene–nitrogen bond and (c) the methylene–pyrenyl bond. In the present work we have determined the X-ray structure of tweezer-molecule **7** in the *absence* of bound substrate, and although the two amide units adopt the near-coplanar *syn*-conformation required for tweezer-type binding, the relationship between the pyrenyl "arms" in the crystal is found to be closer to perpendicular than it is to the parallel arrangement required for tweezer-binding (Fig. 1).

As shown in Fig. 1, one carbonyl oxygen, O(3), of a second tweezer-molecule (from symmetry element $0.25 - y, x - 0.25, z - 0.25$) interacts with the first molecule by forming two intermolecular hydrogen bonds to N(2) and N(11), with distance H···O(3), angle N–H···O(3), and distance N···O(3) of 2.24 Å, 155°, 3.043(6) Å and 2.41 Å, 163°, 3.241(7) Å respectively. These interactions are repeated throughout the crystal such that each successive molecule along the *c* axis is encapsulated by the previous one, so forming a 1-dimensional molecular chain linked by pairs of hydrogen bonds. There is also evidence for π–π stacking between a pyrenyl arm of one molecule and the pyridine-2,6-dicarbonyl ring of the next, as these π-systems are essentially parallel in the crystal (Fig. 1), with the heterocyclic ring-atoms lying, on average, *ca.* 3.56 Å from the mean plane of the pyrenyl residue.

In an attempt to generate a structure with a more highly pre-organised binding conformation, we have now designed and synthesised a new type of tweezer-molecule (**12**), derived from 2,2′-iminodibenzoic acid (Scheme 1).

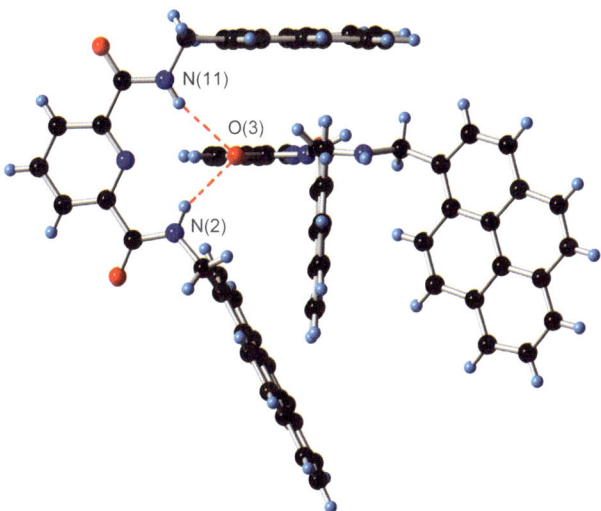

Fig. 1 X-Ray structure of tweezer-molecule **7**. Two molecules of a hydrogen-bonded supramolecular chain are shown (NH···O hydrogen bonds dotted in red).

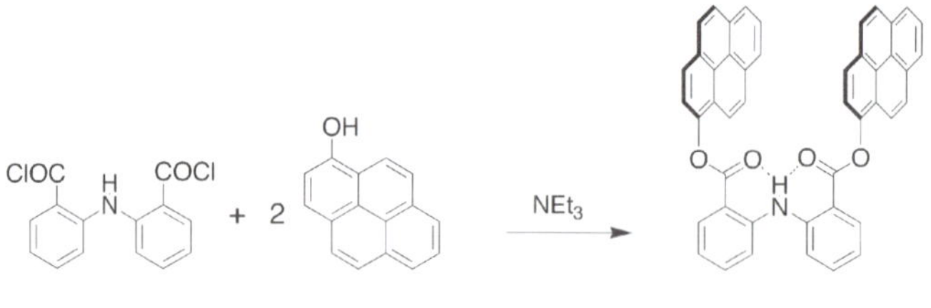

Scheme 1 Synthesis of the new, hydrogen-bonded tweezer-molecule **12**.

Computational modelling (molecular mechanics with charge equilibration) suggested that aromatic esters of this diacid should adopt a conformation in which the carbonyl groups form convergent hydrogen bonds with the bridging iminohydrogen (Scheme 1).

Given the well-established preference for aryl esters to adopt a *trans*-conformation about the ester C–O bond, the proposed intramolecular hydrogen bonding would limit the conformational freedom of the molecule to just two types of rotation, *i.e.* those about the arene–carbonyl and O–pyrenyl bonds. The X-ray structure of tweezer-molecule **12** (Fig. 2) confirmed the presence of the designed, intramolecularly-hydrogen-bonded conformation in the solid state. The flat, pyrenyloxy tweezer arms are arranged anti-parallel to one another (the molecule lies on a crystallographic twofold axis) and are separated by *ca.* 7.12 Å—very close to the separation required for van der Waals contact with an aromatic guest. The space between the tweezer arms is filled by the iminodibenzoyl subunit of an adjacent molecule and this packing motif propagates throughout the crystal.

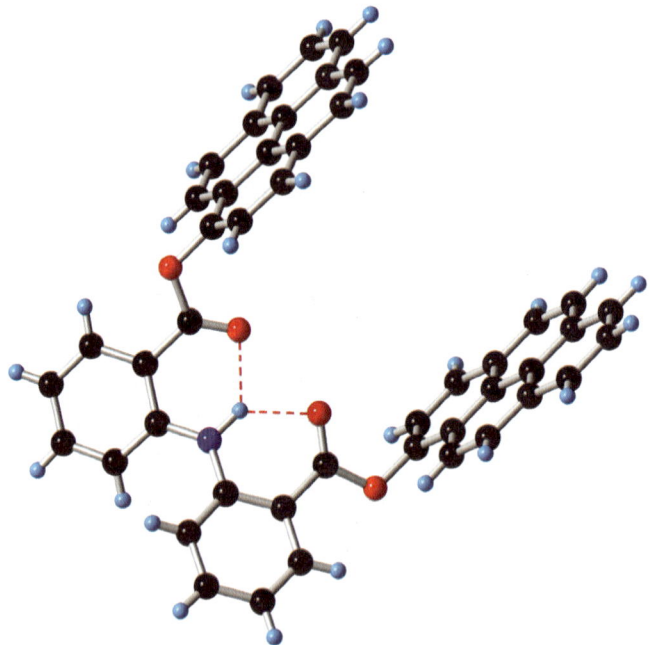

Fig. 2 X-Ray structure of the intramolecularly hydrogen-bonded tweezer-molecule **12**. The molecule has a crystallographic twofold axis. Hydrogen bonds are shown dotted in red [H···O = 2.06(2) Å].

Tweezer molecule **12** showed only very low solubility in chlorinated organic solvents, or in the mixture of chloroform and hexafluoropropan-2-ol (6:1 v/v) successfully employed with tweezer-molecules **6** and **7**. However, it proved reasonably soluble in a mixture of chloroform and trifluoroacetic acid (6:1 v/v) enabling its interactions with a range of different copolyimides to be investigated. Initial studies with the model diimide **11** showed clear evidence for complexation, a solution containing **11** and **12** (1:1) having a deep green colour assignable to a broad charge-transfer absorption at 547 nm. The complexation shifts, $\Delta\delta$, for the diimide protons of **11** in the presence of 50 and 200 mol% of **12** were 0.17 and 0.35 ppm—significantly smaller shifts than those produced by tweezer-molecule **6** (0.41 and 0.90 ppm respectively) under similar conditions. The binding constant for **11** with **12**, measured by the UV-vis dilution method,[15] was 140 M^{-1}.

Crystals of a well-defined complex between the new tweezer-molecule **12** and the naphthalenetetracarboximide model compound **11** were successfully isolated from a solution of the two components (1:1 mole ratio) in chloroform–hexafluoropropan-2-ol (6:1 v/v) by vapour diffusion with hexane. The X-ray structure of complex [**11** + **12**] is shown in Fig. 3. Comparison with the structure of the "free" tweezer (Fig. 2) shows that the internal hydrogen bonding is retained on complexation, but that an improved binding conformation is induced by the ester groups rotating into near-coplanarity with their associated arene rings. This allows the pyrenyloxy arms to overlap one another more effectively, and reduces the mean interplanar distance from 7.12 Å in the "free" tweezer to 6.95 Å in the complex. Both effects lead to enhanced π–π-stacking interactions with the guest diimide, but because of interference from the ester carbonyl groups, the guest molecule does not penetrate completely into the tweezer-cavity. This, together with the absence of hydrogen bonding between host and guest, may well explain the smaller complexation shifts for diimide **11** relative to those observed with tweezer-molecules **6** and **7**.

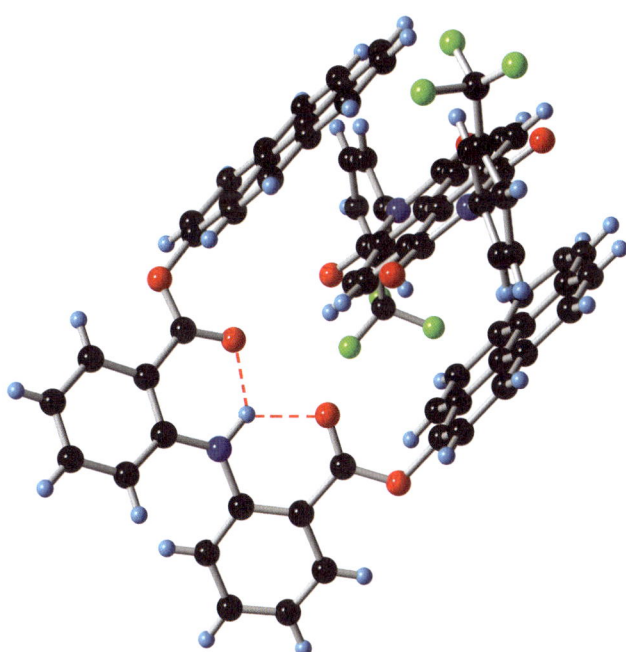

Fig. 3 X-Ray structure of the 1:1 complex between model diimide **11** and tweezer-molecule **12**. Intramolecular hydrogen bonds (NH···O = 1.98, 1.94 Å) are shown dotted in red.

Despite the evidently weak binding of **12** to pyromellitimide residues, the new tweezer-molecule proved startlingly effective at revealing long-range sequence information in chain-folding, aromatic copolyimides. Copolymer **2** was synthesised at high molecular weight (M_n = 68 000; M_w = 109 000) by polycondensation of the six-ring disulfone-diamine **8** with equimolar proportions of pyromellitic dianhydride (**13**) and hexafluoroisopropylidene-diphthalic anhydride (**14**). Designating the diamine residue as "S", and the resulting pyromellitimide and hexafluoroisopropylidene-diphthalimide residues as "I" and "F" respectively, all pyromellitic-centred "triplet" sequences present in the copolymer must be of the form "SIS". In keeping with this, the pyromellitimide protons appear as a sharp singlet at 8.51 ppm in the ¹H

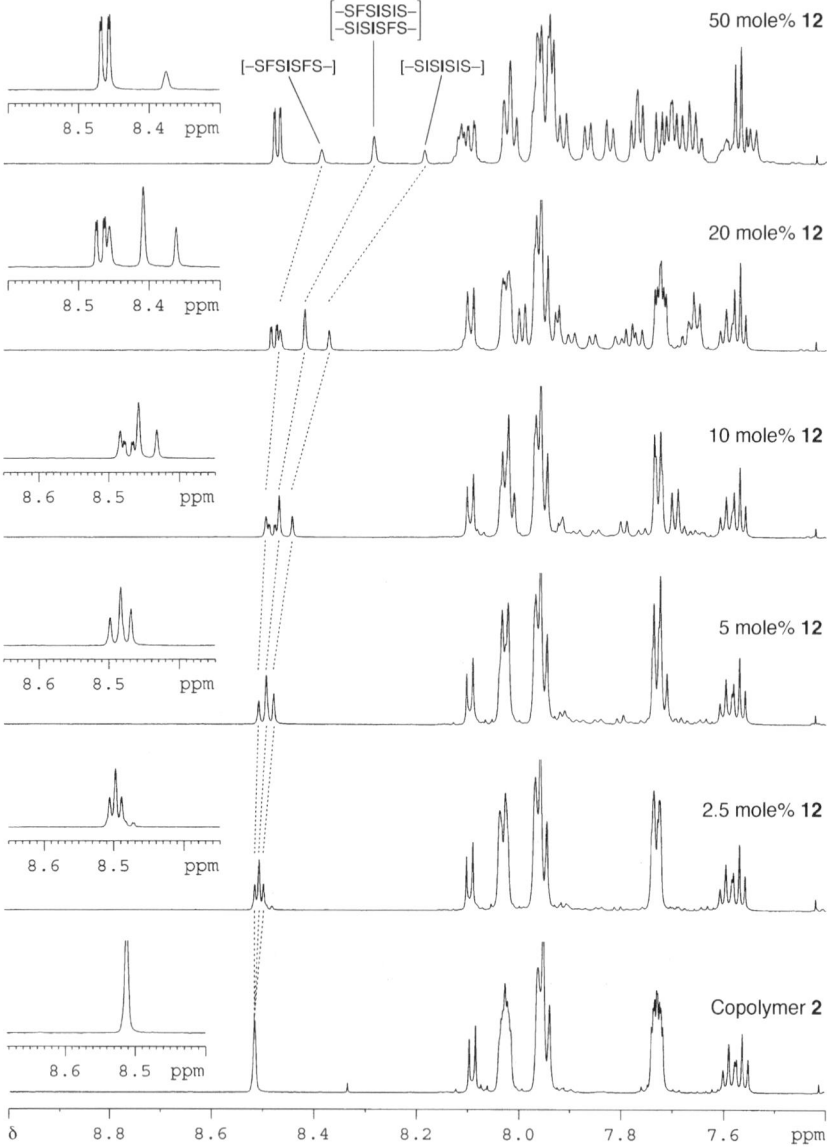

Fig. 4 ¹H NMR spectra (700 MHz) showing the effects of complexation of copolyimide **2** with progressively increasing concentrations (bottom to top) of tweezer-molecule **12**. The weak resonance at 8.48 ppm in the "2.5 mol%" spectrum arises from the added tweezer-molecule.

NMR spectrum of this copolymer (Fig. 4, lowest spectrum). However, addition of as little as 2.5 mol% of tweezer-molecule (relative to pyromellitimide residues) to the polyimide solution produces an immediate splitting of this singlet resonance into a 1:2:1 pattern, and further addition of tweezer results in this group of three resonances moving progressively upfield, with rapidly increasing separation between them. At 50 mol% of tweezer, the splitting reaches *ca.* 0.1 ppm, a very much larger value than that observed with the original tweezer-molecule **6**.[10]

The three newly-resolved diimide resonances can readily be assigned to the three different septet sequences SISISIS, [SFSISIS + SISISFS], and SFSISFS. Here the "observed" pyromellitimide residue (bold **I**) is flanked either by two, one or no pyromellitimide units, permitting the same number of tweezer-molecules to bind at these adjacent positions. For steric and electronic reasons the hexafluoroisopropylidene-diphthalimide residues can have no affinity for tweezer-binding and thus represent "binding-forbidden" sites. As discussed in an earlier publication,[10] splitting of the diimide resonances in the presence of tweezer-type molecules arises from the additional long-range shielding of the central diimide protons generated by adjacently-bound tweezers, over and above the larger degree of shielding resulting from a tweezer-molecule bound directly at the observed site. On this basis, the diimide resonance emerging at highest field can be assigned to the triply-bound sequence SISISIS (Fig. 5); the next-highest field resonance (with twice the intensity) to the directionally-degenerate sequences [SFSISIS + SISISFS], which each have one adjacent binding site "I"; and the resonance at lowest field is assigned to the sequence SFSISFS, which has no adjacent binding sites. The separations between these resonances increase with tweezer-concentration but remain equal, suggesting that the additional magnetic shielding arising from adjacent tweezer-binding increases incrementally with the number of adjacent binding sites (0, 1 or 2). It should be noted that separate resonances associated with (identical) bound and unbound sequences are not observed, indicating that the exchange-kinetics of tweezer binding are fast on the NMR timescale. Also, given the chain-folding conformation which this polymer was designed to adopt, it would be expected that resonances associated with the 4,4′-biphenylenedisulfone unit should also show sequence-selective splitting in the presence of tweezer-molecule **12**. Indeed, in the presence of 10 mol% of **12**, a doublet at *ca.* 7.7. ppm, assignable to the protons *ortho* to the central biaryl linkage of the sequence [–SISIS–], emerges upfield of the main group of resonances associated with this linkage. The latter sequence can bind *two* molecules of **12**, unlike the alternative possible sequences [–SFSIS–], [–SISFS–] and [–SFSFS–], and so is most strongly shifted. Consistent with this assignment, the integral of the upfield resonance is one third of that of the lower field group.

We have also carried out the same tweezer-titration experiment with a copolymer (**3**) in which adjacent binding sites are once again blocked, but now not in the absolute sense represented by the presence of a non-binding hexafluoroisopropylidene-diphthalimide residue. Instead, tweezer-binding is inhibited by the presence of a sterically-hindering methyl substituent on the aromatic ring adjacent to pyromellitimide. Earlier work with tweezer-molecules **6** and **7** has shown that such substituents can reduce the tweezer-binding constant very significantly, though without switching off complexation altogether. Designating the bis(*o*-methyl-substituted) diamine residue as "H" (for "hindered"), the possible "I"-centred triplet sequences are SIS, HIS/SIH and HIH. Indeed, corresponding diimide resonances with relative intensities 1:2:1 are evident in the ^{1}H NMR spectrum of the 1:1 copolymer **3** (M_n = 54 000, M_w = 87 000) even in the *absence* of any tweezer-molecule (Fig. 6, lowest spectrum).

The highest field resonance of the three is assignable to the unhindered sequence SIS, since this resonance shows the most marked upfield shift in the presence of tweezer-molecule **12**. The threefold splitting of this SIS resonance again arises from long-range shielding by adjacently-bound tweezer-molecules, and the possible SIS-centred septet sequences can be represented as SISISIS, [SISISIH + HISISIS],

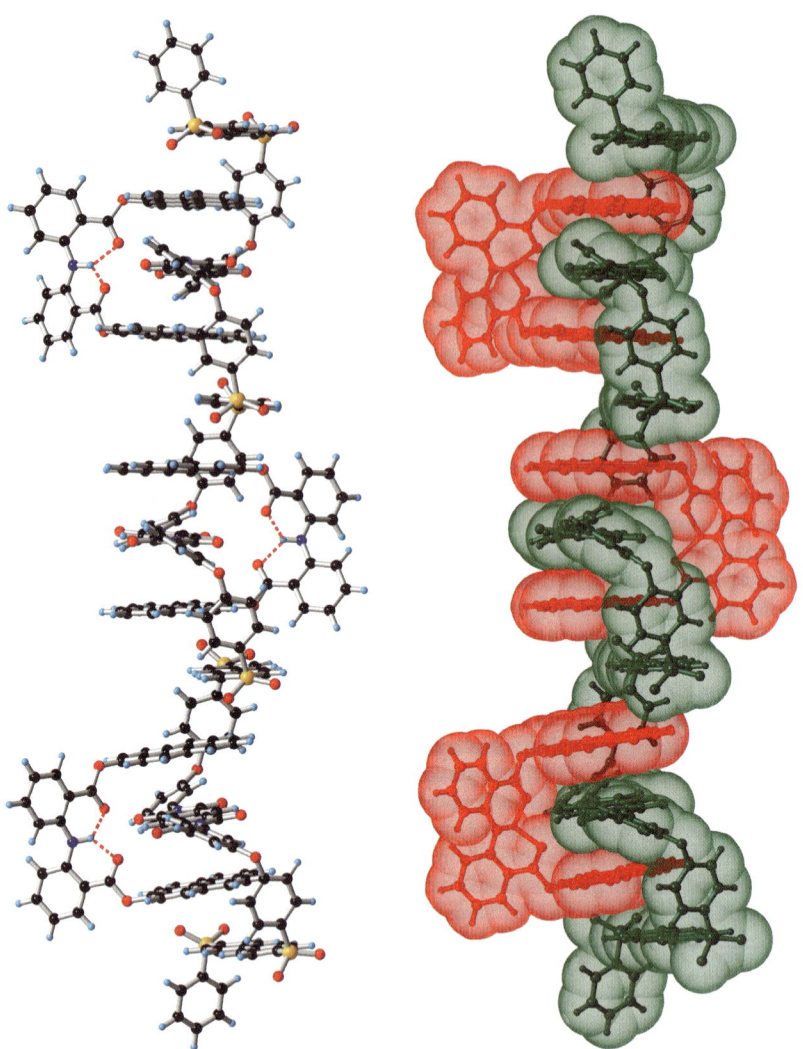

Fig. 5 Computational simulation (molecular mechanics with charge-equilibration) of three tweezer-molecules **12** binding to the monomer sequence [–SISISIS–] in the chain-folding copolyimide **2**, as demonstrated by ^{1}H NMR spectroscopy. Left: ball and stick representation; Right: van der Waals surfaces, showing the high degree of shape-complementarity between the tweezer molecules (red) and the polymer chain (green).

and HISISIH. Again, the 1:2:1 splitting pattern reflects the relative populations of these three sequences in the random 1:1 copolymer **3**. It is noticeable however that the separation between the three resonances—at any given tweezer-molecule to diimide ratio—is very much less (by about a factor of two) than the corresponding separation observed for the hexafluoroisopropylidene di-imide copolymer **2**, although the chemical shift of the resonance assigned to SISISIS is the same for both systems. It is clear then that additional ring-current shielding from adjacently (albeit weakly) bound tweezer-molecules at the sequence -HIS- must occur in the methyl substituted copolymer **3**, thereby narrowing the separations relative to those observed with copolymer **2**, where the probability of tweezer-binding at an adjacent hexafluoroisopropylidene-diphthalimide residue is zero.

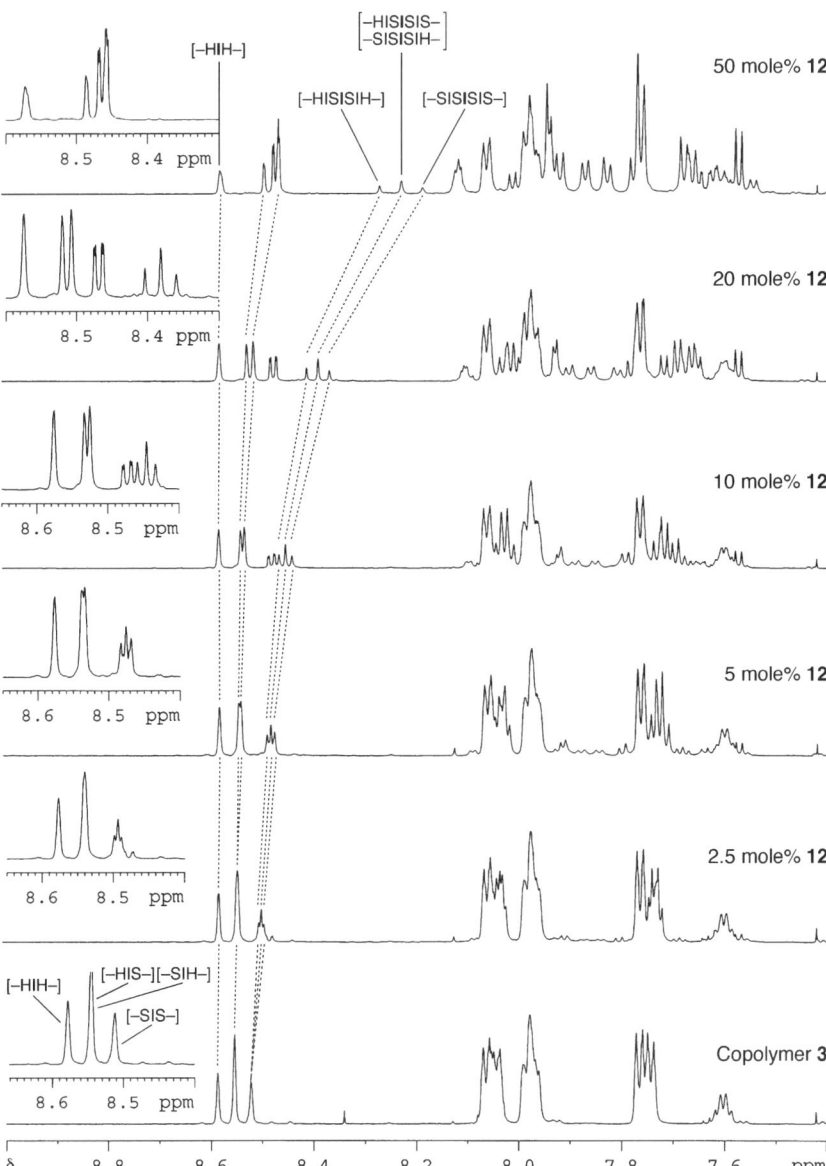

Fig. 6 ^{1}H NMR spectra showing the effects of sequence-specific complexation of copolyimide **3** with progressively increasing concentrations (bottom to top) of tweezer-molecule **12**. The weak resonance at 8.48 ppm in the "2.5 mol%" spectrum arises from the added tweezer-molecule.

So far we have concentrated on sequences centred on the strongly-bound triplet "SIS", but sequence-related effects are also evident for other resonances in the ^{1}H NMR spectrum of copolymer **3**. Thus the resonance assigned to HIS (plus SIH) splits in the presence of tweezer-molecule to give two resonances of equal intensity, the splitting increasing incrementally with increasing concentration of tweezer (Fig. 6 and 7). Enumeration of the different septet sequences centred on HIS shows that these fall into two groups: (i) SIHISIS and HIHISIS (plus their directionally-degenerate reverse sequences), which both have a strongly binding SIS triplet

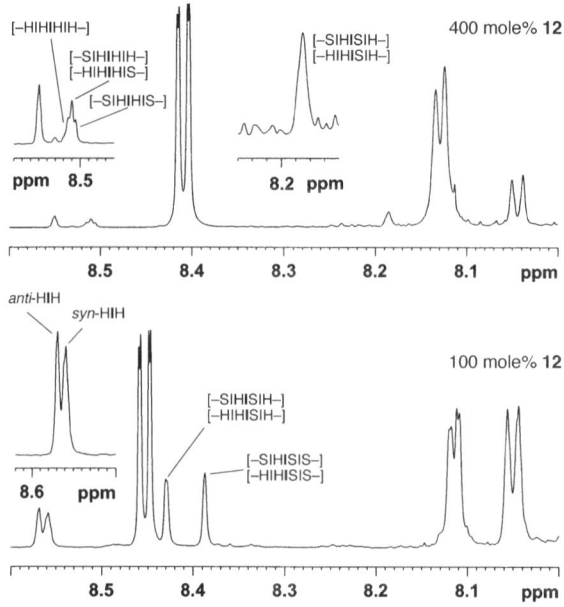

Fig. 7 ^{1}H NMR spectra showing the effects of complexation of copolyimide **3** with high concentrations (100 and 400 mol% relative to diimide) of tweezer-molecule **12**.

adjacent to the central, observed diimide residue, and (ii) SIHISIH and HIHISIH (plus their reverse sequences), neither of which can strongly bind a tweezer-molecule at an adjacent site. We therefore assign the splitting of the original HIS resonance in the presence of the tweezer to the presence (or absence) of ring-current shielding associated with strong tweezer-binding at an *adjacent* site, even when the *observed* site is itself only weakly bound. It is however noticeable that the separation of HIS resonances is, for a given tweezer-concentration, only about 60% of the separation seen for SIS-based sequences. As both splittings depend on the presence of an adjacent, strongly-binding SIS sequence, this observation suggests that the magnetic shielding of a dimide proton produced by an *adjacently* bound tweezer molecule may be **amplified** by the presence (at least in the dynamic sense implied by fast exchange) of a *proximally* bound tweezer.

Finally, the septet sequences centred on HIH can also be analysed in terms of the presence or absence of adjacent binding sites. There is no possibility of a strongly binding SIS sequence lying adjacent to HIH, but weakly-binding HIS sequences are certainly present. The HIH-centred septets thus fall into three groups: (i) SIHI-HIS, with two adjacent HIS triplets, (ii) HIHIHIS and SIHIHIH, each with one adjacent HIS group, and (iii) HIHIHIH, with no adjacent HIS sequences. On this basis, very high concentrations of tweezer-molecule **12** might be expected to split the HIH resonance into a 1:2:1 pattern. In the presence of 100 mol% of **12** however, this peak splits into two resonances of equal intensity, and only with 400 mol% of tweezer does *one* of these signals resolve into a 1:2:1 resonance pattern (Fig. 7).

We have previously shown that the initial resolution of two HIH-type resonances arises from the presence of two non-interconverting conformers (*syn* and *anti*) of the *N,N'*-diarylpyromellitimide unit—since the *o*-methyl groups impose restricted rotation about the N–Ar bonds.[16] It is not certain at this stage which of the two conformers interacts more strongly with the tweezer. Perhaps the *syn* conformer, with one unhindered face, is able to stack with one arm of the tweezer, and this very weak interaction is sufficient to differentiate (just) the various HIH-centred septets. The fact that the non-complexing conformer shows no resonance-splitting

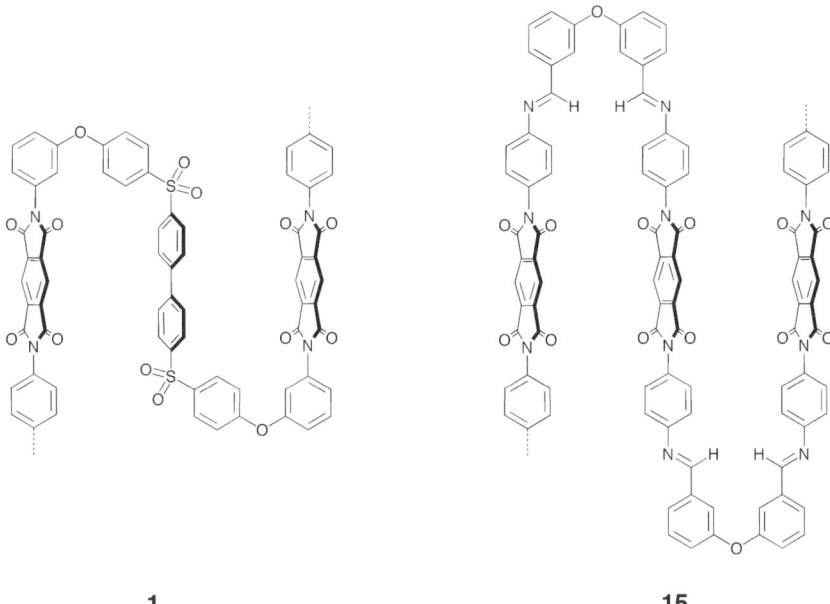

Chart 3 Polyimide-sulfone **1** and hypothetical polyimide-imine **15**, showing the chain-folded, tweezer-binding conformation in each case.

at all is once again consistent with the proposed amplification of magnetic shielding effects between adjacently bound tweezer molecules.

The Future

Up till now we have focused on the use of small molecules for *reading* sequence-information, but we are also developing ideas for *writing* such information to polymer chains. Achieving this will require a more labile type of polymer than our current, all-aromatic polyimide-sulfone system, to enable the tweezer-molecule to impose a preferred sequence of some type on the polymer. For example, the hypothetical polyimide-imine **15**, shown in Chart 3, can in principle have a very similar conformation to that of the strongly tweezer-binding polyimide-sulfone **1**. However, in **15** there are two potential geometric isomers at every imine linkage, so that what appears to be a homopolymer would normally contain a random sequence of imine units with either *syn* or *anti* geometry about the double bonds. It is conceivable that low-temperature photochemical equilibration of this polymer in the presence of tweezer-molecules could drive adjacent imine linkages from an initially random geometric relationship towards the preferred *anti*, *anti*, tweezer-binding arrangement. In this way a random copolymer sequence might be overwritten by the tweezer molecules to give a more regular one. Alternatively, random copolymer sequences arising from the incorporation of sterically hindered comonomers, as described in the present paper, might also be susceptible to tweezer-driven rearrangements through acid-catalysed imine-exchange.

Conclusions

A novel type of tweezer-molecule, designed specifically for binding to sterically- and electronically-complementary sequences in synthetic copolyimides, has been successfully synthesised and structurally characterised. This molecule maintains a pre-organised binding conformation through the formation of a pair of convergent, intramolecular hydrogen bonds, as shown by single crystal X-ray analyses of both the "free" tweezer and its complex with a model aromatic diimide. The new tweezer-molecule shows very high sensitivity to long-range sequence effects in terms of the complexation shifts it produces in the ^{1}H NMR spectra of copolyimides. Evidence from complexation shifts associated with tweezer-binding suggests that ring-current shielding from *adjacent* tweezer-molecules is significantly enhanced by tweezer-binding at the "observed" site, and hence that amplification of ring-current effects may be occurring between adjacently-bound tweezer-molecules.

Acknowledgements

This work was supported by the Royal Society (Leverhulme Senior Research Fellowship to HMC) and by EPSRC (grant numbers EP/C533526/1, EP/E00413X/1, EP/F013663/1 and EP/G026203/1). The assistance of Claire Murray in preparing crystallographic data for publication is gratefully acknowledged

References

1 J. D. Watson and F. H. C. Crick, *Nature*, 1953, **171**, 964.
2 F. H. C. Crick, *Symp. Soc. Exp. Biol.*, 1958, **12**, 138.
3 (*a*) M. Mansuripur and P. Khulbe, in *Advances in Information Recording, (DIMACS Series in Discrete Mathematics, vol. 73)*, American Mathematical Society, Providence, 2008, pp. 63–77; (*b*) D. Branton, D. W. Deamer, A. Marziali, H. Bayley, S. A. Benner, T. Butler, M. Di Ventra, S. Garaj, A. Hibbs, X. H. Huang, S. B. Jovanovich, P. S. Krstic, S. Lindsay, X. S. S. Ling, C. H. Mastrangelo, A. Meller, J. S. Oliver, Y. V. Pershin, J. M. Ramsey, R. Riehn, G. V. Soni, V. Tabard-Cossa, M. Wanunu, M. Wiggin and J. A. Schloss, *Nat. Biotechnol.*, 2008, **26**, 1146; (*c*) S. Matysiak, A. Montesi, M. Pasquali, A. B. Kolomeisky and C. Clementi, *Phys. Rev. Lett.*, 2006, **96**, 118103; (*d*) P. K. Khulbe, M. Mansuripur and R. Gruener, *J. Appl. Phys.*, 2005, **97**, 104317.
4 C. R. Dawkins, *The Blind Watchmaker*, Longmans, London, 1986.
5 See for example: H. Lodish, A. Berk, C. A. Kaiser, M. Krieger, M. P. Scott, A. Bretscher, H. Ploeghand P. T. Matsudaira, *Molecular Cell Biology*, Freeman, New York, 6th edn, 2007.
6 L. E. Orgel, *Nature*, 2006, **439**, 915.
7 L. E. Orgel, *The Origins of Life: Molecules and Natural Selection*, Wiley, New York, 1973.
8 H. M. Colquhoun and Z. Zhu, *Angew. Chem., Int. Ed.*, 2004, **43**, 5040.
9 H. M. Colquhoun, Z. Zhu, C. J. Cardin and Y. Gan, *Chem. Commun.*, 2004, 2650.
10 H. M. Colquhoun, Z. Zhu, C. J. Cardin, Y. Gan and M. G. B. Drew, *J. Am. Chem. Soc.*, 2007, **129**, 16163.
11 Leading references to tweezer-type molecules:(*a*) M. Harmata, *Acc. Chem. Res.*, 2004, **37**, 862; (*b*) F.-G. Klärner and B. Kahlert, *Acc. Chem. Res.*, 2003, **36**, 919; (*c*) C. W. Chen and H. W. Whitlock, *J. Am. Chem. Soc.*, 1978, **100**, 4921; (*d*) S. C. Zimmerman and C. M. VanZyl, *J. Am. Chem. Soc.*, 1987, **109**, 7894; (*e*) H. Kurebayashi, T. Haino, S. Usui and Y. Fukazawa, *Tetrahedron*, 2001, **57**, 8667; (*f*) V. Balzani, H. Bandmann, P. Ceroni, C. Giansante, U. Hahn, F.-G. Klärner, U. Müller, W. M. Müller, C. Verhaelen, V. Vicinelli and F. Vögtle, *J. Am. Chem. Soc.*, 2006, **128**, 637; (*g*) F.-G. Klärner, B. Kahlert, A. Nellesen, J. Zienau, C. Ohsenfeld and T. Schrader, *J. Am. Chem. Soc.*, 2006, **128**, 4831; (*h*) V. K. Potluri and U. Maitra, *J. Org. Chem.*, 2000, **65**, 7764; (*i*) A. J. Goshe, I. M. Steele, C. Ceccarelli, A. L. Rheingold and B. Bosnich, *Proc. Natl. Acad. Sci. U. S. A.*, 2002, **99**, 4823; (*j*) F. C. Krebs and M. Jørgensen, *J. Org. Chem.*, 2001, **66**, 6169; (*k*) X.-X. Peng, H.-Y. Lu and C.-F. Chen, *Org. Lett.*, 2007, **9**, 895.
12 *Polyimides*, ed. D. Wilson, H. D. Stenzenberger and P. M. Hergenrother, Blackie, London, 1990.
13 H. M. Colquhoun, D. J. Williams and Z. Zhu, *J. Am. Chem. Soc.*, 2002, **124**, 13346.
14 H. M. Colquhoun, Z. X. Zhu and D. J. Williams, *Org. Lett.*, 2003, **5**, 4353.
15 M. B. Nielsen, J. O. Jeppesen, J. Lau, C. Lomholt, D. Damgaard, J. P. Jacobsen, J. Becher and J. F. Stoddart, *J. Org. Chem.*, 2001, **66**, 3559.
16 K. D. Shimizu, T. M. Dewey and J. Rebek, *J. Am. Chem. Soc.*, 1994, **116**, 5145.

PAPER

DNA self-assembly: from 2D to 3D

Chuan Zhang,[a] Yu He,[a] Min Su,[b] Seung Hyeon Ko,[a] Tao Ye,[a] Yujun Leng,[b] Xuping Sun,[a] Alexander E. Ribbe,[a] Wen Jiang[b] and Chengde Mao*[a]

Received 16th March 2009, Accepted 6th April 2009
First published as an Advance Article on the web 27th July 2009
DOI: 10.1039/b905313c

This paper describes our recent efforts on the self-assembly of three-dimensional (3D) DNA nanostructures from DNA star motifs (tiles). DNA star motifs are a family of DNA nanostructures with 3, 4, 5, or 6 branches; they are named as 3-, 4-, 5-, 6-point-star motifs, respectively. Such motifs are programmed to further assemble into nanocages (regular polyhedra or irregular nanocapsules) with diameters ranging from 20 nm to 2 μm. Among them, DNA nanocages derived from 3-point-star motif consists of a group of regular polyhedra: tetrahedra, hexahedra (or cubes), dodecahedra and buckyballs (containing 4, 8, 20, and 60 units of the 3-point-star motif, respectively). An icosahedron consists of twelve 5-point-star motifs and is similar to the shapes of spherical viruses. 6-point-star motifs can not assemble into regular polyhedra; instead, some sphere-like or irregular cages with diameters about 1–2 μm will form. Similar large cages can also assemble from the 5-point-star motif when the DNA concentrations are higher than those for assembling regular icosahedra. In our study, we have identified several important factors for assembly of well-defined 3D nanostructures, including the concentration, the flexibility, and the arm length of the DNA tiles and the association strength between the DNA tiles.

1. Introduction

Beyond the genetic interest, DNA has been shown as a superb molecular system in self-assembly towards bottom-up nanofabrication.[1–3] It has many attractive properties: the excellent capability of molecular recognition, the well-defined secondary structure (duplex), reasonable chemical stability, and readily commercial availability. These properties together endow DNA molecules with a great potential to serve as "building blocks" for the preparation of nanostructures.

In structural DNA nanotechnology, DNA motifs (tiles) are designed and investigated as building blocks to fabricate nanostructures. For example, a range of DNA motifs including double crossover (DX) and multiple-crossover motifs, star motifs, triangle and DX triangle motifs, and parallelogram motifs, have been introduced to construct various DNA two-dimensional (2D) arrays and non-periodic 2D patterns.[4–18] However, the investigation of DNA 3D nanostructures is quite limited.[19–27] In the early 90s, DNA 3D nanostructures with the connectivities of a cube[19] and of a truncated octahedron[20] were constructed by Seeman and his coworkers. Limited by characterization methods (denaturing gel electrophoresis), such 3D structures had to be covalently linked, which required the authors to use

[a]*Department of Chemistry, Purdue University, West Lafayette, Indiana, 47907, USA. E-mail: mao@purdue.edu*
[b]*Markey Center for Structural Biology and Department of Biological Sciences, Purdue University, West Lafayette, Indiana, 47907, USA*

stepwise synthesis *via* enzymatic ligation. Successive steps of ligation and purification, unfortunately, greatly increased the workload and resulted in a very low overall synthetic yield. With the introduction of new characterization tools [cryogenic electron microscopy (cryoEM) and atomic force microscopy (AFM)], one was able to visualize non-covalently associated DNA structures by direct imaging. Starting from 2004,[21] strategies of one pot self-assembly were used to assemble DNA 3D nanostructures. For example, Turberfield and his coworkers used four oligonucleotides to assemble DNA tetrahedra;[22] the resulting structures were characterized by AFM. In another elegant example, Shih and his coworkers folded a long single DNA strand into pre-designed, branched, secondary structures, which further folded into an octahedon through intra-complex paranemic interactions with the help of a number of short DNA helper strands.[21] Essentially, this is related to the so-called "DNA origami" approach.[16] The resulting octahedral structures were clearly shown by cryoEM characterization. This work together demonstrated that DNA polyhedra could be readily assembled from synthetic DNA molecules.

Though tetrahedra, cubes, octahedral, and truncated octahedra are highly symmetric, those reported DNA 3D objects do not really possess these symmetries if viewed at the level of DNA sequences. Consequently, every object requires many unique DNA strands. As the objects become larger and more complicated, the number of DNA strands with unique sequences quickly increases. It is not easy to apply the above mentioned methods to construct complicated polyhedra, which require too many unique DNA strands. Thus, while these studies are elegant and exciting, they do not provide a general and simple route to fabricate DNA 3D nanostructures. In biology, many complex nanostructures with various biological functions assemble in different ways. For instance, the outer protein shells (capsids) of spherical viruses have highly symmetric, icosahedral structures. They are composed of many copies of identical protein subunits, which associate with each other *via* non-covalent bonding. The complexity and robustness of viral capsids indicate

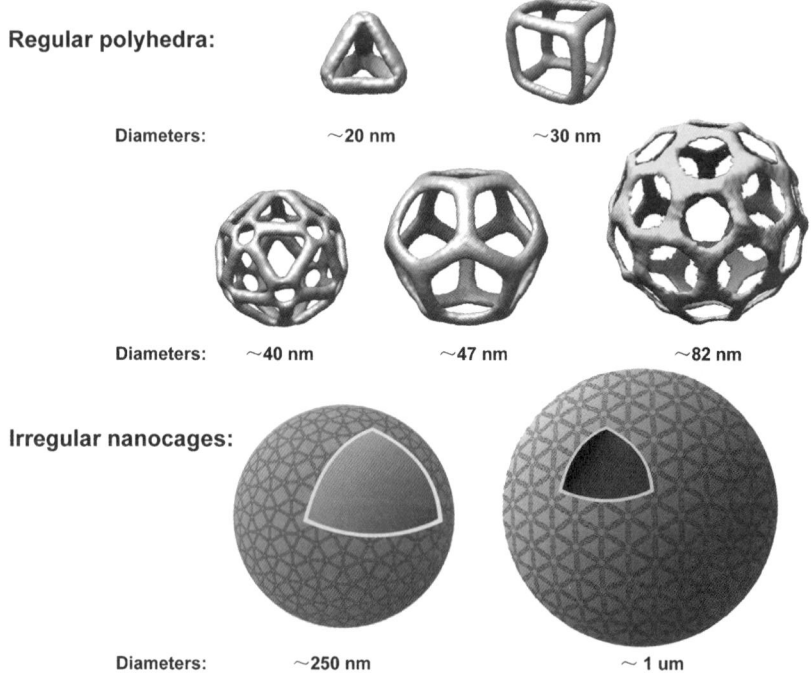

Fig. 1 DNA nanocages self-assembled from DNA star motifs.

that this architectural strategy is very effective. Despite the difficulty to fully understand and mimic the self-assembly of viral capsids, it should be possible to rationally design DNA nanomotifs (tiles) that can behave in a similar way to capsid component proteins to self-assemble into highly symmetric DNA 3D nanostructures. Recently we have successfully used symmetric DNA 3-point-star motifs to assemble three DNA polyhedra (tetrahedra, dodecahedra, and buckyballs) and proved their structures by cryoEM and single particle 3D reconstruction.[28] Since then, a number of other DNA polyhedra and DNA cages have been assembled from a series of DNA star motifs (Fig. 1).

In this paper, we will discuss our efforts on the self-assembly of discrete, 3D DNA nanostructures from symmetric DNA star motifs. The DNA 3D assemblies will be classified by their component units, the nanomotifs. We will start from a 3-point-star motif and discuss the design and parameter controls for fabrication of different types of DNA nanocages; 5- and 6-point-star DNA cages will be discussed in the same way. At the end, we will present our thoughts on future directions.

2. DNA nanocages assembled from 3-point-star motifs

From the view of geometry, a number of regular polyhedra can be considered as assemblies of 3-branched tiles and these polyhedra differ from each other by the number of the component building blocks (tiles). Such polyhedra include tetrahedra, hexahedra (cubes), dodecahedra and buckyballs with increasing sizes. Each 3-branched motif can be represented by a DNA 3-point-star motif. Similar to the self-assembly of DNA 2D arrays, the assembly of DNA nanocages can be carried out by a one-pot hierarchical self-assembly process (Fig. 2): as the solution temperature decreases, individual single DNA strands assemble into sticky-ended, 3-point-star motifs (tiles), and then the tiles further assemble into 3D nanocages through sticky-end association between the tiles.

DNA tetrahedra, dodecahedra, and buckyballs were the first three examples assembled by 3-point-star motifs with component numbers of 4, 20 and 60.[28] This work has clearly demonstrated this biomimetic, hierarchical assembly strategy. The design is shown in Fig. 2 and the key parameters that determine the final structures are the flexibility and the concentration of the 3-point-star DNA tiles.

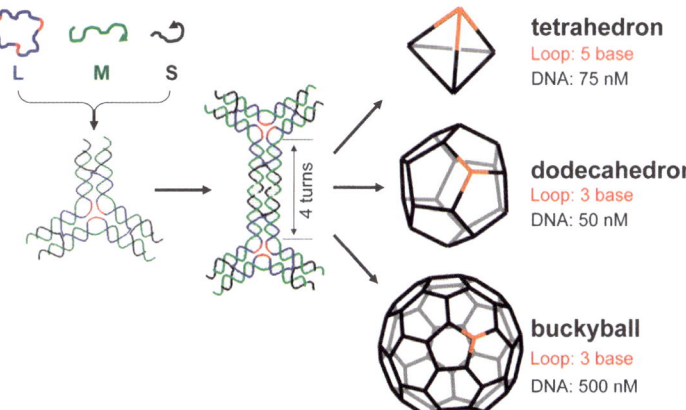

Fig. 2 Self-assembly of DNA polyhedra. Three different types of DNA single strands (**L**, **M**, and **S**) stepwise assemble into symmetric 3-point-star motifs (tiles) and then into polyhedra in a one-pot process. Note that there are three single-stranded loops (red) in the center of the 3-point-star tile. The final structures (polyhedra) are determined by the loop length and the DNA concentration. In each of the final polyhedral structures, one of the component 3-point-star tiles is highlighted in gold.

In general, DNA tiles with less flexibility and at higher concentrations will result in larger assemblies. The three-point-star motif contains a three-fold rotational symmetry and consists of 7 strands: a long repetitive central strand (**L**), three identical medium strands (**M**), and three identical short peripheral strands (**S**). Three single-stranded loops (colored red) are introduced at the center of the DNA tile. The tile flexibility can be easily adjusted by varying the loop length. Increasing the loop length increases the tile flexibility. At the termini of each branch of the tile, there are two self-complementary, 4-base-long, single-stranded overhangs, or sticky ends. Sticky-ends association (base-pairing) between the tiles allow finite numbers of tiles to assemble into 3D nanostructures.

Agarose gel electrophoresis was used for the initial characterization of those DNA cages and for estimation of the assembly yields. It was found that small structures generally had higher assembly yields than the larger cages did. For example, the assembling yield for the tetrahedra was over 90% as estimated from the gel by an image processing software, ImageJ (developed at the National Institute of Health, NIH), but it became lower and lower for the dodecahedra and the buckyballs. The yield of the properly assembled DNA buckyballs was only ~60%. Non-uniform growth and deformation of the larger assemblies might be responsible for the observed, low assembly yields.

Tetrahedra

A tetrahedron is the smallest closed structure assembled from the 3-point-star DNA tiles. Each vertex is one 3-point-star tile and each edge comprises two associated branches from two tiles. The tiles in a tetrahedron adopt a significantly bent conformation when compared with the free, flat tiles. To accommodate such bending, the tiles have to be quite flexible; such flexibility can be provided by using long central loops (5 bases long for the red segments). At a sufficiently low DNA concentration (75 nM), tetrahedra are the dominating complexes assembled. Dynamic light scattering (DLS) shows that the apparent radius of the DNA complexes is $\sim 10.3 \pm 0.6$ nm, agreeing well with the radius (10.9 nm) of the circumscribed sphere of the DNA tetrahedral model if taking the parameters of the standard B-DNA conformation (pitch: 0.33 nm base pair^{-1}; diameter: 2 nm). The DNA assemblies also appear to be uniformly-sized, triangular particles when imaged by AFM in air. The observed structures in the AFM images are probably collapsed DNA tetrahedra due to the strong interaction between the DNA samples and the substrates.

The tetrahedral structure is more clearly shown by direct cryoEM imaging (Fig. 3, top panel). In cryoEM images, most visible particles have tetrahedral shapes of the expected size. The observed edges are 16 nm long, nicely matching with the designed model (16.2 nm). With experimentally observed particles, a DNA tetrahedral structure can be modeled at 2.8 nm resolution through a well-established technique of single particle 3D reconstruction.

Dodecahedra

A dodecahedron consists of twenty 3-point-star tiles, 12 pentagonal faces, and 30 edges. Compared to those in a tetrahedron, the 3-point-star tiles in a dodecahedron are less bent and require much less flexibility. Thus, the length of the central single-stranded loop is reduced to 3 bases long, which stiffens the star tiles. The concentration of DNA tiles is also critical to dodecahedron assembly. The dodecahedral complex is the main product only at low DNA concentration (50 nM). The apparent radius of the assembled objects in DLS measurement is $\sim 24.0 \pm 1.8$ nm, agreeing with the calculated value (23.6 nm) from the designed structural model. Under AFM imaging (in air), the assembled objects show circular features and uniform sizes. Pentagonal structures are visible at the center of the top layer, representing a reasonable, collapsed 2D projection of a dodecahedron.

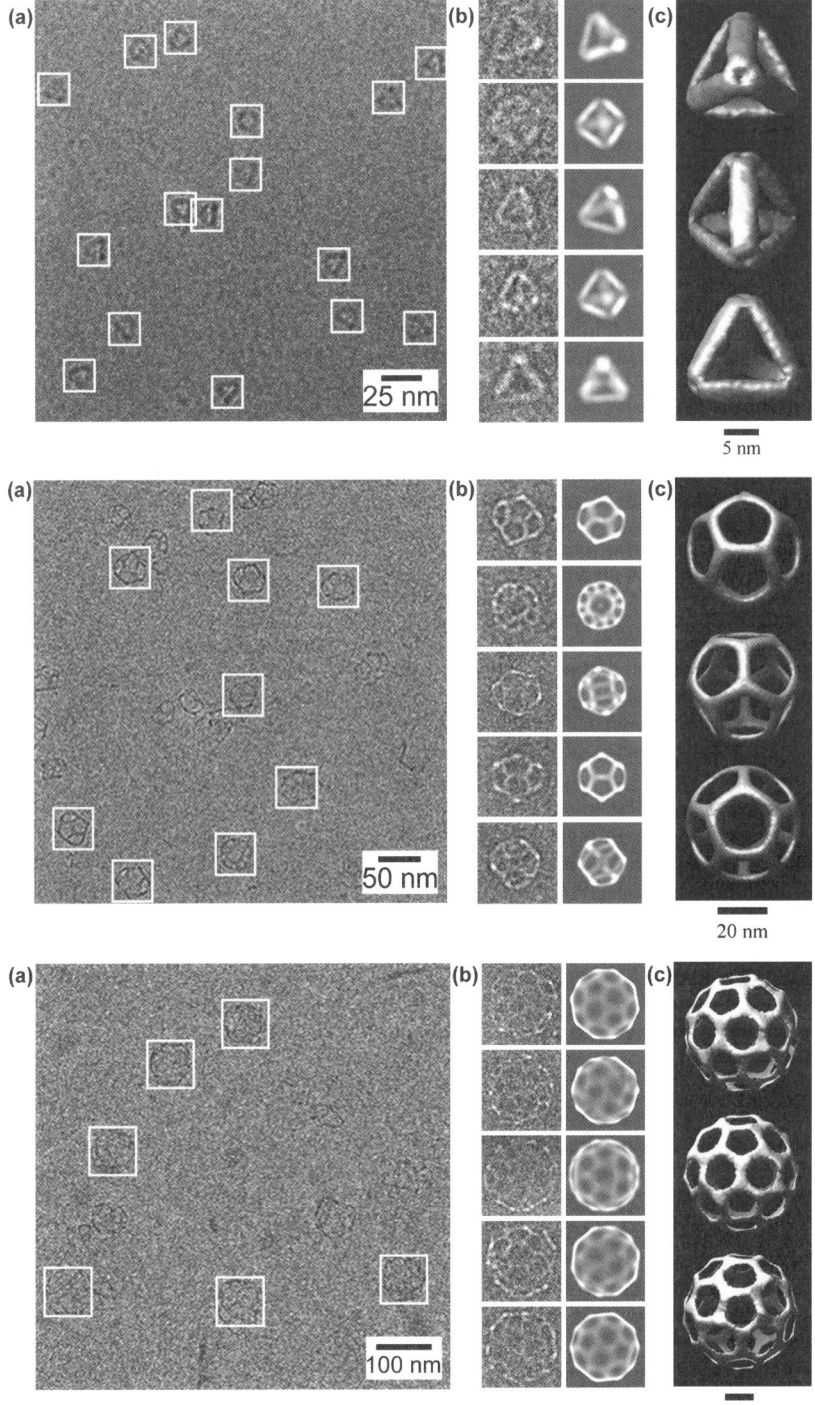

Fig. 3 Visualization of the self-assembled DNA polyhedra by cryogenic transmission electron microscopy (cryoTEM). (top panel) tetrahedra, (middle panel) dodecahedra, and (bottom panel) buckyballs. (a) A representative image. White boxes indicate the DNA particles. (b) Comparison between raw images of individual particles (left) and the corresponding

CryoEM (Fig. 3, middle panel) showed dodecahedral-shaped objects of the expected size (in the boxed area). All particles with the right size were picked up for single particle 3D reconstruction. The icosahedral symmetry was imposed during the reconstruction, resulting in a well-defined dodecahedral structure (at a resolution of 2.5–3.0 nm). The individual, raw cryoEM particles matched with the computer-generated model projections very well.

Buckyballs

3-point-star tiles can also assemble into Buckminsterfullerenes (or buckyballs), a type of highly symmetric polyhedron. A buckyball contains 60 vertexes, 90 edges, and 32 faces (12 pentagons and 20 hexagons). Assembling DNA buckyballs and dodecahedra require exactly the same DNA strands. To selectively assemble these structures, the DNA concentration has to be carefully controlled. At a high DNA concentration (500 nM), the 3-point-star DNA tiles (the central, single-stranded loops are 3 bases long) readily assemble into the large buckyball structures rather than small dodecahedra. The DLS measurement indicates that the DNA assemblies have an apparent hydrodynamic radius of 42.2 ± 4.0 nm, close to the calculated radius of the buckyball model (41 nm). However, the polydispersity of the assembled DNA buckyballs is significantly higher than those of DNA tetrahedra and dodecahedra. Under AFM imaging, the DNA assemblies appear as uniform-sized, round, collapsed DNA nanostructures with a diameter of ~110 nm. Round-like DNA structures can be seen in the raw CryoEM images (Fig. 3, bottom panel) and a buckyball structure has been obtained from reconstruction based on hand-picked particles. Compared to tetrahedra and dodecahedra, the size distribution of the particles in the cryoEM images is much worse than that from the AFM images. This phenomenon is likely due to larger assemblies being easier to deform or break up, which lowers the quality of the reconstructed map.

The first three 3D objects have a common feature in their design. Each pseudo-continuous DNA duplex of the edge in the final polyhedra is 4 turns (42 bases) long after sticky end hybridization. In such a design, all component DNA tiles are facing the same direction and any intrinsic curvatures of the tiles would add up toward the same direction to promote the formation of closed structures. Meanwhile, the DNA concentration plays a key role during the self-assembly. Self-assembly relies on inter-tile interactions. High DNA concentrations favor large assemblies in the same design such as buckyballs, and low DNA concentrations will promote small assemblies such as dodecahedra. The desired products dominate in self-assembly only in a certain range of DNA concentration.

Cubes (or hexahedra)

When carefully balancing the flexibilities and the rigidities of the DNA motifs and controlling the DNA concentrations, 3-point-star motifs can selectively assemble into tetrahedra, dodecahedra, or buckyballs. However, the cube structure, a simple polyhedron that contains eight 3-point-star tiles, has never been found in the study discussed above. It implies that cube structures cannot be obtained by simply changing the concentration and the flexibility of the DNA tiles.

To overcome this problem, we introduced another strategy to the self-assembly of 3D DNA nanostructures.[30] This strategy exploits the helical nature of the DNA double helix structure. When being separated by an odd number of half helical turns, two objects along a DNA duplex will be on the opposite sides of the DNA duplex and are related by a two-fold rotational symmetry. To assemble cubes, each edge

---

computer-generated model projections (right). (c) Three views of the DNA polyhedral structure reconstructed from cryoTEM images. (Replicated from ref. 28).

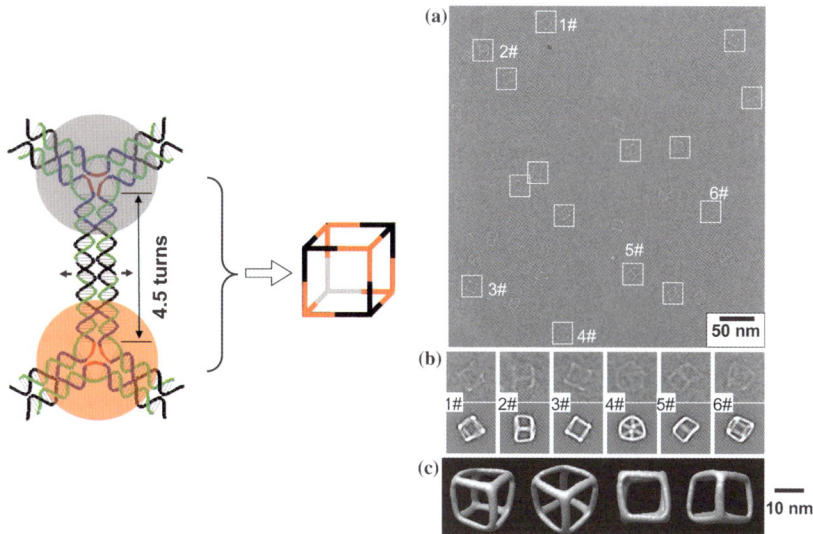

Fig. 4 Self-assembly of DNA cubes from 3-point-star tiles. (left) Assembly scheme. Note that the separation of any two adjacent tiles is 4.5 DNA helical turns. Any two interacting tiles (shadowed with different colors) are related by a two-fold rotational axis, indicated by arrowed black lines. (right) cryoEM characterization of the DNA cubes. (a) a raw cryo-EM image of the DNA sample; (b) close-up view of several representative raw particles from (a) (upper row) and their corresponding computer-generated model projections (lower row); (c) the 3D reconstruction maps of the DNA cube, reconstructed by imposing a tetrahedron symmetry. (Replicated from ref. 30).

is designed to be 4.5 turns long (Fig. 4, left). The odd number of DNA helical turns leads to any two adjacent tiles to face to different sides of the DNA tile plane. Thus, any closed rings must contain an even number of tiles: half of the tiles face inward and the other half face outward. The smallest polyhedron that meets this requirement is a cube; which suggests that the 3-point-star motif at sufficiently low concentrations will assemble into cubes.

The DNA cubes were assembled at a DNA concentration of 50 nM. Most of the DNA tiles were incorporated into a large, well-defined, molecular complex, which appeared as a sharp band in polyacrylamide gel electrophoresis. The assembly yield was ~82% as estimated from the gel. DLS experiment indicated that the DNA complex had an apparent hydrodynamic radius of 16.0 ± 2.8 nm. This value agreed well with the radius of the circumscribed sphere of the DNA cube (radius: 16.5 nm).

Under cryoEM imaging (Fig. 4, right), most observed particles were consistent with the 2D projections of cubes at the expected size (the cube edge is ~15 nm long). With experimentally observed particles, a DNA cube structure at ~2.9 nm resolution was revealed by single particle 3D reconstruction in which tetrahedron symmetry was imposed. By comparing the 2D projections of the reconstructed cube model and the raw images with similar orientations, clear similarities confirmed that the self-assembled DNA complex indeed had a cube structure.

3. DNA nanocages assembled from 5-point-star motifs

The planar tiling theory predicts that objects with 5-fold rotational symmetry can not assemble into regular (2D) lattices to completely tile a plane. So in previous studies of 2D self-assembly, 5-point-star motifs are avoided. However, the obstacle for 2D crystal growth might be a benefit for 3D self-assembly. The 5-point-star motifs have been shown to be excellent building blocks for DNA 3D nanocages in

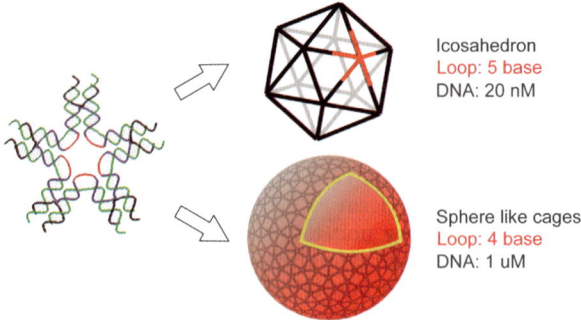

Fig. 5 Self-assembly of 5-point-star motifs. (upper) DNA icosahedra. Each vertex in the icosahedra is a 5-point-star tile; one of them is highlighted in gold. (lower) Large DNA nanocages with diameters around 250 nm.

our recent work.[29] Flexibility and concentration control can also be applied to the self-assembly of 5-point-star motifs (Fig. 5).

Icosahedra

An icosahedron is the complementary structure of a dodecahedron. They share the same icosahedral symmetry. An icosahedron contains 12 vertices, 20 faces and 30 edges, which is the smallest closed structure that can be assembled from 5-point-star motifs. The assembled DNA icosahedra require that each motif has 5-base-long central loops (red segment) to provide enough flexibility. Otherwise it can not be assembled even at extremely low concentrations. Such DNA tiles at a low concentration (20 nM) can readily assemble into icosahedra, which appear as a dominant band in polyacrylamide gel electrophoresis. DLS studies reveal that the DNA complex has an apparent hydrodynamic radius of 20.0 ± 2.8 nm, agreeing well with the radius of the circumscribed sphere of the DNA icosahedral model (19.5 nm). In AMF images, round particles with uniform size are the dominant species.

In CryoEM images (Fig. 6), most observed particles have diameters ~40 nm, nicely matching the diameter of the designed icosahedron (39.0 nm). With experimentally observed particles, a DNA icosahedral structure has been reconstructed at a resolution of 2.8 nm. Such a model is strongly supported by comparison between the computed projections from the model and the class averages of raw particle images with similar views. They match with each other very well. In reconstruction, an icosahedral symmetry is imposed. Though the same symmetry is imposed in reconstruction for both DNA dodecahedra and icosahedra, the final 3D maps are dramatically different. Such a contrast validates that the single particle 3D reconstruction method is reliable and the final structural difference is indeed due to their intrinsic structures. When lower symmetries are imposed during the reconstructions, we obtain similar 3D maps at poorer resolution.

Large cages

With increased DNA concentrations and reduced DNA tile flexibility (with 4-base-long central loops), large DNA nanocages can be assembled. Those cages are not uniform in sizes and shapes. For such large and non-uniform sized DNA complexes, the technique of 3D single particle reconstruction is not applicable. Thus we turn to AFM characterization (Fig. 7).

Besides the flexibility and concentration of the DNA motifs, the strength of the sticky-end association also influences the DNA self-assembly. The stronger the sticky ends cohesion is, the more likely the DNA motifs assemble into smaller closed

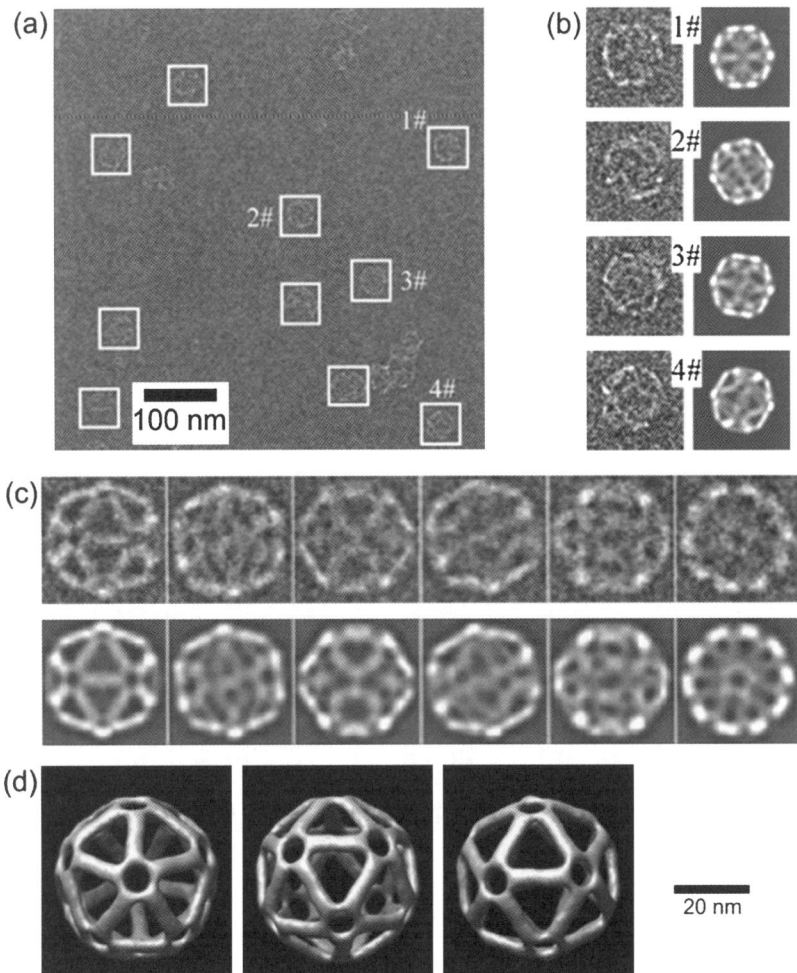

Fig. 6 Cryo-EM analysis of DNA icosahedra. (a) A representative raw cryo-EM image. White boxes indicate the DNA particles. (b) Comparison of raw images (left) of individual particles at a high magnification and the corresponding computer-generated model projections (right). (c) Comparison of class average of particle images with similar views (upper panel) and the corresponding computer-generated model projections (lower panel). (d) Three views of the DNA icosahedron structure reconstructed from cryo-EM images. (Replicated from ref. 29).

structures. It is presumably due to some kinetic traps in the assembly process. If the sticky end cohesion is strong enough to overcome the strain energy penalty of twisting neighboring DNA motifs, the DNA assemblies are prone to close themselves, thus preventing further lateral growth. In the 5-point-star tile, there are a pair of 4-base-long, single-stranded sticky ends locating at the end of each arm. This pair of sticky ends is complementary to each other. The five identical pairs of sticky ends can readily associate among themselves. To vary the strength of the sticky-end association, we have varied the composition of the sticky-ends, from containing 4 GC pairs to containing only 1 GC pair. With the same length, DNA sticky-ends with higher GC contents will provide a stronger association than those with lower GC contents will.

Varying the strength of the sticky-end association brings great differences in the morphology of DNA assemblies. With 4-GC-basepair sticky-ends, the DNA

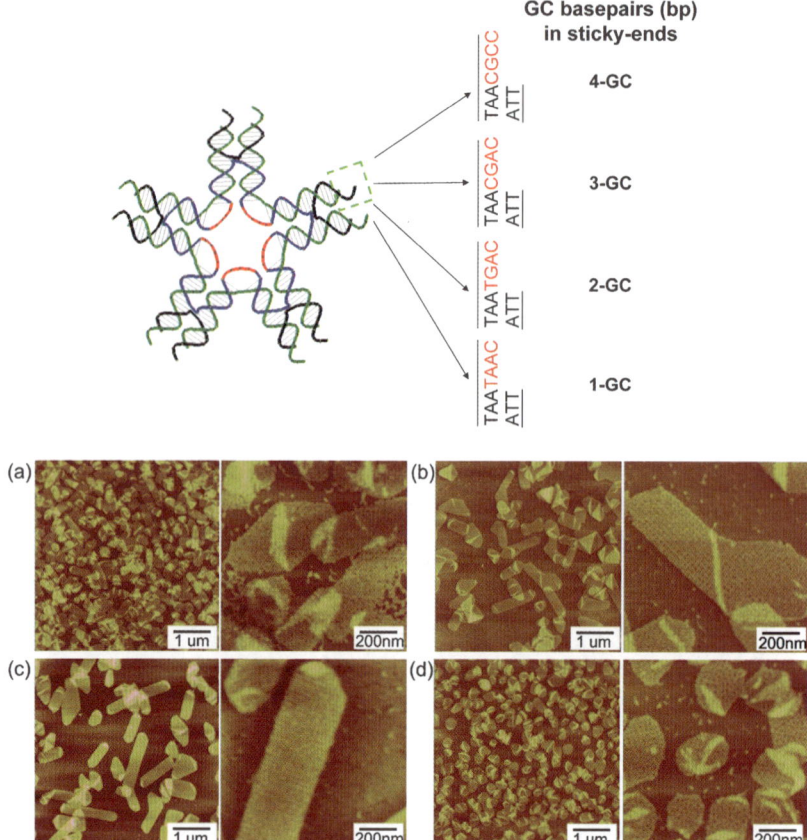

Fig. 7 AFM characterization of DNA nanocages assembled from 5-point-star tiles with different strengths of sticky-end association. (a) 1-GC-basepair sticky ends; (b) 2-GC-basepair sticky ends; (c) 3-GC-basepair sticky ends; (d) 4-GC-base pair sticky ends. AFM images are taken in air. Each pair of AFM images shows the DNA structures at two imaging scales.

nanocages are round spheres with diameters of 200–300 nm. When such DNA cages land onto AFM substrate (mica surfaces), they collapse into double-layer structures due to strong interactions between DNA and mica surfaces. After several successive AFM scans, the top layers are occasionally scratched out by the AFM probes and only the monolayers are left. For such monolayer samples, the DNA assemblies show regular, periodic arrays, which have a tetragonal symmetry. With 3-GC-basepair sticky-ends, most DNA assemblies are capped nanotube-like structures, indicating an anisotropic growth during the self-assembly process. The origin of the anisotropy is unclear. One speculation is that different bending orientations of the five arms break the growth symmetry. The complexes with 2-GC-basepair sticky-ends show similar morphologies to the complexes with 3-GC-basepair sticky-ends, but with a shorter tube length. With 1-GC-basepair sticky-ends, the DNA complexes are not as stable as others because of the weak association. After washing, cracks can be found in the nanocages under AFM imaging.

The strength of sticky-end association clearly influences the stabilities of the DNA nanocages as well. The DNA nanocages are stable at low temperatures and disassociate at high temperatures. This process can be monitored by DLS measurement (Fig. 8). The hydrodynamic radii of the DNA nanocages are much larger than those

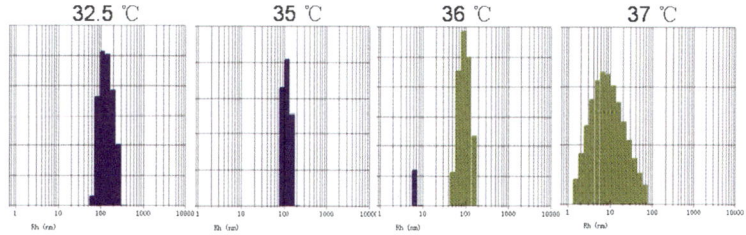

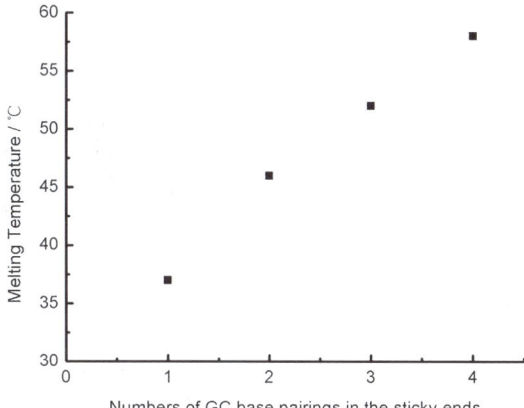

Fig. 8 DLS monitoring of the dissociation of DNA nanocages. (top) The DLS data at different temperatures for DNA cages assembled from 5-point-star tiles with 1-GC sticky-ends. (bottom) The melting temperatures of DNA nanocages assembled from different 5-point-star tiles determined by DLS. DNA concentration: 1 μM.

of the individual tiles. As shown in Fig. 8, higher GC contents in the sticky-ends results in more stable DNA nanocages and higher melting temperatures.

4. DNA nanocages assembled from 6-point-star motifs

A 6-point-star motif is the most branched star motif investigated so far. It can associate into nanocages with sphere-like or irregular shapes. Their sizes range from several hundred nanometers to ~2 μm as determined by AFM imaging (Fig. 9).

The sticky-end effect exists for the self-assembly of 6-point-star tiles as that of 5-point-star tiles. Under the same assembly conditions, the 3-GC tiles assemble into stable nanocages that collapse into double-layer discs (less than 2 μm wide) upon deposition onto mica surfaces. For 2-GC tiles, less cage structures are found. Instead, large 2D monolayers (>10 μm) can be found under AFM imaging. When further reducing the GC content to 1-GC tiles, mica surfaces are fully covered by small monolayer 2D arrays with domain sizes typically less then 2 μm.

5. Conclusions

In this paper, we have discussed all the DNA nanocages derived from DNA star tiles in our recent research. The DNA cages have sizes ranging from 20 nm–2 μm. In such 3D self-assembling processes, balancing the rigidity and the flexibility of the DNA tiles is the key idea to make nanostructures in the desired size. The DNA

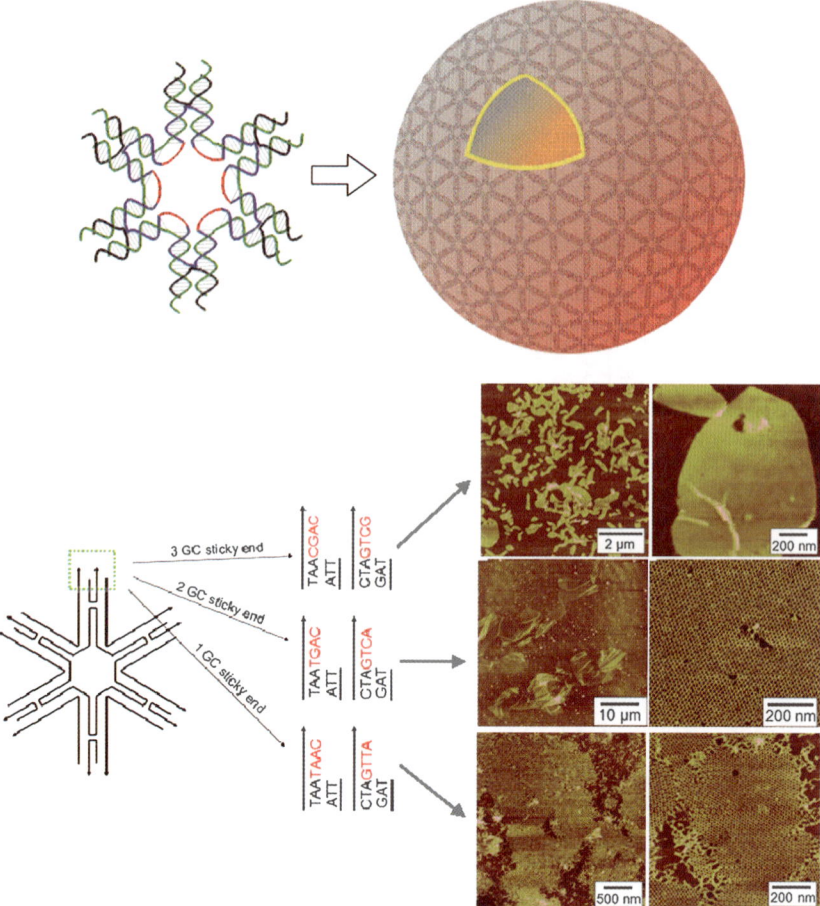

Fig. 9 Self-assembly of DNA nanocages from 6-point-star motifs. (top) Scheme of the DNA self-assembly. (bottom) AFM images of the DNA nanocages with different association strengths.

concentration is another important factor that influences the assembly kinetics and sometimes determines the final morphologies of the assemblies.

Future research might include several aspects:

(1) To organize nanoparticles or proteins in 3D space using DNA polyhedra as templates;

(2) To encapsulate drugs in the DNA cages for delivery or capsulate biomacromolecules for single particle studies;

(3) To develop a new strategy to fabricate complicated 3D structures and develop responsive DNA cages;

(4) To quantitatively study 3D DNA self-assembly in terms of both thermodynamics and kinetics.

Acknowledgements

This work was supported by the National Science Foundation (CCF-0622093). DLS studies were carried out in the Purdue Laboratory for Chemical Nanotechnology (PLCN). The cryo-EM images were taken in the Purdue Biological Electron

Microscopy Facility and the Purdue Rosen Center for Advanced Computing (RCAC) provided the computational resource for the 3D reconstructions.

References

1 N. C. Seeman, *Nature*, 2003, **421**, 427–431.
2 C. Lin, Y. Liu, S. Rinker and H. Yan, *ChemPhysChem*, 2006, **7**, 1641–1647.
3 F. A. Aldaye, A. L. Palmer and H. F. Sleiman, *Science*, 2008, **321**, 1795–1799.
4 E. Winfree, F. Liu, L. A. Wenzler and N. C. Seeman, *Nature*, 1998, **394**, 539–544.
5 T. H. LaBean, H. Yan, J. Kopatsch, F. Liu, E. Winfree, J. H. Reif and N. C. Seeman, *J. Am. Chem. Soc.*, 2000, **122**, 1848–1860.
6 C. Mao, W. Sun and N. C. Seeman, *J. Am. Chem. Soc.*, 1999, **121**, 5437–5443.
7 H. Yan, S. H. Park, G. Finkelstein, J. H. Reif and T. H. LaBean, *Science*, 2003, **301**, 1882–1884.
8 D. Liu, M. Wang, Z. Deng, R. Walulu and C. Mao, *J. Am. Chem. Soc.*, 2004, **126**, 2324–2325.
9 B. Ding, R. Sha and N. C. Seeman, *J. Am. Chem. Soc.*, 2004, **126**, 10230–10231.
10 P. W. K. Rothemund, N. Papadakis and E. Winfree, *PLoS Biol.*, 2004, **2**, 2041–2053.
11 J. Malo, J. C. Mitchell, C. Venien-Bryan, J. R. Harris, H. Wille, D. J. Sherratt and A. J. Turberfield, *Angew. Chem., Int. Ed.*, 2005, **44**, 3057–3061.
12 F. Mathieu, S. Liao, J. Kopatsch, T. Wang, C. Mao and N. C. Seeman, *Nano Lett.*, 2005, **5**, 661–665.
13 A. Chworos, I. Severcan, A. Y. Koyfman, P. Weinkam, E. Oroudjev, H. G. Hansma and L. Jaeger, *Science*, 2004, **306**, 2068–2072.
14 N. Chelyapov, Y. Brun, M. Gopalkrishnan, D. Reishus, B. Shaw and L. Adleman, *J. Am. Chem. Soc.*, 2004, **126**, 13924–13925.
15 Y. He, Y. Tian, A. E. Ribbe and C. Mao, *J. Am. Chem. Soc.*, 2006, **128**, 15978–15979.
16 P. W. K. Rothemund, *Nature*, 2006, **440**, 297–302.
17 R. Jungmann, T. Liedl, T. L. Sobey, W. Shih and F. C. Simmel, *J. Am. Chem. Soc.*, 2008, **130**, 10062–10063.
18 S. H. Park, G. Finkelstein and T. H. LaBean, *J. Am. Chem. Soc.*, 2008, **130**, 40–41.
19 J. H. Chen and N. C. Seeman, *Nature*, 1991, **350**, 631–633.
20 Y. W. Zhang and N. C. Seeman, *J. Am. Chem. Soc.*, 1994, **116**, 1661–1669.
21 W. M. Shih, J. D. Quispe and G. F. Joyce, *Nature*, 2004, **427**, 618–621.
22 R. P. Goodman, I. A. T. Schaap, C. F. Tardin, C. M. Erben, R. M. Berry, C. F. Schmidt and A. J. Turberfield, *Science*, 2005, **310**, 1661–1665.
23 F. A. Aldaye and H. F. Sleiman, *J. Am. Chem. Soc.*, 2007, **129**, 13376–13377.
24 N. Jonoska and R. Twarock, *J. Phys. A: Math. Theor.*, 2008, **41**, 304043.
25 J. Zimmermann, M. P. J. Cebulla, S. Monninghoff and G. von Kiedrowski, *Angew. Chem., Int. Ed.*, 2008, **47**, 3626–3630.
26 F. F. Andersen, B. Knudsen, C. L. P. Oliveira, R. F. Frøhlich, D. Krüger, J. Bungert, M. Agbandje-McKenna, R. McKenna, S. Juul, C. Veigaard, J. Koch, J. L. Rubinstein, B. Guldbrandtsen, M. S. Hede, G. Karlsson, A. H. Andersen, J. S. Pedersen and B. R. Knudsen, *Nucleic Acids Res.*, 2008, **36**, 1113–1119.
27 D. Bhatia, S. Mehtab, R. Krishnan, S. S. Indi, A. Basu and Y. Krishnan, *Angew. Chem., Int. Ed.*, 2009, **48**, 4134–4137.
28 Y. He, T. Ye, M. Su, C. Zhang, A. E. Ribbe, W. Jiang and C. Mao, *Nature*, 2008, **452**, 198–201.
29 C. Zhang, M. Su, Y. He, X. Zhao, P. Fang, A. E. Ribbe, W. Jiang and C. Mao, *Proc. Natl. Acad. Sci. U. S. A.*, 2008, **105**, 10665–10669.
30 C. Zhang, S. H. Ko, M. Su, Y. Leng, A. E. Ribbe, W. Jiang and C. Mao, *J. Am. Chem. Soc.*, 2009, **131**, 1413–1415.

PAPER | www.rsc.org/faraday_d | Faraday Discussions

Self-assembly of liquid crystal block copolymer PEG-*b*-smectic polymer in pure state and in dilute aqueous solution

Bing Xu,[†][a] Rafael Piñol,[‡][a] Merveille Nono-Djamen,[§][ab] Sandrine Pensec,[b] Patrick Keller,[a] Pierre-Antoine Albouy,[c] Daniel Lévy[a] and Min-Hui Li[*][a]

Received 30th January 2009, Accepted 23rd March 2009
First published as an Advance Article on the web 1st July 2009
DOI: 10.1039/b902003a

A series of amphiphilic LC block copolymers, in which the hydrophobic block is a smectic polymer poly(4-methoxyphenyl 4-(6-acryloyloxy-hexyloxy)-benzoate) (PA6ester1) and the hydrophilic block is polyethyleneglycol (PEG), were synthesized and characterized. The self-assembly of one of them in both the pure state and the dilute aqueous solution was investigated in detail. Nano-structures in the pure state were studied by SAXS and WAXS on samples aligned by a magnetic field. A hexagonal cylindrical micro-segregation phase was observed with a lattice distance of 11.2 nm. The PEG blocks are in the cylinder, while the smectic polymer blocks form a matrix with layer spacing 2.4 nm and layer normal parallel to the long axis of the cylinders. Faceted unilamellar polymer vesicles, polymersomes, were formed in water, as revealed by cryo-TEM. In the lyotropic bilayer membrane of these polymersomes, the thermotropic smectic order in the hydrophobic block is clearly visible with layer normal parallel to the membrane surface.

1. Introduction

Nanostructures formed by block copolymers composed of immiscible blocks have been extensively studied theoretically and experimentally.[1,2] It is now well established that these block copolymers can phase separate into a variety of organized structures (lamellae, cylinder, double gyroid, spherical *etc.*) in the solid state,[3] and form a variety of micellar aggregates (spherical and cylindrical micelles, vesicles *etc.*) when dissolved in a solvent selective for one of the blocks.[4] These materials could find potential applications in various fields of nanotechnology, such as nano

[a]Institut Curie, Centre de Recherche, CNRS, UMR 168, Université Pierre et Marie Curie, 26 rue d'Ulm, 75248 Paris CEDEX 05, France. E-mail: min-hui.li@curie.fr; Fax: +33 1 4051 0636; Tel: +33 1 5624 676
[b]Université Pierre et Marie Curie, CNRS UMR7610, Laboratoire de Chimie des Polymères, Case 185, 4, Place Jussieu, 75252 Paris CEDEX 05, France
[c]Université Paris-Sud, CNRS UMR8502, Laboratoire de Physique des Solides, 91405 Orsay CEDEX, France

† Present address: Shanghai Record Pharmaceuticals Co. Ltd, 799 Dun-Huang Road, Shanghai 200331, China (E-mail: xubing97@hotmail.com).
‡ Present address: Instituto de Ciencia de Materiales de Aragón, Fisica de la Materia Condensada, Facultad de Ciencias, Universidad de Zaragoza-CSIC, Plaza San Francisco s/n, 50009-Zaragoza, Spain.
§ Present address: Laboratoire Polymères, Colloïdes, Interfaces, UMR CNRS 6120, Université du Maine, 1 Av Olivier Messiaën, 72085 Le Mans Cedex 9, France.

soft-lithography, nanoreactors and drug delivery systems.[5] These applications require, however, that these materials possess additional properties beyond those inherent to their intrinsic nanostructures. Perfect periodic nanodomain ordering is necessary, for example, for nano soft-lithography, while response to stimuli is desirable for nanoreactors and drug delivery systems.

In the past decade, electric fields,[6] temperature gradients,[7] directional solidification,[8] modification of substrate surfaces,[9] shearing,[10] solvent evaporation,[11] and roll casting[12] have been explored to induce orientation and/or long-range order in the nanostructures of classical coil–coil block copolymers in the solid state. A perfect long-range pattern of such nanostructures by non-contacted and remote approaches still remains however a challenge. Recent progress in synthetic strategies simplifies the molecular design of block copolymers. Block copolymers with a variety of chemical compositions and architectures can be prepared by synthetic tools developed in macromolecular chemistry, including living and controlled polymerization techniques. One of the ideas is to develop block copolymers which possess intrinsic responses to external remote triggers. These responses should help achieve nanostructure control. Liquid crystalline (LC) rod–coil block copolymers are good candidates because they respond to multiple stimuli including electric and magnetic fields, temperature and light (if the mesogen is a chromophore). LC rod–coil block copolymers exhibit hierarchical order in the solid state: microphase segregation in the block copolymer may occur on the 10–100 nm length scale, while LC ordering is observed on the length scale of mesogen (typically 2–4 nm). These two levels of order, in the block copolymer and liquid crystalline mesophases, interplay with each other.[13–15] Non-contacted and remote approaches, such as magnetic fields[16–18] and light,[19,20] have recently been used to align the nanostructures of LC block copolymers.

When the coil block is hydrophilic, an amphiphilic LC rod–coil block copolymer results representing a new class of amphiphilic copolymers. These copolymers form interesting micellar aggregates in aqueous solution, including vesicles and nanofibers.[21,22] Much effort has been made recently to develop stimuli-responsive polymer micelles and vesicles[23,24] because of their potential applications in nanoreactors and drug delivery systems. The key challenge is the controlled release of the active substances in the micellar aggregates, both in space and time. Most of the developed systems make use of chemical stimuli, which require the addition of chemical reagents, and are not always compatible with the application environments. In contrast, LC block copolymers are a promising system for the development of responsive micelles and vesicles sensitive to non-contacted and remote physical stimuli. We have shown recently that the use of light-sensitive LC nematic block copolymers allows the creation of polymer vesicles which burst under UV illumination.[25]

As a part of our endeavour to develop well-defined and stimuli-responsive nanostructures made from LC block copolymers, we report in this paper the synthesis and characterization of a series of amphiphilic LC block copolymers, in which the hydrophobic block is a smectic polymer and the hydrophilic block is polyethyleneglycol (PEG). The self-assembly of one of the LC block copolymers in the pure state and in dilute aqueous solution was then investigated in detail. A well aligned cylindrical nanostructure was obtained in the solid state using a magnetic field to induce orientation in the LC block. In water, faceted unilamellar polymer vesicles (polymersomes) with smectic order in the lyotropic bilayer were formed.

2. Experimental

2.1. Materials and analytical methods

For monomer synthesis, 4-hydroxybenzoic acid, 6-chlorohexanol, KOH, KI, acrylic acid, *p*-toluenesulfonic acid, hydroquinone, 4-methoxyphenol, 4-pyrrolidinopyridine and dicyclohexylcarboxydiimide of analytical grade were purchased from Aldrich and used without further purification. Anhydrous CH_2Cl_2 and $CHCl_3$, as well as ethanol and isopropanol, were purchased from Carlo Erba-SDS. For the polymer synthesis, methoxy poly(ethylene)glycol (MPEG2000, M_n = 2000, DP = 45, from Fluka) was dried by azeotropic distillation with toluene before use, and traces of residual toluene were removed under vacuum. CuBr was purified by stirring with acetic acid for several hours, then filtering, washing with acetic acid, ethanol and diethyl ether in succession and storing under vacuum. 4,4'-Di(n-nonyl)-2,2'-bipyridine (from Aldrich) was recrystallized twice from ethanol. Other reagents (ethyl 2-bromo-2-methylpropionate, 2-bromopropionyl bromide and triethylamine from Aldrich) and solvents of analytical grade were used as received without further purification.

The molecular structures of all products were analyzed by ^{1}H-NMR using a Bruker HW300MHz spectrometer. Molecular weight distributions (M_w/M_n) of polymers were evaluated by size exclusion chromatography (SEC) calibrated with polystyrene standards.[26] For SEC, we used Waters Styragel HR5E columns and a Waters 410 differential refractometer with THF as eluent at a flow rate of 1.0 mL min^{-1} at 40 °C. Molecular weights of the LC homopolymer and the diblock copolymers were calculated from the NMR signals, using the following equations. For the LC homopolymer,

$$M_n = 195 + n \times 398$$

$$n = \frac{3}{8} \times \frac{I_{aryl}}{I_{1.22}}$$

where I_{aryl} denotes the integration of peaks of aryl hydrogens from 6.91 to 8.14 ppm and $I_{1.22}$ that of the methyl hydrogens in the chain end (see Scheme 1). n is the degree of polymerization (DP) of the LC polymer. The molecular weight of the monomer is 398 and that of the initiator 195. For block copolymers,

$$M_n = M_n \text{(MPEG-Br)} + n \times 398$$

$$n = \frac{3}{8} \times \frac{I_{aryl}}{I_{3.38}}$$

where $I_{3.38}$ denotes the integration of peaks of terminal methyl hydrogens in MPEG.

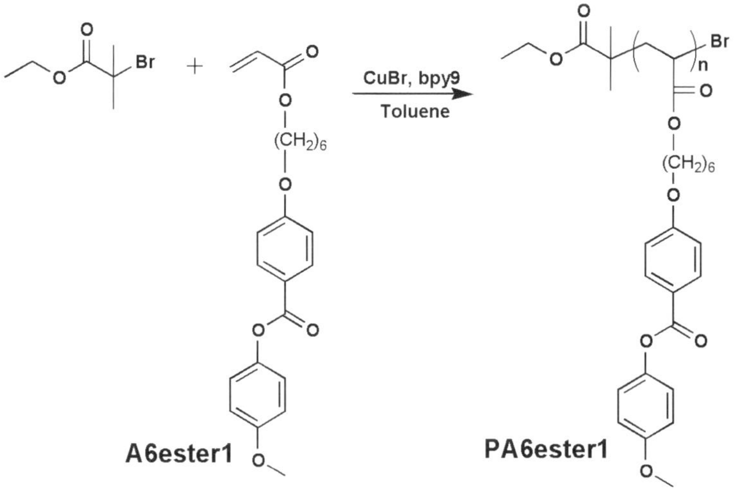

Scheme 1

2.2. Synthesis of the LC monomer and homopolymer

The LC acrylate monomer, 4'-methoxyphenyl 4-(6''-(acryloyloxy)hexyloxy)benzoate (A6ester1) was synthesized from 4-hydroxybenzoic acid by a three-step procedure as described in reference 27. The monomer A6ester1, purified by recrystallization from ethanol (5 times), was obtained as white crystals ready for polymerization. ^{1}H NMR (300 MHz, CDCl$_3$): δ 1.47–1.86 (m, 8H, –OCH$_2$(CH$_2$)$_4$CH$_2$O–), δ 3.82 (s, 3H, –OCH$_3$), δ 4.02–4.06 (t, 2H, –COOCH$_2$CH$_2$–), δ 4.16–4.20 (t, 2H, –CH$_2$CH$_2$OC$_6$H$_4$–), δ 5.80–5.83 (d, 1H, –CH=HCH), δ 6.07–6.17 (m, 1H, –CH=CH$_2$), δ 6.37–6.43 (d, 1H, –CH=HCH), δ 6.91–8.14 (m, 8 H arom.)

The homopolymer was synthesized by atom transfer radical polymerization (ATRP). A general ATRP procedure is as follows:

A Schlenk flask with a magnetic stir bar was charged with Cu^{I}Br (14.34 mg, 0.1 mmol), 4,4'-di(n-nonyl)-2,2'-bipyridine (bpy9, ligand) (81.76 mg, 0.2 mmol), ethyl 2-bromo-2-methylpropionate (I, initiator) (19.51 mg, 0.1 mmol) and monomer A6ester1 (M) (0.4776 g, 1.2 mmol). The flask was degassed by four vacuum-argon cycles. Toluene (1 mL), degassed by argon bubbling for 30 min, was then introduced into the flask using a syringe purged with argon. The flask was then immersed in an oil bath held at 80 °C by means of a thermostat. After reacting for 24 h, the mixture was cooled to room temperature. The resulting polymer solution was poured into a large volume of diethyl ether. The precipitated polymer was purified thrice by dissolution in a small amount of dichloromethane and precipitation into a large volume of diethyl ether. The purified polymer was dried under vacuum at room temperature for 3 days. Yield: 0.2g (42%). ^{1}H NMR (300 MHz, CDCl$_3$): δ 1.11–1.14 (6H, –C(CH$_3$)$_2$), δ 1.21 (t, 3H, –CH$_2$CH$_3$), δ 1.32–1.88 (8nH, –OCH$_2$(CH$_2$)$_4$CH$_2$O–), δ 2.33 (nH, –CH$_2$CH(COO)–), δ 3.80 (3nH–OCH$_3$), δ 3.91–4.18 (4nH + 2H, –COOCH$_2$CH$_2$–, –CH$_2$CH$_2$OC$_6$H$_4$– and –COOCH$_2$CH$_3$), δ 6.82–8.15 (8nH arom.) (n is DP of LC polymer calculated by ^{1}H NMR spectrum). M_n of PA6ester1 was 3300 (NMR) and $M_w/M_n = 1.17$ (SEC).

2.3. Synthesis of block copolymers PEG-b-smectic polymer

The synthesis of the macroinitiator MPEG2000-Br (I) was first carried out as previously described.[21] Yield 58%. ^{1}H NMR (300 MHz, CDCl$_3$): δ 1.81(d, $J = 6$Hz, 3H, BrCH(CH$_3$)–), 3.36(s, 3H, CH$_3$OCH$_2$CH$_2$O–), 3.51~3.87(m, 181H, –OCH$_2$CH$_2$O–), 4.30(t, $J = 3$Hz, 2H, –CH$_2$CH$_2$OCO–), 4.38(q, $J = 9$Hz, 1H, BrCH(CH$_3$)–).

The block copolymers PEG-b-PA6ester1 were synthesized by ATRP by the same procedure and conditions used for homopolymer PA6ester1, except that the macroinitiator MPEG2000-Br was used here instead of ethyl 2-bromo-2-methylpropionate. The molar amount of the initiator was kept to be [I] = 0.1 mmol, and the molar ratio [I]/[bpy9]/[CuBr] = 1/2/1. The monomer molar ratio was varied as [M]/[I] = 5/1, 10/1, 15/1, 20/1 and 30/1, in order to obtain different lengths for the LC block (see Table 1). For a typical synthesis of copolymer CP3 (Table 1), 597.7 mg monomer A6ester1 (1.5 mmol) and 213.5 mg MPEG2000-Br (0.1 mmol) were polymerized according to the procedure described above, yielding 578.7 mg diblock copolymer (71.4%). ^{1}H NMR (300 MHz, CDCl$_3$): δ 1.29–2.02 (8n H, –CH$_2$(CH$_2$)$_4$CH$_2$O–), δ 2.31 (n H, –CH$_2$CH(COO)–), δ 3.37 (3H, –OCH$_3$), δ 3.57–3.57 (181 H–OCH$_2$CH$_2$O–), δ 3.77 (3n H –OCH$_3$), δ 3.89–4.15 (4nH + 2H, –COOCH$_2$CH$_2$–, –CH$_2$CH$_2$OC$_6$H$_4$– and –COOCH$_2$CH$_3$), δ 6.79–8.13 (8n H, arom.) (n is the DP of the LC block determined by the ^{1}H NMR spectrum) M_n(CP3) = 7900 (NMR) and $M_w/M_n = 1.13$ (SEC).

2.4. Physical characterization

The mesomorphic properties of the diblock copolymers in bulk were studied by thermal polarizing optical microscopy (POM) using a Leitz Ortholux microscope equipped with a Mettler FP82 hot stage, and differential scanning calorimetry using a Perkin-Elmer DSC7.

Table 1 Molecular characterization of the PA6ester1 and PEG-b-PA6ester1 series (CP1–CP5)

Polymer	[M/[I]^a	M_n (Da) (NMR)	M_w/M_n (SEC)	Hydrophilic/Hydrophobic weight ratio (NMR)	DP of PA6ester1 block$n$ (NMR)
PA6ester1	12/1	3300	1.17	—	8
CP1	5/1	4000	1.07	50/50	5
CP2	10/1	6000	1.09	33/67	10
CP3	15/1	7900	1.13	25/75	15
CP4	20/1	10000	1.20	20/80	20
CP5	30/1	14200	1.32	14/86	31

^{a} Molar concentration ratio of monomer to ATRP initiator. Reaction temperature: 80 °C, reaction time: 24h.

The self-assembled phases of the diblock copolymers in solid state were studied by X-ray scattering using CuKα radiation ($\lambda = 1.54$ Å) from a 1.5 kW rotating anode generator. The diffraction patterns were recorded on photosensitive imaging plates. The experiments were performed on fibre samples drawn from molten polymer for LC homopolymer and on samples contained in glass capillaries (1 mm diameter) for block copolymers. WAXS and SAXS were performed in two separate apparati. WAXS experiments examined the wave vector domain ($q = 4\pi\sin\theta/\lambda$) from 1.83–32 nm^{-1} and SAXS experiments probed the range 0.36–2.96 nm^{-1}. For the diblock copolymers, samples were first aligned by slowly cooling (0.1 °C min^{-1}) from the isotropic phase ($T = 85$ °C) to the smectic phase ($T = 40$ °C) in a magnetic field of 1.7 T. The magnetic field (**H**) was perpendicular to the long axis of the capillary containing the block copolymers. WAXS experiments on aligned samples were then made *in-situ* in the smectic phase (40 °C), followed by SAXS experiments on the same sample cooled to room temperature. The X-ray was transverse to the plane formed by the magnetic field (**H**) and the capillary axis.

The preparation of polymer vesicles and the turbidity measurements were performed according to published procedures.[21] The diblock copolymers were first dissolved in dioxane, which is a good solvent for both polymer blocks, at a concentration of 1.0 wt%. Deionized water was then added very slowly to the solution (2–3 μL of water per minute to 1 mL of polymer solution) under slight shaking. After each addition of water, the solution was left to equilibrate for 10 or more minutes until the optical density was stable. The optical density (turbidity) was measured at a wavelength of 650 nm using a quartz cell (path length: 2 cm) with a Unicam UV/Vis spectrophotometer. The cycle of water addition, equilibration and turbidity measurement was continued until the turbidity reaches a plateau. The solution was then dialyzed against water for 3 days to remove dioxane using a Spectra/Por regenerated cellulose membrane with a molecular weight cut-off of 3500. The morphological analysis of the turbid polymer solutions was performed by cryo-TEM on samples flash frozen in liquid ethane. Images were recorded in low dose conditions using a Philips CM120 electron microscope equipped with a Gatan SSC 1K × 1K CCD camera. The calibration of the microscope was performed with a purple membrane leading to 0.386 nm pixel^{-1} at 45 K magnification.

3. Results and discussion

3.1. Synthesis

LC homopolymer and block copolymers were prepared by ATRP as detailed in the Experimental section (see Scheme 1 and Scheme 2). The same hydrophilic macroinitiator (MPEG2000-Br) was used to synthesize a series of block copolymers. By varying the monomer to macroinitiator molar ratio [M]/[I] from 5/1 to 30/1, one obtains

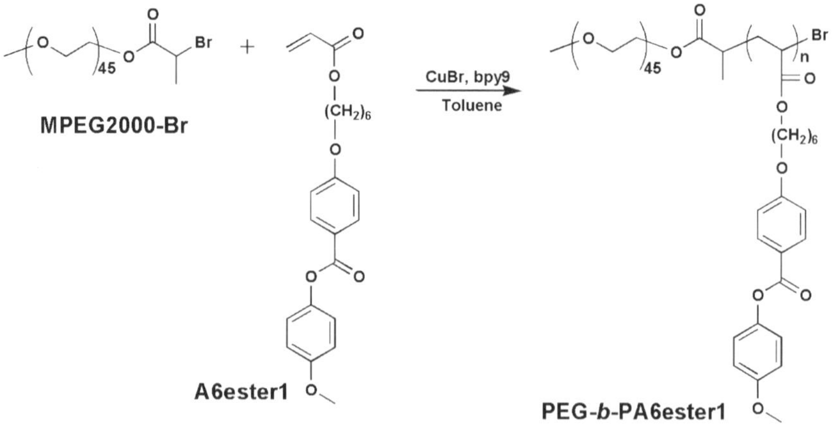

Scheme 2

LC hydrophobic blocks with different DP. Block copolymers with hydrophilic/hydrophobic weight ratios from 50/50 to 14/86 were synthesized. The SEC measurement showed narrow molecular distributions for all polymer samples (M_w/M_n = 1.07–1.32). In Table 1 we summarize the molecular characteristics of the LC homopolymer and block copolymers analysed by ^{1}H NMR and SEC. In this paper we will focus on the self-assembling properties of one of the block copolymers, PEG$_{45}$-b-PA6ester1$_{20}$ (CP4, hydrophilic/hydrophobic weight ratio of 20/80) in both the solid state and dilute aqueous solution.

3.2. Mesomorphic properties

Before discussion of the self-assembly of block copolymers, we examine the individual mesomorphic properties of the LC monomer, homopolymer and copolymer CP4 respectively. The LC monomer A6ester1 has a monotropic nematic (N) phase with the phase sequence: Cr-61.9°C-I on heating, and I-47.7°C-N 35.4°C-Cr on cooling, as observed optically with a temperature change rate of 5 °C min^{-1} (Cr represents the crystalline phase and I the isotropic phase). Typical Schlieren-type textures were observed in the nematic phase by POM. Once polymerized, the nematic A6ester1 gave a homopolymer (PA6ester1) with a richer polymorphism. PA6ester1 has a nematic phase at high temperature and a smectic phase at low temperature, in agreement with results published previously for the homopolymer with the same monomer structure, but much higher molecular weight.[27] Fig. 1a shows nematic Schlieren-type textures taken at T = 84.8 °C and Fig. 1b shows fan-shape textures, typical for a smectic A (SmA) phase, taken at T = 75.0 °C. The mesomorphic properties of PA6ester1, as determined by DSC, are shown in Fig. 2 and Table 2. The SmA phase is further confirmed by SAXS. Fig. 3 shows the SAXS pattern obtained on an aligned fiber sample of PA6ester1. The signals along the meridian give a period p = 2.52 nm for this smectic A phase. The extended molecular length of the side group is estimated to be 2.35–2.45 nm. A one-layer antiparallel packing is therefore the most probable arrangement for the SmA mesophase.[28]

For the block copolymer sample CP4 the nematic and smectic mesophases seem to be preserved in the LC block, as suggested by the DSC curves where the same number of peaks were detected for CP4 and PA6ester1 (Fig. 2). The transition domains broaden and transition temperatures drop because of the presence of the amorphous PEG block.[21,26] The POM textures are Schlieren-type between 74 and 60 °C when cooling from the isotropic phase, indicating a nematic phase

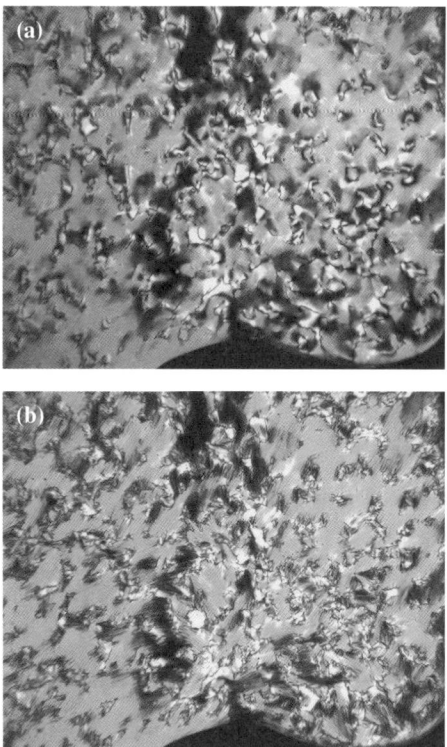

Fig. 1 Textures of the homopolymer PA6ester1 observed under a polarizing optical microscope. (a) nematic textures at $T = 84.8\ °C$. (b) SmA textures at $T = 75.0\ °C$.

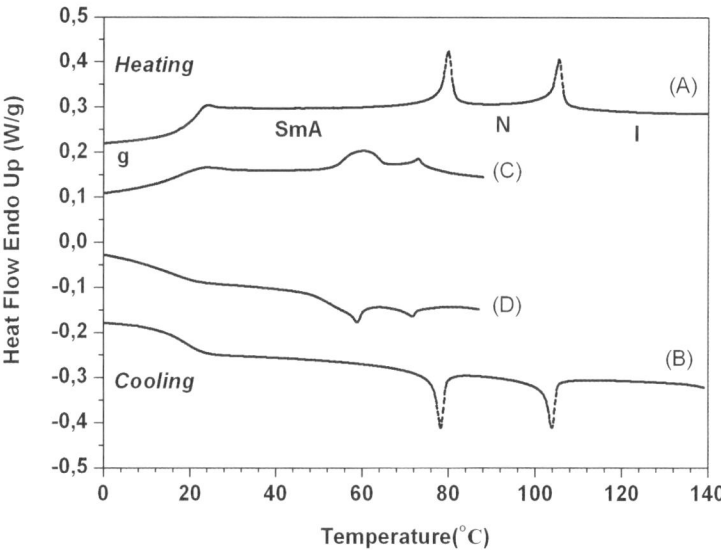

Fig. 2 DSC thermograms taken at 10 °C min^{-1}: (A) and (B) the homopolymer PA6ester1 on heating and on cooling, (C) and (D) the block copolymer CP4 on heating and on cooling. The curves correspond to the first heating and cooling scans for samples without any thermal treatment.

Table 2 Transition temperatures taken as the peak temperatures in DSC thermograms (°C) and enthalpies (in brackets) (J g^{-1}) obtained by DSC analysis on heating at 10 °C min^{-1}

Sample	Glass (g)	Smectic A (SmA)	Nematic (N)	Isotropic (I)
PA6ester1	•[a]	20 •	79.9 (2.1) •	105.4 (1.6) •
CP4	•	15 •	60.8 (1.5) •	73.9 (0.3) •

[a] The • symbol means the phase exists.

in this temperature range. At lower temperatures the fan-type textures are observed only after a long annealing time. The final smectic phase assignment at $T < 60$ °C is made by WAXS on samples aligned by a magnetic field.

The driving force for the magnetic field alignment of mesophases composed of typical aromatic LC mesogens is the collective anisotropy of the mesophase diamagnetic susceptibility, $\chi_\alpha = \chi_\parallel - \chi_\perp$, with respect to the direction of the long axis of the LC mesogen (director n).[29] This anisotropy causes a free energy difference between the state of randomly oriented polydomains and the state of parallel- or perpendicular-oriented monodomains. For typical aromatic LC mesogens this difference is sufficiently large compared to typical thermal energy, kT, that the system forms a monodomain mesophase as prescribed by the diamagnetic anisotropy. The mesophase with positive anisotropy ($\chi_\alpha > 0$) will have the director n aligned parallel to the field direction H, while the mesophase with negative anisotropy ($\chi_\alpha < 0$) will have the director perpendicular to H. In our system the smectic A phase of PA6ester1 block has a positive diamagnetic anisotropy, and so the mesogens' long axes are oriented parallel to the magnetic field H.

Fig. 4a shows the WAXS pattern obtained at $T = 40$ °C on a CP4 sample aligned by a magnetic field. A typical aligned SmA structure is obtained. Small-angle Bragg reflections (I) are due to the smectic layers, and wide-angle reflections (II) are associated with molecular arrangements of mesogenic side groups within the smectic layers. Reflection at wide-angles (II) is preferentially located near the equator and indicates the liquid-like arrangement of the mesogens aligned macroscopically in a direction parallel to the magnetic field. The order parameter is estimated to be $S = 0.74$, through the angular intensity profile of wide angle reflection according

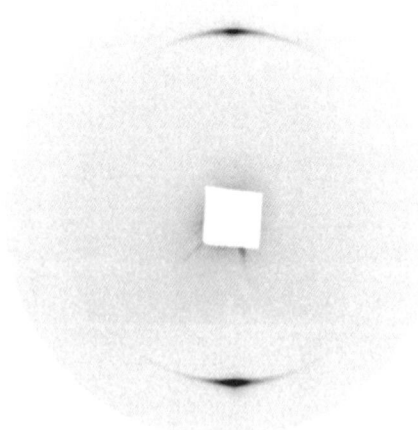

Fig. 3 SAXS pattern of the homopolymer PA6ester1 at room temperature in the glassy SmA phase. The fibre sample was drawn from melted polymer, the long axis of the fibre being along the vertical direction. The smectic period is measured as $p = 2.52$ nm.

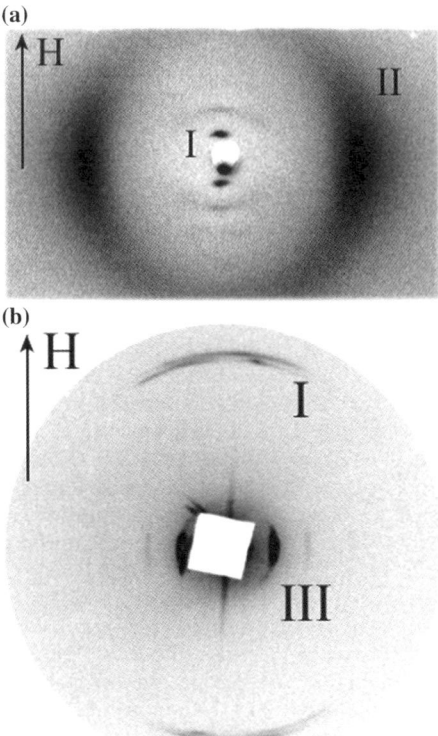

Fig. 4 X-Ray diffraction patterns of the block copolymer CP4. (a) WAXS pattern taken at 40 °C *in-situ* after cooling from isotropic phase (85 °C) at 0.1 °C min⁻¹ in a magnetic field of 1.7 T. (b) SAXS pattern taken at room temperature for the same sample in the absence of magnetic field. *H* is the direction of the applied magnetic field.

to the reported method.[30] The average distance between side groups is estimated to be 0.43 nm. Three orders of Bragg reflections (I) are visible along the meridian (parallel to the magnetic field). The smectic layer spacing associated with the Bragg reflections is $p = 2.37$ nm. This value is very close to that measured for the homopolymer ($p = 2.52$ nm).

In conclusion, even though the LC monomer is a monotropic nematic molecule, the LC homopolymer PA6ester1 possesses both nematic and smectic phases with the phase sequence: g-20 °C-SmA-79.9 °C-N-105.4 °C- I. The block copolymer PEG$_{45}$-*b*-PA6ester1$_{20}$ (CP4) preserves the nematic and smectic phases in its LC block, with the phase sequence: g-15 °C-SmA-60.8 °C-N-73.9 °C-I. The SmA phases in the homopolymer and in the block copolymer have very similar periods: p(PA6ester1) = 2.52 nm and p(CP4) = 2.37 nm. One-layer of anti-parallel packing is found for the SmA structure. Besides this basic molecular organization related to the mesomorphic properties of the LC block, what other type of self-assembled nanostructures, reflecting the incompatibility between the PEG block and the LC block, are formed in this copolymer? In the following section we will examine this issue using SAXS at larger length scales.

3.3. Self-assembly of PEG$_{45}$-*b*-PA6ester1$_{20}$ (CP4) in the pure state

The sample aligned in a magnetic field was then studied by SAXS in another apparatus at room temperature in the absence of any magnetic field. The diffraction pattern is shown in Fig. 4b. Bragg reflections (I) correspond to the first order of reflections (I) (Fig. 4a) in a WAXS experiment and are due to smectic layers. The smaller angle reflections (III) are associated with the nanostructures of the microphase segregation of block copolymers. The smectic layer spacing by reflections (I) is $p = 2.4 \pm 0.1$ nm, in good agreement with that found in WAXS. The signals (I) here are less oriented along the meridian than those for $T = 40\,^\circ$C in a magnetic field. The absence of the magnetic field causes partial loss of the smectic alignment. Nevertheless, the block copolymer nanostructure is well aligned at room temperature as shown by the signals (III). Up to three orders of reflections are visible along the equator in Fig. 4b. These orders are associated with distances $d_1 = 9.73$ nm and $d_1/d_2/d_4 = 1/3^{1/2}/7^{1/2}$, which correspond to the 1st, 2nd and 4th orders of a hexagonal phase. The hexagonal lattice parameter is $a_{\text{hex}} = 2d_1/3^{1/2} = 11.24$ nm. When the copolymer composition is considered, the hexagonal structure must comprise cylinders of PEG and a matrix of smectic polymer. The orthogonal orientation of reflection signals (I) from the smectic layers with respect to those (III) from the hexagonally packed cylinder planes indicates that the mesogens are parallel to the intermaterial dividing surface (IMDS) between the cylindrical nanodomains and the matrix. We infer then a homogeneous (planar) anchoring of mesogens at the IMDS. In Fig. 5 we show a schematic representation of the aligned self-assembled nanostructure of CP4 in a magnetic field. The degree of parallelism of the cylinders' long axes related to the magnetic field is $10 \pm 1^\circ$, which was evaluated from the angular profile of the diffraction signals III (*i.e.* their HWHM: half width at half maximum in Fig. 4b).

The anchoring state (homogeneous or homeotropic) of the mesogens at IMDS and the sign of the diamagnetic anisotropy of the mesophase χ_a, discussed in the preceding section, are two crucial parameters controlling the orientation of the copolymer nanostructures. These two parameters themselves are dictated by the chemical structure of the mesogens and the architecture of the block copolymers. The end-on side-chain block copolymer CP4 studied here combines positive diamagnetic anisotropy with homogeneous anchoring. The PEG cylinders consequently orient parallel to the field lines (1.7 T), yielding a monodomain with uniaxial symmetry, as shown in Fig.5.

Hamley *et al.*[14] reported an end-on side-chain LC block copolymer (Scheme 3) where the smectic mesophase also showed a positive χ_a, but with the difference that the mesogens had homeotropic anchoring at the IMDS of polystyrene (PS) cylinders. The application of a magnetic field (1.8 T) failed to align the morphology of the PS cylinders. This was ascribed to the nucleation of defects around the PS nano-cylinders in the LC matrix. In contrast Thomas *et al.*[16] described another end-on side-chain block copolymer (Scheme 4) with a negative χ_a for mesophase

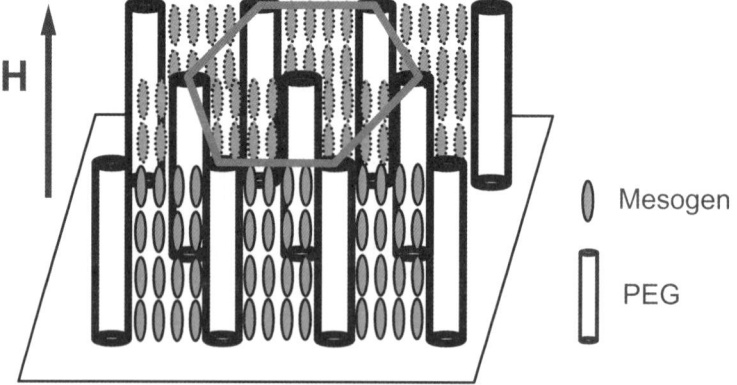

Fig. 5 Schematic representation of the self-assembled nanostrucure of CP4 aligned in a magnetic field. The hexagonal lattice parameter is $a_{\text{hex}} = 11.2$ nm, and smectic period is $p = 2.4$ nm.

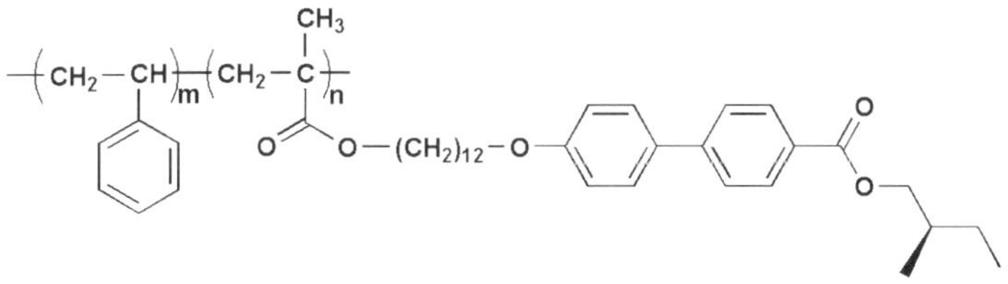

Scheme 3

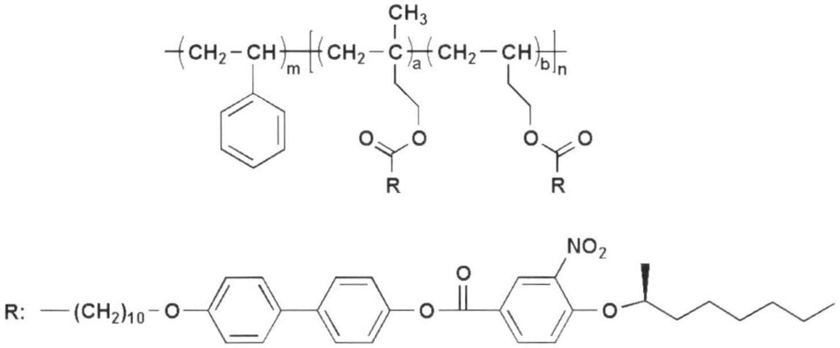

Scheme 4

and homogeneous anchoring of mesogens at the IMDS of polystyrene (PS) cylinders were found. A rather high magnetic field (9 T) still failed to produce long range order in the hexagonal cylindrical phase. As a matter of fact, the combination of homogeneous anchoring (mesogens parallel to the long axes of the cylinders) and the negative χ_a (mesogens aligned perpendicular to H) results in a degeneracy in the orientation of the cylinders with respect to the magnetic field lines. Clearly, there are an infinite number of in-plane orientations in which the long axes of the cylinders can lie perpendicular to the magnetic field.

In conclusion, our block copolymer with the combination of positive diamagnetic anisotropy and homogeneous anchoring at the IMDS offers a good system for the formation of a long range ordered hexagonal cylindrical nanostructure. In the present study a capillary of 1mm diameter was used as a container for the block copolymer sample. The magnetic field for alignment was perpendicular to the capillary's long axis. The circular surface geometry of the capillary is not very favourable for the perpendicular alignment of mesogens along the magnetic field. Nevertheless, the orientation of the cylindrical phase and the smectic structure in the LC matrix was still clearly observed even at room temperature. We believe the alignment will be improved further for plane geometry where the block copolymer is in the form of a thin film on a substrate. As shown in Fig. 5, PEG cylinders are perpendicular to the surface along the magnetic field. If we change the magnetic field such that it is parallel to the surface, we would expect the cylinders to reorient parallel to the surface. This would constitute a magnetic field-triggered alignment of a self-assembled nanostructure.

3.4. Self-assembly of PEG$_{45}$-b-PA6ester1$_{20}$ (CP4) in dilute aqueous solution

Not only are the two blocks, PEG and PA6ester1, of the block copolymer incompatible in the pure state, but they also have very different affinities for water. The PA6ester1 block is hydrophobic, while the PEG block is hydrophilic. Self-assembly studies of CP4 in dilute water solution at room temperature were performed with the aid of the co-solvent dioxane. Fig. 6 shows the turbidity diagram when water is added progressively to dioxane solutions of the copolymer. The jump

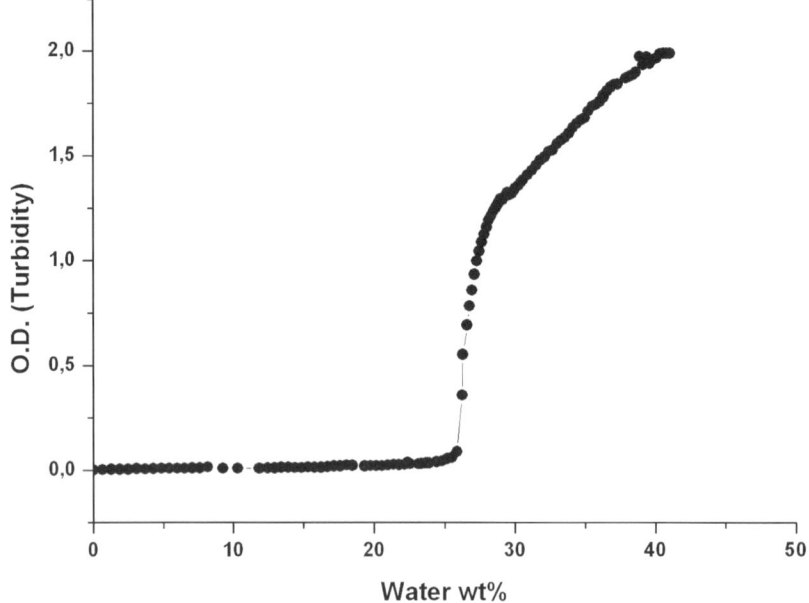

Fig. 6 Turbidity diagrams of 1.0 wt% CP4 in dioxane upon addition of water.

in turbidity upon the addition of water corresponds to the formation of particles which scatter light. The turbid mixtures at the end of the measurement (at around 40 wt% of water added) were dialysed against water to remove the dioxane and the particles suspended in pure water were then analysed by cryo-TEM.

Fig. 7 and Fig. 8 show typical cryo-TEM images of nanoparticles of CP4 embedded in ice. They are faceted vesicles with polygonal shapes, some of them exhibiting complex sharp-edge contours in several directions of the space (Fig. 7a). The vesicles are clearly unilamellar, with total sizes ranging from 100 to 800 nm. The measured thickness of the hydrophobic part of the membrane is of the order 10 nm. More interestingly, a careful analysis revealed a striped structure in some parts of the membrane (see Fig. 8 and the Fourier transform in the inset). The period measured is $p = 2.5 \pm 0.1$ nm, in good agreement with that of the

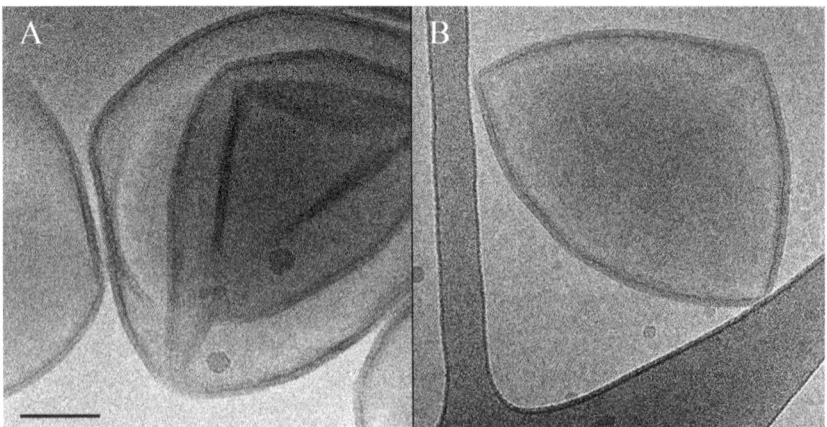

Fig. 7 Cryo-transmission electron micrographs of polymer vesicles of CP4 in water, showing complex sharp-edge contour (A) and triangle contour (B). Scale bar = 100 nm.

Fig. 8 Cryo-transmission electron micrographs of polymer vesicles of CP4. Scale bar = 100 nm. The inset at higher left is an enlargement of the upper left area of the vesicle. The inset at lower right is the Fourier transform of the image, diffraction spots corresponding to a period of 2.5 ± 0.1 nm.

SmA phase measured in the pure sample of homopolymer and block copolymer CP4. We can conclude that the SmA structure of the PA6ester1 block is preserved in some parts of the vesicle membrane. Fig. 9 shows a schematic representation of the smectic structure within a cross-section of the membrane where the stripes are visible.

Faceted vesicles have already been reported for small-molecular weight amphiphiles,[31–38] but not for polymer amphiphiles. It is known that planar bilayers in aqueous surroundings, or vesicle bilayers formed by small-molecular weight

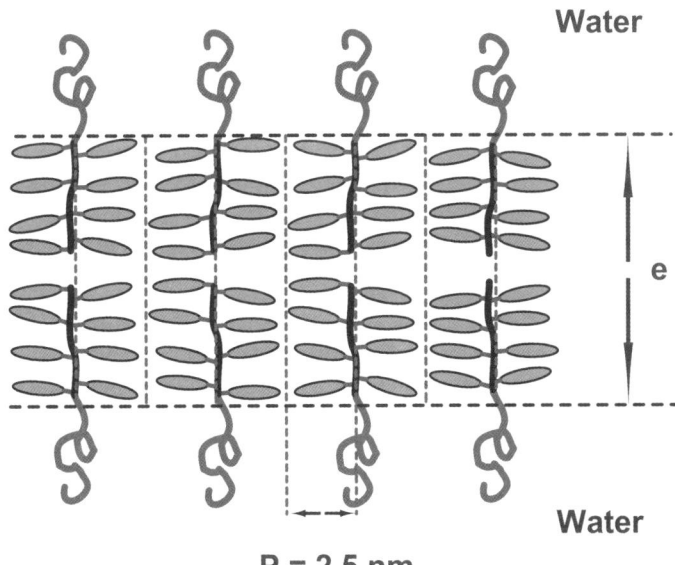

Fig. 9 Schematic representation of the smectic structure within a cross-section of the membrane of CP4 polymer vesicles. The mesogens are represented by small elongated ellipsoids.

amphiphiles can exhibit several thermotropic substates. Above a certain melting transition, T_m, the amphiphile has fluid disordered alkyl chains and the bilayers appear in the fluid L_α phase. Below this temperature, upon the crystallization of the chains, gel-like phases (such as $P_{\beta'}$, $L_{\beta'}$ or L_β) are observed in the bilayers. Consequently, vesicles typically exhibit spheroidal shapes above T_m, while below T_m non-spheroidal aggregates (*e.g.*, disks, planar fragments, lens-shaped vesicles, regular polygon-shaped vesicles, and irregularly faceted vesicles),[31–35] can be formed because of the curvature constraints imposed by chain crystallization. In the case of faceted vesicles due to chain crystallization, reheating restores the spheroidal shape. In two-component catanionic vesicles or polymer associated catanionic vesicles, segregation has also been shown to occur in the frozen state and be responsible for the observed polygonal-shaped or faceted vesicles.[36,37] Only spheroidal shapes, however, have been observed up to now for polymer vesicles. Most of the polymer amphiphiles reported are coil–coil block copolymers, such as poly(ethylene oxide)-*b*-polybutadiene (PEO-*b*-PBD) or polyacrylic acid-*b*-polystyrene (PAA-*b*-PS). Rod–coil block copolymers with nematic LC hydrophobic block were used to form polymer vesicles in our recent research.[21] We also observed spheroidal shaped vesicles for them. In the present work, the smectic LC structure of the hydrophobic block could be responsible for the faceted polymer vesicle shape. Work is in progress to study the vesicle shape evolution as a function of temperature (above the SmA-N transition and the N–I transition). Detailed investigations of the membrane structure and the topological defects of vesicles will also be carried out and described in a further paper.

4. Conclusions

In this paper we describe the synthesis and characterization of a series of amphiphilic LC block copolymers, in which the hydrophobic block is a smectic polymer poly(4-methoxyphenyl 4-(6-acryloyloxy-hexyloxy)-benzoate) (PA6ester1) and the hydrophilic block is polyethyleneglycol (PEG). The focus is on the self-assembly of one of the copolymers, PEG_{45}-*b*-$PA6ester1_{20}$ (CP4), in the pure state and in dilute aqueous solution.

In water, faceted unilamellar vesicles were formed by the copolymer CP4 as revealed by cryo-TEM. This is the first example of faceted shapes observed in polymer vesicles. In the lyotropic bilayer membrane, the thermotropic smectic order in the hydrophobic block is clearly visible in some places. The SmA layer normal is parallel to the membrane surface. Further investigations are necessary to elucidate the relationship between the smectic order in the membrane and the faceted shape of polymer vesicles.

Nano-structures in the pure state were studied by SAXS and WAXS. A hexagonal cylindrical micro-segregation phase was observed with a lattice distance of 11.2 nm. PEG blocks are located in cylinders, while the smectic polymer blocks constitute the matrix with a SmA layer spacing of 2.4 nm. The SmA phase has a positive diamagnetic anisotropy and the mesogens exhibit a homogeneous (planar) anchoring at the IMDS of PEG cylinders. The application of a magnetic field aligns the SmA layer normal and the PEG cylinders parallel to the field lines, yielding a monodomain with uniaxial symmetry. This is a good system for forming long-range-ordered hexagonal cylindrical nanostructures. By changing the magnetic field direction, a magnetic field-triggered change in orientation of the nanostructure may be possible. We believe this system reveals important clues for the design of block copolymers with applications in nanolithography, high-density information storage media and organic photovoltaics.

Acknowledgements

MHL thanks Jacques Prost, Mark Bowick and Qiang Fang for fruitful discussions. We thank Aurélie Di Cicco for performing the cryo-TEM images and Lin Jia for his help in drawing the figures. We acknowledge financial support from the CNRS, the Institut Curie and the Agence Nationale de la Recherche (ANR-08-BLAN-0209-01).

References

1. I. W. Hamley, *The Physics of Block Copolymers*, Oxford University Press, Oxford, 1998.
2. P. Alexandridis and B. Lindman, *Amphiphilic Block Copolymers: Self-Assembly and Applications*, Elsevier Science B.V., Amsterdam, 2000.
3. (a) L. Leibler, *Macromolecules*, 1980, **13**, 1602; (b) F. S. Bates, *Science*, 1991, **251**, 898; (c) G. H. Fredrickson and F. S. Bates, *Annu. Rev. Mater. Sci.*, 1996, **26**, 501.
4. I. W. Hamley, *Block Copolymers in Solution: Fundamentals and Applications*, John Wiley & Sons, Chichester, Sussex, 2005.
5. M. Lazzari, G. Liu, S. Lecommandoux, *Block Copolymers in Nanoscience*, Wiley-VCH Verlag GmbH, Weinheim, 2006.
6. (a) T. L. Morkved, M. Lu, A. M. Urbas, E. E. Ehrichs, H. M. Jaeger, P. Mansky and T. P. Russell, *Science*, 1996, **273**, 931; (b) T. Thurn-Albrecht, J. Schotter, G. A. Kästle, N. Emley, T. Shibauchi, L. K. Elbaum, K. Guarini, C. T. Black, M. T. Tuominen and T. P. Russell, *Science*, 2000, **290**, 2126.
7. J. Bodycomb, Y. Funaki, K. Kimishima and T. Hashimoto, *Macromolecules*, 1999, **32**, 2075.
8. C. D. Rosa, C. Park, E. L. Thomas and B. Lotz, *Nature*, 2000, **405**, 433.
9. (a) R. A. Segalman, H. Yokoyama and E. J. Kramer, *Adv. Mater.*, 2001, **13**, 1152; (b) J. Y. Cheng, C. A. Ross, E. L. Thomas, H. I. Smith and G. J. Vansco, *Appl. Phys. Lett.*, 2002, **81**, 3657; (c) J. Y. Cheng, A. M. Mayes and C. A. Ross, *Nat. Mater.*, 2004, **3**, 823; (d) S. O. Kim, H. H. Solak, M. P. Stoykovich, N. J. Ferrier, J. J. de Pablo and P. F. Nealey, *Nature*, 2003, **424**, 411.
10. (a) D. E. Angelescu, J. H. Waller, D. H. Adamson, P. Deshpande, S. Y. Chou, R. A. Register and P. M. Chaikin, *Adv. Mater.*, 2004, **16**, 1736; (b) D. E. Angelescu, J. H. Waller, R. A. Register and P. M. Chaikin, *Adv. Mater.*, 2005, **17**, 1878.
11. S. H. Kim, M. J. Misner, T. Xu, M. Kimura and T. P. Russell, *Adv. Mater.*, 2004, **16**, 226.
12. (a) C. C. Honeker, E. L. Thomas, R. J. Albalak, D. A. Hajduk, S. M. Gruner and M. C. Capel, *Macromolecules*, 2000, **33**, 9395; (b) C. C. Honeker and E. L. Thomas, *Chem. Mater.*, 1996, **8**, 1702.
13. H. Fischer, S. Poser, M. Arnold and W. Frank, *Macromolecules*, 1994, **27**, 7133.
14. I. W. Hamley, V. Castelletto, Z. B. Lu, C. T. Imrie, T. Itoh and M. Al-Hussein, *Macromolecules*, 2004, **37**, 4798.
15. E. Verploegen, L. C. McAfee, L. Tian, D. Verploegen and P. T. Hammond, *Macromolecules*, 2007, **40**, 777.
16. C. Osuji, P. J. Ferreira, G. Mao, C. K. Ober, J. B. Vander Sande and E. L. Thomas, *Macromolecules*, 2004, **37**, 9903.
17. M.-H. Li, J.-Y. Yang, P. Keller and P.-A. Albouy, *Adv. Mater.*, 2004, **16**, 1922.
18. Y. Tao, H. Zohar, B. D. Olsen and R. A. Segalman, *Nano Lett.*, 2007, **7**, 2742, DOI: 10.1021/nl0712320.
19. Y. Morikawa, S. Nagano, T. Watanabe, K. Kamata, T. Iyoda and T. Seki, *Adv. Mater.*, 2006, **18**, 883.
20. H. F. Yu, T. Iyoda and T. Ikeda, *J. Am. Chem. Soc.*, 2006, **128**, 11010.
21. (a) J. Yang, D. Lévy, W. Deng, P. Keller and M.-H. Li, *Chem. Commun.*, 2005, 4345; (b) J. Yang, R. Piñol, F. Gubellini, P.-A. Albouy, D. Lévy, P. Keller and M.-H. Li, *Langmuir*, 2006, **22**, 7907.
22. R. Piñol, L. Jia, F. Gubellini, D. Lévy, P.-A. Albouy, P. Keller, P. A. Cao and M.-H. Li, *Macromolecules*, 2007, **40**, 5625.
23. C. J. F. Rijcken, O. Soga, W. E. Hennink and C. F. van Nostrum, *J. Controlled Release*, 2007, **120**, 131.
24. M.-H. Li and P. Keller, *Soft Matter*, 2009, **5**, 927.
25. E. Mabrouk, D. Cuvelier, F. Brochard-Wyart, P. Nassoy and M.-H. Li, *Proc. Natl. Acad. Sci. U. S. A.*, 2009, **106**, 7294.
26. M.-H. Li, P. Keller and P.-A. Albouy, *Macromolecules*, 2003, **36**, 2284.
27. M. Portugall, H. Ringsdorf and R. Zentel, *Makromol. Chem.*, 1982, **183**, 2311.
28. V. P. Shibaev and N. A. Platé, in *Advances in Polymer Science, Liquid Crystal Polymers II/III*, vol. 60/61, ed. M. Gordon and N. A. Platé, Springer-Verlag, Berlin, 1984, p. 173.
29. P.-G. De Gennes and J. Prost in *The Physics of Liquid Crystals* (Oxford University Press, New York, 2nd edn, 1993) (ch. 3).
30. P. Davidson, D. Petermann and A.-M. Levelut, *J. Phys. II*, 1995, **5**, 113.
31. M. Andersson, L. Hammarstrom and K. Edwards, *J. Phys. Chem.*, 1995, **99**, 14531.

32 B. Klösgen, in *Lipid Bilayers: Structure and Interactions*, ed. J. Katsaras and T. Gutberlet, Springer-Verlag, Berlin, 2001, pp. 47–88.
33 J. Cocquyt, U. Olsson, G. Olofsson and P. Van der Meeren, *Langmuir*, 2004, **20**, 3906.
34 P. Oliger, M. Schmutz, M. Hebrant, C. Grison, P. Coutrot and C. Tondre, *Langmuir*, 2001, **17**, 3893.
35 C. Vautrin, T. Zemb, M. Schneider and M. Tanaka, *J. Phys. Chem. B*, 2004, **108**, 7986.
36 M. Dubois, B. Deme, T. Gulik-Krzywicki, J.-C. Dedieu, C. Vautrin, S. Desert, E. Perez and T. Zemb, *Nature*, 2001, **411**, 672.
37 M. Dubois, V. Lizunov, A. Meister, T. Gulik-Krzywicki, J. M. Verbavatz, E. Perez, J. Zimmerberg and T. Zemb, *Proc. Natl. Acad. Sci. U. S. A.*, 2004, **101**, 15082.
38 F. E. Antunes, R. O. Brito, E. F. Marques, B. Lindman and M. Miguel, *J. Phys. Chem. B*, 2007, **111**, 116.

A novel self-healing supramolecular polymer system

Stefano Burattini, Howard M. Colquhoun,* Barnaby W. Greenland and Wayne Hayes*

Received 15th January 2009, Accepted 31st March 2009
First published as an Advance Article on the web 23rd July 2009
DOI: 10.1039/b900859d

Utilising supramolecular π–π stacking interactions to drive miscibility in two-component polymer blends offers a novel approach to producing materials with unique properties. We report in this paper the preparation of a supramolecular polymer network that exploits this principle. A low molecular weight polydiimide which contains multiple π-electron-poor receptor sites along its backbone forms homogeneous films with a siloxane polymer that features π-electron-rich pyrenyl end-groups. Compatibility results from a complexation process that involves chain-folding of the polydiimide to create an optimum binding site for the π-electron-rich chain ends of the polysiloxane. These complementary π-electron-rich and -poor receptors exhibit rapid and reversible complexation behaviour in solution, and healable characteristics in the solid state in response to temperature. A mechanism is proposed for this thermoreversible healing behaviour that involves disruption of the intermolecular π–π stacking cross-links as the temperature of the supramolecular film is increased. The low T_g siloxane component can then flow and as the temperature of the blend is decreased, π–π stacking interactions drive formation of a new network and so lead to good damage-recovery characteristics of the two-component blend.

Introduction

Stimuli-responsive polymers have become a major field of investigation over recent years. Materials with precise and dynamically tuneable physical properties have opened up new fields of research, including those of drug and gene delivery,[1] liquid crystal displays[2] and functional polymeric gels.[3] One of the most recent additions to this field are healable polymers, which are characterised by the ability to recover their mechanical strength[4] or surface appearance[5] after suffering a fracture or scratch.

The first class of self-healing polymers were based on a composite design whereby bulk polymers were impregnated with either microcapsules[6] or fibres[7] containing a healing agent, for example, a liquid monomer[8] or simply a solvent.[9] When the material was fractured, reactive monomers were released into the void and subsequently polymerised, restoring, at least in part, the mechanical strength of the original pre-fractured material.

Studies by Wudl et al. have investigated a different approach to afford healable polymer networks.[10] In these investigations, a crosslinked polymer was synthesised that contained latent dienes (furans) and dienophiles (maleimides) residues. These components are able to participate in thermally-reversible covalent bond-forming

Department of Chemistry, University of Reading, Whiteknights, Reading, UK RG6 6AD.
E-mail: w.c.hayes@reading.ac.uk

reactions (*i.e.* Diels–Alder reactions). When a crack formed in the bulk resin, the system could be thermally annealed thus promoting the formation of new covalent bonds between dienyl and maleimide moieties on either side of the crack, enabling partial restoration of the mechanical properties of the pristine material. This approach to healable polymer networks has since been expanded upon and a range of reversible covalent bond forming reactions have been studied with often impressive healing characteristics.[11] For example, Broekhuis *et al.* synthesised[12] a covalently cross-linked, formally thermoset polymer, that can nevertheless be reformed into a single bulk material, even after being shattered into many fragments.

The two approaches to healable polymer networks outlined above rely upon the production of new covalent bonds to effect healing in the material. However, a recent study in 2008 by Leibler *et al.* reported that it is possible to use *non-covalent* supramolecular interactions to produce materials with healable properties.[13] It should be noted, however, that prior to Leibler's publication, Meijer and co-workers patented[14] and commented on[15] a healable supramolecular polymer based on their well-documented ureido-pyrimidone hydrogen bonded motif. Supramolecular materials generally comprise of low molecular oligomers which contain receptor units that are designed to enable the oligomers to assemble into higher order structures through dynamic, reversible, non-covalent interactions.[16] The resulting supramolecular materials frequently display physical properties such as high solution viscosity and mechanical strength that are more akin to conventional high molecular weight covalent polymers.[17]

In a supramolecular polymer, a fracture will tend to propagate *via* scission of the weak, supramolecular bonds between each oligomer, rather than by breaking the stronger, covalent bonds within the oligomers. Upon fracture of the supramolecular polymer, application of an appropriate stimulus (*e.g.* heat or light) will disrupt the non-covalent interactions, and allow the material to flow with the low viscosity of its low molecular weight oligomeric components. This process will allow rapid filling of the fracture void and, upon removal of the applied stimulus, reformation of the supramolecular network, with the original strength of the material. Depending on the thermodynamic stability of the supramolecular polymer, the healed region may possess a different local structural composition when compared to the bulk. This situation was found in the self-healing supramolecular rubber described by Leibler *et al.*[13] and resulted in a reduction in strength of the healed elastomer when compared to the original, pristine material.

We have investigated the supramolecular interactions between pyrenyl derivatives and polymers containing diimide residues.[18] Model studies on the complexes formed between these species showed them to adopt folded conformations in order to maximise the number of attractive face-to-face interactions between the electronically complementary components of the complex.[18,19] We envisaged that it should be possible to exploit these π–π stacking interactions—rather than hydrogen bonding interactions—to provide the basis for a novel, thermoresponsive, supramolecular material.

As a prelude to the synthesis of complex supramolecular π–π stacked blends, we carried out a detailed study into the feasibility of producing an easily accessible π-electron-poor receptor motif which would allow us to evaluate the binding mode of the receptor at the molecular level.[20] The result of this preliminary investigation was receptor **1** (Scheme 1) which contains two π-electron-poor diimide residues separated by 2,2′-(ethylenedioxy)bis(ethylamine) (the inclusion of the 2-ethylhexyl groups in **1**, is a synthetic necessity to maintain solubility, and is not thought to be integral to complex formation). Computational modelling of **1** led us to believe that the 2,2′-(ethylenedioxy)bis(ethylamine) spacer unit would possess sufficient conformational flexibility to allow the diimide moieties to encapsulate a π-electron-rich pyrenyl derivative, thus forming a two-component π-stacked complex.[20] A blend of the π-electron-poor receptor molecule **1** and pyrenyl derivative **2** was found to form a bright red solution, indicative of a π–π stacked complex.

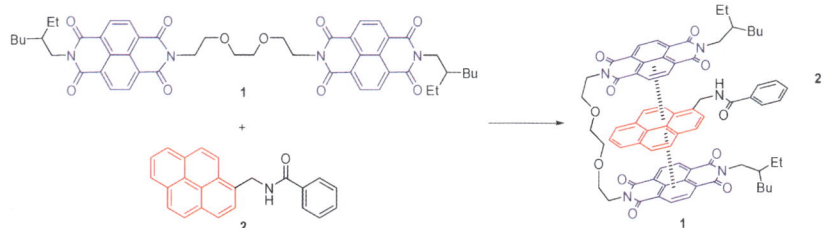

Scheme 1 Proposed conformation[20] of the supramolecular complex formed between **1** and **2**

The complex had a binding constant of constant of 130 M^{-1} and the adoption of a chain-folded conformation for **1** in solution was proposed on the basis of extended spectroscopic and computational studies.[20]

The chemical nature of the supramolecular motif enables it to be readily incorporated into a diverse range of polymers using simple and scalable synthetic strategies. Herein we describe the synthesis of a novel supramolecular polymer blend featuring a polydiimide **3** and a component featuring pyrene end-groups with a predominantly inorganic backbone (**4**) and the investigation of the interactions between the two polymers in solution and in the solid state.

General experimental

Materials

Reagents were purchased from Aldrich: 1-pyrenebutyric acid N-hydroxysuccinimide ester (95%), 2,2′-(ethylenedioxy)bis-(ethylamine) (98%), 2-ethylhexylamine (98%), bis-3-aminopropyl terminated polydimethylsiloxane (M_w = 2200), 1,4,5,8-naphthalene tetracarboxylic dianhydride (95%), and all were used as received. Triethylamine (99%) was distilled from CaH$_2$ under nitrogen prior to use. Solvents were used as supplied, with the exception of chloroform (CHCl$_3$), which was distilled under nitrogen from calcium hydride. Anhydrous dimethylacetamide (DMAc, 99.8%) was purchased from Aldrich and used as received.

Instrumentation

NMR spectra were recorded on a Bruker AC250 spectrometer operating at 250 MHz and 62.5 MHz for ^{1}H and ^{13}C nuclei, respectively. Infrared (IR) spectroscopic analyses were performed on a Perkin–Elmer 1720-X Infrared Fourier Transform spectrometer using thin polymer films cast onto KBr disks. Matrix-assisted laser desorption–ionization time-of-flight mass spectra (MALDI-TOF MS) were obtained using a Scientific Analysis Instruments LT3 LaserTof (Manchester, UK) spectrometer using dithranol as the matrix. A typical sample preparation is described as follows: 3 μL of a solution of the analyte in chloroform (1–10 mg mL^{-1}) was combined with 10–20 μL of the freshly prepared matrix (40 mg mL^{-1} in CHCl$_3$) in a mini-vial, and from the mixture was taken an aliquot, which was transferred onto a sample plate and left to air dry prior to analysis. UV/vis absorption measurements were performed in 1 cm path length quartz cuvettes on a double-beam Perkin–Elmer Lambda 25 spectrophotometer (interfaced with UV WinLab software) with a slit width of 1 nm. Spectra were recorded using samples dissolved in analytical grade chloroform mixed 6:1 (v/v) with hexafluoroisopropanol (HFIP) and blank-corrected for the possible absorption of the solvent. DSC analysis were performed with a Mettler-Toledo DSC823^e at a scan rate of 20 °C min^{-1} with 40 μL aluminium crucibles. Scanning electron micrographs were obtained with a FEI Quanta FEG 600 Environmental Scanning Electron Microscope (ESEM) equipped with a hot stage. Fluorescence measurements were performed on a Perkin-Elmer LS50B luminescence spectrometer. Solutions of the analyte in chloroform were excited at 345 nm and the emission spectra recorded over the stated range of wavelengths.

Synthesis of polydiimide 3

To a stirring suspension of 1,4,5,8-naphthalene tetracarboxylic dianhydride **5** (2.50 g, 9.33 mmol) in DMAc (50 mL) at 100 °C was added 2,2'-(ethylenendioxy)-bis-(ethylamine) **6** (1.24 g, 1.22 mL, 8.40 mmol). Over a period of 20 min the reaction became homogeneous at which point 2-ethylhexylamine **7** (0.24 g, 0.30 mL, 1.9 mmol) was added to the dark brown solution. After a further 20 min, toluene (5 mL) was added and the solution heated to 135 °C while water was removed azeotropically using toluene. After 18 h, the solution was cooled to room temperature and the polymer was precipitated from MeOH (200 mL), collected by filtration and washed with MeOH (2 × 50 mL). The dark brown polymer was subjected to Soxhlet extraction with MeOH (150 mL) for 24 h before being dried under vacuum to give polyimide **3** (3.1 g, 86%) as a grey/brown solid. ^{1}H NMR (CDCl$_3$/HFIP 250 MHz) δ = 8.78–8.64 (m, Ar*H*), 3.90 (m, NC*H$_2$*CH$_2$O), 3.66 (m, NCH$_2$C*H$_2$*O), 3.39 (m, OC*H$_2$*CH$_2$O), 1.47–1.28 (m, CC*H$_2$*C), 1.00–0.84 (m, CH$_3$); ^{13}C NMR (CDCl$_3$, 62.5 MHz) δ = 163–162, 131–130, 126–125, 69.6, 68.0, 39.5, 37.9, 30.6, 28.5, 23.9, 22.9, 13.9, 10.4; IR (KBr) ν = 3365, 3080, 2960, 2929, 2873, 1785 (C=O), 1709 (C=O), 1666 (C=O), 1580, 1453, 1425, 1373, 1335, 1245, 1099, 912; DSC T_g 89 °C, mp 201 °C; M_w = 7400; M_n = 3200.

Synthesis of pyrene end-capped polysiloxane 4

Bispropylamine-terminated polydimethylsiloxane **8** (nominal M_w = 2200) (0.74 g, 0.30 mmol) was dissolved in dry chloroform (100 mL) with dry triethylamine (0.149 g, 0.205 mL, 1.475 mmol). To this solution was added 1-pyrenebutyric acid *N*-hydroxysuccinimide ester **9** (0.250 g, 0.649 mmol) and the mixture then heated under reflux overnight under an argon atmosphere. The solution was concentrated *in vacuo*, diluted with chloroform–ethanol 5:1 (v/v, 100 mL), and then filtered through Celite®. The filtrate was dried and a dark yellow wax was obtained (0.67 g, 83%). ^{1}H NMR (CDCl$_3$, 250MHz,) δ = 8.25–7.7(m, Ar-*H*), 5.37 (m, N–*H*), 3.21–3.03 (m, Ar-C*H$_2$* & NHC*H$_2$*), 2.23–2.11 (m, COC*H$_2$*), 1.46–1.21 (m, CC*H$_2$*C), 0.47–0.40 (m, SiC*H$_2$*), 0.01–0.11 (m, SiC*H$_3$*); ^{13}C NMR (CDCl$_3$, 62.5 MHz) δ = 171.8, 134.8–122.3, 74.5, 72.3, 44.9, 41,5, 35.1, 31.7, 26.4, 22.4, 16.3, 14.3, 7.5, 0.0; IR (KBr) ν = 3305, 2963, 2924, 2901, 2850, 1651 C=O, 1537, 1418, 1259, 1093, 1020; MALDI-TOF-MS (M + Na)$^+$ *m/z* 811.3, 885.1, 960.6, 1034.2, 1108.6, 1181.6, 1257.3, 1331.2, 1404.6, 1478.1, 1554.7, 1627.0, 1704.5, 1780.1, 1849.0; M_w = 24500, M_n = 6800.

Results and discussion

Synthesis

A polymeric analogue of model compound **1** that was capable of chain-folding[21] was accessed readily *via* a one-pot procedure by the condensation of diamine **6** with 1,4,5,8-naphthalene tetracarboxylic dianhydride **5** in the presence of a small quantity of the end-capping mono amine **7**, to aid solubility of the product and control the molecular weight (Scheme 2).

The pyrenyl end-capped polysiloxane component **4** was synthesised by reaction of 1-pyrenebutyric acid *N*-hydroxysuccinimide ester **9** with the commercially available amine-end-capped polydimethylsiloxane **8**. The desired pyrenyl substituted polysiloxane **4** (M_n = 6800) was afforded as a yellow wax in 83% yield (Scheme 3).

The MALDI-TOF-MS of pyrenyl end-capped polysiloxane **4** revealed high intensity signals corresponding to either the sodium or potassium adducts of the desired

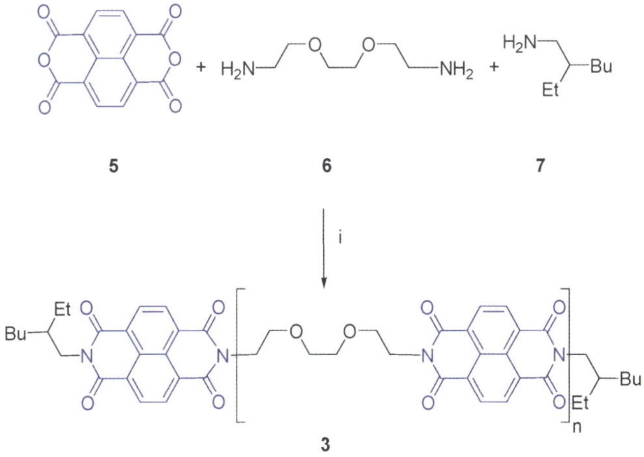

Scheme 2 Synthesis of the chain-folding polydiimide **3**

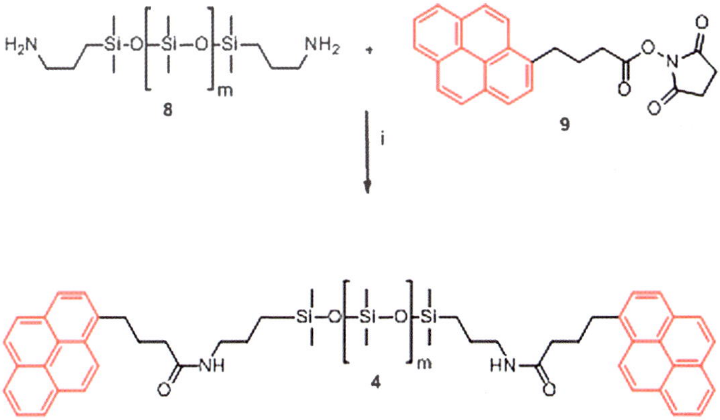

i) DMAc, Toluene, 135 °C, 18 hours, 86%

Scheme 3 Synthesis of the pyrenyl end-capped polysiloxane **4**

i) CHCl$_3$, Et$_3$N, overnight, 83%

bis-end-capped polysiloxane **4** (Fig. 1). Mass ions corresponding to the polysiloxane **8** or mono-end-capped analogue of **4** were not evident in the MALDI-TOF mass spectrum. The ^{1}H NMR spectrum of the pyrene-end-capped polysiloxane (**4**) featured a resonance at 3.0 ppm, characteristic of methylene protons adjacent to the newly formed amide bond whilst proton resonances at 2.6 ppm corresponding to the methylene units adjacent to the terminal amine functionalities of **8** were not apparent. These data, taken together, clearly suggest that amide bond formation to the terminal amine residues of polysiloxane **4** proceeds with better than 95% efficiency.

Solution phase complexation studies

Blending (in an equimolar ratio based on the number of pyrene end-groups in **4** relative to the number of potential chain-folded receptor units in the polydiimide **3**)

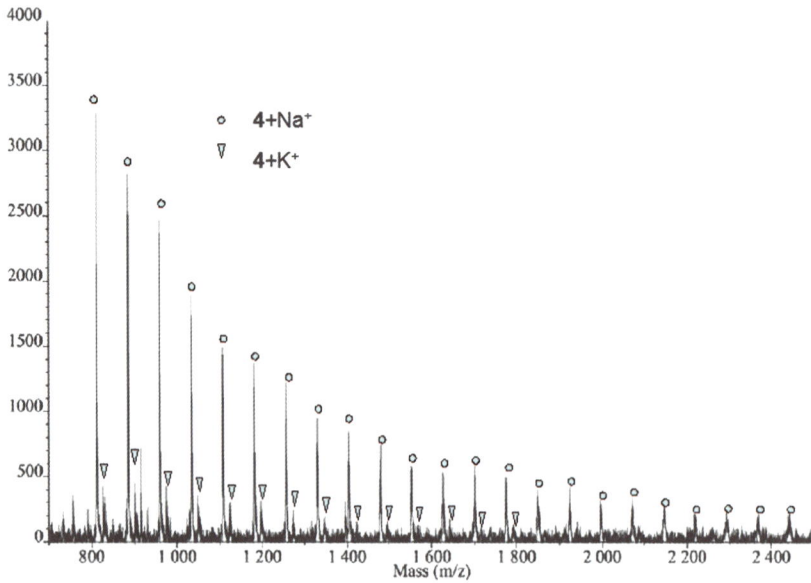

Fig. 1 MALDI-TOF-MS of pyrenyl end-capped siloxane **4**.

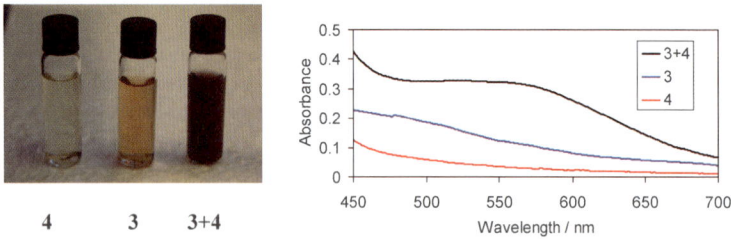

Fig. 2 (Left) Solutions of **4**, **3** and a blend of **3** and **4** (all 4 mg mL^{-1}) under ambient light conditions. (Right) UV/vis absorption spectra of solutions of **3** (0.001 M chain-folding motif), **4** (0.001 M pyrenyl groups), solution blend where the concentration of the chain-folding motif = concentration of the pyrenyl groups = 0.001 M (Solvent system in all cases: 6:1 v/v CHCl$_3$/HFIP).

a pale yellow solution of polyimide **3** with an almost colourless solution of pyrene end-capped siloxane **4** lead to the instantaneous formation of a deep cherry-red solution (see photograph in Fig. 2, right hand vial). This colour is attributed to the presence of a new absorption band in the blend, centred on 540 nm (Fig. 2). An absorption at this wavelength is characteristic of a charge-transfer band associated with π–π-stacked complex formation between π-electron-poor species such as diimides and π-electron-rich species such as pyrene.[22] This absorption may also be compared with the UV-visible spectrum obtained previously for the complex between model compounds **1** and **2** (Scheme 1), which was found to have a charge-transfer absorption centred at 530 nm.[20]

In addition to the change in UV/vis absorbance of the blend relative to its components (either **3** or **4** individually), there is a dramatic change in the fluorescence characteristics of the polymers upon mixing. Pyrenyl groups are known to exhibit strong fluorescence when exposed to UV radiation.[23] This property has led to their exploitation as fluorescent tags in numerous chemical[24] and biological applications[25] and, more recently, as potential sensors for the detection of explosives.[26]

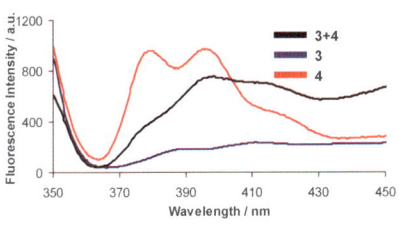

Fig. 3 (Left) fluorescence spectra for solutions of **3** (5×10^{-6} M chain-folding motif), **4** (8×10^{-9} M pyrenyl end groups) and a blend of these solutions of **3** and **4** mixed 1:10 (v/v). (Right) Solutions (all 4 mg mL^{-1}) of **4**, **3** and a blend of **3** and **4**, irradiated with UV light (348 nm). (Solvent system in all cases: 6:1 v/v CHCl$_3$/HFIP).

Fig. 3 (left) shows the emission spectra from solutions of **3**, **4** and a blend of **3** and **4** following excitation at 345 nm. The signals at 380 nm and 396 nm in the fluorescence spectra of **4** correspond to an isolated pyrenyl species. Addition of a solution of the polyimide **3**, which itself shows little emission at 380 nm, results in substantial quenching of the emission from the pyrenyl end-groups in this spectral region. This phenomenon can even be observed by eye by placing solutions of the polymers **3** and **4** (4 mg mL^{-1}) and their blend under a standard laboratory UV lamp emitting at 348 nm (Fig. 3, right).

In order to achieve a useful, solid-state polymer blend that can be thermally healed on a short timescale, it is necessary that the complex formed between the two polymers should be rapidly reversible over a suitable temperature range. If the complex formed between the pyrenyl chain ends of the polysiloxane **4** and the diimide polymer **3** remained intact at high temperatures, or was slow to associate or dissociate, then the effectiveness of healing of this type of polymer blend by the proposed mechanism would be severely compromised. Heating a solution of polymer **3** and **4** to *ca.* 65 °C produced an almost instantaneous reduction in the intensity of the cherry-red colour, implying a significant reduction in the strength and number of interactions between the two polymers. The efficiency of the polymer–polymer binding process was investigated by placing a heated cuvette (held at *ca.* 65 °C) containing a solution of the polymer blend in a UV/vis spectrometer (at ambient temperature) and measuring the change in absorbance over the range 490 nm to 600 nm as the system cooled under ambient conditions (Fig. 4 (top)). A plot of this spectroscopic data clearly reveals the progressive reappearance of the charge transfer band at 540 nm associated with complex formation. Analysing the intensity of the absorption of the solution blend at 540 nm specifically, it can be seen that the signal due to complex formation reaches a maximum after only ~280 s (Fig. 4 (bottom)).

As a final probe of the solution-state interaction between polydiimide **3** and pyrene end-capped polysiloxane **4**, the ^{1}H NMR spectrum of the blend (**3**:**4** 1:1 w/w, 20 mg mL^{-1}, 10% TFA in CDCl$_3$ (v/v)) was compared to the spectra obtained for the isolated components (Fig. 5). The resonances associated with the aromatic protons in both the π-electron-rich pyrenyl groups and π-electron-poor diimide residues experience a significant upfield shift (~0.35 ppm) when blended. In addition, the signals associated with the 'naphthyl' protons in polymer **3** split into two distinct populations on mixing. The complex pattern of naphthyl proton resonances observed may arise as a consequence of the range of different binding sequences in the polydiimide **3**, some strongly binding, some less so, occurring between the intercalating pyrene end groups of **4** with the chain-folding polydiimide backbone.[19] The broad nature of the proton resonances for both of the polymeric receptors indicates that association under a slow exchange regime may be occurring (with respect to the timescale of the ^{1}H NMR spectroscopic analysis).

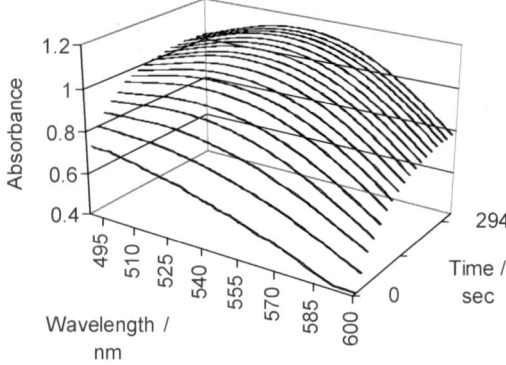

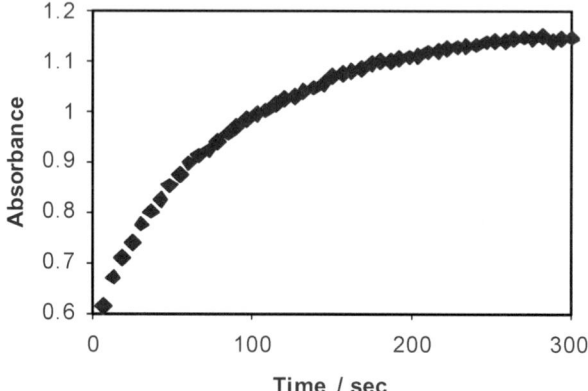

Fig. 4 (Top) time dependent UV/vis spectra of a blend between **3** and **4**; (Bottom) Time dependent UV/vis spectroscopic data showing the magnitude of the absorbance at 540 nm for the blend between **3** and **4**. In both cases the temperature of the blend at $t = 0 \approx 65\ °C$ and $t = 300 \approx 20\ °C$.

Solid state complexation studies

A film cast from a solution blend of polymers **3** and **4** (equimolar ratio of **3**:**4** (based on the number of pyrene end-groups in **4** relative to the number of potential chain-folded receptor units in the polydiimide **3**)) generated a deep red homogeneous solid. In contrast, attempts to cast films from either **3** or **4** solely proved unsuccessful, leading to the formation of a pale tan powder (**3**) or a soft pale yellow wax (**4**): neither of these latter two materials could be peeled from the Teflon casting surface or were self-supporting in nature. The deep red colour of the film cast from the blend between **3** and **4** demonstrates that complexation between the pyrenyl end-groups of **4** and the diimide units in **3**, which had been observed in the solution studies, is maintained in the solid state. In addition, complexation between the two polymer components appears crucial to the generation of a smooth, visually homogeneous film.

Differential scanning calorimetric analyses between −20 °C and 220 °C, for the individual polymers **3** and **4** and the blend of **3** and **4** are shown in Fig. 6. Polysiloxane **4** did not reveal any significant thermal transitions over the temperature range studied, whereas the polydiimide **3** exhibited a glass transition temperature (T_g) at approximately 97 °C and a well-defined melting point at *ca.* 200 °C. This melting was not evident in the polymer blend, although the new material exhibited a

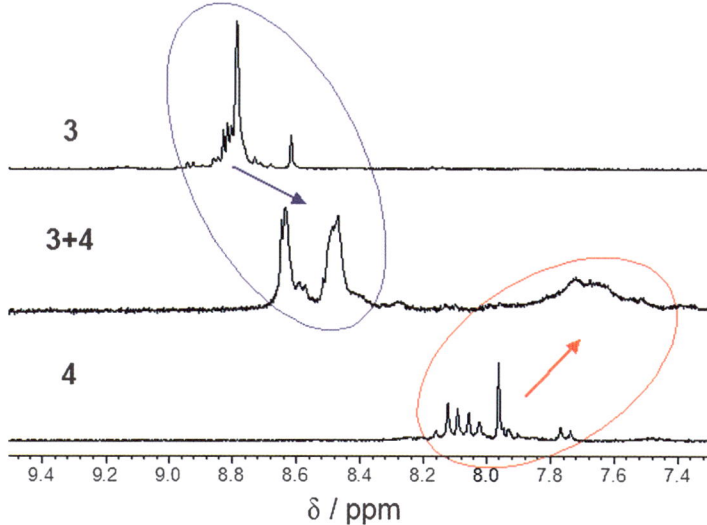

Fig. 5 Partial ^{1}H NMR spectra of: (top) polydiimide **3**, (centre) 1:1 w/w blend of **3** and **4**, (bottom) pyrene end-capped polysiloxane **4**.

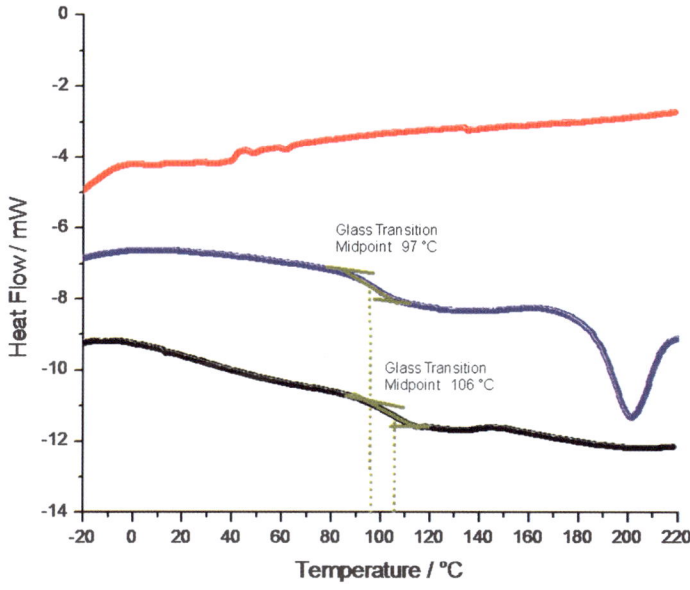

Fig. 6 DSC traces of polymers **3** (blue) and **4** (red) and the blend of polymers **3** and **4** (black).

T_g (106 °C) at a higher temperature (+9 °C) than that of **4**. This analysis demonstrates the miscibility of the two polymers in the blend and the fact that complex formation between these two components serves to reduce the mobility of the polydiimide backbone.

It is clearly important to verify that the apparently homogeneous nature of the blend represents a thermodynamic minimum for two polymers rather than being an unstable mixture of the components that is kinetically trapped during the film casting process. Accordingly, the heat flow of the freshly prepared solid-state blend

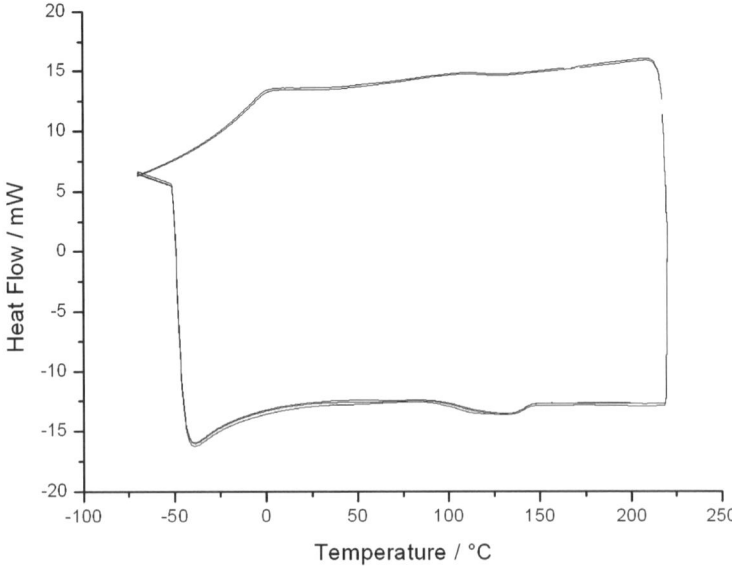

Fig. 7 DSC traces for the polymer blend (**3** and **4**) recorded over three consecutive heat–equilibrate–cool cycles.

was measured over the same temperature range as previously (−20 °C and 220 °C), but on this occasion it was allowed to equilibrate at 220 °C (20 °C above the T_m of **3**) for 10 min prior to returning the sample to −20 °C. The complete heat–equilibrate–cool cycle was repeated, in total, three times. The results of the heat flow recorded over these three cycles are presented in Fig. 7. The close agreement in heat flow results recorded for each cycle clearly demonstrates that phase separation does not occur over the temperature range and timescales investigated, and that the homogeneous blend of these two structurally disparate polymers represents the thermodynamic minimum for the system.

Fig. 8 (column a, top image) shows environmental scanning electron microscopy (ESEM) images of a sample of the film cast from a blend between **3** and **4** which has been cut with a scalpel (*ca.* 60 μm cut-width). The undamaged area of the polymer can be seen to be entirely homogeneous at this magnification (×200). This homogeneity contrasts with the surface of a control film cast from a blend between the polyimide **3** and the non-pyrenyl end-capped polysiloxane **8**, which exhibited macroscopic phase separation under these conditions (Fig. 8, column b, top image). Thus, again, it is evident that complexation between the two polymers is of crucial importance in driving miscibility of the components.

To investigate the potential for this type of material to show healable characteristics, the damaged film from **3** and **4** was heated to 115 °C (heating rate: 5 °C min^{-1}) in the ESEM instrument, whilst images were acquired in real time (Fig. 8 column a). For comparative purposes the inhomogeneous material cast from **3** and **8** was treated in an identical manner (Fig. 8 column b).

It can be seen that the surface of the film cast from the polymers containing complementary recognition motifs (**3** and **4**) remained homogeneous over the entire temperature range studied. In addition, at around 90 °C the cut made in the material was seen to close rapidly, and for that at 115 °C the damage is barely visible.

As a control experiment, a film was cast consisting of chain folding diimide **3** and an analogue of **4** which contained benzyl end-groups rather than pyrenyl end-groups (**10**). The surface of this material cast from a blend of **3** and **10**, remained inhomogeneous over the entire temperature range studied. Furthermore, the cracks in the surface, which were produced whilst transferring the material into the instrument,

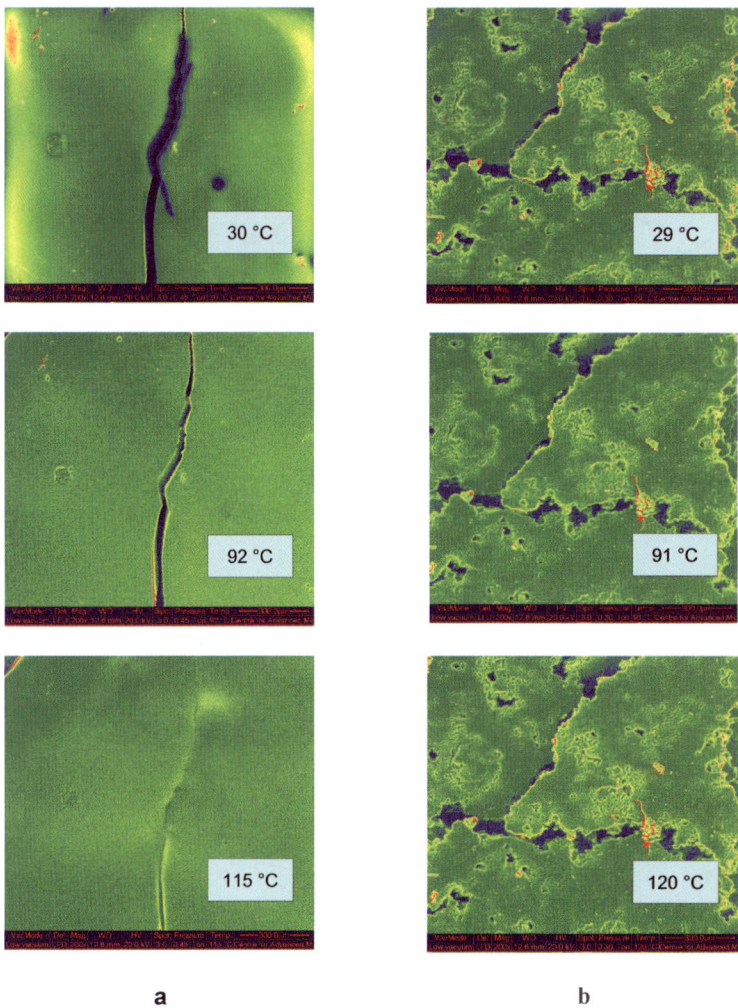

Fig. 8 ESEM images (magnification ×200) of (a) the fractured film of the blend between **3** and **4** as at increasing temperature; (b) the film of the blend formed between **3** and **9** that does not feature the π-electron rich pyrene receptor end-groups at increasing temperature.

remain essentially unaltered throughout the experiment, with no evidence of transfer of material into the voids.

A separate set of healing experiments were carried out using a visible-light microscope by heating (to *ca.* 100 °C) a fractured sample of the polymer blend on a glass slide. Upon cooling, it was observed that the material filling the fractured void retained the deep red colour of the complex. Although the exact composition of the material in the void could not be quantified in this simple test, the signature red colour of the complex suggests that it contains both components of the blend, despite polydiimide **3** displaying a T_m of over 200 °C. Thus, in agreement with the DSC results, it is apparent that at elevated temperatures, interactions between the components still permit the blend to flow homogeneously, rather than separating into its constituents.

From the ESEM data and healing experiments reported above, we propose that the supramolecular polymer network recovers its physical form by a mechanism that involves partial dissociation and subsequent regeneration of numerous

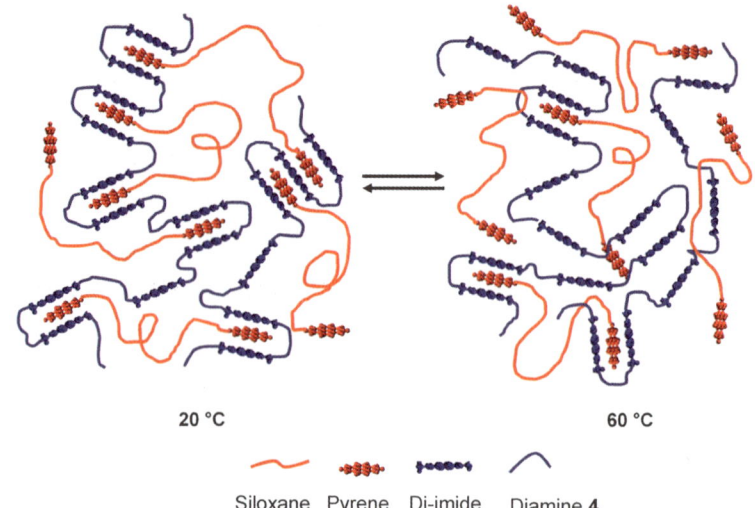

Scheme 4 Schematic representation of the proposed self-healing mechanism for the blend comprising the chain-folding polydiimide **3** and the pyrene end-capped siloxane **4**

π–π stacking and hydrogen bonding interactions between the chain ends of the polysiloxane **4** and the chain-folding 'tweezer' sequences of the polydiimide **3** (see Scheme 4). It is also apparent that the complementary π–π stacking interactions between the components of the blend serve a dual rôle in this system. As the temperature of the supramolecular film is increased, the intermolecular π–π stacking cross-links are disrupted (but not entirely—the intense red colour of the film does not fade completely) and this disengagement of the pyrene end-groups enables the polysiloxane component **4** to flow in unison with polydiimide **3** at temperatures below its T_m. Thus, even at temperatures where flow is observed, there remains a sufficient number of complementary interactions to maintain the homogeneity of the blend. As the temperature of the blend is reduced, increasing levels of π–π stacking can occur, thus forming a new network, and, in the context of healing processes, enabling the annealed supramolecular material to regain its physical form and properties.

Conclusions

This paper reports the synthesis of two very simple, relatively low molecular weight polymers—a polydiimide **3** that is capable of adopting a chain-folded conformation, thereby generating π-electron deficient 'tweezer-type' receptor units and a linear polysiloxane **4** that features complementary π-electron rich pyrene end-groups. Casting a blend of these two polymers from solution affords a thermodynamically stable material that is capable of autonomous healing when exposed to temperatures above ambient. In contrast to the supramolecular hydrogen bonded self-healing system reported by Leibler *et al.* the healable two component polymer blend described here is not reliant upon plasticising additives for repair to occur. The low T_g value of the polysiloxane component and the weaker nature of the π–π stacking interactions involved enable efficient healing to occur at temperatures of *ca.* 100 °C.

The healing studies presented here were conducted on films positioned horizontally in the ESEM instrument. To date, we have not conducted healing studies on the blend in the vertical plane. However, given the viscosity characteristics of this material at the healing temperature (*ca.* 100 °C) it is not unreasonable to postulate that healing would still occur (provided the film were not held under tension) if the sample was annealed over the fracture site to allow the flow of material into the void

Preliminary solution-phase UV-vis, fluorescence and NMR spectroscopic investigations have shown that the two polymers bind to each other via π–π stacking interactions. An intense charge-transfer absorption band was evident at 540 nm in the UV-vis spectra of a 1:1 equimolar solution of **3** and **4** (the molar ratios used were based upon the number of pyrene end-groups in **4** relative to the number of potential chain-folded receptor units in the polydiimide **3**). Furthermore, the fluorescence observed from the solution of the pyrenyl end-capped polymer **4** was strongly quenched upon addition of polydiimide **3** to the solution, thus demonstrating intimate group-specific contact between the two polymers in solution. Complementing this spectroscopic data, ^{1}H NMR spectroscopic titrations have shown that the resonances assigned to the aromatic protons in both the π-electron-rich pyrenyl end-groups and π-electron-poor diimide residues experience a significant upfield shift when the two components are blended together in solution—this data is similarly consistent with the formation of π–π stack complexes between these two aromatic residues.

Casting a solution containing equimolar quantities of the π-electron-rich and π-electron-deficient receptor units (again, the molar ratios used were based upon the number of pyrene end-groups in **4** relative to the number of potential chain-folded receptor units in the polydiimide **3**), at ambient temperatures gives a dimensionally-stable, flexible, homogeneous film that evidently constitutes a non-covalently cross-linked network polymer. In stark contrast, the two individual components of the blend are a brown/grey incoherent solid (**3**) and a yellow wax (**4**), respectively. The physical form of the 1:1 blend of **3** and **4** represents a more robust material in comparison to the individual constituent components in isolation. DSC analysis of the blend of **3** and **4** revealed a dramatic change in the thermal characteristics of these individual components—a composite T_g value in between the T_g values of the individual polymers was not observed as is evident in many conventional polymer blends. Indeed a higher T_g value (106 °C) was evident in comparison to the T_g of **3** (97 °C) indicating that complex formation between these two components serves to reduce the mobility of the polydiimide backbone.

In summary, this paper reports a new supramolecular polymer blend which assembles via π–π stacking interactions and is capable of healing when fractured. This is the first report of the use of π–π stacking interactions in the generation of healable supramolecular polymer materials.

Acknowledgements

This work was supported by EPSRC (EP/D0743471/1) (post-doctoral research fellowship for BWG and post-graduate studentship for SB) and (EP/G026203/1) (post-doctoral research fellowship for BWG). We wish to thank Dr John Baum for technical assistance with the time-dependent IR experiments.

References

1 C. D. H. Alacon, S. Pennadam and C. Alexander, *Chem. Soc. Rev.*, 2005, **34**, 276.
2 For example see:Y. Kang, J. J. Walish, T. Gorishnyy and E. L. Thomas, *Nat. Mater.*, 2007, **6**, 957.
3 S. K. Ahn, R. M. Kasi, S. C. Kim, N. Sharma and Y. X. Zhou, *Soft Matter*, 2008, **4**, 515; R. V. Ulijn, N. Bibi, V. Jayawarna, P. D. Thornton, S. J. Todd, M. J. Mart, A. M. Smith and J. E. Gough, *Mater. Today*, 2007, **10**, 40.
4 R. P. Wool, *Soft Matter*, 2008, **4**, 400; S. D. Bergman and F. J. Wudl, *J. Mater. Chem.*, 2008, **18**, 41; D. Y. Wu, S. Meure and D. Solomon, *Prog. Polym. Sci.*, 2008, **33**, 479.
5 S. Ghosh and M. W. Urban, *Science*, 2009, **323**, 1458–1460.
6 S. R. White, N. R. Sottos, P. H. Geubelle, J. S. Moore, M. R. Kessler, S. R. Sriram, E. N. Brown and S. Viswanathan, *Nature*, 2001, **409**, 794.
7 J. W. C. Pang and I. P. Bond, *Compos. Sci. Technol.*, 2005, **65**, 1791.
8 E. N. Brown, S. R. White and N. R. Sottos, *J. Mater. Sci.*, 2004, **39**, 1703.

9 M. M. Caruso, D. A. Delafuente, V. Ho, N. R. Sottos, J. S. Moore and S. R. White, *Macromolecules*, 2007, **40**, 8830; M. M. Caruso, B. J. Blaiszik, S. R. White, N. R. Sottos and J. S. Moore, *Adv. Funct. Mater.*, 2008, **18**, 1898.
10 X. Chen, M. A. Dam, K. Ono, A. Mal, H. Shen, S. R. Nutt, K. Sheran and F. Wudl, *Science*, 2002, **295**, 1698.
11 X. Chen, A. Mal, H. Shen, S. R. Nutt and F. Wudl, *Macromolecules*, 2003, **36**, 1802; E. B. Murphy, E. Bolanos, C. Schaffner-Hamann, F. Wudl, S. R. Nutt and M. L. Auad, *Macromolecules*, 2008, **41**, 5203; B. J. Adzima, H. A. Aguirre, C. J. Kloxin, T. F. Scott and C. N. Bowman, *Macromolecules*, 2008, **41**, 9112.
12 Y. Zhang, A. B. Broekhuis and F. Picchioni, *Macromolecules*, 2009, **42**, 1906–1912.
13 P. Cordier, F. Tournilhac, C. Soulié-Ziakovic and L. Leibler, *Nature*, 2008, **451**, 977.
14 See for example: G. M. L. Hoorne-van-Gemert, H. M. Janssen, E. W. Meijer and A. W. Bosman, PCT Pat. Appl. WO2008/063057 A2, 2008, to Suprapolix B.V.
15 A. W. Bosman, R. P. Sijbesma and E. W. Meijer, *Mater. Today*, 2004, **7**, 34–39.
16 *Supramolecular Polymers*, ed. A. Ciferri, Marcel Dekker, New York, 2000, pp. 511–529; C. Schmuck and W. Wienand, *Angew. Chem., Int. Ed.*, 2001, **40**, 4363; L. Brunsvel, B. J. B. Folmer, E. W. Meijer and R. P. Sijbesma, *Chem. Rev.*, 2001, **101**, 4071.
17 S. H. M. Söntjens, R. P. Sijbesma, M. H. P. van Genderen and E. W. Meijer, *Macromolecules*, 2001, **34**, 3815; A. T. ten Cate, H. Kooijman, A. L. Spek, R. P. Sijbesma and E. W. Meijer, *J. Am. Chem. Soc.*, 2004, **126**, 3801; C. L. Elkins, K. Viswanathan and T. E. Long, *Macromolecules*, 2006, **39**, 3132; T. Park and S. C. Zimmerman, *J. Am. Chem. Soc.*, 2006, **128**, 11582.
18 H. M. Colquhoun and Z. X. Zhu, *Angew. Chem., Int. Ed.*, 2004, **43**, 5040.
19 H. M. Colquhoun, Z. X. Zhu, C. J. Cardin and Y. Gan, *Chem. Commun.*, 2004, 2650; H. M. Colquhoun, Z. X. Zhu, C. J. Cardin, Y. Gan and M. G. B. Drew, *J. Am. Chem. Soc.*, 2007, **129**, 16163.
20 B. W. Greenland, S. Burattini, W. Hayes and H. M. Colquhoun, *Tetrahedron*, 2008, **64**, 8346.
21 R. S. Lokey and B. L. Iverson, *Nature*, 1995, **375**, 303.
22 R. Foster, *Organic Charge Transfer Complexes*, Academic Press, London, 1969, pp. 33–93.
23 F. M. Winnik, *Chem. Rev.*, 1993, **93**, 587.
24 For example see: J. Duhamel, *Acc. Chem. Res.*, 2006, **39**, 953.
25 M. Manoharan, K. L. Tivel, M. Zhao, K. Nafisi and T. L. Netzel, *J. Phys. Chem.*, 1995, **99**, 17461.
26 H. Bai, C. Li and G. Q. Shi, *Sens. Actuators, B*, 2008, **130**, 777; S. Burattini, H. M. Colquhoun, B. W. Greenland, W. Hayes and M. Wade, *Macromol. Rapid Commun.*, 2009, **30**, 459.

General discussion

Dr Titmuss opened the discussion of the paper by Professor Mao: I would ask you to clarify the statement you made in the 'outlook' part of your presentation regarding the production of synthetic viruses. Is your aim to provoke an immune response or to avoid an immune response? If your aim is to avoid an immune response, I would suggest that making something resemble a virus in size and shape is not a smart strategy, as the immune system is primed to recognize nanoparticles, and viruses in particular.

Professor Mao replied: Provoking and suppressing an immune response are related. If we can control the spacing between the antigens, I believe that we can purposely promote either of them.

Professor Reinhoudt enquired: Do you have a qualitative idea about the kinetics of formation/dissociation rates? Do you see imperfect structures next to the desired product and if so, do you know what they are? How stable are these structures in solution and do you see changes as a function of time?

Professor Mao replied: We have not studied the kinetics yet. It would be an interesting study.

We don't have a way to study the imperfect structure. The current characterization method, cryoEM-3D reconstruction, is essentially an average method: it would average all objects. To study the imperfect structures, we need to study each individual object. Currently, there is no method that allows such studies.

The DNA structures are relatively stable and no change has been observed over at least two weeks.

Professor Ulijn asked: Can you comment on the mechanical properties of the resulting materials (honeycomb etc.)?

Professor Mao responded: We haven't really characterize the mechanical propoerties of the DNA nanostructures. Our guess is that they are moderately strong.

Professor Ulijn opened the discussion of the paper by Professor Matile: Can you comment on the difference in ion capture when comparing Lys and Arg—is this exclusively a pKa effect?

Professor Matile responded: I like to believe so and thought this interpretation is wonderfully supported in the present paper with the charge-reversed situation, where DNA is inactive and polyE is active. Support from other lines of research includes the anion selectivity of K-rich pores and the cation-selectivity(!) of R-rich pores, phase transfer data, *etc.* However, it is understood that it is difficult to get hard evidence for these dynamic systems, and I wonder about other explanations as you do.

Mr Aufderhorst-Roberts continued the discussion of the paper by Professor Mao: Firstly, the paper reports the self-assembly of 3-point star motifs into ordered polyhedra, with the largest ordered structure reported being a buckyball containing 60 units of the 3-point star. Can you envisage the self-assembly of larger structures of this type and do you believe there is an upper limit to the size of such ordered structures?

Secondly, through using a higher volume fraction of the DNA motifs reported would you expect the formation of 3-dimensional networks?

Thirdly, in the case of larger nanocages, does disorder occur above a threshold size?

Professor Mao responded: Firstly, I don't think that there is a clear-cut upper limit. More likely, we can build large (but more heterogeneous) structures if we choose the right conditions. In fact, we have observed structures at least several times larger than buckyball structures.

Secondly, yes we would expect such networks. However, we have no experimental evidence so far.

Thirdly, large nanocages have higher heterogeneities, we are not sure whether there is a threshold in terms of size.

Professor Ikkala asked: Professor Mao, could you discuss the merits of your approach in comparison to the DNA-origami method based on long single-stranded DNA and short staple strands.[1]

1 P. W. K. Rothemund, *Nature*, 2006, **440**, 297–302.

Professor Mao answered: The two approaches are complementary to each other. Both are effective, but not perfect. At the current stage, origami can construct arbitrarily designed nanostructures, but require many (>200) different synthetic DNA strands. Our approach only requires very few (*e.g.* 3) different DNA strands, but can only build highly symmetric objects.

Professor Huskens remarked: All your structures are fully endgroup-saturated/closed, so enthalpy is optimally used regardless of the size of your structures. This means that selectivity between the formation of different structures is governed by entropy differences, *e.g.* the basic number of building blocks used. The theory for this is given in polymer chemistry, mainly for macrocyclization. Have you considered using such theory for predicting the preferred structures and the ratios between them?

Professor Mao responded: Excellent point! As experimental exploration goes, we certainly would like to understand self-assembly in terms of theory and enhance our prediction power. Thanks for pointing this out. In this particular case, I am not sure whether entropy is the governing factor. The system contains DNA and small molecules (water, salt, *etc.*). When thinking about entropy, we cannot only look at the number of DNA units and ignore those small molecules. We need to look at the related experimental data in the literature more carefully. At the end, we might find that entropy is indeed the governing factor.

Professor Huskens returned to the discussion of the paper by Professor Matile: What is the role of the counterion size in the exchange dynamics for your counterion-activated polyion systems? Would there be an interest in using divalent counterions (both truly divalent ones such as Ca^{2+}, or synthetic ones like R_3N^+–R–N^+R_3) for changing the dynamics and membrane crossing?

Professor Matile answered: Exchange dynamics are unknown, presumably faster than we can measure in our assays. This could be an interesting question to address for experts in the field. We have studied di- to tetravalent counterions with amphiphilic calixarenes; increasing multivalency usually gave better activity, although the final result depends on many parameters.

Professor Sen asked: How much of the observed counterion effect is due to the conformational change of the polymer, which may affect cell uptake?

Professor Matile answered: Conformation is presumably less important, although there are conflicting reports in the literature concerning CPPs. DNA/RNA and counteractions are known to collapse into compact lipoplexes. In vesicles, the hydrophobicity, multivalency and nature of the ion pair are more important.

Professor Whitesides opened the discussion of the paper by Professor Colquhoun: What is the proper measure of information in a biological system? The Shannon approach based on binary information transmission can't be right for biology. The number of bits needed to specify the amino acids in a 300 amino acid proteins (= 900 bases $\approx$ 4000 bits) can't be enough to specify the structure and function of an enzyme active site can it? Or maybe it can?

Professor Colquhoun replied: The Shannon information content of a digital string is the maximum amount of information that can, in principle, be represented by that string. It says nothing about the *actual* information content, which is entirely context-dependent and generally very much less than the Shannon content. For example, the molecular operation of the genetic code requires amino acid sequences to be templated by triplets of nucleotides rather than by single residues, so that the Shannon information content of a DNA sequence must be at least three times greater than the contextual content. Also, because the coding triplets are consecutive, rather than overlapping, the Shannon information content of the DNA must be increased by a further factor of three in order to specify a given protein sequence. Finally, not all DNA carries genetic information. In eukaryotic cells, for example, a gene will often comprise several widely-separated sections of a DNA molecule. After being copied to RNA, the transcripts must then be spliced together enzymatically to generate the strand of messenger-RNA which is actually translated into protein. For all these reasons, the Shannon information content of DNA is very much greater than the information which actually specifies the protein sequences it encodes. The latter information might be regarded as the *contextual* information content of the DNA.

With regard to your specific question, as to whether the Shannon information content of an enzyme's amino-acid sequence is sufficient to specify the structure of the active site, it has been calculated[1] that the Shannon information content of a protein sequence is generally in the range 2.5–3.0 bits per amino acid residue, and that the information content of the three-dimensional structure of a protein—as measured by its "Kolmogorov information entropy" or "algorithmic complexity"—is less than 1.0 bits per amino acid residue.[2] Consequently, it would seem that there is always enough information (and to spare) in an amino acid sequence to specify the three-dimensional structure of the corresponding protein, and so generate the active site. However, such calculations take the molecular structures and selection of the "standard set" of 20 amino acids as given information. A huge number of other amino acids could, in principle, be incorporated into proteins—and indeed many non-natural amino acids are now being so incorporated by modifying the "context", (*i.e.* the molecular machinery responsible for translation) of the genetic information.[3] If the need to specify the amino acids and their molecular structures (in terms of atomic composition, connectivity and chirality) were taken into account, a vastly greater quantity of Shannon information would of course be required to specify a protein sequence. And if the same information for the nucleotide structures in DNA and RNA were included in the total...

1 B. J. Strait and T. G. Dewey, *BioPhys. J.*, 1996, **71**, 148.
2 T. G. Dewey, *Phys. Rev. E*, 1996, **54**, R39.
3 J. K. Montclare and D. A. Tirrell, *Angew. Chem., Int. Ed.*, 2006, **45**, 4518.

Dr Faul continued the discussion of Professor Matile's paper: Has counterion binding been investigated in terms of cooperativity, as concentration and structure of counterions will both affect cooperativity very drastically?

Professor Matile answered: Hill coefficients are usually larger than 1, but not as large as one would expect for cooperative binding of many counterions. But then the Hill coefficients for membrane lysis with triton X-100 are much lower than one would expect from the formation of the active micelles. Overall, Hill coefficients contain not only information on cooperativity but also on the stability of the active suprastructure.[1] Strong multivalency effects can be seen as support for cooperativity; the efficiency of counterion activators increases with increasing number of monomers that are linked together.

1 S. Bhosale, S. Matile, *Chirality*, 2006, **18**, 849–856.

Professor Woolfson returned to the discussion of Professor Colquhoun's paper: To complete your analogy with storing information in nucleic acids, you will have to "read" the information/sequence of your polymers. Presumably you will have to use multiple tweezers and get these to move along the polymer. Have you thought about this at all?

Professor Colquhoun answered: "Reading" the digital (binary) information represented in AB-type copolymer sequences does not strictly require the information to be expressed in chemical terms as occurs in biological system, although, as I said earlier, we can see ways of doing this in principle. As well as needing "start" and "stop" sequences to be recognised, we would need to design several different tweezer-molecules, with different sequence-selectivities, to act as adaptors between the templating polyimide chain and the monomers being used to produce a "translated" polymer or oligomer chain.

Our current tweezer-polymer "reading" system operates under conditions of fast-exchange on the NMR timescale, so that all the sequences present are "read" simultaneously through rapid and reversible binding of multiple tweezer-molecules. The analogy at the moment is thus probably closer to electronic random-access memory than to the expression of genetic information through protein synthesis.

Professor Woolfson continued the discussion of the paper by Professor Mao: Your beautiful structures show little evidence of stain—*e.g.* the aims of the nanostructures are straight, and the planar assemblies are very flat. Is there any evidence of buckle at all?

Professor Mao replied: It is not clear. The flat 2D arrays shows buckles when imaged by AFM, but we don't know whether the buckles are formed during the deposition of the DNA sample onto a solid surface or the formation in solution. We strongly suspect that the former is the case.

Professor Fery asked Professor Matile: How important is Manning condensation for the observed counterion condensation effects? If this would be the dominating mechanism, one would expect that the effect increases, once the linear charge density falls below the Bjerrum length in the solvent for a strong polyelectrolyte. Do you observe this?

Professor Matile responded: This is not observed; the activity of polyphosphate is in the range of that of calf thymus DNA. The hydrophobicity, multivalency and nature of the ion pair seem more important.

Dr Sporer continued the discussion of the paper by Professor Colquhoun: The type of tweezer molecules Professor Colquhoun used in his experiments should be well suited to perform luminescence experiments. Have you also used fluorescence spectroscopy to study the interaction between the pyrenyloxy tweezer and the aromatic polymer? If yes, what have been the results?

Professor Colquhoun answered: We have carried out only a very limited amount of work on fluorescence-quenching of the pyrene units of our tweezer-molecules on binding to copolyimides, mainly because this technique, though extremely sensitive in terms of detection, is a rather blunt instrument (compared to NMR) when it comes to identifying specific sequences bound by the tweezer-molecules. In a different context, we have however applied fluorescence methods quite successfully to the detection of explosives (polynitroarenes) by polymeric tweezer-type molecules.[1]

1 S. Burattini, H. M. Colquhoun, B. W.Greenland, W. Hayes and M. Wade, *Macromol. Rapid Commun.*, 2009, **30**, 459.

Professor Chau returned to Professor Matile's paper: Can counterion-activated polyions penetrate the cell nucleus?

Professor Matile answered: Results with enhanced green fluorescent protein (EGFP)-tagged cell-penetrating peptides (CPPs) suggest that this is possible. Vesicle studies suggest that internal release can occur together with counterion activators, which then could be reutilized to enter the nucleus, but this is speculation at this point.

Professor Chau then asked Professor Mao: How do you measure the flexibility of the DNA arms in the star structure quantitatively?

Professor Mao replied: One experiment called a ligation-closure experiment is specifically designed to measure DNA flexibility. One polymerize a DNA molecule (by ligation) and then analyze the product (linear oligomers and cyclized oligomers) distribution. The more flexible the DNA molecule is, the more cyclized oligomers are expected.

Professor Hu asked: You have made a nice design using the geometry of DNA construction. There exist many possibilities for misconstruction in the meantime. Do you have any idea of how to purify your designed product?

Professor Mao replied: Misconstructions very likely exist. However, from native gel analysis, DNA samples looks pretty homogeneous.

Dr Channon continued the discussion of the paper by Professor Colquhoun: My question has two parts, the first extends the point raised previously by Professor Woolfson, the second is a more general point. Firstly, the general theme of your presentation suggests that you are interested in the information content of you polymers, and the ability of tweezer parts to recognised the presence of this information; there is an obvious and striking analogy with the role and function of DNA in this respect. My question is, do you have any plans as to how you might go about completing the analogy, as it were? That is, can you have the "sequence" of your polymer coupled to some desirable functional aspect and, *via* the recognition of the tweezers for particular sequences, allow templating of the sequences that perform the function more effectively? This would allow a real evolutionary process to occur in your system.

My second question is: it appears as if the polymeric component of your system should undergo a conformational change on binding of the tweezer component, perhaps becoming more compact. Do you think that this process could be used in a responsive network, of the type described by Professor Jones in his presentation?

Professor Colquhoun answered: We do indeed plan to develop functionalised tweezer-molecules to which polymerisable monomers can be attached (and possibly even activated through their attachment). The tweezers will then act as adaptor-molecules (conceptually analogous to transfer-RNA molecules), allowing a specified sequence of co-monomers to be aligned and then polymerised on the polyimide template. This latter type of chemistry is, however, still some way from being realised, not least because the selection of specific templating sequences from a longer, random sequence would require certain sequences to be recognised as "start" and "stop" messages, as does of course occur in protein synthesis.

On the second point, we do have some indirect evidence for a change of polyimide conformation to a more rigid and extended structure on tweezer-binding, in that the inherent viscosity of the chain-folding homopolyimide **1** in solution increases dramatically (by up to 300%) on addition of tweezer-molecule **6**.

Mr Carew continued the discussion of the paper by Professor Matile: Could your biphilic molecules be tuned to enter certain membranes? For instance a well-known cancer drug delivery strategy is to target drugs to the so-called 'leaky' cancer cell membranes. Could your molecules be used to selectively penetrate such membranes?

Professor Matile answered: The EPR effect is routinely used for liposomal delivery and operates mainly on differences in size. Being much smaller than liposomes, I would expect less significant EPR effects with polyion–counterion complexes. However, other conventional targeting methods such as the attachment of cell-specific ligands (*e.g.* folate) should be applicable without any problems. With counterion-activated CPPs (R_8), objects as large as GFP, NMR probes or quantum dots could be delivered and used *in vitro* (HeLa cells, *etc.*). *In vivo* applications of polyion–counterion complexes appear, however, to be inaccessible so far and would require implementation and optimization of lipoplex-type technologies. From a supramolecular chemistry point of view, I can see much potential in this direction and am tempted to try.

Dr Munz returned to the discussion of the paper by Professor Mao: Firstly, gel electrophoresis is frequently used to characterise the molecular weight distribution of DNA nanostructures. If you applied this technique to the 3D nanostructures, did you observe shifts in the band position as compared to linear molecules of similar molecular weight? Such linear molecules are typically used as a reference. It seems likely that the shape of the 3D nanostructures affects their motion through the gel, thus causing a shape-related shift in the band position.

Secondly, could one employ the sphere-shaped nanostructures as a reference for gel electrophoresis of 3D nanostructures, *i.e.* to complement the linear molecules?

Professor Mao replied: To your first question: yes, we observed different mobilities for the molecules with the same molecular weight but different 3D structures. Secondly, in principle, it is possible, but such a reference might be needed only in some special cases.

Dr Titmuss asked: In Fig. 5 in your paper you suggest that a 250 nm diameter nanocage can be assembled from purely 5-point star motifs. Is this really the case as large viruses incorporate 6-fold/3-fold elements in addition to 5-fold elements to produce a closed curved surface? My specific question is whether the large (250 nm) diameter sphere is composed of purely 5-fold elements that are distorted from planarity to differing amounts (imagine an umbrella in different states of openness) or whether some of the 5-fold DNA stars have actually unbound into 3-fold motifs that then combine with the 5-fold motifs to produce virus like spherical nanocages?

Professor Mao answered: It is very reasonable thinking. I am very eager to know the results. The problem is that we don't have a proper experimental method to distinguish the two possibilities.

Professor Ikkala opened the discussion of the paper by Dr Hayes: You showed interesting self-healing as evidenced by recovered Young's moduli. Do the strength and fracture toughness also self-heal?

Dr Hayes responded: The strength does heal. We have yet to assess the fracture toughness; however, since these initial materials are rubbers this may not be an appropriate parameter to study.

Mr Ahangar remarked: Does your polymer or the healing mature that you use have to be elastic and can it be elastic to the same extent even after healing? Also, what about the effect of high temperatures which you use during the process of healing?

Dr Hayes responded: The self-assembled healing blend that we have developed is elastic in nature as a result of the polymer components used to make them—however, if we elected to use more rigid polymers which possess higher glass transition temperatures then the physical form of the blends would change and would be less elastomeric. The healing studies of the elastomeric blends that we have completed to date indicate that the elongation of the repaired material slightly decreases but not enormously—these results are very encouraging given the infancy of our system.

If we exposed the current reported self-assembled blends to high temperatures (>250 °C) then there is potential for degradation of the covalent integrity since there are ether and amide linkages present in the polymer backbones. However, our healing studies use temperatures below the degradation temperatures of these bonds so the chemical composition should not be affected.

Mr Carew remarked: I have two questions. Firstly, practically speaking, how many times is it possible to break and then heal a single sample of your polymer?

Secondly, how expensive are these materials compared to other bulk plastics like polystyrene?

Dr Hayes responded: First: we have healed structurally related materials to those reported in this paper (these supramolecular blends feature the same π–π stacking motif but different polymeric 'mid-blocks') in excess of 10 times without any noticeable loss in performance.

Second: it is fair to say that these materials are more expensive to produce than polystyrene! That said, both components of the blend are afforded on multigram scales *via* 1-step syntheses using inexpensive commercially available starting materials, so we do not consider that this route to healable materials represents an overly complicated approach.

Dr Harries asked: Have you thought about functionalising the pyrene unit to change the optical properties of the material? Do you think this will have an impact on the π-stacking interaction?

Dr Hayes answered: Alteration of the pyrene or the diimide structures should enable the optical properties of the resultant self-assembled materials to be tailored—at present this material is deep red in colour. Decreasing the π-electron density of the pyrene will affect its ability to bind with the chain-folded polydiimide so may lead to a weaken material. Steric factors also have to be considered in these proposed modifications.

Dr Titmuss remarked: What is the magnitude of the elastic moduli? Do the moduli show any dependence on the length of the mid-blocks? By increasing the length of the mid-blocks could you build in entanglements and in doing so improve the resistance to deformation?

Dr Hayes replied: We have not yet assessed the elastic modulus and in theory, increasing the length of the polymer mid-segment above chain entanglement values should lead to increased toughening and resistance to deformation.

Dr Shaffer commented: Intrinsically, your approach to self-healing relies on breaking supramolecular interactions rather than the polymer backbone; therefore, whichever supramolecular interaction you chose must be relatively weak compared to an intrachain covalent bond. Presumably, also, the density of supramolecular units will always be lower than the number of polymer chains in a conventional matrix, per unit area of crack. Therefore, we might naively expect that supramolecular structures of the type proposed will always suffer from a very much lower strength than a conventional resin; whilst some reduction in strength might be acceptable, any significant weakening would outweigh the benefits of self-healing. Is there any reason to suppose that this simple, pessimistic view is incorrect? As a secondary effect, the intrinsically relatively weak supramolecular interactions, which allow self-healing at low temperatures (even approaching room temperature), must inherently introduce a tendency to creep. Do you have strategies in mind to address these issues?

Dr Hayes answered: These materials, in their current composition, are not especially strong materials as a consequence of the nature of the polymeric 'mid-blocks' that have been used. Altering the chemical composition of the polymeric 'mid-blocks' could, *via* phase separation, offer the ability to produce materials with inherently stronger characteristics. An alternative strategy would be to utilise a composite approach in order to address the strength issues whilst maintaining the ability to heal at lower temperatures (taking combined advantage of (i) the inherent strength of the filler used and (ii) the reversible nature of the relatively weak non-covalent interactions).

The ultimate self-healing material is one that offers high strength and toughness yet is easily healed at room temperature—this is a very tough assignment to address!

Dr Steinke said: After several crack formation–healing cycles what can be said about the chemical composition of the material in the crack region? Given that the healing mechanism relies on flow would one expect a bias towards lower molecular weight polymers at the crack location compared to the composition of the bulk?

Dr Hayes answered: This question is an excellent one. We have yet to analyse the composition of the healed areas—one method to probe the composition could be to cut out the healed zone, dissolve the blended materials in an appropriate deuterated solvent and then subject this solution to NMR spectroscopic analysis. Comparison of the spectroscopic data obtained for the pristine blended material should enable any potential bias this towards lower molecular weight components to be determined. The healing process for these blends does not affect the tensile modulus when compared to the pristine sample—we obtain 100% healing efficiencies for these blends—these observations are encouraging and possibly indicate that the local composition in the healed area is not drastically affected.

Mr Ahangar addressed Dr Hayes and Mr Burattini: What about the time span it takes for healing the fracture? Is there any link between the type of fracture and healing? Do you think that the molecules at the cracks need to be in a particular orientation for healing process to take place?

Dr Hayes responded: The timespan for healing the fractures is dependent upon the components used in the blend. For example, the system described in the paper attains 100% healing in ~5 min by observation. Subsequent derivatives we have developed (but not reported in the paper) take longer to heal. We have not assessed the relationship between the type of fracture and the healing profile.

We have not attempted to orient the molecular direction and study the healing process yet.

Mr Burattini replied: Thanks for the questions. The temperature variable ESEM experiment was conducted at a rate of 5 °C per min and it appeared to heal quickly between 90 and 115 °C. This system has been cycled only a couple of times obtaining same results and other systems showed the possibility to heal the fracture at least 10 times.

If by orientation you mean order, I don't think you need any particular orientation. What you do need is that the guest motif (naphthyl diimide units) assumes a folded conformation around the electron-rich pyrene unit and this driven by thermodynamic effects.

Professor Ikkala commented: Dr. Hayes, I guess one of the aimed applications of self-healing materials could be in engineering, where solvent and moisture insensitivity are important. Could you comment on such properties for your material?

Dr Hayes replied: The relatively low molecular weight polymers used in these healing blends are readily soluble in a range of organic solvents so I anticipate (we have yet to undertake solvent uptake experiments) that exposure to organic solvent may be a factor we have to consider. However, given the molecular design and ease of synthesis of these materials it should not be too difficult to incorporate components that decrease any such effects. We have not seen any loss of physical properties of these healable blends when exposed to moisture—these materials are stable to moisture and the healing efficiencies are not affected.

Professor Huskens opened the discussion of the paper by Dr Piñol: You make intriguing faceted vesicles by using rigid liquid-crystalline vesicle walls. Is there a preference for the angles that are formed between the facets? What does the edge between these facets look like, structurally and molecularly? Do you see any options for designing or tuning the system such that you can control the angles and thus the sizes and numbers of facets?

Dr Piñol replied: There is not really a preference for the angles that are formed between the facets. We have observed angles around 90°, >90° and <90°. The vesicles have not uniform shapes from one to another. The resolution of cryo-TEM suitable for this kind of soft systems doesn't permit us to see the molecular structures around the edge between facets. We suppose disclinations (smectic defects) should exist in these regions. Nano-holes may also be possible in the liquid crystal hydrophobic layer. We can change the hydrophobic and hydrophilic length, as well as hydrophilic/hydrophobic ratio, to change the nano-objects morphology. This study is in course.

Professor Matile continued the discussion of the paper by Dr Hayes: Would self-healing in the sea be as efficient as in a lake? Considering the strongly positive quadruple moments of naphthalenediimides (NDIs), could you comment on the responsiveness toward anion–π and cation–π interactions?

Dr Hayes replied: The materials are hydrophobic in nature and do appear to be affected by atmospheric water. In this regard we do not anticipate any decrease in materials performance in a lake! We have not exposed these supramolecular blends to salt solutions so have not studied anion/cation uptake and the potential effect of this upon the healing characteristics.

Mr Carew said: Two questions: firstly, is there evidence of crystalline structure in your materials?

Secondly, would placing bonding units pendant to the main polymer chain improve the strength of your materials?

Dr Hayes replied: With respect to your first question, in the blend there is no evidence of crystallinity. Secondly, in principle a comb polymer in which the recognition units are appended from the polymer could afford a stronger material; we have already made materials of this type for sensor applications.[1]

1 S. Burattini, H. M. Colquhoun, B. W. Greenland, W. Hayes and M. Wade, *Macromol. Rapid Commun.*, 2009, **30**, 459–463.

Dr Titmuss continued the discussion of the paper by Dr Piñol: The facets of the unilamellar vesicles you observe are essentially planar structures (smectic forming). By analogy with structures formed by lipids/surfactants, and the Israelachvili model for self-assembly of surfactant/lipid structures in terms of molecular curvature, could you induce a preferred curvature into the lamellae by using PEG tail groups of differing lengths? For example, a longer PEG tail group would play the same role as a lipid/surfactant with a larger head group area and so tend to form a curved interface.

Dr Piñol responded: Yes. My answer to this question relies on another smectic block copolymer system. By changing the PEG length, we can change the morphology from smectic vesicles to smectic nanofibers. In nanofibers, the preferred curvature is higher than in the vesicles (please see ref. 1 and 2).

In these systems, no faceted vesicles, but ellipsoidal vesicles were formed. Work also in progress to understand these different smectic polymersomes.

Additionally, the situation of block copolymer amphiphiles is more complicated than that of small molecular surfactants. For example, the temperature also plays an important role in self-assembling, because the PEG conformation in water is temperature-dependent.

1 L. Jia, A. Cao, D. Lévy, B. Xu, P.-A. Albouy, X. Xing, M. J. Bowick and M.-H. Li, *Soft Matter*, 2009, DOI: 10.1039/B907485F.
2 R. Piñol, L. Jia, F. Gubellini, D. Lévy, P.-A. Albouy, P. Keller, A. Cao and M.-H. Li, *Macromolecules*, 2007, **40**, 5625–5627.

Dr Channon returned to the discussion of the paper by Dr Hayes: The materials that you have presented appear to perform their "healing" function very well. In addition to healing, biological materials (such as bone and tendon) have the capacity to increase their ability to bear stress in areas that are subject to higher stresses. Have you considered how you might extend your materials to encompass such a function? This could perhaps be achieved through the incorporation of fibrous nano-structures to form a composite, in which alignment is induced in the fibres by stress on the material, so increasing the strength of the material in the direction of the applied stress?

Dr Hayes replied: We are aware of the current limitations of our self-healing systems in terms of their strength characteristics—however, these materials are very much 'proof of concept' prototypes that have been designed to test the concept of using π–π stacking interactions between two different polymer components to generate blends that have healing properties. Your suggestion of investigating a composite structure that incorporates fibres and our self-healing material in order to improve the ability of the system to cope with higher stresses is a very good idea and a logical route for this work to follow—in our polymer group meetings at Reading we have discussed this type of composite but the immediate focus of our research has been to understand the basic assembly/disassembly process so that we can tailor the healing characteristics accordingly.

Mr Carew asked: Are your polymers conductive? Perhaps if doped with ions?

Dr Hayes responded: We have not examined the conductive nature of these polymer blends. However, if they were, it might offer another healing mechanism!

Dr Munz asked: My first question is: considering the self-healing properties of such supramolecular polymers, they seem to be promising as a sizing material for the fibres in fibre-reinforced polymers. Do you think that such an application would make sense?

My second question: frequently, failure of the fibre/polymer interface is due to high shear stresses. Could you give us some idea of the fracture behaviour of supramolecular polymer networks under shear loading, as in mode II or mode III testing? Do you expect similar healing effects as for the case of normal loading?

Dr Hayes responded: In response to the first question, I agree with the suggestion made here—supramolecular polymers of this type may prove to be very good sizing systems for fibre-reinforced polymers.

As to the second, we have yet to test our supramolecular blends under shear stress assays so cannot provide an indication of how they are likely to behave. These tests will be conducted in the near future and the results made available in publications or *via* our websites.

Dr Titmuss commented: The π–π bonding interaction presumably involves some degree of (partial) charge transfer between the π-donor and π-acceptor groups. When you apply a stress that breaks the material across the weakest interface (presumably between π-acceptor and π-donor) is there an electrical discharge caused by the separation of charge across the interface?

Dr Hayes answered: We do not know! We will need to look into this phenomenon.

Mr Ahangar asked: Does your method involve extensive study of bonding at the surface and modifying chemistry of surfaces?

If we look at the mechanism of healing of skin when it cracks, can we learn something from that and apply the same strategy to improve your system? Are there any other alternative forms which nature uses to heal the surfaces and which can be applied to your concept?

Dr Hayes replied: Our studies are at a very early stage so we have not studied in detail the nature of the bonding within the blend specifically at surfaces—however, we realise that the way in which these supramolecular blends behave at surface interfaces is an important parameter to understand. In terms of your point regarding 'improving' our system, we are not sure that we need to improve it as we have not studied the physical properties of sufficient number of analogues in order to rank the physical properties of these derivatives. Nature has many healing processes which can and should be studied—skin, bone and key organs are all regenerated and each of these has its own perfected repair mechanisms.

Quantitative approaches to defining normal and aberrant protein homeostasis

Michele Vendruscolo* and Christopher M. Dobson*

Received 23rd March 2009, Accepted 17th April 2009
First published as an Advance Article on the web 28th July 2009
DOI: 10.1039/b905825g

Protein homeostasis refers to the ability of cells to generate and regulate the levels of their constituent proteins in terms of conformations, interactions, concentrations and cellular localisation. We discuss here an approach in which physico-chemical properties of proteins and their environments are used to understand the underlying principles governing this process, which is crucial in all living systems. By adopting the strategy of characterising the origins of specific diseases to inform us about normal biology, we are bringing together methods and concepts from chemistry, physics, engineering, genetics and medicine. In particular, we are using a combination of *in vitro*, *in silico* and *in vivo* approaches to study protein homeostasis through the analysis of the effects that result from its perturbation in a select group of specific proteins, from either amino acid mutations, or changes in concentration and solubility, or interactions with other molecules. By developing a coherent and quantitative description of such phenomena, we are finding that it is possible to shed new light on how the physical and chemical properties of the cellular components can provide an understanding of the normal and aberrant behaviour of living systems. Through such an approach it is possible to provide new insights into the origin and consequences of the failure to maintain homeostasis that is associated with neurodegenerative diseases, in particular, and the phenomenon of ageing, in general, and hence provide a framework for the rational design of therapeutic approaches.

Introduction

One of the essential characteristics of living systems is the ability of their molecular components to self-assemble into functional structures.[1,2] Equally important, however, is the way in which the processes leading to this organisation are balanced within the cellular environment through the mechanism of homeostasis.[3-7] Of central importance in the study of this mechanism is to focus specifically on proteins, since these are the molecules that enable, regulate and control essentially all chemical processes on which life depends. In order to function, the large majority of our proteins need to fold into a specific three-dimensional structure.[4,8-10] Indeed, the wide variety of highly specific structures that results from protein folding, and which serve to bring key functional groups into close proximity, has enabled living systems to develop astonishing diversity and selectivity in their underlying chemical processes by using a common set of just twenty building blocks—the amino acids.[11] Much research has addressed the fundamental mechanism of protein folding through a combination of *in vitro* and *in silico* studies, and we now have considerable understanding at a molecular level of the

Department of Chemistry, University of Cambridge, Lensfield Road, Cambridge, UK CB2 1EW. E-mail: mv245@cam.ac.uk; cmd44@cam.ac.uk

fundamental principles underlying this complex process.[8,12–14] The next challenge is to relate this information to events occurring in living systems.

As well as simply generating biological activity, however, we now know that protein folding in living systems takes place in a complex environment, and as a polypeptide chain emerges from the ribosome on which it has been synthesised, it interacts with a wide range of ancillary molecules including molecular chaperones.[4,15,16] Much less is known about such events at a molecular level and a primary objective in protein science is to extend the studies of folding from the test tube to the cell, and to understand how this process takes place in the cellular environment. Moreover, it is clear that protein folding and unfolding are closely coupled to many other biological processes ranging from the trafficking of molecules to specific cellular locations to the regulation of the growth and differentiation of cells.[7,10] In addition, only correctly folded proteins have the ability to remain soluble in crowded biological environments and to interact selectively with their natural partners.[10,15] The manner in which proteins are able to maintain such homeostasis is a subject of central interest in molecular biology.

Given the tremendous importance of protein folding, it is not surprising that the failure of proteins to fold correctly, or to remain correctly folded, is the origin of a wide variety of pathological conditions, including cystic fibrosis, α1-antitrypsin deficiency and Alzheimer's disease.[10,17–20] In many of these diseases proteins self-assemble in an aberrant manner into large molecular aggregates, including amyloid fibrils. Considerable attention has been devoted to exploring the nature and origin of such disorders from a structural viewpoint and to understanding the manner in which the balance between normal and aberrant conformational transitions can be perturbed.[20] Several studies have involved *in vitro* studies coupled with computer simulations,[20–23] and many others have been concerned with the goal of relating processes studied in atomic detail in the test tube to their quantitative effects in living systems.[24–26] Moreover, recent findings suggest that further developments in this area could have much more general relevance to understanding the way in which well-established physical and chemical principles can provide new insights into the apparent complexity of biology.[27]

The discovery of the common existence of amyloid and amyloid-like states is of unique importance in understanding the nature of biological systems because it reveals that there is an alternative stable and highly ordered state, accessible essentially to all proteins, in addition to the native one;[10,28,29] this observation has profound implications in diverse fields ranging from medicine to materials science. Because the structural interactions within the amyloid state and the native state are similar—although the latter are largely intramolecular whilst the former also include strong intermolecular contributions—the stability of the native and amyloid states can be comparable.[30] There is thus a competition between the two states that results in normal or aberrant biological behaviour depending on whether the native or the aggregated state is populated.[10,28,29] More generally, the maintenance of the correct balance in the populations of different states of proteins, one facet of protein homeostasis, is of great significance, as even marginal alterations in such populations can result in disease in the long term.[7,27] Indeed, it has been recently realised that the limit to the safe concentration of proteins in living systems is likely to be reached when the amyloid state becomes more stable than the native state.[27]

It is therefore of great importance to complement the well-established characterisation of the structure, folding and stability of native states with a similar analysis of the structure, assembly and stability of other states—ranging from unfolded and partially folded species, including natively unfolded states, to aggregated species such as amyloid fibrils. This is one of the main thrusts of our own work, together with the exploration of the effects of the balance between the normal and aberrant states of proteins in living organisms such as the *Drosophila* model system, which we believe will inform us on the origins of amyloid-related disease and hence more generally on the mechanism of protein homeostasis.

A conceptual framework for understanding protein homeostasis

It is becoming clear that the interplay between the various states of different protein molecules creates a highly complex system, whose behaviour determines whether a living organism functions in a normal or aberrant manner, and yet, as with other complex systems,[31,32] may be determined by the combination of relatively simple underlying processes.[3,6,33] This complexity[6] underlies the phenomena now often referred to as protein homeostasis or "proteostasis"[7] perhaps in a similar manner that, for example, individual organisms interact[34] in an ecosystem. The investigation of this particular class of biological molecules could therefore potentially shed a great deal of light on more general questions of the design and evolution of biological molecules and the environments in which they function. Such information lies at the heart both of understanding the molecular aspects of the phenomenon of life and of rational approaches to molecular medicine.

In this paper we present a strategy for describing and understanding in a coherent manner the behaviour of proteins in living systems, including their folding, misfolding and assembly processes. Our approach is primarily based on five technical and conceptual developments that have recently been made in protein science:

(1) The ability to describe quantitatively, by a combination of experimental and computational approaches, the often disordered and dynamic structures of the multiple states of proteins on which their biological behaviour depends.[35–38]

(2) New ideas about protein aggregation,[10,28] including the finding that the ability to assemble into stable and highly organised structures (*e.g.* amyloid fibrils) is not an unusual feature exhibited by a small group of peptides and proteins with special sequence or structural properties, but rather a property shared by most, if not all, proteins;

(3) The discovery that specific aspects of protein behaviour, including their aggregation propensities[21,23,39,40] and the cellular toxicity associated with the aggregation process,[24,41] can be predicted with a remarkable degree of accuracy from the knowledge of their amino acid sequences;

(4) The realisation that a wide variety of techniques originally devised for applications in nanotechnology can be used to probe the nature of protein aggregation and assembly and of the structures that emerge;[30,42–44] and

(5) The development of powerful approaches using model organisms for probing the origins and progression of misfolding diseases by linking concepts and principles emerging from *in vitro* studies to *in vivo* phenomena such as neurodegeneration.[24]

An analysis of these results, which span across a wide range of subjects from neuroscience to nanoscience, reveals that the ability to keep proteins in their soluble form is absolutely central for the maintenance of cell homeostasis.

Protein solubility and biological complexity

Considerable advances have been made in recent years in the search for the chemical and physical principles that underlie the complexity of biological phenomena. We believe that it is possible to describe, for example, processes such as folding and aggregation, at least in outline, in generic terms and link them to well-established and quantifiable concepts of chemistry and physics. In this context, it should then be possible to study in depth a relatively small number of carefully chosen proteins, and yet extract general principles from such studies. In particular, we believe that there are many common features underlying the diseases associated with amyloid formation. Thus, much of our research is focused on the development of this theme using amyloid-related diseases as a paradigm of the way the ideas of chemistry and physics can provide fundamental insight into both normal and aberrant behaviour and suggest novel therapeutic strategies.

Biological systems have evolved to be efficient by achieving astonishing levels of molecular crowding within cells and extracellular space, which is of the order of

300 g l⁻¹.[10] Information can then be transmitted rapidly, largely by molecular diffusion, between different components enabling them to function efficiently.[3] Molecules such as proteins remain soluble and able to avoid interaction with all but a relatively small and specific selection of other molecules, yet are composed of chemical species that are often extremely hydrophobic and prone to self-assemble. We are beginning to think that the ability to maintain the solubility of its component molecules is of much more general significance in biology than previously imagined.[27] Thus the observation that the sequence determines the solubility of proteins[45] could be just as important as the fact that it determines their structure and the ability to fold. As we understand in increasing detail how sequences define solubility, it is becoming possible to predict aspects of biology in ways that were previously unsuspected.

Multiple forms of protein structure

Much is understood about the nature of globular protein structures and about the principles by which isolated denatured polypeptide chains are able to achieve such states.[4,8-10] A more complete knowledge of the behaviour of proteins in the cell has, however, been limited by the challenges involved in defining the structures of native states in complex environments and of the highly dynamic structural ensembles that describe most of the additional forms of proteins that are now known to be of biological importance, including natively unfolded states and partially folded states involved in folding and in aggregation.

A detailed structural description of native and non-native states lies at the heart of our ability to describe in a quantitative manner the complex behaviour of proteins within a cell. Powerful techniques are being developed to complement more established methods to overcome the challenges posed by the task of providing such a description.[30,42-44,46] Our own approach is based primarily on methods that directly combine experimental and computational techniques.[5,6,35] These procedures involve the use of experimental data, largely derived from NMR spectroscopy, as restraints in computer simulations.[36,37] We have already used in this way a range of different types of experimental data, for example distance measurements from paramagnetic probes introduced by mutagenesis,[47,48] and structural and dynamical information from hydrogen–deuterium exchange experiments.[49-51] But a major breakthrough has come recently through the discovery[37,52-55] of ways in which chemical shift data can be used in this approach to generate structures of native states to an accuracy comparable to that of conventional methods (Fig. 1). The use of chemical shifts requires only a resolved and assigned spectrum and thus renders unnecessary the measurement of large numbers of additional parameters such as interatomic distances derived from nuclear Overhauser effects (NOEs). The latter measurements are very challenging (and in practice virtually impossible) in highly dynamical systems and for the conformationally heterogeneous states populated, for example, along the folding and misfolding pathways.

These chemical shift techniques, particularly in conjunction with the measurement of residual dipolar couplings (RDCs), can be used to generate structural ensembles of non-native states for which these parameters are often the only ones that can be readily measured.[56,57] These methods can be used to study a wider range of protein states, including highly dynamical ones such as low-populated conformations involved in the folding and misfolding processes. Indeed we have already provided a proof of principle that the use of chemical shift restraints, here derived from relaxation dispersion techniques, can be used to characterise transient species.[58] We believe that these computational approaches, as well as advances in experimental techniques that enable the systematic measurements of chemical shifts in non-native states of proteins,[56] will enable us to increase very considerably the resolution to which this type of non-native structure can be determined.

This approach should enable the characterisation of proteins whose structures have proved elusive by conventional means, but which are crucial to understanding

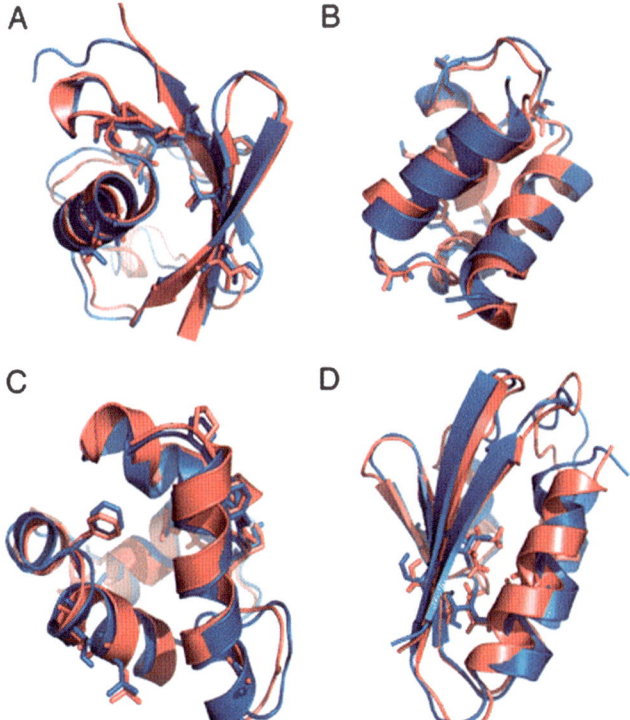

Fig. 1 Comparison between protein structures determined by X-ray crystallography (pink) and by the technique that we have recently introduced that uses only NMR chemical shift information (blue).[37] Despite being only at the initial stages of development of the method, we are already able to generate structures of globular proteins of up to 120 residues in length that agree with the those determined by conventional methods with RMSD values of 1.2–1.8 Å for the backbone atoms and 2.1–2.6 Å for the side-chain atoms.

the process of protein folding. Two recent studies[59,60] have already demonstrated this approach by using as test cases calmodulin,[59] a protein that plays an essential role in signal transduction pathways, and ubiquitin,[60] a protein that is a key component in degradation pathways, that the use of molecular dynamics simulations with NMR restraints[36] enables the changes in structure and dynamics upon binding to be described at nearly atomic level resolution, thus enabling analysis of the molecular mechanisms responsible for binding to be carried out. These studies have provided strong support for the "equilibrium shift" model,[61,62] according to which the conformations that proteins adopt in the bound state are already present, although with low statistical weights, in the unbound states in solution. This mechanism enables proteins such as calmodulin and ubiquitin to interact with large numbers of other proteins in a selective and efficient manner.

Another crucial area in which the inclusion of experimental measurements in molecular dynamics simulations is having a major impact is in the investigation of the structures of protein complexes, where even weak interactions can be detected by NMR methods.[63] In a first study,[53] we have already established, using the case of the structure of the cytotoxic endonuclease domain from bacterial toxin colicin (E9) in complex with its cognate immunity protein (Im9),[64] that chemical shifts enable the determination of protein complexes even when the complexes themselves exhibit significant dynamics and the component proteins undergo conformational rearrangements upon binding. It is also possible to determine the structure of protein–protein complexes using chemical shift information when the chemical shift changes upon binding are relatively small and hence particularly difficult to compute accurately. This result is a consequence of the well-known fact in NMR spectroscopy that the availability of a large number of restraints—in this case derived from chemical shifts—can provide enough information for high-resolution structure determination, even if they are not accurately known individually.[65] We have developed a computer code called CamDock to enable the structures of protein–protein complexes to be determined by combining advanced docking methods[53] with the information provided by chemical shifts[53] and residual dipolar couplings.[66]

It is then of very great importance to be able to relate the principles that emerge from studies in the test tube to analogous events occurring in the cell. Interesting work has been done using NMR spectroscopy in environments designed to mimic the cellular milieu,[67] but our ambition is to go further and ultimately to explore processes taking place in the cellular environment itself.

One of the most fundamental, and yet so far elusive, aspects of protein folding concerns the way in which this process is initiated during or following biosynthesis on the ribosome, *i.e.* the manner in which folding occurs in the cellular environment.[4,9,15,16] We have recently shown the feasibility of applying advanced NMR techniques to obtain detailed structural insights into the conformations of nascent proteins during the process of their synthesis on the ribosome.[38] In collaboration with Dr John Christodoulou (UCL) we generated ribosome–nascent chain complexes (RNCs) by arresting RNA translation, and used this technique in the first instance to study a tandem immunoglobulin (Ig) domain repeat (Ig2) of an actin-binding protein.[38] Analysis of the spectra of these RNCs selectively ^{15}N/^{13}C-labelled in the nascent chains reveals that the first Ig domain of the translation-arrested nascent chain is able to fold to a native-like state that remains tethered to the ribosome by the second highly disordered Ig domain. This study is now being extended by studying nascent chains of different lengths to probe the progressive development of structure in a growing nascent chain. We believe that this approach can open the door to descriptions at an atomistic level of detail of the process of co-translational folding, so as to characterise in detail the process by which proteins fold as the nascent chain emerges from the ribosomal exit tunnel.

In the immediate future, there are also great opportunities provided by the inclusion of NMR observables, including chemical shifts,[37,52–54] residual dipolar couplings[68] and interatomic distances obtained by paramagnetic relaxation enhancement experiments,[47,48] with molecular dynamics simulations to probe in detail the crucial processes by which cellular components such as molecular chaperones, including Hsp70 and trigger factor, interact with nascent chains and help to promote correct folding and inhibiting misfolding and aggregation.[4,15] This work has the potential of opening up a vast range of new opportunities to explore the study of the fundamental mechanism of folding in environments directly relevant to living systems.

Molecular basis of protein aggregation

The structures, dynamics and interactions that stabilise protein aggregates are difficult to study, since these species are often insoluble and resist crystallisation, thus making it very challenging to apply standard solution NMR spectroscopy and X-ray crystallography techniques.[20,69] Interdisciplinary approaches appear to be particularly suitable to address the problem of describing the structures of a variety of protein assemblies.

Very considerable progress has been recently made to characterise quantitatively the physical properties of fully formed amyloid fibrils and of their partially ordered protofibrillar precursors by bringing together solid-state NMR spectroscopy, cryo-electron microscopy and techniques of nanoscience. Advances in solid-state NMR methods are making it possible to obtain interatomic distance information in states that are insoluble and non-crystalline such as amyloid fibrils.[69] In addition to information about the structure of the amyloid fibrils formed by an 11-residue fragment of human transthyretin, these approaches have provided great insight into the structures of amyloid fibrils formed by several other peptides and proteins, including Het-S and Aβ.[70–72]

Together with the strategy mentioned above, nanoscience techniques are also emerging as powerful tools that can provide insight into the factors that stabilise amyloid fibrils.[30,42,44] In an initial study, by using atomic force microscopy (AFM) imaging we described the changes in the distribution of inter- *versus* intra-molecular bonding interactions associated with the transition of proteins from their native globular structures into ordered supramolecular assemblies. This work reveals that hydrogen bonded arrays of polypeptide chains exhibit material properties intermediate between those of hard materials such as steel and carbon nanotubes and softer biological fibres such as tubulin and actin (Fig. 2).[30] AFM and other techniques also generate information about the dimensions of the cross-β structural core of amyloid fibrils and of the less structured regions flanking them. These measurements provide unique insight into the kinetic and thermodynamic factors responsible for the stability of amyloid fibrils.

Much of our current understanding of the process of aggregation has been obtained by light scattering and fluorescence measurements of the kinetics of their growth.[20,29] As the spectroscopic signals in these techniques do not exclusively arise from amyloid fibrils but also from other types of aggregates that may be present in solution, and because of the non-linear relationship between aggregate abundance and signal intensity, the results can be difficult to interpret in a highly quantitative manner. In order to overcome such problems we have shown that very accurate measurements of growth rates can be obtained by a strategy in which real-time monitoring of the increase in mass of the fibrils themselves is carried out by measuring the variation in the frequency of a quartz crystal oscillator.[42] The application of this quartz crystal microbalance technique for monitoring the kinetics of aggregation of a series of proteins and under a variety of conditions is enabling a systematic analysis of the factors that can influence protein aggregation, particularly the mechanism by which molecular chaperones can inhibit fibril growth.[42]

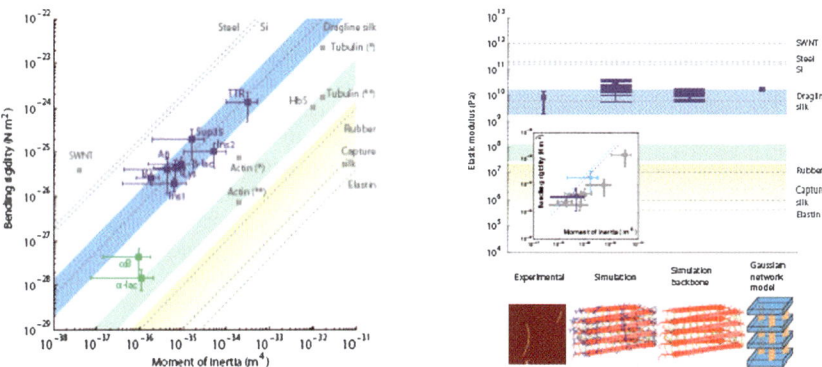

Fig. 2 A clear relationship between bending rigidity and moment of inertia has revealed the existence of universal mechanical properties of amyloid fibrils.[30] Left: illustration for a set of proteins of the values determined by atomic force microscopy (AFM) for the bending rigidities as a function of their cross-sectional moments of inertia (blue rectangles). For comparison, values for single-wall carbon nanotubes (SWNT), steel, actin, tubulin, rubber and elastin are also shown. The green shaded region shows the range of elastic moduli for materials held together by amphiphilic interactions. Right: comparison of the elastic moduli of fibrils from (blue squares, left to right) AFM measurements, full atom simulations, contribution of backbone alone, and results from the Gaussian network theory, in which the stability of a structure is estimated from the contributions of its hydrogen bonds.

Origins and consequences of aggregation in living systems

As we have discusses above, a wide range of human diseases is associated with protein misfolding and aggregation.[10,17–20] A powerful approach that has recently been proposed to study these phenomena is based on the combination of *in silico*, *in vitro* and *in vivo* studies to investigate the factors responsible for the abnormal assembly of proteins into insoluble aggregates and the effects that these conformational species have once released in the cellular environment.

Considerable progress has been made in characterising the major physicochemical factors that promote the aggregation of polypeptide chains, and increasingly sophisticated computational approaches have been developed[21,23,39,40] that enable predictions to be made about a variety of features of the process of aggregation of peptides and proteins. On this basis a method of predicting "aggregation propensity profiles" has been established that enables the identification of regions with a high intrinsic propensity for aggregation,[21,23,39,40] providing a platform for further development of this approach.

In the case of fully folded proteins, we have shown that it is possible to take into account the fact that the most amyloidogenic regions are protected from aggregation since they are located in the structural core of the native state and hence they are unable to form inter-molecular interactions without at least a degree of unfolding.[23] In essence, given the amino acid sequence of a protein, we have shown how it is possible to combine the predictions of the intrinsic aggregation propensity profiles with those for folding into stable structures to determine new aggregation propensity profiles of structured or partially structured proteins that account for the influence of the structural context. We have provided a initial demonstration of the potential of this approach through its application to the prediction of aggregation profiles for a range of peptides and proteins whose aggregation propensities have been characterized experimentally in particular detail, including the human prion protein, which is involved in sporadic, inherited and infectious forms of Creutzfeld–Jakob disease.

We anticipate that, in addition to its relevance for understanding misfolding diseases, the insight provided by these studies will in time represent a significant contribution to improving the biotechnological production of therapeutic peptides and proteins, in drug discovery initiatives and for antibody production. The ability to design rationally, and with increasing reliability, specific amino acid substitutions capable of altering significantly the aggregation propensities of peptides and proteins will enable us to investigate the physico-chemical factors responsible for the formation of amyloid fibrils and their oligomeric precursors.[20,21]

A range of strategies is being developed to combine *in vivo* approaches with *in vitro* and *in silico* methods to obtain a quantitative understanding of the molecular basis of neurodegenerative and other misfolding diseases. In an initial study we have demonstrated the potential of this approach in the case of the Aβ peptide, by showing that the relative toxicity in *Drosophila* of its mutational variants can be predicted with a remarkable 83% accuracy from their amino acid sequences (Fig. 3).[24] The advantage of using *Drosophila* models for such studies is that the brevity of their lifecycle, the power of the associated genetic tools, and the ease with which a range of toxicity-related phenotypes may be measured allows us to quantify the links between the *in vivo* toxicity of protein aggregates and the fundamental chemical properties of peptides and proteins.[24,41,73,74]

The results discussed above were obtained by developing a method for predicting the rate of formation of protofibrillar aggregates based on the physico-chemical properties of the amino acids comprising the sequences of the mutational variants of the Aβ42 peptide that we have investigated. It is also remarkable that, despite the fact that the intrinsic aggregation propensities of typical protein sequences vary by at least five orders of magnitude, we have been able to achieve profound alterations in the pathogenic effects of Aβ42 by increasing or decreasing its propensity to aggregate by less than 15%. This result suggests that proteins implicated in

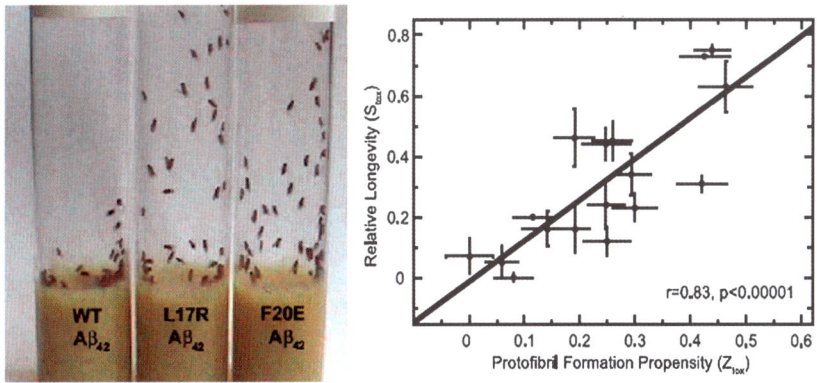

Fig. 3 Rational design of the toxic effects of Aβ42 mutants in a transgenic *Drosophila* model of Alzheimer's disease.[24] The relative longevity (S_{tox}, *y*-axis) of flies expressing a range of Aβ42 variants is predicted accurately by a score (Z_{tox}, *x*-axis) for the propensity to form protofibrillar aggregates ($r = 0.83$, $p < 0.00001$).

misfolding diseases are likely to be extremely close to the limit of their solubility under normal physiological conditions and consequently the small alterations in their concentration, environment or sequence, such as those that occur with genetic mutations or with increasing age, are likely to be the fundamental origin of these highly debilitating and increasingly common conditions.

The approach that we are developing is already enabling us to obtain accurate quantitative measurements of the relationships between the manifestations of neuronal dysfunction in a complex organism, such as locomotor defects and reduced lifespans, and the fundamental physico-chemical factors that determine the propensities of peptides and proteins to aggregate into oligomeric species and protofibrils. Our research is aimed at demonstrating that, despite the presence within the cell of multiple regulatory mechanisms such as molecular chaperones and degradation systems, it is the intrinsic, sequence-dependent propensity of the polypeptide chains to aggregate to form protofibrillar aggregates that is the primary determinant of its pathological behaviour in living systems.

Thus, by using quantitative *in vivo* and *in vitro* techniques we are exploring the links between various conformational states and pathological effects. Our strategy is to use the results of *in vitro* biophysical methods, including NMR spectroscopy, fibril formation assays and amyloid staining, to deduce the events occurring *in vivo*. We have pioneered this strategy in the case of the Aβ peptide to differentiate the effects of the propensities to form either fibrillar or protofibrillar aggregates, by rationally designing mutations that alter either the fibrillar or the protofibrillar propensities.[24]

Protein homeostasis in normal and aberrant biology

The fundamental connection between two aspects of proteins in the cell—their abundance and their solubility—is increasingly evident. A direct characterisation of this relationship is the very high correlation (97%) observed between the *in vitro* aggregation rates of a series of human proteins and the corresponding *in vivo* mRNA expression levels.[27] Thus, even relatively small alterations in protein abundance and *in vivo* solubility can be linked to human disease, as described below.

The existence of a close relationship between mRNA expression levels and protein aggregation rates (Fig. 4) provides a new perspective on the phenomenon of protein aggregation and of its connections with misfolding diseases.[27] In essence, these results suggest that the amino acid sequences of proteins determine not only their

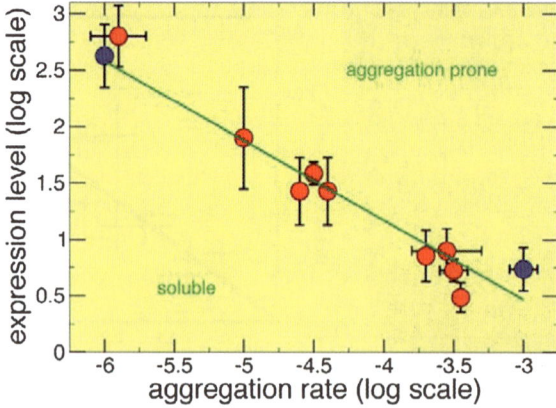

Fig. 4 Relationship between mRNA expression levels measured *in vivo* through microarray technologies and the aggregation rates of the corresponding proteins measured *in vitro*. We have considered all the 11 human proteins, either involved in disease (red circles) or not (blue circles), for which protein aggregation rates have been measured under near-physiological conditions.[27]

folding behaviour, but also their aggregation propensities and ultimately the susceptibility of an organism to contracting aggregation-related diseases.

The strong degree of anticorrelation between expression levels and aggregation rates suggests that the aggregation propensities of the proteins needed by the cell are precisely tuned to levels that enable them to be functional at the concentrations required for optimally efficient performance. It also indicates that protein molecules have co-evolved with their cellular environments to be sufficiently soluble for their biological roles, but no more so, and hence that aggregation can result from even minor changes in the chemistry and in the regulation of otherwise harmless proteins. Indeed, expression at higher levels than those found naturally is likely to result in enhanced clearance or in deposition of the proteins involved, both of which can generate disease. The intimate relationship between expression levels and aggregation rates is the net result of the opposing effects of an evolutionary pressure to remain soluble at the concentrations needed by the cell and random mutational processes that tend to increase their aggregation propensity.[27] Thus, over time, evolutionary selection generates proteins able to resist aggregation just at the levels required. Such a relationship provides dramatic evidence for the generic ideas about aggregation[10,28]—specifically that the propensity of proteins to revert to the amyloid state is the ultimate origin of amyloid-related diseases.

We are just beginning to explore in detail the hypothesis that proteins have evolved to have low enough aggregation propensities to enable an organism to function optimally, but with almost no scope for dealing with any situation where these levels increase further. We are now investigating the way in which variations in the concentration of proteins influence their behaviour in the cell. These studies are enabling us to understand the interplay of the physico-chemical properties of proteins and of the quality control mechanisms present in the cell to regulate the way in which they act. The approach that we are developing will enable us to obtain accurate measurements of the relationships between manifestations of neuronal dysfunction in a complex organism, such as locomotor defects and reduced life-spans, and the fundamental chemical factors that determine the propensities of peptides and proteins to aggregate into protofibrils.[24,75] We aim to understand how, despite the presence within the cell of multiple regulatory mechanisms such as molecular chaperones and degradation systems to avoid protein deposition, the intrinsic sequence-dependent propensity of polypeptide chains to aggregate remains a major determinant of the pathologies associated with misfolding and aggregation in living systems. In addition to demonstrating that rational mutagenesis can be used to alter systematically the toxicity of peptides and proteins, the transgenic *Drosophila* models that we are developing are enabling us to perform a quantitative analysis of the effects on the lifespan and mobility of *Drosophila* of other factors likely to be relevant to the *in vivo* aggregation process, in particular molecular chaperones, small therapeutic molecules and antibodies, or antibody-like species, by co-expression techniques.

Towards a quantitative biology based on physico-chemical principles

Complex regulatory networks, involving primarily nucleic acids and proteins, orchestrate the cellular functions required to maintain protein homeostasis.[3,33] These same cellular functions are also, however, dependent on the basic chemistry of the molecules taking part in them. Therefore, the "chemical" and the "cellular" views of cell biology are closely related, as is revealed, for example, by the high correlation between expression levels and aggregation propensities. By this statement, we do not mean that gene regulation itself is not important, as there is a huge amount of evidence that demonstrates its key role in protein homeostasis.[76] What we are saying is that very significant advances can be made by considering the "chemical view" of protein homeostasis. Studies are under way to explore the validity of the hypothesis that we have formulated according to which the necessity of avoiding aggregation plays a major evolutionary role in ensuring that proteins can remain soluble in the cell at the concentrations required for their function.

As an initial example, we discuss here the case of gene expression, which is the process through which the information contained in the DNA sequence of an organism is converted into functional proteins.[76] In response to the requirements of a cell, each step in the process of gene expression is regulated by complex cellular mechanisms, from the transcription of DNA into mRNA to the post-translational modification of proteins. The conversion of the information stored in DNA into proteins takes place through several phases that are highly regulated in response to the functional requirements of proteins by the cell. A detailed knowledge of the mechanisms of regulation can be used to rationalise and ultimately predict the

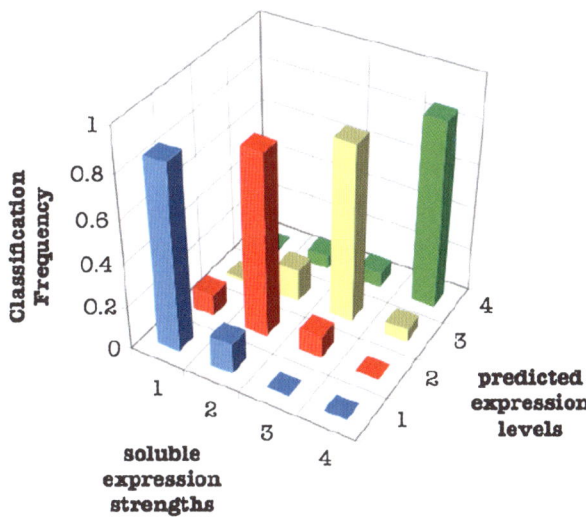

Fig. 5 Prediction of mRNA expression levels of recombinant human proteins in *E. coli*. When soluble expression strengths are compared with our predictions the overall accuracy is above 90%.[45]

outcome of the gene expression process. For example, the study of *cis*-regulatory motifs encoded in DNA sequences has reported an accuracy of about 70% for the prediction of expression patterns, while the correlation between the frequency of translational codons and gene expression levels is estimated to be around 60%.[77]

As an alternative to the strategy of exploiting the knowledge of the cellular regulatory processes to predict gene expression, we have proposed an approach, which has been prompted by the observation that proteins, once expressed, must remain soluble and avoid misfolding and aggregation in order to function optimally and to avoid cellular damage.[27] Since, as discussed above, we have established quantitative methods for predicting aggregation rates of proteins from the knowledge of the chemical properties of their sequences, we are in a position to investigate the relationship between these properties and the levels of expression of the genes. Our results indicate that it is possible to predict mRNA expression levels in *E. coli* with an accuracy of 90% or better from the knowledge of the sequences of the corresponding proteins (Fig. 5).[45]

Conclusions

Quantitative methods in molecular biology are providing unprecedented insights into the molecular mechanisms by which protein homeostasis is maintained in the cell. By drawing primarily on results from our own research we have discussed a variety of strategies based on the physico-chemical properties of proteins that appear to be particularly promising for increasing our ability to describe their behaviour *in vivo* and to suggest rational approaches to modulate it.

Acknowledgements

The results that we have described from our own research have been generated through a series of interdisciplinary collaborations. We would like in particular to acknowledge the contributions from Dr Andrea Cavalli, Dr Damian Crowther, Dr Anthony Fitzpatrick, Dr Bob Griffin, Dr Thomas Jahn, Dr Tuomas Knowles, Dr David Lomas, Dr Leila Luheshi, Dr Rinaldo Montalvao, Dr Gian Gaetano Tartaglia, Dr Mark Welland and Dr Helen Saibil. Our research is supported by the Royal Society, the Wellcome Trust, the Leverhulme Trust, the Alzheimer's Research Trust, the European Community, the European Molecular Biology Organisation, the Biotechnology and Biological Sciences Research Council and the Medical Research Council.

References

1 P. Aloy, B. Bottcher, H. Ceulemans, C. Leutwein, C. Mellwig, S. Fischer, A. C. Gavin, P. Bork, G. Superti-Furga, L. Serrano and R. B. Russell, Structure-based assembly of protein complexes in yeast, *Science*, 2004, **303**, 2026–2029.
2 C. V. Robinson, A. Sali and W. Baumeister, The molecular sociology of the cell, *Nature*, 2007, **450**, 973–982.
3 L. H. Hartwell, J. J. Hopfield, S. Leibler and A. W. Murray, From molecular to modular cell biology, *Nature*, 1999, **402**, C47–C52.
4 F. U. Hartl and M. Hayer-Hartl, Protein folding—Molecular chaperones in the cytosol: from nascent chain to folded protein, *Science*, 2002, **295**, 1852–1858.
5 M. Vendruscolo, J. Zurdo, C. E. MacPhee and C. M. Dobson, Protein folding and misfolding: a paradigm of self-assembly and regulation in complex biological systems, *Philos. Trans. R. Soc. London, Ser. A*, 2003, **361**, 1205–1222.
6 M. Vendruscolo and C. M. Dobson, Towards complete descriptions of the free-energy landscapes of proteins, *Philos. Trans. R. Soc. London, Ser. A*, 2005, **363**, 433–450.
7 W. E. Balch, R. I. Morimoto, A. Dillin and J. W. Kelly, Adapting proteostasis for disease intervention, *Science*, 2008, **319**, 916–919.
8 A. R. Fersht, *Structure and Mechanism in Protein Science: A Guide to Enzyme Catalysis and Protein Folding*, W.H. Freeman, New York, 1999.

9 M. J. Gething and J. Sambrook, Protein folding in the cell, *Nature*, 1992, **355**, 33–45.
10 C. M. Dobson, Protein folding and misfolding, *Nature*, 2003, **426**, 884–890.
11 C. M. Dobson, Chemical space and biology, *Nature*, 2004, **432**, 824–828.
12 K. A. Dill and H. S. Chan, From Levinthal to pathways to funnels, *Nat. Struct. Biol.*, 1997, **4**, 10–19.
13 C. M. Dobson, A. Sali and M. Karplus, Protein folding: A perspective from theory and experiment, *Angew. Chem., Int. Ed.*, 1998, **37**, 868–893.
14 J. N. Onuchic and P. G. Wolynes, Theory of protein folding, *Curr. Opin. Struct. Biol.*, 2004, **14**, 70–75.
15 J. Frydman, Folding of newly translated proteins *in vivo*: The role of molecular chaperones, *Annu. Rev. Biochem.*, 2001, **70**, 603–647.
16 B. Bukau, J. Weissman and A. Horwich, Molecular chaperones and protein quality control, *Cell*, 2006, **125**, 443–451.
17 R. W. Carrell and D. A. Lomas, Conformational disease, *Lancet*, 1997, **350**, 134–138.
18 B. Caughey and P. T. Lansbury, Protofibrils, pores, fibrils, and neurodegeneration: Separating the responsible protein aggregates from the innocent bystanders, *Annu. Rev. Neurosci.*, 2003, **26**, 267–298.
19 C. Haass and D. J. Selkoe, Soluble protein oligomers in neurodegeneration: lessons from the Alzheimer's amyloid beta-peptide, *Nat. Rev. Mol. Cell Biol.*, 2007, **8**, 101–112.
20 F. Chiti and C. M. Dobson, Protein misfolding, functional amyloid, and human disease, *Annu. Rev. Biochem.*, 2006, **75**, 333–366.
21 F. Chiti, M. Stefani, N. Taddei, G. Ramponi and C. M. Dobson, Rationalization of the effects of mutations on peptide and protein aggregation rates, *Nature*, 2003, **424**, 805–808.
22 A. M. Fernandez-Escamilla, F. Rousseau, J. Schymkowitz and L. Serrano, Prediction of sequence-dependent and mutational effects on the aggregation of peptides and proteins, *Nat. Biotechnol.*, 2004, **22**, 1302–1306.
23 G. G. Tartaglia, A. Pawar, S. Campioni, F. Chiti and M. Vendruscolo, Prediction of aggregation-prone regions of structured proteins, *J. Mol. Biol.*, 2008, **380**, 425–436.
24 L. M. Luheshi, G. G. Tartaglia, A.-C. Brorsson, A. P. Pawar, I. E. Watson, F. Chiti, M. Vendruscolo, D. A. Lomas, C. M. Dobson and D. C. Crowther, Systematic *in vivo* analysis of the intrinsic determinants of amyloid beta pathogenicity, *PLoS Biol.*, 2007, **5**, e290.
25 R. L. Wiseman, E. T. Powers, J. N. Buxbaum, J. W. Kelly and W. E. Balch, An adaptable standard for protein export from the endoplasmic reticulum, *Cell*, 2007, **131**, 809–821.
26 T. W. Mu, D. S. T. Ong, Y. J. Wang, W. E. Balch, J. R. Yates, L. Segatori and J. W. Kelly, Chemical and biological approaches synergize to ameliorate protein-folding diseases, *Cell*, 2008, **134**, 769–781.
27 G. G. Tartaglia, S. Pechmann, C. M. Dobson and M. Vendruscolo, Life on the edge: A link between gene expression levels and aggregation rates of human proteins, *Trends Biochem. Sci.*, 2007, **32**, 204–206.
28 C. M. Dobson, Protein misfolding, evolution and disease, *Trends Biochem. Sci.*, 1999, **24**, 329–332.
29 T. R. Jahn and S. E. Radford, Folding *versus* aggregation: Polypeptide conformations on competing pathways, *Arch. Biochem. Biophys.*, 2008, **469**, 100–117.
30 T. P. J. Knowles, A. W. Fitzpatrick, H. R. Mott, S. Meehan, M. Vendruscolo, C. M. Dobson and M. E. Welland, Mechanical properties reveal the dominance of backbone interactions in stabilising amyloid fibrils, *Science*, 2007, **318**, 1900–1903.
31 A. L. Barabási and R. Albert, Emergence of scaling in random networks, *Science*, 1999, **286**, 509–512.
32 H. Jeong, B. Tombor, R. Albert, Z. N. Oltvai and A. L. Barabasi, The large-scale organization of metabolic networks, *Nature*, 2000, **407**, 651–654.
33 A. C. Gavin, P. Aloy, P. Grandi, R. Krause, M. Boesche, M. Marzioch, C. Rau, L. J. Jensen, S. Bastuck, B. Dumpelfeld, A. Edelmann, M. A. Heurtier, V. Hoffman, C. Hoefert, K. Klein, M. Hudak, A. M. Michon, M. Schelder, M. Schirle, M. Remor, T. Rudi, S. Hooper, A. Bauer, T. Bouwmeester, G. Casari, G. Drewes, G. Neubauer, J. M. Rick, B. Kuster, P. Bork, R. B. Russell and G. Superti-Furga, Proteome survey reveals modularity of the yeast cell machinery, *Nature*, 2006, **440**, 631–636.
34 I. Volkov, J. R. Banavar, S. P. Hubbell and A. Maritan, Patterns of relative species abundance in rainforests and coral reefs, *Nature*, 2007, **450**, 45–49.
35 M. Vendruscolo, E. Paci, C. M. Dobson and M. Karplus, Three key residues form a critical contact network in a protein folding transition state, *Nature*, 2001, **409**, 641–645.
36 K. Lindorff-Larsen, R. B. Best, M. A. DePristo, C. M. Dobson and M. Vendruscolo, Simultaneous determination of protein structure and dynamics, *Nature*, 2005, **433**, 128–132.
37 A. Cavalli, X. Salvatella, C. M. Dobson and M. Vendruscolo, Protein structure determination from NMR chemical shifts, *Proc. Natl. Acad. Sci. U. S. A.*, 2007, **104**, 9615–9620.

38 S. T. D. Hsu, P. Fucini, L. D. Cabrita, H. Launay, C. M. Dobson and J. Christodoulou, Structure and dynamics of a ribosome-bound nascent chain by NMR spectroscopy, *Proc. Natl. Acad. Sci. U. S. A.*, 2007, **104**, 16516–16521.
39 K. F. Dubay, A. P. Pawar, F. Chiti, J. Zurdo, C. M. Dobson and M. Vendruscolo, Prediction of the absolute aggregation rates of amyloidogenic polypeptide chains, *J. Mol. Biol.*, 2004, **341**, 1317–1326.
40 A. P. Pawar, K. F. DuBay, J. Zurdo, F. Chiti, M. Vendruscolo and C. M. Dobson, Prediction of "aggregation-prone" and "aggregation-susceptible" regions in proteins associated with neurodegenerative diseases, *J. Mol. Biol.*, 2005, **350**, 379–392.
41 D. C. Crowther, R. Page, D. Chandraratna and D. A. Lomas, A *Drosophila* model of Alzheimer's disease, *Methods Enzymol.*, 2006, **412**, 234–255.
42 T. P. J. Knowles, W. M. Shu, G. L. Devin, S. Meehan, S. Auer, C. M. Dobson and M. E. Welland, Kinetics and thermodynamics of amyloid formation from direct measurements of fluctuations of fibril mass, *Proc. Natl. Acad. Sci. U. S. A.*, 2007, **104**, 10016–10021.
43 T. P. J. Knowles, J. F. Smith, A. Craig, C. M. Dobson and M. E. Welland, Spatial persistence of angular correlations in amyloid fibrils, *Phys. Rev. Lett.*, 2006, **96**, 238301.
44 J. F. Smith, T. P. J. Knowles, C. M. Dobson, C. E. MacPhee and M. E. Welland, Characterization of the nanoscale properties of individual amyloid fibrils, *Proc. Natl. Acad. Sci. U. S. A.*, 2006, **103**, 15806–15811.
45 G. G. Tartaglia, S. Pechmann, C. M. Dobson and M. Vendruscolo, A relationship between mRNA expression levels and protein solubility in E. coli, *J. Mol. Biol.*, 2009, **388**, 381–389.
46 H. J. Dyson and P. E. Wright, Intrinsically unstructured proteins and their functions, *Nat. Rev. Mol. Cell Biol.*, 2005, **6**, 197–208.
47 M. M. Dedmon, K. Lindorff-Larsen, J. Christodoulou, M. Vendruscolo and C. M. Dobson, Mapping long-range interactions in alpha-synuclein using spin-label NMR and ensemble molecular dynamics simulations, *J. Am. Chem. Soc.*, 2005, **127**, 476–477.
48 K. Lindorff-Larsen, S. Kristjansdottir, K. Teilum, W. Fieber, C. M. Dobson, F. M. Poulsen and M. Vendruscolo, Determination of an ensemble of structures representing the denatured state of the bovine acyl-coenzyme a binding protein, *J. Am. Chem. Soc.*, 2004, **126**, 3291–3299.
49 R. B. Best and M. Vendruscolo, Determination of protein structures consistent with NMR order parameters, *J. Am. Chem. Soc.*, 2004, **126**, 8090–8091.
50 R. B. Best and M. Vendruscolo, Structural interpretation of hydrogen exchange protection factors in proteins: Characterization of the native state fluctuations of CI2, *Structure*, 2006, **14**, 97–106.
51 J. Gsponer, H. Hopearuoho, S. B. M. Whittaker, G. R. Spence, G. R. Moore, E. Paci, S. E. Radford and M. Vendruscolo, Determination of an ensemble of structures representing the intermediate state of the bacterial immunity protein Im7, *Proc. Natl. Acad. Sci. U. S. A.*, 2006, **103**, 99–104.
52 Y. Shen, O. Lange, F. Delaglio, P. Rossi, J. M. Aramini, G. H. Liu, A. Eletsky, Y. B. Wu, K. K. Singarapu, A. Lemak, A. Ignatchenko, C. H. Arrowsmith, T. Szyperski, G. T. Montelione, D. Baker and A. Bax, Consistent blind protein structure generation from NMR chemical shift data, *Proc. Natl. Acad. Sci. U. S. A.*, 2008, **105**, 4685–4690.
53 R. W. Montalvao, A. Cavalli, X. Salvatella, T. L. Blundell and M. Vendruscolo, Structure determination of protein–protein complexes using NMR chemical shifts: Case of an endonuclease colicin–immunity protein complex, *J. Am. Chem. Soc.*, 2008, **130**, 15990–15996.
54 P. Robustelli, A. Cavalli and M. Vendruscolo, Determination of protein structures from solid-state NMR chemical shifts, *Structure*, 2008, **16**, 1764–1769.
55 D. S. Wishart, D. Arndt, M. Berjanskii, P. Tang, J. Zhou and G. Lin, CS23D: a web server for rapid protein structure generation using NMR chemical shifts and sequence data, *Nucleic Acids Res.*, 2008, **36**, W496–W502.
56 P. Vallurupalli, D. F. Hansen and L. E. Kay, Structures of invisible, excited protein states by relaxation dispersion NMR spectroscopy, *Proc. Natl. Acad. Sci. U. S. A.*, 2008, **105**, 11766–11771.
57 A. De Simone, B. Richter, X. Salvatella and M. Vendruscolo, Toward an accurate determination of free energy landscapes in solution states of proteins, *J. Am. Chem. Soc.*, 2009, **131**, 3810–3811.
58 D. M. Korzhnev, X. Salvatella, M. Vendruscolo, A. A. Di Nardo, A. R. Davidson, C. M. Dobson and L. E. Kay, Low-populated folding intermediates of Fyn SH3 characterized by relaxation dispersion NMR, *Nature*, 2004, **430**, 586–590.
59 J. Gsponer, J. Christodoulou, A. Cavalli, J. M. Bui, B. Richter, C. M. Dobson and M. Vendruscolo, A coupled equilibrium shift mechanism in calmodulin-mediated signal transduction, *Structure*, 2008, **16**, 736–746.

60. O. F. Lange, N. A. Lakomek, C. Fares, G. F. Schroder, K. F. A. Walter, S. Becker, J. Meiler, H. Grubmuller, C. Griesinger and B. L. de Groot, Recognition dynamics up to microseconds revealed from an RDC-derived ubiquitin ensemble in solution, *Science*, 2008, **320**, 1471–1475.
61. D. D. Boehr, D. McElheny, H. J. Dyson and P. E. Wright, The dynamic energy landscape of dihydrofolate reductase catalysis, *Science*, 2006, **313**, 1638–1642.
62. M. Vendruscolo and C. M. Dobson, Dynamic visions of enzymatic reactions, *Science*, 2006, **313**, 1586–1587.
63. E. R. P. Zuiderweg, Mapping protein–protein interactions in solution by NMR spectroscopy, *Biochemistry*, 2002, **41**, 1–7.
64. C. Kleanthous, U. C. Kuhlmann, A. J. Pommer, N. Ferguson, S. E. Radford, G. R. Moore, R. James and A. M. Hemmings, Structural and mechanistic basis of immunity toward endonuclease colicins, *Nat. Struct. Biol.*, 1999, **6**, 243–252.
65. K. Wuthrich, Protein structure determination in solution by nuclear magnetic resonance spectroscopy, *Science*, 1989, **243**, 45–50.
66. A. Grishaev, J. Wu, J. Trewhella and A. Bax, Refinement of multidomain protein structures by combination of solution small-angle X-ray scattering and NMR data, *J. Am. Chem. Soc.*, 2005, **127**, 16621–16628.
67. P. Selenko, Z. Serber, B. Gade, J. Ruderman and G. Wagner, Quantitative NMR analysis of the protein G B1 domain in *Xenopus laevis* egg extracts and intact oocytes, *Proc. Natl. Acad. Sci. U. S. A.*, 2006, **103**, 11904–11909.
68. G. M. Clore and C. D. Schwieters, How much backbone motion in ubiquitin is required to account for dipolar coupling data measured in multiple alignment media as assessed by independent cross-validation?, *J. Am. Chem. Soc.*, 2004, **126**, 2923–2938.
69. M. Baldus, ICMRBS founder's medal 2006: Biological solid-state NMR, methods and applications, *J. Biomol. NMR*, 2007, **39**, 73–86.
70. C. Wasmer, A. Lange, H. Van Melckebeke, A. B. Siemer, R. Riek and B. H. Meier, Amyloid fibrils of the HET-s(218–289) prion form a beta solenoid with a triangular hydrophobic core, *Science*, 2008, **319**, 1523–1526.
71. A. T. Petkova, W. M. Yau and R. Tycko, Experimental constraints on quaternary structure in Alzheimer's beta-amyloid fibrils, *Biochemistry*, 2006, **45**, 498–512.
72. N. Ferguson, J. Becker, H. Tidow, S. Tremmel, T. D. Sharpe, G. Krause, J. Flinders, M. Petrovich, J. Berriman, H. Oschkinat and A. R. Fersht, General structural motifs of amyloid protofilaments, *Proc. Natl. Acad. Sci. U. S. A.*, 2006, **103**, 16248–16253.
73. M. B. Feany and W. W. Bender, A *Drosophila* model of Parkinson's disease, *Nature*, 2000, **404**, 394–398.
74. J. Bilen and N. M. Bonini, *Drosophila* as a model for human neurodegenerative disease, *Annu. Rev. Genet.*, 2005, **39**, 153–171.
75. L. M. Luheshi, D. C. Crowther and C. M. Dobson, Protein misfolding and disease: from the test tube to the organism, *Curr. Opin. Chem. Biol.*, 2008, **12**, 25–31.
76. M. Levine and R. Tjian, Transcription regulation and animal diversity, *Nature*, 2003, **424**, 147–151.
77. H. J. Bussemaker, B. C. Foat and L. D. Ward, Predictive modeling of genome-wide mRNA expression: From modules to molecules, *Annu. Rev. Biophys. Biomol. Struct.*, 2007, **36**, 329–347.

PAPER

Evolving nanomaterials using enzyme-driven dynamic peptide libraries (eDPL)

Apurba K. Das,[a] Andrew R. Hirst[b] and Rein V. Ulijn*[ab]

Received 30th January 2009, Accepted 22nd April 2009
First published as an Advance Article on the web 1st July 2009
DOI: 10.1039/b902065a

This paper describes the application of dynamic combinatorial libraries (DCL) towards the discovery of self-assembling nanostructures based on aromatic peptide derivatives and the continuous enzymatic exchange of amino acid sequences. Ultimately, the most thermodynamically stable self-assembling structures will dominate the system. In this respect, a library of precursor components, based on N-fluorenyl-9-methoxycarbonyl (Fmoc)–amino acids (serine, S and threonine, T) and nucleophiles (leucine, L–; phenylalanine, F–; tyrosine, Y–; valine, V–; glycine, G–; alanine, A–OMe amino-acid esters) were investigated to produce Fmoc–dipeptide esters, denoted Fmoc–XY–OMe. Upon exposure to a protease (thermolysin), which catalyses peptide bond formation and hydrolysis under aqueous conditions at pH 8, dynamic libraries of self-assembling gelator species were generated. Depending on the molecular composition of the precursors present in the library different behaviours were observed. Single components, Fmoc–SF–OMe and Fmoc–TF–OMe, dominated over time in Fmoc–S/(L+F+Y+V+G+A)–OMe and Fmoc–T/(L+F+Y+V+G+A)–OMe libraries. This represented >80% of all peptide formed suggesting that a single component molecular structure dominates in these systems. In a competition experiment between Fmoc–(S+T)/F–OMe, conversions to each peptide corresponded directly with ratios of starting materials, implying that a bi-component nanostructure, where Fmoc–TF–OMe and Fmoc–SF–OMe are incorporated equally favourably, was formed. Several techniques including HPLC, LCMS and fluorescence spectroscopy were used to characterize library composition and molecular interactions within the self-selecting libraries. Fluorescence spectroscopy analysis suggests that the most stable peptide nanostructures show significant π–π intermolecular electronic communication. Overall, the paper demonstrates a novel evolution-based approach with self-selection and amplification of supramolecular peptide nanostructures from a complex mixture of amino acid precursors.

Introduction

Self-assembling nanomaterials may have applications in various areas, including 3D cell culture and tissue engineering,[1,2,3] biosensing,[4] templating[5] and supramolecular optical and electronic materials.[6] Despite significant advances in synthetic[7,8] and biomolecular self-assembly,[9–11] design rules are not always understood and many new self-assembling structures are discovered by serendipity, rather than by rational

---

[a] WestCHEM/Department of Pure & Applied Chemistry, University of Strathclyde, Glasgow, UK G1 1XL. E-mail: Rein.Ulijn@strath.ac.uk; Tel: +44 (0) 141 548 2110
[b] School of Materials & Manchester Interdisciplinary Biocentre (MIB), University of Manchester, Manchester, UK M1 7HS

design. Dynamic combinatorial libraries (DCLs) are molecular component libraries that continuously exchange their components towards a thermodynamic equilibrium.[12,13] In this regard, DCLs hold considerable promise as a tool for self-assembled material discovery. Recently, this DCL approach has been successfully employed to identify molecular binding and folding events through thermodynamically driven component selection[14–17] but as yet is under-represented in materials discovery.[18]

Peptides and their derivatives are particularly interesting as building blocks for designed nanomaterials, due to their rich chemistry (involving 20 amino acid building blocks covering a range of chemical functionality), ease of synthesis and relevance to and inherent compatibility with biological systems. The attraction of DCLs for peptide nanomaterials discovery is obvious: one need only consider the very large number of structurally diverse materials that could be developed based on the combination of the 20 gene-coded amino acids (plus an infinite number of synthetic ones). For example, even a short peptide based on five naturally occurring amino acids generates a library of 20^5 or 3.2 million structurally diverse species. In this respect, testing all these sequences would be a major synthetic and analytical effort and, of the enormous numbers of possible sequences, only comparatively few are expected to fold into stable structures. In the context of biological systems these have emerged through evolution, *i.e.* by DNA-mutation-induced amino acid exchange. A peptide-based DCL would mimic such a system and allow for new nanostructures to evolve *in vitro*.

To successfully exploit dynamic combinatorial chemistry the component exchange reactions must be fully reversible (*i.e.* close to equilibrium under relevant conditions) and ideally should not display kinetic bias (*i.e.* all building blocks should be interconverted at equal rates, therefore possessing similar activation energies).

Peptide-based dynamic combinatorial libraries have been developed through the introduction of reversible bonds in a range of reactions including disulfide exchange,[19,20] and metal–peptide coordination.[21] However, the selection of different species that reversibly exchange amide bonds from the same library is still in its infancy. An early example involved the use of protease catalysed scrambling of short peptides and proteins.[22] This research showed that, as expected, under dilute aqueous conditions hydrolysis was favoured with only small amounts of peptide formed, hampering analysis of this system. This problem could be overcome by introducing a chemical re-activation step to re-introduce peptide components in protease controlled dipeptide libraries, a system that is a pseudo-DCL as it does not operate under thermodynamic control.[23]

Effective peptide based dynamic combinatorial libraries would therefore rely on methods to shift amide synthesis/hydrolysis equilibria away from hydrolysis toward synthesis of peptides. Equilibrium shifts of this type are known to be feasible by replacing water with organic solvents,[24] eutectic mixtures of substrates[25] or ionic liquids.[26,27] However, these systems involve changes in experimental parameters that move away from physiological and aqueous conditions. They therefore have limited compatibility with identification of aqueous-based self-assembling peptide structures that usually rely on a balance between hydrophobic effects, ionic interactions and hydrogen bonding. A number of approaches that also shift enzymatic reactions, yet operate under physiological conditions in aqueous buffers, have been described. For example, highly concentrated suspensions of amino acid precursors have been utilized in so-called solid-to-solid or precipitation-driven reactions.[28,29] In addition, polymer-supported reactions allow significant equilibrium shifts to allow protease driven production of peptide derivatives in aqueous media.[30,31,32] We recently demonstrated a third method that allowed for protease catalysed amide synthesis in aqueous media, which involved stabilization of peptides by molecular self-assembly of gel-phase materials.[33] This ability to affect the equilibrium position of amide synthesis/hydrolysis should allow for conditions to be identified where fully reversible amide exchanges are possible, opening up opportunities

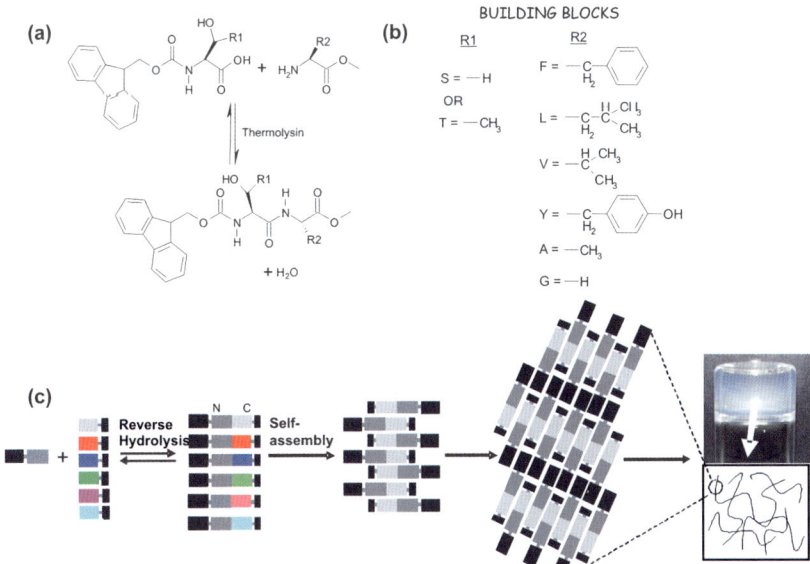

Fig. 1 (a) Reversible amide synthesis/hydrolysis catalysed by thermolysin. (b) Building blocks used for enzyme-driven dynamic peptide libraries. (c) Example of peptide library where the most stable sequence is preferentially produced, forming a self-supporting gel (libraries I and II).

for enzyme-driven dynamic peptide libraries (eDPLs) that self select for the most stable self-assembling species in component mixtures.

Our recent proof-of-concept work demonstrated a first example of eDPL for identification of self-assembling peptide nanostructures based on peptide length.[34] In that paper, we demonstrated that the self-assembly of aromatic short peptide derivatives (Fmoc–peptides) could provide a driving force that enables a protease enzyme to produce building blocks in a reversible and spatially confined manner. We demonstrated that this system allowed for fully reversible self-assembly under thermodynamic control and allowed for identification, *via* self-selection, of the most stable molecular self-assembling peptide derivatives.

In this paper, we provide further examples of eDPL *via* enzymatic amide exchange. In this evolution-based approach selection and amplification of self-assembling species from a complex mixture of non-assembling precursors is investigated. We expand upon the previous work[34] in three important ways. (i) Instead of selecting for peptide length, we show selection for sequence libraries for Fmoc–dipeptide esters, where six amino acid esters are competitively linked to either Fmoc–serine (Fmoc–S, **library I**) or Fmoc–threonine (Fmoc–T, **library II**) (Fig. 1). (ii) We then study the most stable self-assembling structure from the above two libraries in direct competition and show that 50 : 50 mixtures of two components dominate, suggesting that *bi-molecular assemblies* form preferentially (**library III**). (iii) We also provide evidence that π–π stacking interactions play a key role in the most stable assemblies, suggesting that this method may provide a means of identifying materials that show electronic communication.

Materials and methods

All reagents were purchased from commercial sources (Bachem, Germany and Sigma-Aldrich) at the highest purity (≥98%) and were used as supplied, unless stated otherwise in the experimental procedures.

Enzyme-driven peptide libraries

Fmoc–amino acids and amino acid methyl esters were supplied by Sigma-Aldrich and Bachem, Germany. Peptide libraries were prepared *via* reverse hydrolysis using thermolysin as a catalyst. 20 mmol L^{-1} of Fmoc–amino acid and 80 mmol L^{-1} of amino acid methyl ester hydrochloride were solubilised in 2 mL of 0.1 mol sodium phosphate buffer at pH 8. Thermolysin was supplied by Sigma Aldrich in the form of a lyophilised powder from *Bacillus thermoproteolyticus rokko* at 40 units per mg. No gel formation was observed prior to enzyme addition. Upon addition of 1 mg of thermolysin, gelation was observed a few hours after formation of Fmoc–dipeptide–methyl ester at room temperature (25 °C).

HPLC

A Dionex P680 HPLC pump was used to quantify conversions to peptide derivatives. A 20 µL sample was injected onto a Macherey-Nagel C18 column of 250 mm length with an internal diameter of 4.6 mm and 5 µm fused silica particles at a flow rate of 1 mL min^{-1} (linear gradient of 20% (v/v) acetonitrile in water for 4 min, gradually rising to 80% (v/v) acetonitrile in water at 35 min). This concentration was kept constant until 40 min when the gradient was decreased to 20% (v/v) acetonitrile in water at 42 min. Sample preparation involved mixing 100 µL of gel with acetonitrile–water (900 µL, 50 : 50 mixture) containing 0.1% trifluoroacetic acid. The samples were then filtered through a 0.45 µm syringe filter (Whatman, 150 units, 13 mm diameter, 2.7 µm pore size) prior to injection. The purity of each identified peak was determined by the UV detection at 265 nm. The peak retention times and peak areas were compared with known standards.

Fluorescence spectroscopy

Fluorescence emission spectra were measured on a Jasco FP-6500 Spectrofluorometer with light measured orthogonally to the excitation light at a scanning speed of 100 nm min^{-1}, with excitation at 280 nm and emission data range between 300 nm and 600 nm. The spectra were measured with a bandwidth of 5 nm with a medium response and a 1 nm data pitch.

Results and discussion

Our self-assembly strategy involves short aromatic peptide derivatives (generally five amino acids, or less) that contain aromatic functionality, either as amino acid side chains or as synthetic appended ligands. It was first highlighted in the 1990s that certain Fmoc–dipeptides can self-assemble to form gel-phase materials[35] and more recently Xu *et al.*,[36,37] Gazit *et al.*[38] and ourselves[39,40] have started to explore these materials in a number of biotechnological applications. A detailed study of one of these peptide derivatives (*N*-fluorenyl-9-methoxycarbonyl diphenylalanine, Fmoc–FF) showed that in this particular system, self-assembly was governed by a combination of β-sheet formation and π–π interactions. Additionally, formation of antiparallel beta sheets forces the Fmoc fluorenyl groups to be presented at alternating sides of each sheet. Finally, multiple sheets are interlocked through (antiparallel) π-stacking interactions (a novel molecular architecture that we termed *π-interlocked β-sheets* or *π-β structures*). The resulting arrays of sheets may form nanoscale sheet-like structures[34] or, in cases where curvature allows both edges of an array of beta-sheets to lock together, hollow tubes[41] or fibrous structures.[36–40,42,43] Under certain conditions, macroscopic gelation can be triggered in an on-demand fashion. Typical examples include dilution from organic solvents,[38] sequential pH changes[39,40] and more recently use of enzymes.[33,36,41] In the latter approach, enzymes convert non-assembling precursors into self-assembling components (or *vice versa*). We previously demonstrated that fully reversible amide exchange may be achieved in protease driven self-assembling Fmoc–peptide systems.[34]

Enzyme-assisted self-assembly *via* reversed hydrolysis

We first studied the dynamics of peptide and structure formation of one system consisting of Fmoc–S and F–OMe to form Fmoc–SF–OMe. As before,[33,34,41] thermolysin was used as the enzyme of choice. This enzyme has been used extensively in reversed hydrolysis systems[30,33] and is suitable for eDPL due to its relatively relaxed substrate specificity. Although it does preferentially synthesise/hydrolyse peptide bonds with hydrophobic amino acids in the P1′ position (*i.e* the position directly adjacent to the N end of the scissile bond), given enough time it will attack most bonds (with the exception of proline residues in the position at the C terminus of the affected amide bond). Therefore, kinetic bias towards hydrophobic amino acids can be expected at an early stage of the experiments but equilibrium distribution should eventually be reached.

As shown in Fig. 2a, the formation of Fmoc–SF–OMe from Fmoc–S and F–OMe was monitored over time using HPLC. It is clear that a constant conversion (98%) was achieved after approximately 24 h. This high conversion to peptide correlates well with the formation of a self supporting, transparent gel-phase material. During this same time period, photoluminescence experiments showed an initial red-shift and increased emission intensity, suggesting formation of more brightly emitting fluorenyl excimer species (Fig. 2b).[40,44,45] Over time, a broader feature emerged at approximately 365 nm, which eventually dominated the system. This feature is thought to represent formation of a J-aggregate, where a large number of fluorenyl groups stack together in antiparallel fashion, forming extended electronic communication paths along the nanostructures. Over time, the intensity of this peak decreases, which may relate to scattering of fibrous structures that are formed, or may indicate self-quenching as order increases further. Within approximately 250 min this system forms a self-supporting hydrogel as observed by tube inversion.

The above experiments clearly demonstrate that Fmoc–peptide esters can be formed in high yield from Fmoc–amino acid and amino acid ester building blocks at physiological conditions driven by molecular self-assembly. In addition, it is shown that formation of a peptidic gelator structure corresponds with self-assembly resulting in significant changes in fluorescence emission, highlighting that π–π stacking interactions play a key role in the self-assembly process. Next, this system was exposed to competing nucleophiles with different properties, both in parallel and in direct competition (eDPL).

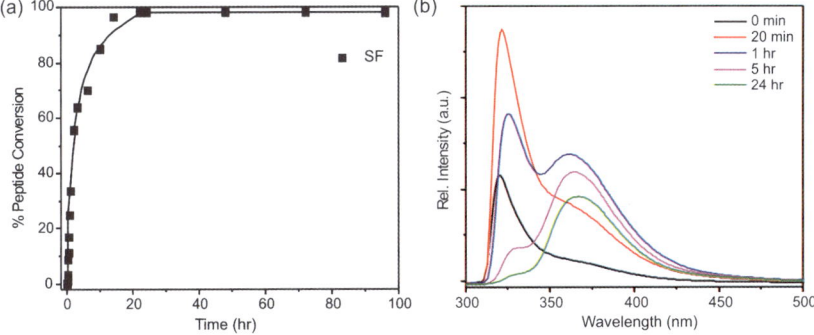

Fig. 2 Enzyme-assisted self-assembly of Fmoc–SF–OMe: (a) % peptide conversion as a function of time followed by HPLC and (b) observed changes in emission spectra monitored using photoluminescence spectroscopy.

The self-assembly precursors consisted of N-(fluorenyl-9-methoxycarbonyl)-protected amino acids (serine, S, or threonine, T) with a fourfold excess of nucleophile (G–, A–, V–, Y–, L–, F–OMe amino acid esters) (Fig. 1b). First, each Fmoc–T/X–OMe combination was tested in isolation (Fig. 3a). In each case, the corresponding Fmoc–peptide–ester was detected by HPLC, indicating that the selectivity of the enzyme allowed for the synthesis to take place. A constant conversion was obtained after 96 h with final yields of Fmoc–TF–OMe (TF, 96%), Fmoc–TL–OMe (TL, 84%), Fmoc–TV–OMe (TV, 39%), Fmoc–TY–OMe (TY, 14%), Fmoc–TA–OMe (TA, 0.7%) and Fmoc–TG–OMe (TG, 0.43%). Only L and F peptide systems giving rise to formation of self-supporting hydrogels as confirmed by tube inversion, with T_{gel} values of 78.1 and 80.9 °C, respectively. The low yields obtained for the other peptide esters relate to their less favourable self-assembly, *i.e.* these systems lack the thermodynamic driving force to produce high yields of peptide.[33] Similar results were obtained when Fmoc–S was used as the acyl donor (Fig. 3c). A constant conversion was also obtained after 96 h with final yields of Fmoc–SF–OMe (SF, 98%), Fmoc–SL–OMe (SL, 85%), Fmoc–SV–OMe (SV, 56%), Fmoc–SY–OMe (SY, 27%), Fmoc–SA–OMe (SA, 0.77%) and Fmoc–SG–OMe (SG, 0.01%). Only SF, SL and SV peptide systems self-assembled to form self-supporting hydrogels.

The sequence of yields appeared to correlate with the hydrophobicity (expressed as the partition coefficient between water and octanol, log P) of the nucleophile precursor, as demonstrated in Fig. 3, with the clear exception of Fmoc–Y peptide esters that assemble less favourably compared to what the hydrophobicity might suggest. These results indicate that design rules for hydrogelators are not straightforward, emphasizing the need for screening methods such as the one described here.

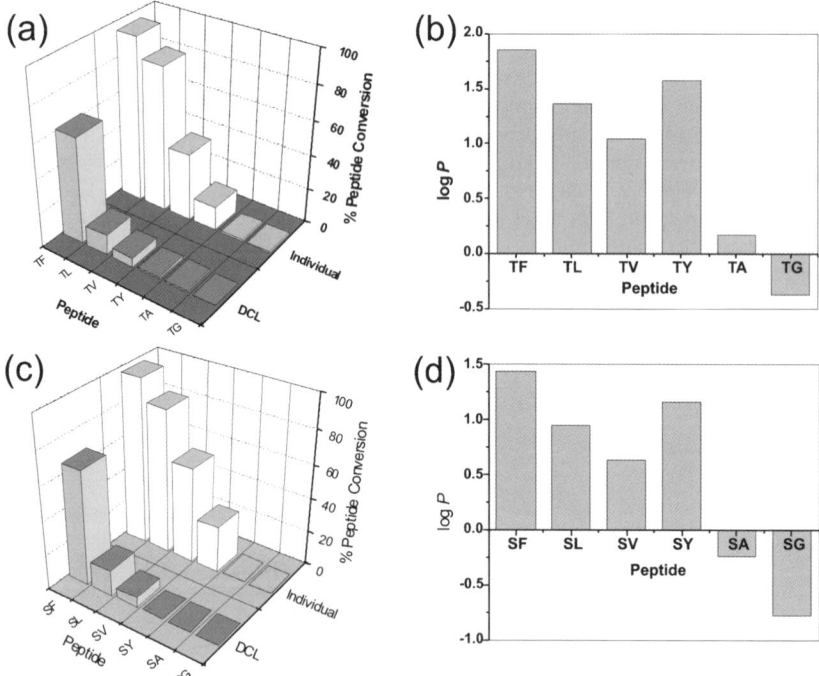

Fig. 3 (a) Formation of peptide structures based on Fmoc–T and X–OMe in isolation and as part of a DCL system. (b) and (d) The corresponding hydrophobicites (log P) of the peptidic structures as determined using HyperChem. (c) Formation of peptide structures based on Fmoc–S and X–OMe in isolation and as part of a DCL system.

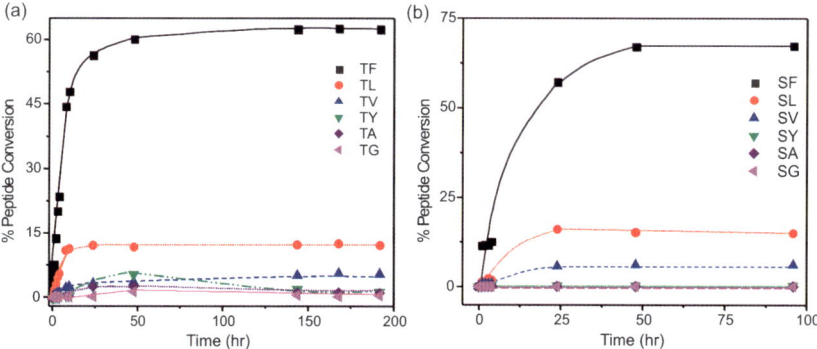

Fig. 4 Time course of Fmoc–TX–OMe library as measured by HPLC showing kinetic bias at early time points (a) and time course for Fmoc–SX–OMe library (b).

Next, Fmoc–S or Fmoc–T were presented with a mixture of the same six amino acid esters (**Libraries I** and **II**) (Fig. 1 and 4). When carried out in competition, the same sequence of yields was eventually obtained, however in this case, it was clear that a single component (Fmoc–TF–OMe and Fmoc–SF–OMe) was preferentially produced (61 and 65% conversions, respectively; corresponding to >80% of the total peptide yield), indicating that this peptide ester represents the lowest free-energy well in the free-energy hypersurface (Fig. 4). To investigate any enzyme-induced kinetic bias, the evolution of both mixtures was followed over time. For **Library II** at 50 h a correlation between yields and log P (representing molecular hydrophobicity) was apparent, in accordance with the enzyme's preference for hydrophobic amino acids as the nucleophiles in the position on the N terminus of the affected amide bond. After nearly 200 h the distribution has changed to reflect the equilibrium composition. In **Library I** this kinetic bias was not observed, however it could have been missed due to the infrequence of sampling. These experiments clearly demonstrate the ability of this system to identify the most stable structure in component mixtures.

Having shed some light on the impact of the nucleophile on the self-assembly process, we next turned our attention to the role of the acyl donor. In order to determine which of the two Fmoc amino acids (Fmoc–T and Fmoc–S) would give rise to the most stable self-assembling structure a direct competition experiment was performed. In this case, we used Fmoc–T and Fmoc–S in the presence of the nucleophile F–OMe (*i.e.* the most favoured structures formed in component **Libraries I** and **II** are now in competition to form **Library III**). Recall that in isolation Fmoc–SF–OMe formed in 98% yield while Fmoc–TF–OMe formed in 96% yield. If hydrophobicity was the main driving force then we would expect a higher conversion yield of Fmoc–TF–OMe as Fmoc–T is more hydrophobic than Fmoc–S (as observed in **Libraries I** and **II**, where L and F–OMe peptides are formed in high yield in isolation, and in competition the F–OMe containing peptides dominate five-fold). Instead, a 50 : 50 mixture emerged, which suggests to us that bi-component nanostructures are produced preferentially, with Fmoc–TF–OMe and Fmoc–SF–OMe components that are interchangeable within the structure. Indeed, heterodimers and trimers have previously been described, which gave rise to more stable structures compared to the single component assemblies.[20,46]

To confirm that the bi-component mixture represents the equilibrium state (*i.e.* represented by the free energy diagram given in Fig. 5d), two sequential experiments were performed. In these experiments, at the initial starting point, only Fmoc–T or Fmoc–S was employed in the presence of a four-fold excess of F–OMe. Once the equilibrium conversions were attained the competing acyl donor was introduced. In both cases the same 50 : 50 mixture was obtained over time. These observations provide evidence that the bi-component structure is more stable compared to the

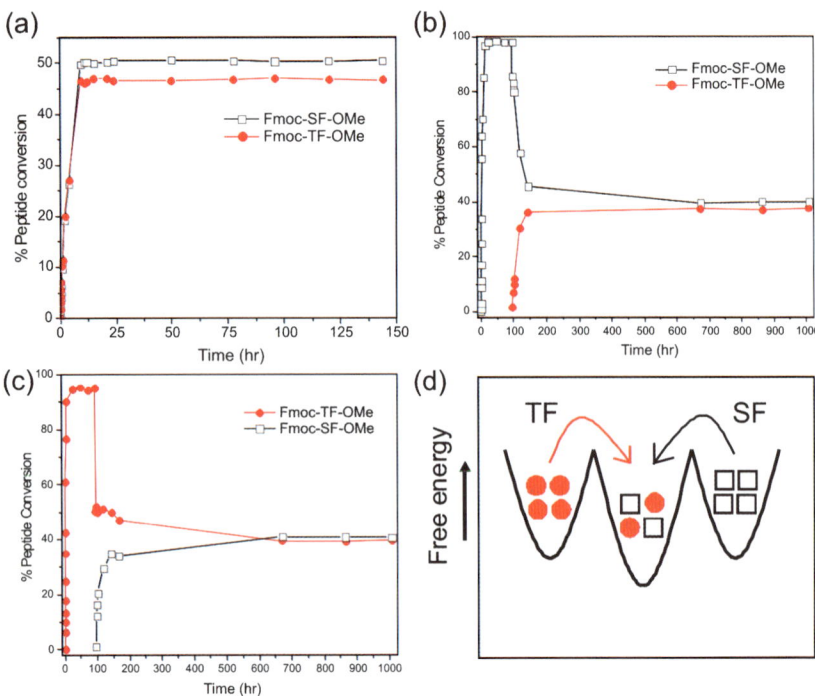

Fig. 5 (a) Time course of Fmoc–S and Fmoc–T in the presence of a four-fold excess of F–OMe showing an equal distribution of species under equilibrium conditions. (b) Time course of Fmoc–S in the presence of a four-fold excess of F–OMe followed by addition of Fmoc–T after 100 h showing that under equilibrium conditions a bi-molecular self-assembled state is obtained. (c) Time course of Fmoc–T in the presence of a four-fold excess of F–OMe followed by addition of Fmoc–S after 100 h showing that under equilibrium conditions a bi-molecular self-assembled state is obtained. (d) Diagram showing the free energy profile of enzyme-assisted self-assembly in which the formation of Fmoc–SF–OMe and Fmoc–TF–OMe are equally favoured leading to the formation of a bi-molecular self-assembled state.

single component ones, suggesting a free energy hypersurface as drawn schematically in Fig. 5d.

We hypothesise that, in addition to H-bonding interactions between peptide fragments which should be similar for each Fmoc–XY–OMe, it is likely that differences in stability arise from the induced positioning of Fmoc groups, resulting in differences in π-stacking interactions. Therefore, fluorescence spectra were acquired for Fmoc–(T/S)(L/F)–OMe peptides, *i.e.* the most stable candidates that emerged from **Libraries I** and **II**. Prior to enzyme addition, an emission maximum centered at 320 nm was observed in each case, with a similar overall emission spectra observed in each case. At 96 h after addition of enzyme different behaviours were observed for each combination. For Fmoc–SF–OMe (see also Fig. 2), the broad peak at 365 nm confirms a higher order J-aggregate of aromatic moieties stabilized by π-stacking interactions, and dominates the system (Fig. 6a). For Fmoc–SL–OMe, an emission maximum at 324 nm with a shoulder at 375 nm was observed due to the fluorenyl excimer formation prior to enzyme addition. Fluorescence emission maxima at 325 nm and an enhanced emission around 455 nm suggest the rearrangement of molecules into a higher order J-aggregate of fluorenyl rings for SL after 96 h (Fig. 6b), although significant non-associated fluorenyl groups were apparent. Fluorescence spectra for Fmoc–TF–OMe and Fmoc–TL–OMe systems taken 96 h after enzyme addition showed emission maxima shifted from 320 nm to 330 nm and

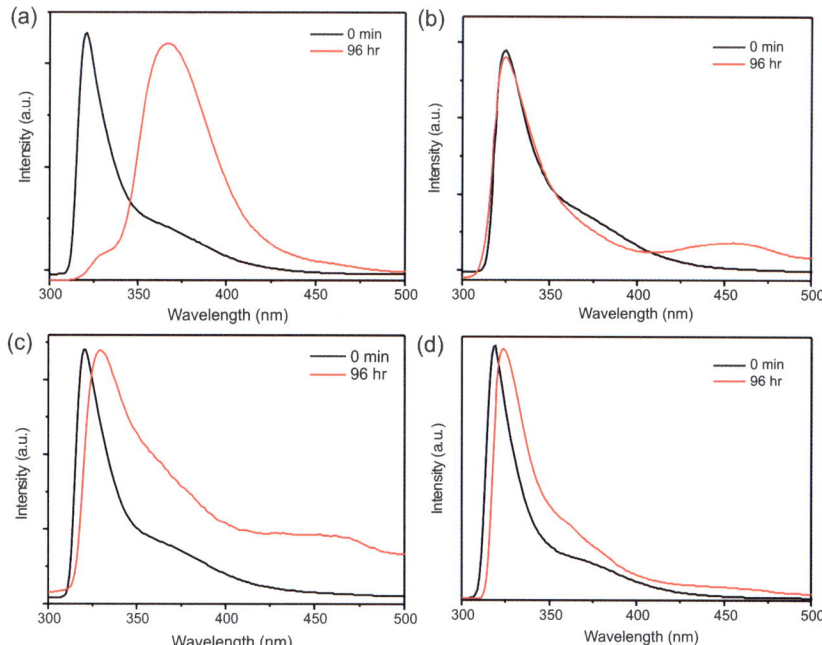

Fig. 6 Normalized fluorescence spectra were taken for (a) SF, (b) SL, (c) TF and (d) TL systems before enzyme addition (0 min, black line) and at 96 h (red line) after enzyme addition. All spectra at 96 h suggest structural changes occur after enzyme addition.

another enhanced peak at 460 nm appeared for Fmoc–TF–OMe system at 96 h after enzyme addition (Fig. 6c). For the Fmoc–TL–OMe system, the emission maximum peak is red shifted and an enhanced peak at 455 nm is observed at 96 h after enzyme addition (Fig. 6d).

Although complete interpretation of the emission spectra will require further study, it is clear that the more stable peptide self assemblies as identified by eDPL show the most significant overall shifts in emission spectra, which may suggest that eDPL of Fmoc–XY–OMe libraries selects for highly conjugated π-stacking among molecular components, *i.e.* it may provide a tool for discovery of electronically conductive peptide-based nanomaterials.

Conclusions

The work presented in this paper demonstrates that eDPL is a feasible method to identify the most stable self-assembling peptides from component mixtures. When comparing experiments individually, or in direct competition, the same distribution of species emerged. We show that the enzyme selected for this study has an inherent preference for more hydrophobic nucleophiles leading to a initial kinetic bias in species distribution. However, over time the final distribution of species reflected the thermodynamic equilibrium composition.

Future work will focus on enhancing the rates, applying the system to the identification of binders and exploring the limitation of peptide length in this type of self-assembling DCL system. Furthermore, identification of different enzymes with less kinetic bias and other near-equilibrium reactions will be investigated. We envisage enzymatic DPL to have broad ranging applications in nanomaterials and ligand discovery as a complementary technique to phage display.[47,48]

Acknowledgements

We thank EPSRC and The Leverhulme Trust for funding.

References

1. S. Zhang, *Nat. Biotechnol.*, 2003, **21**, 1171–1178.
2. G. A. Silva, C. Czeisler, K. L. Niece, E. Beniash, D. A. Harrington, J. A. Kessler and S. I. Stupp, *Science*, 2004, **303**, 1352–1355.
3. L. Haines-Butterick, K. Rajagopal, M. Branco, D. Salick, R. Rughani, M. Pilarz, M. S. Lamm, D. J. Pochan and J. P. Schneider, *Proc. Natl. Acad. Sci. U. S. A.*, 2007, **104**, 7791–7796.
4. S. Kiyonaka, K. Sada, I. Yoshimura, S. Shinkai, N. Kato and I. Hamachi, *Nat. Mater.*, 2004, **3**, 58–64.
5. M. Reches and E. Gazit, *Nat. Nanotechnol.*, 2006, **1**, 195–200.
6. (*a*) A. P. H. J. Schenning and E. W. Meijer, *Chem. Commun.*, 2005, 3245–3258; (*b*) A. R. Hirst, B. Escuder, J. F. Miravet and D. K. Smith, *Angew. Chem., Int. Ed.*, 2008, **47**, 8002–8018.
7. H. Cui, Z. Chen, S. Zhong, K. L. Wooley and D. J. Pochan, *Science*, 2007, **317**, 647–650.
8. P. Jonkheijm, P. van der Schoot, A. P. H. J. Schenning and E. W. Meijer, *Science*, 2006, **313**, 80–83.
9. M. G. Ryadnov and D. N. Woolfson, *Nat. Mater.*, 2003, **2**, 329–332.
10. R. M. Capito, H. S. Azevedo, Y. S. Velichko, A. Mata and S. I. Stupp, *Science*, 2008, **319**, 1812–1816.
11. R. V. Ulijn and A. M. Smith, *Chem. Soc. Rev.*, 2008, **37**, 664–675.
12. R. F. Ludlow and S. Otto, *Chem. Soc. Rev.*, 2008, **37**, 101–108.
13. M. M. Rozenman, B. R. McNaughton and D. R. Liu, *Curr. Opin. Chem. Biol.*, 2007, **11**, 259–268.
14. N. Sreenivasachary and J. M. Lehn, *Proc. Natl. Acad. Sci. U. S. A.*, 2005, **102**, 5938–5943.
15. E. Buhler, N. Sreenivasachary, S. J. Candau and J. M. Lehn, *J. Am. Chem. Soc.*, 2007, **129**, 10058–10059.
16. N. Sreenivasachary and J. M. Lehn, *Chem.–Asian J.*, 2008, **3**, 134–139.
17. S. Otto, R. L. E. Furlan and J. K. M. Sanders, *Science*, 2002, **297**, 590–593.
18. P. T. Corbett, J. Leclaire, L. Vial, K. R. West, J.-L. Weitor, J. K. M. Sanders and S. Otto, *Chem. Rev.*, 2006, **106**, 3652–3711.
19. Y. Krishnan-Ghosh and S. Balasubramanian, *Angew. Chem., Int. Ed.*, 2003, **42**, 2171–2173.
20. E. H. C. Bromley, R. B. Sessions, A. R. Thomson and D. N. Woolfson, *J. Am. Chem. Soc.*, 2009, **131**, 928–930.
21. H. J. Cooper, M. A. Case, G. L. McLendon and A. G. Marshall, *J. Am. Chem. Soc.*, 2003, **125**, 5331–5339.
22. P. G. Swann, R. A. Casanova, A. Desai, M. M. Frauenhoff, M. Urbancic, U. Slomczynska, A. J. Hopfinger, G. C. Le Breton and D. L. Venton, *Biopolymers*, 1996, **40**, 617–625.
23. A. D. Corbett, J. D. Cheeseman, R. J. Kazlauskas and J. L. Gleason, *Angew. Chem., Int. Ed.*, 2004, **43**, 2432–2436.
24. M.-Y. Lee and J. S. Dordick, *Curr. Opin. Biotechnol.*, 2002, **13**, 376–384.
25. R. López-Fandiño, I. Gill and E. N. Vulfson, *Biotechnol. Bioeng.*, 1994, **43**, 1016–1023.
26. R. Sheldon, *Chem. Commun.*, 2001, 2399–2407.
27. M. Erbeldinger, A. J. Mesiano and A. J. Russell, *Biotechnol. Prog.*, 2000, **16**, 1129–1131.
28. R. V. Ulijn and P. J. Halling, *Green Chem.*, 2004, **6**, 488–496.
29. M. Erbeldinger, U. Eichhorn, P. Kuhl and P. J. Halling, *Methods Biotechnol.*, 2001, **15**, 471–477.
30. R. V. Ulijn, B. Baragana, P. J. Halling and S. L. Flitsch, *J. Am. Chem. Soc.*, 2002, **124**, 10988–10989.
31. R. V. Ulijn, N. Bisek, P. J. Halling and S. L. Flitsch, *Org. Biomol. Chem.*, 2003, **1**, 1277–1281.
32. P. J. Halling, R. V. Ulijn and S. L. Flitsch, *Curr. Opin. Biotechnol.*, 2005, **16**, 385–392.
33. S. Toledano, R. J. Williams, V. Jayawarna and R. V. Ulijn, *J. Am. Chem. Soc.*, 2006, **128**, 1070–1071.
34. R. J. Williams, A. M. Smith, R. Collins, N. Hodson, A. K. Das and R. V. Ulijn, *Nat. Nanotechnol.*, 2009, **4**, 19–24.
35. R. Vegners, I. Shestakova, I. Kalvinsh, R. M. Ezzell and P. A. Janmey, *J. Pept. Sci.*, 1995, **1**, 371–378.
36. Z. Yang, H. Gu, D. Fu, P. Gao, J. K. Lam and B. Xu, *Adv. Mater.*, 2004, **16**, 1440–1444.
37. Y. Zhang, H. Gu, Z. Yang and B. Xu, *J. Am. Chem. Soc.*, 2003, **125**, 13680–13681.

38 A. Mahler, M. Reches, M. Rechter, S. Cohen and E. Gazit, *Adv. Mater.*, 2006, **18**, 1365–1370.
39 V. Jayawarna, M. Ali, T. A. Jowitt, A. F. Miller, A. Saiani, J. E. Gough and R. V. Ulijn, *Adv. Mater.*, 2006, **18**, 611–614.
40 A. M. Smith, R. J. Williams, C. Tang, P. Coppo, R. F. Collins, M. L. Turner, A. Saiani and R. V. Ulijn, *Adv. Mater.*, 2008, **20**, 37–41.
41 A. K. Das, R. Collins and R. V. Ulijn, *Small*, 2008, **4**, 279–287.
42 Z. Yang, K. Xu, Z. Guo, Z. Guo and B. Xu, *Adv. Mater.*, 2007, **19**, 3152–3156.
43 Z. Yang, G. Liang, L. Wang and B. Xu, *J. Am. Chem. Soc.*, 2006, **128**, 3038–3043.
44 T. Forster, *Angew. Chem., Int. Ed. Engl.*, 1969, **8**, 333–343.
45 J. P. Pinion, F. L. Minn and N. Filipescu, *J. Lumin.*, 1971, **3**, 245–252.
46 V. Gauba and J. D. Hartgerink, *J. Am. Chem. Soc.*, 2008, **130**, 7509–7515.
47 R. R. Naik, S. J. Stringer, G. Agarwal, S. E. Jones and M. O. Stone, *Nat. Mater.*, 2002, **1**, 169–172.
48 S. R. Whaley, D. S. English, E. L. Hu, P. F. Barbara and A. M. Belcher, *Nature*, 2000, **405**, 665–668.

Rational design of peptide-based building blocks for nanoscience and synthetic biology

Craig T. Armstrong,[a] Aimee L. Boyle,[a] Elizabeth H. C. Bromley,[a] Zahra N. Mahmoud,[a] Lisa Smith,[a] Andrew R. Thomson[a] and Derek N. Woolfson[*ab]

Received 28th January 2009, Accepted 23rd March 2009
First published as an Advance Article on the web 21st July 2009
DOI: 10.1039/b901610d

The rational design of peptides that fold to form discrete nanoscale objects, and/or self-assemble into nanostructured materials is an exciting challenge. Such efforts test and extend our understanding of sequence-to-structure relationships in proteins, and potentially provide materials for applications in bionanotechnology. Over the past decade or so, rules for the folding and assembly of one particular protein-structure motif—the α-helical coiled coil—have advanced sufficiently to allow the confident design of novel peptides that fold to prescribed structures. Coiled coils are based on interacting α-helices, and guide and cement many protein–protein interactions in nature. As such, they present excellent starting points for building complex objects and materials that span the nano-to-micron scales from the bottom up. Along with others, we have translated and extended our understanding of coiled-coil folding and assembly to develop novel peptide-based biomaterials. Herein, we outline briefly the rules for the folding and assembly of coiled-coil motifs, and describe how we have used them in *de novo* design of discrete nanoscale objects and soft synthetic biomaterials. Moreover, we describe how the approach can be extended to other small, independently folded protein motifs—such as zinc fingers and EF-hands—that could be incorporated into more complex, multi-component synthetic systems and new hybrid and responsive biomaterials.

1. Introduction

Nature provides considerable inspiration for what might be achieved through self-assembly in water. It uses lipids, nucleic acids, carbohydrates and polypeptides as building blocks for self-assembling systems, though nucleic acids and polypeptides form the best-defined biosupramolecular (tertiary and quaternary) structures.[1] Taking such inspiration, nanoscientists and synthetic biologists are achieving much in synthetic self-assembly using nucleic-acid building blocks.[2,3] Arguably, however, polypeptides present better long-term building materials as they form more-diverse examples of self-assembled structures and functions in biology; they are stable, or can be stabilised over a broad range of conditions; and they can be produced in large quantities from renewable sources using recombinant DNA technologies and gene expression in a variety of hosts. For these reasons, we have adopted Nature's favoured macromolecules and use peptides and proteins to design and

[a]*School of Chemistry, University of Bristol, Bristol, UK BS8 1TS. E-mail: D.N.Woolfson@bristol.ac.uk*
[b]*Department of Biochemistry, University of Bristol, Bristol, UK BS8 1TD*

engineer new self-assembled systems and biomaterials that span the nano-to-meso scales.

This choice of peptides and proteins as building blocks carries a problem, however: for nucleic acids the double helical structure of DNA, and the various structures formed by RNA are dictated, and can largely be predicted, by the underlying base-pairing;[4,5] for peptides and proteins, however, there are no, or at least very few, direct rules that relate polypeptide sequence to 3D structure. This is the so-called protein-folding problem, which despite considerable effort, remains largely unsolved. That is not to say that protein structure cannot be predicted from a sequence; in certain cases, this can be done through both homology (evolution-based) methods and *ab initio* predictions.[6] However, these tend to look at sequences, or indeed multiple sequence alignments, holistically to make predictions; that is, rather than identifying specific motifs or patterns of residues that dominate folding, they rest on identifying common general features between protein sequences of known structure and those under interrogation.

Ideally, for rational peptide design, we would like to identify clear rules for folding and that can be readily distilled from natural sequences. These could then be applied in *de novo* sequences in which a small number of specified positions dictate folding, and the remainder can be filled initially with "innocuous" amino acids (such as alanine and glutamine) with the option of adding function; here, we refer to such template sequences as *vanilla* and give them the prefix "v", (Fig. 1). Through such an approach, we aim (i) to tackle fundamental problems in protein structure prediction and design—by garnering rules for protein folding—and (ii) to generate a number of basic designed peptide components. Indeed, we see the protein folding and design problems as intimately linked, as successful protein design provides the acid test of our understanding of sequence-to-structure relationships in proteins.

Furthermore, we aim to combine the basic peptide components to generate more complex, multi-component, self-assembling systems such as discrete nanoscale objects and biocompatible biomaterials. In turn, such assemblies could provide the basis for applications in bionanotechnology and synthetic biology. Applications might include materials for cell and tissue growth and assemblies for sensing and delivery. An important consideration in this approach is size: peptides of up to ~30 amino acids in length are amenable to rational design using *in Biro* (writing straight down on paper using a small and straightforward set of design rules) and *in silico* approaches, and they are readily accessible to modern solid-phase synthesis. Nonetheless, to achieve our goal of building supramolecular objects and materials with designed peptides, they must harbour information for both autonomous folding and higher-order self-assembly, which is challenging in a short peptide

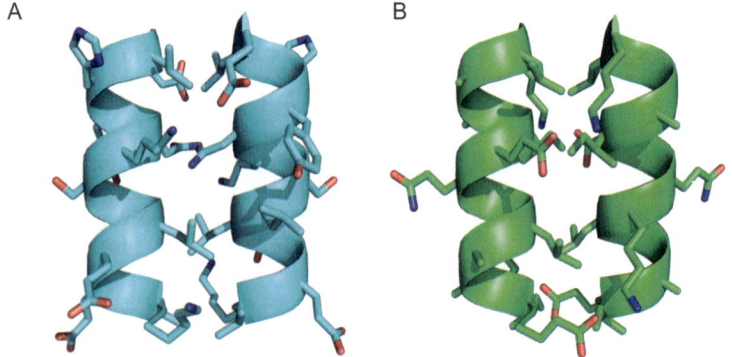

Fig. 1 (A) Two heptad repeats of the leucine zipper from GCN4 (PDB identifier 2ZTA) compared with (B) two heptads of a vanilla dimeric coiled coil.

frame. *N.B.* Larger proteins can be made using recombinant DNA technology, though the complexity of interactions usually puts these beyond current rational design, with notable exceptions.[7]

Given these objectives and constraints, what peptide structures have been designed to date? The answer is depressingly few; though many *de novo* designs have been reported, these tend to centre on a small number of protein-folding motifs, which we highlight with the following examples.

Zinc-finger motifs have been successfully designed over the last few decades.[8,9] The folding of these ~30 residues sequences is driven by four residues (usually combinations of thiol-containing cysteine and the imidazole side chain of histidine) that coordinate the metal (Fig. 2A). The zinc finger is one of the few motifs that has been progressed further along the vanilla-design philosophy that we outline here, and—along with the Ca^{2+}-binding EF hand, (Fig. 2B)—the C2H2 class of zinc finger is one of our target components described in the next section.

The two examples above might be classed as discrete protein-folding motifs. However, one of our aspirations is to design and engineer extended biomaterials for such building blocks.[10] Until recently, this area of biomaterials design has been dominated by designed β-strand or β-sheet-forming peptides that assemble into extended amyloid-like configurations. The field is too large to review here;[11–14] however, peptides tend to form β-strands that hydrogen bond to form extended sheets, which bundle to form fibres with the cross-beta structure in which the strands run perpendicular to the long-axis of the fibres, (Fig. 2C). Related to these studies and also in the biomaterials area, recent successes have broadened to self-assembling collagen sequences.[15–17] These bring a further set of problems associated with proline-rich sequences, notably: potentially problematic chemical synthesis, and inefficient folding due to slow *cis–trans* isomerisation of peptide bonds. Nonetheless, a number of groups have reported successful designs of peptides that make freestanding collagen triple helices (Fig. 2D),[18,19] and others that assemble further to form extended nano-to-micron scale fibres, some of which exhibit collagen-like features.[16,20,21] In all of these studies the basic peptide repeat is Gly-Pro-Hyp, which is embellished with charged residues at some of the proline and hydroxyproline (Hyp) positions to foster self-assembly to the discrete and extended structures.

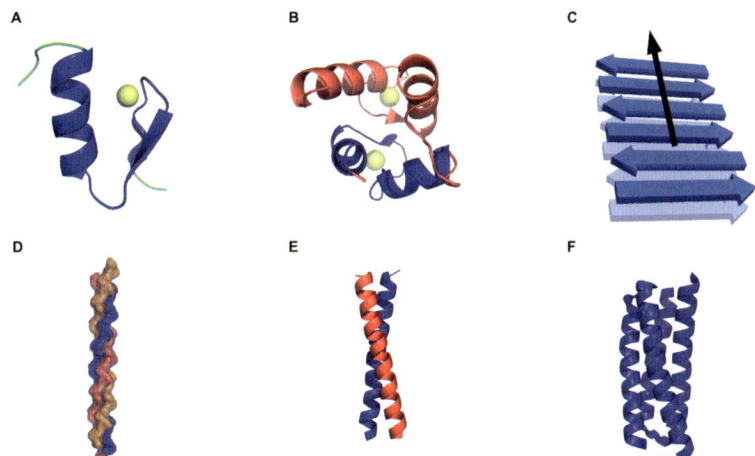

Fig. 2 (A) The C2H2-type zinc-finger structure shown coordinating a zinc ion (PDB code 1ARD). (B) Two covalently linked EF-hand domains (distinguished in red and blue) coordinating calcium ions (PDB code 1B1G). (C) Cartoon of a β-strand-based assembly of amyloid-like fibrils, with the black arrow pointing along the long-fibre axis. (D) Three polypeptides (red, orange and blue) associating to form a collagen triple helix (PDB code 1K6F). (E) The coiled coil domain (PDB code 2ZTA). (F) The four-helix bundle (PDB code 1ROP).

In contrast to the above studies, our focus has been on peptides that adopt α-helical structures, and moreover, those that associate specifically and in a well-defined manner known as the coiled-coil structure,[22–24] (Fig. 2E). Many others work in this area, and the field is reviewed more fully elsewhere.[25,26] Alongside the zinc-finger and collagen motifs described above, sequence-to-structure relationships in coiled coils are perhaps the most clear and well understood at this time. Moreover, the rules for coiled-coil assembly appear to be particularly robust and transferable between different coiled-coil contexts. This, together with the ingenuity of peptide chemists and designers has lead to an impressive array of coiled-coil-based designs for supramolecular, discrete, and materials systems, as well as peptide-based switches and functional units.[10,27]

Significant progress towards vanilla sequences has also been made for the related motif of four-helix bundles[28] (Fig. 2F). Here complex sequences with novel functions are built up from minimal originals termed maquettes. This approach has been used to successfully design a four-helix bundle with cytochrome functionality.[29]

Excepting the above examples, the majority of protein-design work focuses on designing a motif purely to fold to the specified conformation. To be useful in synthetic biology and biomaterials technology we must move beyond this and design sequences that not only fold to the prescribed discrete structures, but which also possess sites available, for instance, for functionalization, conformational switching and further self-assembly.

In this paper, we outline a general route to other vanilla peptide sequences that could provide the basis for more-complex designs. We illustrate how a combined bioinformatics and experimental approach can be used to deliver new peptide-based components with different structures and properties. We give three examples to illustrate our approach: firstly, the zinc finger—a tractable target on which much work has already been published; secondly the coiled coil—a more complex target that shows great potential as a tecton for synthetic biology,[1,30] but for which more sequence-to-structure rules are required to define better partner and oligomer-state selection; and, finally, the EF hand—as an example of applying our philosophy to a general small folding domain. We also describe how these different components are being used, and might be used in the future to create new peptide-based switches, nanoscale objects and biomaterials of increasing complexity.

2. Materials and methods

2.1 Computational analysis

The seed file from Pfam A 23.0[31] was searched using a PERL script for entries fitting the description of domains less than 100 amino acids long with five or more examples of sequences, and one or more example of structures deposited in the Protein Data Bank (PDB).[32] Corresponding structures were visualized using Pymol[33] and examined by eye. Those domains deemed likely to fold independently—either the domain alone was in the PDB entry, or it was suitably exposed even in the context of a larger protein—were considered as potential design targets.

2.2 Peptide synthesis

Peptides were synthesised using standard solid-phase Fmoc chemistry on a CEM Liberty automated synthesizer using HBTU (Advanced Chemtech) activation. All Fmoc-protected amino acids and resins were purchased from Novabiochem. Peptide-grade N,N-dimethylformamide (DMF) and piperidine were purchased from Rathburn chemicals. All other reagents and solvents were obtained from Aldrich. Acetylation of the peptide N-termini was achieved on the resin by treating the Fmoc-deprotected peptides with acetic anhydride (0.5 ml) and pyridine (0.75 ml) in DMF (5 ml). Peptides were cleaved from the resin supports using trifluoroacetic acid (TFA) (5 ml) containing 2.5% triisopropylsilane and 2.5% water. Crude peptides were obtained by precipitation of the TFA mix in ice-cold diethyl ether (80 ml) followed by centrifugation and drying of the resultant solid material under vacuum.

HPLC was carried out using a 250 × 10 mm, C18 reversed-phase column (Kromatek), eluting with a water–acetonitrile gradient on a Jasco HPLC system. Peptide identity was confirmed using matrix-assisted laser desorption/ionisation–time-of-flight (MALDI-ToF) mass spectrometry (Applied Biosystems ABI 4700) and purity by analytical high-pressure liquid chromatography (HPLC).

2.3 Biophysical characterisation

2.3.1 Circular dichroism (CD).
CD measurements were made using a JASCO J-815 spectropolarimeter fitted with a Peltier temperature controller. Peptide solutions were prepared in phosphate buffered saline (PBS; 137 mM NaCl, 2.7 mM KCl, and 10 mM phosphate buffer) adjusted to pH 7.0, and examined in 1 mm quartz cuvettes. Spectra were recorded at 5 °C using 1 nm interval, 1 nm bandwidth and 2 s response times. After baseline correction, ellipticities in mdeg were converted to molar ellipticities (mdeg cm^2 dmol^{-1}) by normalizing for the concentration of peptide bonds and pathlength. Thermal unfolding curves were recorded at 222 nm through 1 °C min^{-1} ramps using a 1 nm bandwidth, averaging the signal for 8 s every 1 °C interval.

2.3.2 Dynamic light scattering (DLS).
DLS measurements were made using a Malvern Zetasizer Nanoseries instrument. 100 μM samples were prepared in PBS and centrifuged to remove any large particulate material. Measurements were made at 20 °C using auto-evaluated settings. The data were analyzed using the associated Malvern software. Predictions for the apparent hydrodynamic diameters were made using Hydropro.[34] Model coiled-coil atomic coordinate files were produced using the MAKECCSC program written by Offer based on previously published work.[35]

2.3.3 Sedimentation equilibrium experiments by analytical ultracentrifugation (AUC).
These were conducted at 20 °C in a Beckman-Optima XL-I analytical ultracentrifuge fitted with an An-60 Ti rotor. Peptide solutions were prepared in the concentration range 50–150 μM in PBS and spun at speeds in the range 25 000 to 55 000 rpm. Data were fitted simultaneously assuming a single ideal species model using Ultrascan.[36] The partial specific volumes of the peptide combinations (0.7777, 0.7783, 0.7766 for vCC-pIL, vCC-pII and vCC-pLI, respectively) and the density of the solvent (1.00442 g ml^{-1}) were calculated using Sednterp.[37]

3. Results and discussion

3.1. Previous studies towards vanilla zinc-finger motifs

Since the 1980s, researchers have aimed to unravel the determinants of both folding and DNA-binding specificity of zinc fingers. These ββα motifs[38] (Fig. 2A) offer a simple, robust scaffold for studies into protein folding, and would serve as an ideal addition to the synthetic biologist's toolkit. Two notable efforts geared towards elucidating folding rules for zinc fingers have come out of the Berg laboratory, which has studied both a consensus and a minimalist zinc finger, both designed from multiple sequence alignments[39,40] such as that depicted as a Weblogo (Fig. 3A).[41] This work showed the zinc finger to be a remarkably robust fold, with the consensus sequence binding zinc with an unprecedented affinity, providing evidence that much, if not all, of the information needed for specifying the zinc finger fold can be extracted in a straightforward manner from linear sequence information. The minimalist zinc finger is an example of a vanilla sequence, with many of the non-conserved residues mutated to alanine. This peptide folded upon the addition of

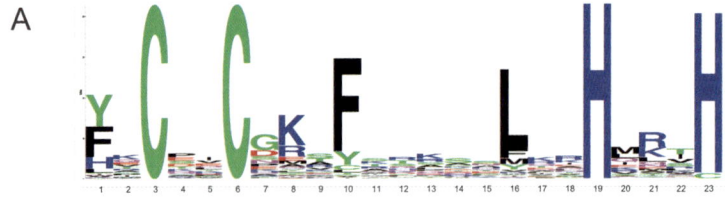

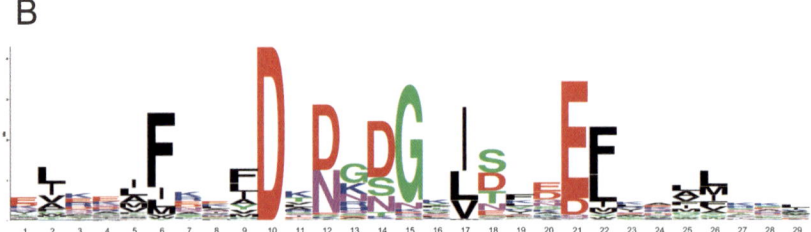

Fig. 3 Weblogos showing the key, structure-determining residues for (A) the zinc-finger, and (B) the EF-hand.

zinc, although there was evidence for some dimerization, suggesting that one or more of the mutated residues had played a role in ensuring zinc fingers fold as monomers.

Since the discovery that DNA binding by zinc fingers is modular,[42] a vast body of work has been carried out to determine their DNA-binding specificity.[43] The advent of so-called zinc-finger nucleases[44]—namely, zinc-finger domains attached to FokI restriction endonucleases, able to cleave DNA at points specified by the zinc fingers—is encouraging in that once the DNA-recognition code is cracked, potential new research areas may open in medicine and molecular biology, with the nucleases being touted as potential therapeutics for SCID and HIV.[45,46] Indeed, one group has even designed a zinc finger able to cut DNA without the need for an ancillary nuclease domain.[8]

In summary the zinc finger, with its robust design rules, is already in widespread use in designed proteins. Not only is it amenable to the vanilla design approach, it has already had novel function incorporated in the form of new DNA binding specificities. Therefore, it offers promise as a component for synthetic biology.

3.2 Towards new vanilla coiled-coil components

Coiled coils provide excellent components for synthetic biology because they occur in a range of oligomerization states and topologies, which have clear potential as spacers and general building blocks, and as hubs to co-localize and orient other functional domains.[1,10] They are attractive to the vanilla peptide design philosophy because all the interactions are in the core leaving the external face potentially free for exploitation and functionalization, (Fig. 4). At the most basic level of design most coiled-coil sequences display a *heptad* pattern of hydrophobic (*H*) and polar (*P*) residues, *HPPHPPP*. In such repeats, hydrophobic residues are spaced alternately 3 and 4 residues apart.

This spacing closely matches the 3.6 residues per turn of the α-helix. Thus, when configured into an α-helix, heptad repeats lead to amphipathic structures with hydrophobic and polar faces (Fig. 4B). In turn, the hydrophobic faces of two or more such helices can combine to give different helical bundles. However, in itself, the hydrophobic interaction is not specific. Thus, one of the challenges in the coiled-coil structure and design fields is to understand how the very many different coiled-coil architectures and topologies that are possible, and which are indeed

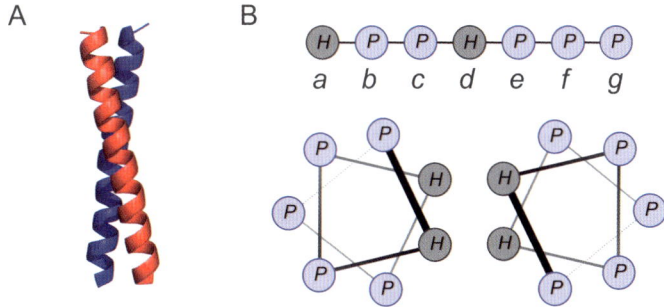

Fig. 4 (A) A dimeric coiled-coil structure, comprising two α-helices wrapped around each other with a left-handed supercoil (PDB code 2ZTA). (B) The heptad pattern of hydrophobic (*H*) and polar (*P*) residues typically found in coiled-coil sequences (top), and the burial of hydrophobic residues that drives coiled-coil formation (bottom).

observed, are distinguished at the sequence level.[24,47] Next, we describe how the heptad pattern can be embellished to bring different numbers of helices together. For our purposes sequences that span 3 or 4 heptads (21–28 residues) are sufficient to specify stable autonomously folded units.

The aforementioned heptad repeat is a good starting point for design, but it is only that. Unlike the zinc-finger case the majority of coiled-coil interactions occur across the hydrophobic interface and are therefore not only pairwise, but intermolecular. Fortunately, the problem of specifying coiled-coil structures appears to be simplified somewhat as follows: first, natural sequence do not use all hydrophobic residues equally at coiled-coil interfaces, with the aliphatic side chains (Ala, Ile, Leu, Met and Val) being preferred over the bulky aromatics.[48,49] Secondly, seminal research by Harbury and co-workers led to what we term the Harbury rules for oligomer-state selection.[50,51] To explain these, it is best to consider the *HPPHPPP* repeat as *abcdefg* to highlight that the two *H* sites are different in this odd-numbered repeat (Fig. 4B). In essence, Harbury showed that different hydrophobic patterns using the isomers Ile and Leu give different oligomers. For instance, a = Ile + d = Leu directs dimer formation; a = d = Ile, trimer; and a = Leu + d = Ile, tetramer. Harbury was able to rationalize his findings through crystal structures of the three different states, which revealed different packing arrangements in the cores that were best satisfied by the above combinations; rather like the tight and specific packing of pieces in a jigsaw puzzle.

Other issues in coiled-coil assembly that are relevant to design, as will become apparent below, include specifying helix orientation (*i.e.* parallel and antiparallel topologies) and partner selection (*i.e.* homo *versus* hetero-oligomers). Considerable work has been done in this area, particularly for dimeric coiled coils.[25] The rules that can be distilled from these are: (i) interhelical pairings of oppositely charged residues at *e* and *g* sites that bridge the hydrophobic interface (Fig. 4B), can be used to influence both helix orientation and partner selection; and (ii) certain polar residues can be inserted at *a* and *d* sites to help specify helix orientation, oligomer state and partner selection (Fig. 5). Though the above, distilled rules including the Harbury rules are apparent in many natural sequences,[49] they were determined from mutational studies of one natural coiled-coil motif, namely the leucine-zipper region from the yeast transcriptional activator GCN4.[52] GCN4-p1, as it is called, has proved to be one of the best models for protein folding. Nonetheless, it is clearly a very plastic structure that can form other oligomer states and topologies upon mutation,[53] and we wondered how transferable the Harbury rules would be to other constructs, specifically to vanilla-peptide constructs.

To this end, we designed three peptides based on a standard vanilla homo-oligomeric heptad repeat of Glu-*H*-Ala-Ala-*H*-Lys-Gln (with the revised assignment *gabcdef*)

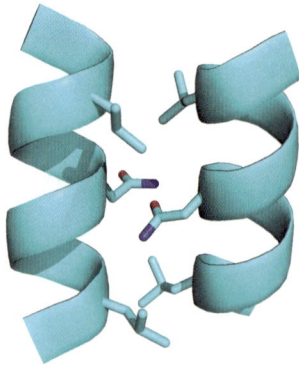

Fig. 5 A buried polar–polar pair in the core of a dimeric coiled-coil structure. This particular example is a pair of asparagine residues at complementary *a* sites in the GCN4 leucine zipper, flanked by leucines at two *d* sites (2ZTA). The inclusion is thermodynamically destabilizing, but specifying as the side chain–side chain hydrogen bond can only be made in the parallel, in-register dimer.

(Table 1). If the Harbury rules are context independent then the sequences would oligomerize as predicted, and the vanilla peptides would provide a basis set that could be adapted for future applications in synthetic biology and bionanotechnology.

Initially, the peptides were characterized by circular dichroism (CD) spectroscopy, which indicated that all three sequences folded extensively as α-helices, but that the folding was reduced for vCC-pIL over vCC-pII with vCC-pLI being the least folded (Fig. 6A). The variation of CD signal with temperature was also measured to explore the co-operativity of thermal denaturation (Fig. 6B). These data gave sigmoidal

Table 1 The Harbury rules translated into a vanilla homo-oligomeric coiled-coil background. The peptides were synthesized using standard solid-phase Fmoc/HBTU-based peptide synthesis, purified by reverse-phase HPLC and confirmed by mass spectrometry. Key: ac, acetyl; am, amide; otherwise standard one-letter codes are used for the amino acids; red, blue, orange and bold text signify acidic, basic, aromatic (UV chromophores) and hydrophobic residues, respectively; the remaining residues are vanilla, alanine or glutamine

Peptide	Sequence	Target oligomer
	fgabcdefgabcdefgabcdefg	
vCC-pIL	ac-GEIAALKKEIAALKYEIAALKYG-am	2
vCC-pII	ac-GEIAAIKKEIAAIKWEIAAIKQG-am	3
vCC-pLI	ac-GELAAIKYELAAIKKELAAIKQG-am	4

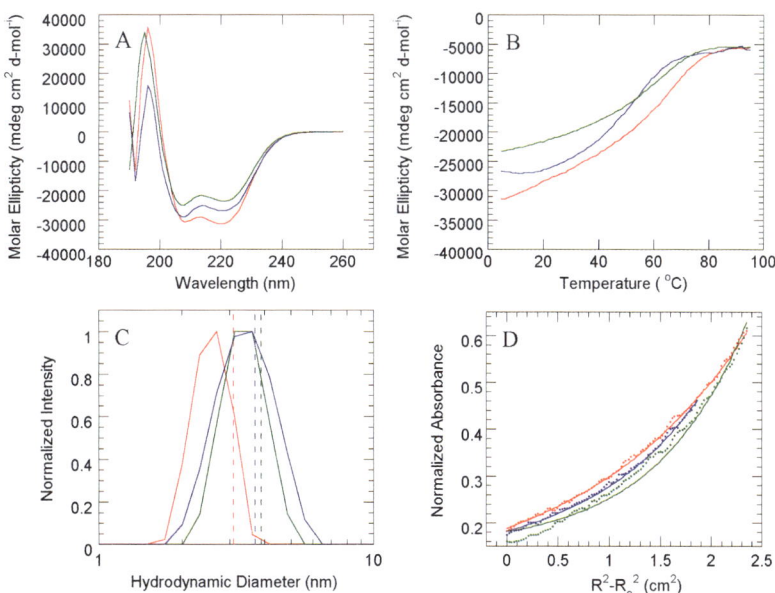

Fig. 6 Biophysical characterization of the coiled-coil basis set peptides. (A) CD spectra at 5 °C of 100 μM peptide in PBS. (B) CD signal at 222 nm as a function of temperature. (C) Hydrodynamic diameter as measured by DLS. The dotted lines indicate the predicted hydrodynamic diameters as calculated using the program Hydropro.[34] (D) Representative sedimentation equilibrium AUC traces for 50 μM peptide at 20 °C and at a speed of 46000 rpm. Key: vCC-pIL, red; vCC-pII, blue; vCC-pLI, green; the solid black lines show calculated curves for dimer, trimer and tetramer models.

unfolding curves largely consistent with co-operatively and uniquely folded structures. Although, the unfolding of vCC-pIL was preceded by a linear decrease in helicity, which is likely associated with fraying of the helices.[54] Peptide vCC-pII exhibited a sharper unfolding phase than vCC-pLI, as would be expected for a higher oligomerization state. Peptide vCC-pLI, however, unfolded with a broader transition that may indicate either a lower oligomerization state than targeted, less folding, or a mixture of states each with different melting temperatures.

The peptides were also examined by DLS to estimate the size of the assemblies in solution (Fig. 6C). Peptides vCC-pIL, vCC-pII and vCC-pLI all gave hydrodynamic diameters somewhat less than those predicted by the software Hydropro[34] for 3-heptad dimeric, trimeric and tetrameric coiled-coil models, respectively. The predictions match qualitatively, however, and the quantitative discrepancy is likely due to the aforementioned fraying of the helices and/or our ability to model hydration *in silico*, reducing the apparent sizes of the structures. Therefore, we subjected the peptides to AUC to gain more detailed information on the molecular weight in solution (Fig. 6D). The peptides were centrifuged over a range of speed and concentrations. The resulting data were fitted globally using Ultrascan[36] assuming single, ideal species in solution. These returned molecular weights of vCC-pIL, 4790 Da (expected dimer 5010 Da); vCC-pII, 6450 Da (expected trimer 7579 Da); and vCC-pLI 5660 Da (expected tetramer 9880 Da). These data show that while vCC-pIL appeared to be dimeric, vCC-pII gave a mass somewhat less than predicted, which we interpret as a monomer–trimer equilibrium, consistent with the reduced folding by CD spectroscopy. A much lower mass was found for vCC-pLI than that predicted for a tetramer, and, in combination with the CD data, this confirms that vCC-pLI does not form a stable tetrameric state. Further work is required to determine the range of species present.

In summary, we have demonstrated that in two out of three cases the Harbury rules for coiled-coil oligomer-state selection translate from a natural to a vanilla background. However, those for the tetramer do not appear to transfer as well. Studies are under way to improve the design towards the targeted tetramer.

3.3 A new target: the Ca^{2+}-binding EF-hand motif

We aim to add new domains to the field of peptide-based self-assembly. Protein–protein interaction motifs other than coiled coils would, for example, aid in the design of mutually exclusive interactions, and catalytic or metal-binding domains could be incorporated into assemblies to give rise to designed functional biomaterials. To determine the scope of domains that might be used in future designs, the online resource of protein domains, Pfam,[55] was searched for potential new targets for synthetic biology, as outlined in materials and methods. The *ca.* 30 domains returned from this search included protein–protein interaction motifs such as UBA domains[56,57] and the WW domain,[58] DNA binding motifs such as the ribbon–helix–helix domain[59,60] and the Fis family helix–turn–helix domain,[61] as well as the family of small enzymes related to 4-oxalocrotonate tautomerase.[62] Below, we focus on the EF-hand to illustrate our approach of rational vanilla design.

In many ways the EF-hand motif is similar to the zinc finger; it is small (*ca.* 30 amino acids), requires a metal ion for folding (in this case, calcium), and is found in a wide range of proteins; this time those which respond to Ca^{2+} signals. Structurally, EF-hands consist of two helices linked by a calcium-binding loop.[63] Upon binding calcium, many EF-hands undergo a conformational rearrangement, thereby transmitting information encoded in calcium signals into changes in protein activities. Some however undergo little or no change upon binding metal, and are thought to act as buffers.[64] The EF-hand offers a very different challenge to the zinc finger in terms of design—much like coiled coils, they are capable of oligomerizing, although often with EF-hands found on the same polypeptide chain.[64] The observation that isolated EF-hands are capable of pairing specifically with their intended partners in solution[65] has sparked some interest in determining the rules of EF-hand dimer assembly,[66–68] but the picture is by no means complete. Designing EF-hands from profiles such as those depicted in Fig. 3 has the obvious drawback that no pairwise information is contained, hence any attempt to design specific interactions between EF-hands must rely on more sophisticated methods. Of course, there are many ways of doing this, from the *in Biro* methods described above, to modeling methods employed by many protein designers.[69,70] A potentially powerful method for finding the determinants of such protein–protein interactions however, stems from sequence alignments alone. Recently one group has managed to switch the specificities of histidine kinase–response regulator pairings by changing a few amino acids selected by studying covarying residues in multiple sequence alignments.[71] Approaches such as this might give valuable insights into the design rules for more complicated systems, and indeed have also been used to increase the success rate of designing monomeric WW domains when compared to using flat profiles alone.[72] Once the key residues in the EF-hand have been recognized, the remaining task will be to create the vanilla scaffold upon which to graft them. The choice of residues at these non-essential positions should reflect both the backbone dihedral angles, and solvent accessibilities observed at the corresponding positions in structures of the domain. Further to this, these residues can be selected to fine-tune properties such as the isoelectric point.

Whilst we argue that the EF-hand offers the most immediate potential as a component for synthetic biology, the philosophy for designing vanilla versions of any other small domain remains the same. Establishing a robust protocol for gleaning the necessary rules for the design of these targets would vastly expand the repertoire of domains which the synthetic biologist might use in the design of complex systems or materials, and therefore is a key goal in the field.

4. Conclusion

We have outlined an approach to the design of peptide motifs that involves (i) gleaning those sequence-to-structure relationships most critical to directing and stabilizing the folding of the motifs, and (ii) incorporating the rules delivered into otherwise vanilla sequences. Our ultimate aims are: (A) to use these basic units as scaffolds upon which other structural and functional motifs can be added; and, in turn, (B) that these will provide routes to components or building blocks for synthetic biology and biomaterials design, and thence more complex, multi-component, peptide-based systems. We recognize that this borrows much from our own past work and that of many others in the field. However, we hope that by presenting it in this way it will focus minds, and a clear approach to the rational design of peptide-folding motifs and basic components will emerge.

To finish with biomaterials design: as hinted in the introduction, many peptide-based fibrous and gel materials are being developed and these have potential applications ranging from templating electronic circuits through to acting as synthetic extracellular matrices to scaffold the assembly of new tissue from stem cells.[11,73,74] However, most of these systems are currently just bare scaffolds comprising one, or, at most, two peptide components. To develop this field, and specifically to engineer responsive, dynamic and functional materials, new peptide components will be needed that assemble in complementary but independent ways to those comprising the main framework of the materials. The approach to new motifs that we describe may provide one route to such components.

We will present this philosophy in protein design, along with examples given herein at the Faraday Discussion Meeting in June 2009, and, hopefully, some new experimental systems that we characterize in the intervening time for discrete folding units and new materials designs based on these.

Acknowledgements

We thank the BBSRC for studentships to CTA and ALB, and grants to DNW (BB/D003016/1 and BB/E022359/1), and the Human Frontiers Science Programme for a grant to DNW and collaborators (RGP0031/2007) that supports ZNM, ART and EHCB. Finally, we are indebted to all members of the DNW group for many helpful discussions, and Dr Kevin Channon for his contribution to Fig. 2.

References

1 E. H. C. Bromley, K. Channon, E. Moutevelis and D. N. Woolfson, *ACS Chem. Biol.*, 2008, **3**, 38–50.
2 L. Jaeger and A. Chworos, *Curr. Opin. Struct. Biol.*, 2006, **16**, 531–543.
3 J. Bath and A. J. Turberfield, *Nat. Nanotechnol.*, 2007, **2**, 275–284.
4 C. M. Erben, R. P. Goodman and A. J. Turberfield, *J. Am. Chem. Soc.*, 2007, **129**, 6992–6993.
5 P. W. K. Rothemund, *Nature*, 2006, **440**, 297–302.
6 Y. Zhang, *Curr. Opin. Struct. Biol.*, 2008, **18**, 342–348.
7 B. Kuhlman, G. Dantas, G. C. Ireton, G. Varani, B. L. Stoddard and D. Baker, *Science*, 2003, **302**, 1364–1368.
8 A. Nomura and Y. Sugiura, *J. Am. Chem. Soc.*, 2004, **126**, 15374–15375.
9 M. Papworth, P. Kolasinska and M. Minczuk, *Gene*, 2006, **366**, 27–38.
10 D. N. Woolfson and M. G. Ryadnov, *Curr. Opin. Chem. Biol.*, 2006, **10**, 559–567.
11 C. E. MacPhee and D. N. Woolfson, *Curr. Opin. Solid State Mater. Sci.*, 2002, 141–149.
12 R. V. Ulijn and A. M. Smith, *Chem. Soc. Rev.*, 2008, **37**, 664–675.
13 K. Rajagopal and J. P. Schneider, *Curr. Opin. Struct. Biol.*, 2004, **14**, 480–486.
14 C. Vepari and D. L. Kaplan, *Prog. Polym. Sci.*, 2007, **32**, 991–1007.
15 W. Kim and V. P. Conticello, *Polym. Rev.*, 2007, **47**, 93–119.
16 F. W. Kotch and R. T. Raines, *Proc. Natl. Acad. Sci. U. S. A.*, 2006, **103**, 3028–3033.
17 V. Gauba and J. D. Hartgerink, *J. Am. Chem. Soc.*, 2007, **129**, 2683–2690.
18 B. Brodsky, G. Thiagarajan, B. Madhan and K. Kar, *Biopolymers*, 2008, **89**, 345–353.

19 V. Gauba and J. D. Hartgerink, *J. Am. Chem. Soc.*, 2008, **130**, 7509–7515.
20 S. Rele, Y. H. Song, R. P. Apkarian, Z. Qu, V. P. Conticello and E. L. Chaikof, *J. Am. Chem. Soc.*, 2007, **129**, 14780–14787.
21 S. E. Paramonov, V. Gauba and J. D. Hartgerink, *Macromolecules*, 2005, **38**, 7555–7561.
22 A. N. Lupas and M. Gruber, *Adv. Protein Chem.*, 2005, **70**, 37–78.
23 O. D. Testa, E. Moutevelis and D. N. Woolfson, *Nucleic Acids Res.*, 2009, **37**, D315–322.
24 E. Moutevelis and D. N. Woolfson, *J. Mol. Biol.*, 2009, **385**, 726–732.
25 D. N. Woolfson, *Adv. Protein Chem.*, 2005, **70**, 79–112.
26 J. M. Mason, K. M. Mueller and K. M. Arndt, *Methods Mol. Biol.*, 2007, 35–70.
27 E. Cerasoli, B. K. Sharpe and D. N. Woolfson, *J. Am. Chem. Soc.*, 2005, **127**, 15008–15009.
28 R. B. Hill, D. P. Raleigh, A. Lombardi and N. F. Degrado, *Acc. Chem. Res.*, 2000, **33**, 745–754.
29 G. Ghirlanda, A. Osyczka, W. X. Liu, M. Antolovich, K. M. Smith, P. L. Dutton, A. J. Wand and W. F. DeGrado, *J. Am. Chem. Soc.*, 2004, **126**, 8141–8147.
30 E. H. Bromley, R. B. Sessions, A. R. Thomson and D. N. Woolfson, *J. Am. Chem. Soc.*, 2009, **131**, 928–930.
31 R. D. Finn, J. Tate, J. Mistry, P. C. Coggill, S. J. Sammut, H.-R. Hotz, G. Ceric, K. Forslund, S. R. Eddy, E. L. L. Sonnhammer and A. Bateman, *Nucleic Acids Res.*, 2007, **36**, D281–288.
32 Protein Data Bank, see http://www.rcsb.org/pdb/home/home.do.
33 W. L. DeLano, *Pymol*, DeLano Scientific, Palo Alto, CA, 2002.
34 J. G. de la Torre, M. L. Huertas and B. Carrasco, *Biophys. J.*, 2000, **78**, 719–730.
35 G. Offer, M. R. Hicks and D. N. Woolfson, in *Workshop on Coiled-Coils, Collagen, and Co-Proteins*, Academic Press Inc., Elsevier Science, Alpbach, Austria, 2001, pp. 41–53.
36 B. Demeler, in *Modern Analytical Ultracentrifugation:Techniques and Methods.*, ed. D. J. Scott, S. E. Harding and A. J. Rowe, Royal Society of Chemistry, Cambridge, UK, 2005, pp. 210–229.
37 T. M. Laue, B. D. Shah, T. M. Ridgeway and S. L. Pelletier, in *Analytical Ultracentrifugation in Biochemistry and Polymer Science*, ed. S. E. Harding, A. J. Rowe and J. C. Horton, The Royal Society of Chemistry, Cambridge, UK, 1992, pp. 90–125.
38 M. S. Lee, G. P. Gippert, K. V. Soman, D. A. Case and P. E. Wright, *Science*, 1989, **245**, 635–637.
39 B. A. Krizek, B. T. Amann, V. J. Kilfoil, D. L. Merkle and J. M. Berg, *J. Am. Chem. Soc.*, 1991, **113**, 4518–4523.
40 S. F. Michael, V. J. Kilfoil, M. H. Schmidt, B. T. Amann and J. M. Berg, *Proc. Natl. Acad. Sci. U. S. A.*, 1992, **89**, 4796–4800.
41 G. E. Crooks, G. Hon, J. M. Chandonia and S. E. Brenner, *Genome Res.*, 2004, **14**, 1188–1190.
42 N. P. Pavletich and C. O. Pabo, *Science*, 1991, **252**, 809–817.
43 S. A. Wolfe, L. Nekludova and C. O. Pabo, *Annu. Rev. Biophys. Biomol. Struct.*, 2000, **29**, 183–212.
44 B. H. Huang, C. J. Schaeffer, Q. H. Li and M. D. Tsai, *J. Protein Chem.*, 1996, **15**, 481–489.
45 F. D. Urnov, J. C. Miller, Y. L. Lee, C. M. Beausejour, J. M. Rock, S. Augustus, A. C. Jamieson, M. H. Porteus, P. D. Gregory and M. C. Holmes, *Nature*, 2005, **435**, 646–651.
46 E. E. Perez, J. B. Wang, J. C. Miller, Y. Jouvenot, K. A. Kim, O. Liu, N. Wang, G. Lee, V. V. Bartsevich, Y. L. Lee, D. Y. Guschin, I. Rupniewski, A. J. Waite, C. Carpenito, R. G. Carroll, J. S. Orange, F. D. Urnov, E. J. Rebar, D. Ando, P. D. Gregory, J. L. Riley, M. C. Holmes and C. H. June, *Nat. Biotechnol.*, 2008, **26**, 808–816.
47 J. Walshaw and D. N. Woolfson, *J. Struct. Biol.*, 2003, **144**, 349–361.
48 J. F. Conway and D. A. D. Parry, *Int. J. Biol. Macromol.*, 1990, **12**, 328–334.
49 D. N. Woolfson and T. Alber, *Protein Sci.*, 1995, **4**, 1596–1607.
50 P. B. Harbury, T. Zhang, P. S. Kim and T. Alber, *Science*, 1993, **262**, 1401–1407.
51 P. B. Harbury, P. S. Kim and T. Alber, *Nature*, 1994, **371**, 80–83.
52 E. K. O'Shea, R. Rutkowski and P. S. Kim, *Science*, 1989, **243**, 538–542.
53 J. Liu, Q. Zheng, Y. Q. Deng, C. S. Cheng, N. R. Kallenbach and M. Lu, *Proc. Natl. Acad. Sci. U. S. A.*, 2006, **103**, 15457–15462.
54 A. I. Dragan and P. L. Privalov, *J. Mol. Biol.*, 2002, **321**, 891–908.
55 Pfam, see http://pfam.sanger.ac.uk.
56 T. Dieckmann, E. S. Withers-Ward, M. A. Jarosinski, C. F. Liu, I. S. Y. Chen and J. Feigon, *Nat. Struct. Biol.*, 1998, **5**, 1042–1047.
57 K. Hofmann and P. Bucher, *Trends Biochem. Sci.*, 1996, **21**, 172–173.
58 P. J. Lu, X. Z. Zhou, M. H. Shen and K. P. Lu, *Science*, 1999, **283**, 1325–1328.
59 E. R. Schreiter and C. L. Drennan, *Nat. Rev. Microbiol.*, 2007, **5**, 710–720.
60 W. S. Somers and S. E. V. Phillips, *Nature*, 1992, **359**, 387–393.

61 H. S. Yuan, S. E. Finkel, J. A. Feng, M. Kaczorgrzeskowiak, R. C. Johnson and R. E. Dickerson, *Proc. Natl. Acad. Sci. U. S. A.*, 1991, **88**, 9558–9562.
62 D. I. Roper, H. S. Subramanya, V. Shingler and D. B. Wigley, *J. Mol. Biol.*, 1994, **243**, 799–801.
63 N. C. J. Strynadka and M. N. G. James, *Annu. Rev. Biochem.*, 1989, **58**, 951–998.
64 Z. Grabarek, *J. Mol. Biol.*, 2006, **359**, 509–525.
65 G. S. Shaw, R. S. Hodges, C. M. Kay and B. D. Sykes, *Protein Sci.*, 1994, **3**, 1010–1019.
66 S. Linse, M. Voorhies, E. Norstrom and D. A. Schultz, *J. Mol. Biol.*, 2000, **296**, 473–486.
67 T. Berggård, K. Julenius, A. Ogard, T. Drakenberg and S. Linse, *Biochemistry*, 2001, **40**, 1257–1264.
68 T. Cedervall, I. Andre, C. Selah, J. P. Robblee, P. C. Krecioch, R. Fairman, S. Linse and K. S. Akerfeldt, *Biochemistry*, 2005, **44**, 13522–13532.
69 J. J. Plecs, P. B. Harbury, P. S. Kim and T. Alber, *J. Mol. Biol.*, 2004, **342**, 289–297.
70 C. Wang, P. Bradley and D. Baker, *J. Mol. Biol.*, 2007, **373**, 503–519.
71 J. M. Skerker, B. S. Perchuk, A. Siryaporn, E. A. Lubin, O. Ashenberg, M. Goulian and M. T. Laub, *Cell*, 2008, **133**, 1043–1054.
72 M. Socolich, S. W. Lockless, W. P. Russ, H. Lee, K. H. Gardner and R. Ranganathan, *Nature*, 2005, **437**, 512–518.
73 T. Scheibel, *Curr. Opin. Biotechnol.*, 2005, **16**, 427–433.
74 T. C. Holmes, S. de Lacalle, X. Su, G. S. Liu, A. Rich and S. G. Zhang, *Proc. Natl. Acad. Sci. U. S. A.*, 2000, **97**, 6728–6733.

The influence of viscosity on the functioning of molecular motors

Martin Klok,[a] Leon P. B. M. Janssen,[b] Wesley R. Browne[a] and Ben L. Feringa[*a]

Received 28th January 2009, Accepted 14th April 2009
First published as an Advance Article on the web 30th July 2009
DOI: 10.1039/b901841g

Light driven molecular motors based on sterically overcrowded alkenes achieve repetitive unidirectional rotation through a sequential series of photochemical and thermal steps. The influence of highly viscous environments on the functioning of unidirectional light driven molecular motors is established in the present report using three distinct media. Liquefied propane, a reference medium due to its low viscosity even at 85 K, and two binary solvent systems that undergo a well-defined glass-transition are used to evaluate the influence of glass-like viscosities on both photochemical equilibria and on thermal helix inversion in the molecular motors. It is found that the greater molecular volume of the stable conformations relative to the unstable conformations is responsible for a shift of the photochemical equilibrium from 20% to 100% in favour of the unstable form when irradiated in high viscosity media. These results demonstrate the critical role excluded solvent volume can play in immobilized photo responsive molecular systems. The volume expansion associated with the thermal reversal to the most stable isomer can be used to determine the maximum possible power output of the rotary cycle.

Introduction

Since the initial report of the first overcrowded-alkene based light-driven unidirectional molecular motors by our group, considerable attention has been focused on acceleration of the rotation process, with a view to applications in nanomechanical devices.[1] A series of reversible photochemical and effectively irreversible thermal steps provide unidirectionality. The rates of the thermal steps are orders of magnitude lower that the rates of the photochemical steps since photo-isomerization proceeds on a picosecond timescale. Hence the focus has been on understanding and controlling the thermal steps. Achievement of MHz rotation rates by reducing the rate of the thermal step to ns-timecales,[2] now requires that the photochemical processes associated with motor rotation receive attention to achieve further improvements in efficiency.

Cis–trans isomerisation, the first step in the unidirectional rotary cycle, is a paradigm photoreaction in molecular photochemistry and indeed within the considerable body of photochemistry literature; medium effects on cis–trans isomerisation have been well-documented.[3,4] As part of our ongoing research program into

---

[a]Laboratory of Organic Chemistry, Stratingh Institute for Chemistry, Zernike Institute for Advanced Materials and Center for Systems Chemistry, University of Groningen, Nijenborgh 4, 9747 AG Groningen, The Netherlands. E-mail: b.l.feringa@rug.nl; Fax: (+) 31 50 363 4278; Tel: (+) 31 50 363 4296
[b]Department of Chemical Engineering, Faculty of Mathematics and Natural Sciences, University of Groningen, Nijenborgh 4, 9747 AG Groningen, The Netherlands

unidirectional rotation in the MHz region,[2] a beneficial effect of organic glasses on the position of the photostationary states was noted. In organic glasses the position of the photostationary state was found to be changed in favour of the unstable state.

Two unidirectional molecular motors were subjected to irradiation in low and high-viscosity media at cryogenic temperatures to study this phenomenon in further detail, in order to gain new insights into the remarkable photochemistry of these compounds. The results show some analogy to stilbene photochemistry, however on key points the behaviour of the molecular motors is opposite to that observed for stilbenes. The origin of these differences is ascribed to the specific substitution pattern of the molecular motors and their effect on the relative molar excluded solvent volumes of the stable and unstable forms of these structures. The differences in excluded solvent volumes allow tuning of the photochemical equilibrium from 20% to 100% by controlling the viscosity of the organic glass at temperatures where thermally activated helix inversion does not take place.

Both viscosity and temperature have been found to have a pronounced effect on the quantum yields for trans → cis and cis → trans photoisomerisation, as well as on fluorescence quantum yields.[3–5] In several notable studies, cis–trans isomerisation of stilbene has been monitored in glassy organic solvent media, displaying pronounced decreases in both the trans → cis as cis → trans quantum yields.[5,6] Although a clear relationship between viscosity and the position of the photostationary state (PSS) was not established, it is apparent that the quantum yield of cis → trans isomerization is less dependent on both temperature and viscosity than the quantum yield of trans → cis isomerization.[4] Furthermore, the cis–trans ratio at photoequilibrium can be altered using different sensitisers.[4,7] However, complete conversion to the higher-energy isomer has not been achieved. In the case of stilbenes, the origin of the shift of the equilibrium has been ascribed to an increase in molecular volume during trans → cis isomerisation.[6,8] The equilibrium could not be shifted completely and thus full control over the photostationary state was not obtained. The rotation cycle for the structures studied in the present contribution are shown in Fig. 1

The compound comprises of a cyclopenta[a]naphthalene upper half (rotor), which is connected to a (thio)xanthyl stator. The central double bond functions as the axis of rotation. As reported earlier these compounds are capable of MHz unidirectional rotation under ideal irradiation conditions through an alternating series of photochemical cis–trans isomerisation and thermal helix inversion (half-life $t_{1/2}$ for the thermal conversion of **1B** to **1A** is 2.3×10^{-6} s, the $t_{1/2}$ for **2B** to **2A** conversion is 1.1×10^{-7} s).[2] Irradiation results in the establishment of a photochemical equilibrium between the stable forms **A** and thermally unstable forms **B**, in which the helicity is inverted. The unstable form **B** has a higher ground state energy due to the greater steric hindrance of the methyl group, at the stereogenic centre, with the stator. The photostationary state (PSS) is governed by the relative rate of forward and reverse photochemical reactions. These are governed by the molar absorptivities

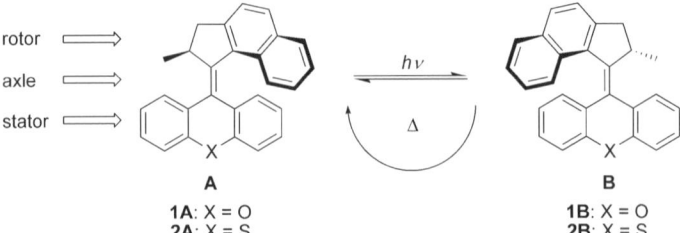

Fig. 1 Rotational cycle for light driven unidirectional molecular motors.

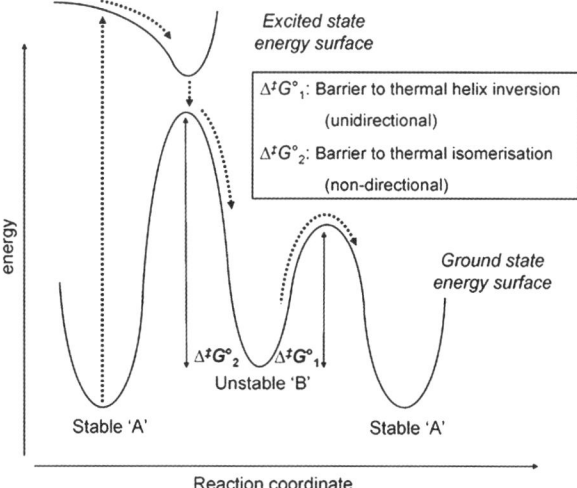

Fig. 2 Potential energy profile for photochemical *cis*–*trans* isomerisation and possible thermal pathways for **1** and **2**.

(ε) of the stable and unstable forms, and the quantum yields of each reaction. The higher energy isomer resulting from irradiation reverts to the most stable isomer by an activated ($\Delta^{\ddagger}G°_1$) helix inversion over the ground state potential energy surface in which the naphthalene part of the rotor unit passes over a phenyl ring of the stator unit (see Fig. 2).

In case of a non-substituted stator one photochemical–thermal cycle results in the formation of the same compound, in which the upper half is rotated 180° along the central double bond relative to the lower half. The direction of rotation is governed by the absolute orientation of the methyl group, whilst full rotation is achieved by consecutive cycles of photochemical *cis*–*trans* isomerisation followed by thermal helix inversion. In case of substituted stators, the pathways of photochemical *cis*–*trans* and *trans*–*cis* isomerisation can be distinguished, as can the pathways for thermal helix inversion over $\Delta^{\ddagger}G°_1$, which is used to study the full rotation cycle.[2]

An alternative thermal pathway with activation energy $\Delta^{\ddagger}G°_2$ is available for these structures, and results in a non-directional thermal *cis*–*trans* isomerisation. However, this process occurs at rates that are orders of magnitude lower due to the higher energy of partial rupture and reformation of the central double bond associated with this process ($\Delta^{\ddagger}G°_1 \ll \Delta^{\ddagger}G°_2$).[2] Therefore the pathway over $\Delta^{\ddagger}G°_2$ can be neglected in the analysis of the photochemical–thermal cycle which is responsible for the rotation.

In the present contribution, the effect of highly viscous environments on the behaviour of molecular motors is examined. High viscosities are reached through the use of binary organic solvent systems that are known to display a glass transition at cryogenic temperatures.[9] As the viscosity–temperature relationship is known in this region, precise control over the temperature provides control over the viscosity in the glass transition region.

Results

Propane gas was the solvent of choice for experiments at low viscosity; sample preparation with propane has been described earlier.[2] Two binary organic glass forming solvent systems with differing viscosity–temperature relationships were employed.

These systems, methylcyclohexane : methylcyclopentane = 1 : 1 (MCHMCP11) and isopentane : methylcyclohexane = 3 : 1 (IPMCH31), are widely used as trapping matrices for unstable species, and for this reason have been studied extensively. An inverse relationship between the viscosity in the glass transition region and temperature has been established by dropping ball viscometry,[9] however one report has described a non-linear inverse relationship on the basis of rotating disk viscosity measurements.[10] Nevertheless the linear relationship has been observed in several studies.[11] The synthesis and photochemical characterisation of compounds **1** and **2** were reported earlier.[2,12]

CD-spectroscopy

CD-spectroscopy of **1** in liquid propane at low solution viscosity indicated that the position of the photochemical equilibrium (PSS) upon irradiation at 365 nm lies in favour of the stable form **A** both at 85 K and at 120 K (Fig. 3). This provides a qualitative indication that the position of the photochemical equilibrium is not affected significantly by temperature within this range. The rate of thermal helix inversion of **1B** was negligible below 115 K, however fast thermal conversion to the stable form **1A** was observed at temperatures above 125 K, as was observed previously in isopentane solutions.[2]

In the organic-glass forming medium MCHMCP11 a different behaviour was observed. The spectrum of the stable form **1A** was not affected significantly by changes in temperature and viscosity. However, irradiation at various temperatures (and hence viscosities), resulted in a shift in the equilibrium position, determined from the change in the spectrum at the PSS relative to the spectrum of the stable form **1A**. At lower temperature, and hence higher viscosity, the spectrum at the PSS was inverted relative to the spectrum of the stable form **1A** more than at higher temperature (lower viscosity). This is a qualitative indication of a shift in the equilibrium position with increasing viscosity in favour of **1B**. Similar results were obtained in IPMCH31 (Fig. 4). The higher the viscosity at a particular temperature, the further the equilibrium lies in favour of the unstable form **1B**. Upon warming the irradiated sample to 150 K and subsequent cooling to reform the glass, a complete restoration of the original spectrum (**1A**) is achieved.

UV/Vis spectroscopy

Irradiation of **1** in IPMCH31 at 100 K (see Fig. 5) resulted in the appearance of a new absorption at longer wavelength. This band disappeared with recovery of the initial spectrum upon warming or by irradiation at $\lambda > 455$ nm. This indicates

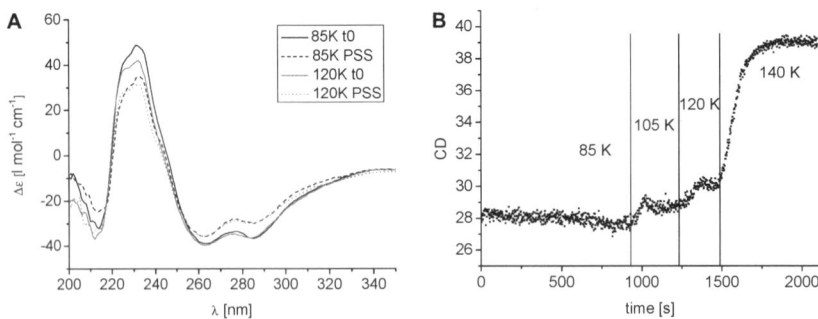

Fig. 3 Irradiation of **1** in liquid propane solution (left), and temperature profile for thermal helix inversion (right). The apparent baseline drift is due to changes in CD-intensity which accompany the change in temperature. Helix inversion takes place between 120 and 140 K. PSS = photostationary state.

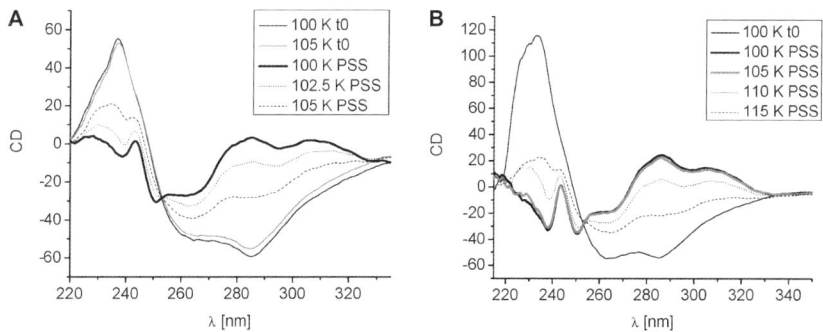

Fig. 4 CD-spectra of the stable isomer of **1** (**A**) and of the irradiated sample mixture (365 nm) at the photostationary state in IPMCH31 (left) and MCHMCP11 (right) at various temperatures.

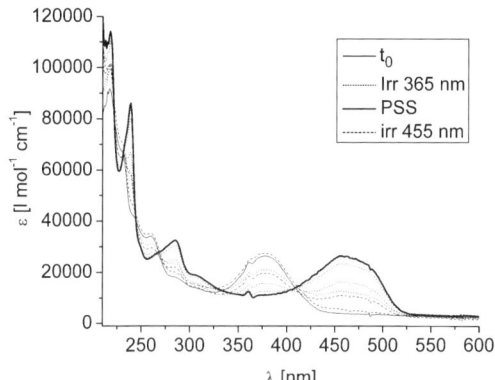

Fig. 5 UV/Vis spectra of the stable form of **1** and the PSS resulting from irradiation of **1** in IPMCH31 at 100 K.

that the unstable form **1B** is formed as part of a photochemical equilibrium upon irradiation. Clear isosbestic points were maintained as expected for a unimolecular process.

Fluorescence spectroscopy

The stable and unstable forms are strongly and weakly fluorescent, respectively. Furthermore, in accordance with the bathochromic shift in absorbance between the stable and unstable forms, the fluorescence of the unstable state is red-shifted with respect to the stable form. The higher energy isomer **1B** does not fluoresce significantly in solutions of low viscosity. At viscosities above 10^5 cP, however, the red-shifted fluorescence of **1B** is observed. This allows the position of the photochemical equilibrium to be determined quantitatively by fluorescence spectroscopy. *In situ* irradiation permits the sample to be maintained in identical positions before and after conversion to the PSS. Hence the decrease in fluorescence intensity provides quantification of the relative change in concentration of the stable and unstable forms. This allows the position of the photostationary state to be determined with reasonable accuracy, dependent on the signal intensity. Identical results were obtained from the decrease in signal intensity at the λ_{max} of the emission of the stable form and by integration of the entire spectrum, due to the near negligible relative intensity of the unstable form in most cases; in the following section the results obtained by integration of the fluorescence spectrum, corrected for the presence of fluorescence of the unstable form are employed.

Despite the weakness of the fluorescence of the stable form **A** in liquid propane at low temperature, irradiation at 365 nm yielded a photostationary state of *ca.* 36 : 64 (stable : unstable) (see Fig. 6). Fluorescence attributable to the unstable form was not observable. The photostationary state in this solvent did not show significant temperature dependency upon irradiation either at 365 nm or with broadband irradiation ($\lambda > 280$ nm). The time required to reach the photostationary state was <15 min under these conditions; the irradiation was carried out under identical conditions throughout the range of temperatures examined.[13]

In the glass forming media MCHMCP11 and IPMCH31 under similar experimental conditions an increased fluorescence intensity and a concomitant increase in irradiation times to reach the photostationary state were observed. Broadband irradiation ($\lambda > 280$ nm) provided quantitative conversion to the unstable isomer **1B**, albeit with irradiation times of up to 8 h to reach the photostationary state.[14] Two examples in which the fluorescence spectra of **1A** and **1B** in MCHMCP11 at relatively high and low viscosity are shown in Fig. 7. At the photostationary state under broadband irradiation ($\lambda > 280$ nm), the fluorescence of **1A** decreases to 82% of the initial intensity (PSS **1A** : **1B** = 82 : 18). Fluorescence from **1B** was not observed even under direct excitation under these conditions ($\lambda_{exc} = 460$ nm, dotted line). By contrast, at high viscosity the fluorescence of **1A** decreases to 3% of the

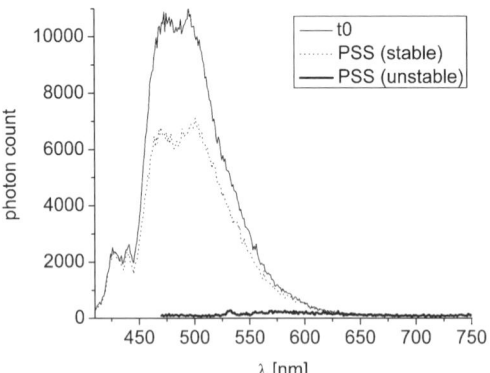

Fig. 6 Fluorescence spectra of **1** in liquid propane at 85 K before irradiation at 365 nm and at the photostationary state (λ_{exc}(**1A**) = 360 nm, λ_{exc}(**1B**) = 460 nm).

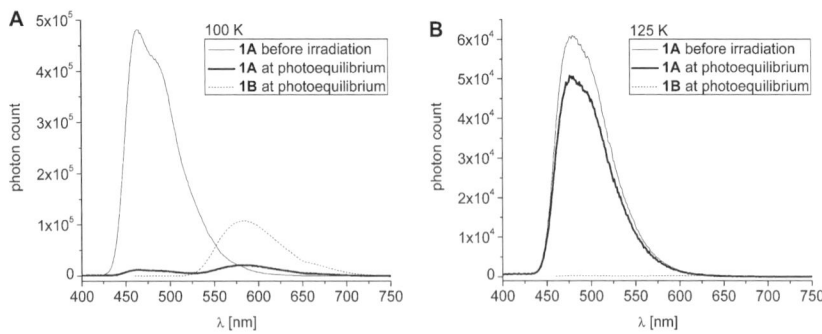

Fig. 7 Fluorescence spectra of pure stable isomer **1A** and at the photostationary state ($\lambda > 280$ nm) of **1A** and **1B** at high (left) and low (right) viscosity in MCHMCP11 (λ_{exc}(stable **1A**) = 350 nm, λ_{exc}(unstable **1B**) = 450 nm).

original intensity (Fig. 7, left), representative of a PSS of **1A** : **1B** = 3 : 97. Under these conditions, fluorescence of **1B** is readily observable (dotted line).

By expressing the position of the photostationary state as $\ln(\phi_B/\phi_A)$,[15] a relationship of the position of the photostationary state (PSS ratio) with temperature and viscosity could be established. A correlation of the position of the photostationary state with temperature was not observed, even where the viscosity remains within the same order of magnitude (Table 1). By contrast, a close correlation was obtained between the position of the photostationary state for both media and viscosity as determined *via* the dropping ball method (Table 2).[9] Therefore, the shift in the photostationary state in favour of the thermally unstable form **1B** is ascribed to the effect of viscosity (Fig. 8). At low viscosity, the data have a higher uncertainty

Table 1 Photostationary states for **1** for various temperatures and irradiation sources in liquid propane as determined by fluorescence spectroscopy

$\lambda > 280$ nm		$\lambda = 365$ nm	
T/K	% **1B**	T/K	% **1B**
90	17	85	47
95	25	109	39
100	17	115	25
105	23	120	31
125	20	125	40

Table 2 PSS positions upon irradiation of **1** in MCHMCP11 at various viscosities

T/K	$\ln(\eta)$	λ_{max} (**1A**)	% **1B**	ϕ_B/ϕ_A	$\ln(\phi_B/\phi_A)$
95	15.66	460	99	90.06	4.50
100	12.66	464	97	38.51	3.65
105	9.90	470	92	11.36	2.43
109	8.29	473	75	2.97	1.09
115	6.45	476	43	0.76	−0.27
120	4.14	478	19	0.24	−1.43
125	2.30	479	18	0.22	−1.51

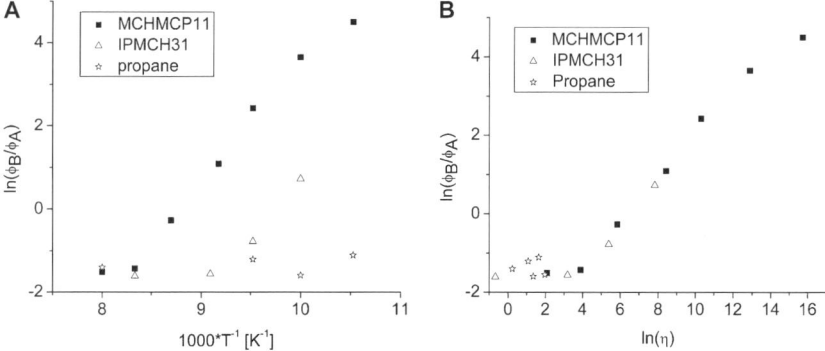

Fig. 8 Position of the photostationary state between **1A** and **1B** upon broadband irradiation ($\ln(\phi_B/\phi_A)$) against reciprocal temperature (left) and against logarithmic viscosity (right) in media with three different viscosity–temperature relationships, as determined by the dropping ball method.

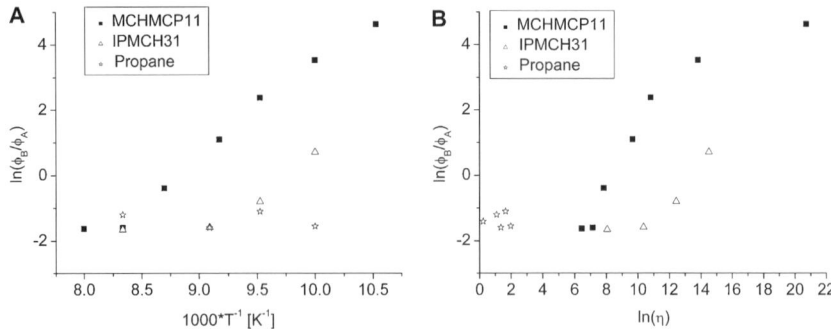

Fig. 9 Position of the photostationary state between **1A** and **1B** upon broadband irradiation ($\ln(\phi_B/\phi_A)$) against reciprocal temperature and logarithmic viscosity in media with three different viscosity–temperature relationships, as determined by the rotating disk method.

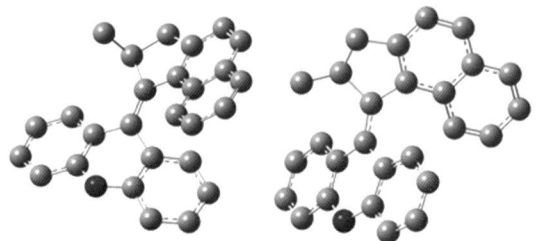

Fig. 10 DFT-optimised structures of stable **1A** (left) and unstable **1B** (right).

due to lower absolute fluorescence, whereas in the high-viscosity regime only the data obtained in MCHMCP11 can be used due to solidification of IPMCH31 at these temperatures.

By contrast, when viscosity values, determined by the rotating disk method for viscosity measurement,[10] were employed an overlap of the curves was not observed (Fig. 9). Although it is difficult to ascertain which method for viscosity determination has greater physical meaning with regard to the present system, it is still possible to conclude that the effect of the shift of the position of the photostationary state is due to the increase in viscosity. The smaller and more flexible isopentane may provide a more 'flexible' microenvironment in comparison to methylcyclopentane, in which the equilibrium is less easily affected. An alternative rationalisation of this difference may lie in the differences between the methods employed for the determination of solvent viscosity; the rotating disk method is performed under high mechanical stresses, whereas dropping ball experiments are likely to leave the glass in a more equilibrated state.

The Conolly solvent-excluded volume was calculated for both **1A** and **1B** on structures optimised (see Fig. 10) with Density Functional Theory (B3LYP 6-31G(d,p)) using Gaussian 03.[16] The most stable conformation of molecular motor **1** (**1A**) has a free volume of 254.9 Å^3, whereas the higher energy isomer **1B** has a free volume of 249.2 Å^3. Hence the transition from the higher energy isomer **1B** to the lowest energy isomer **1A** requires a volume expansion of 2.3%.

Influence of viscosity on the thermal activated helix inversion

Whereas for **1** the thermal conversion of **1B** to **1A** is not observable in the low viscosity (>85 K) solvent propane, the thermal conversion of the related molecular motor **2B** to the stable form **2A** can be observed readily in propane in the temperature range at which the binary solvent mixtures undergo glass–liquid transitions, *i.e.* 85–125 K. The (extrapolated) rate constant for **2B** → **2A** conversion at 100 K is 6.5×10^{-2} s^{-1}, with a corresponding half-life of 11 s. By contrast, the rate constant for thermal conversion of **2B** → **2A** in MCHMCP11 at 100 K was found to be 1.0×10^{-3} s^{-1} ($t_{1/2} = 672$ s). Previously, in low viscosity solvents of varying polarity, the rate of this reaction was found to vary only modestly (within half an order of magnitude).[17] In the present study, the influence of viscosity is much more pronounced. Here, given the close similarity in other solvent characteristics other than viscosity, a much larger influence of the solvent matrix is observed. It is therefore reasonable to assume that the highly viscous solvent matrix provides extra stabilisation of the unstable form **2B** through additional interactions.

The natural logarithm of the rate of the thermal reaction **2B** → **2A** was found to be inversely related to temperature (Fig. 11, left) despite the glass transition that takes place over the same temperature range. As dropping ball methods for the determination of viscosity indicate an inverse linear dependency of viscosity with temperature over the liquid/glass transition, and the Arrhenius relation predicts a linear relationship of $\ln(k)$ with T^{-1} at a constant viscosity, the values for viscosity obtained by the dropping ball method will be used in the present study.

The rate constant for thermal helix inversion can be separated into two components; a temperature dependence and a viscosity dependence. The temperature dependence was determined previously in low viscosity propane solutions;[2] extrapolation of the data to higher temperatures provides rate constants for the temperatures at which rate determinations were performed in MCHMCP11. In the present study, the influence of viscosity on the rate constant for thermal helix inversion can be obtained by subtracting the rate constant at high viscosity from the rate constant at low viscosity. A correction for the change in viscosity of propane over this temperature region[18] has been included in the analysis, although it does not affect the results significantly. The linear relationship obtained is the viscosity dependence of the rate constant for the thermally activated reaction **2B** → **2A**, *where the temperature effect has been excluded* (Fig. 11, right). This relationship establishes the dependency of the rate of thermal helix inversion on the viscosity of the solvent matrix, *i.e.* a mechanically opposing force of known magnitude.

As discussed above, the Conolly-solvent-excluded volume of the unstable isomer of **1** is less than that of the stable isomer. This relative difference in solvent excluded volume is found for **2** also. The unstable form **2B** has a molecular volume of 265.3 Å^3, whereas the most stable form **2A** has a molecular volume of 271.9 Å^3. Hence, thermal relaxation from the unstable form **2B** to the stable form **2A** is accompanied by a volume expansion of 2.4%.

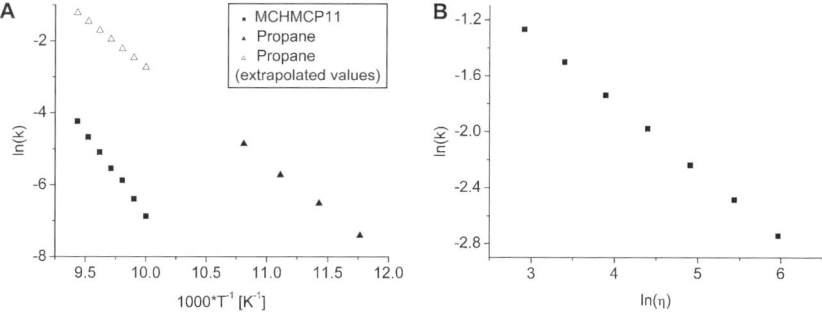

Fig. 11 Rate constants for thermal helix inversion of **2** over the glass transition of MCHMCP11 and in the low-viscosity medium propane (left). The dependence of the rate constant on the glassiness of the solvent environment can be obtained by subtraction of the rate constants at high viscosity from those at low viscosity (right).

Discussion

Viscosity effects on the photochemistry of molecular motors

The photochemical behaviour of alkene based molecular motors shows striking similarities to that of stilbene.[3,4,19–22] Qualitatively, as can be deduced from the increase in time required to reach the photostationary state (from 15 min to 8 h), the photochemical quantum yield for *cis–trans* isomerisation decreases dramatically upon moving from low to high solvent viscosity, as has been observed for stilbenes. Furthermore, whereas the stable form **1A** is moderately fluorescent, the unstable form **1B** is essentially non-fluorescent. This correlates with an energy barrier on the (S^1) excited state surface between the local minima, with the minima of the **1A** form being lowest in energy, as has been observed experimentally for *trans*-stilbene and, more recently, for **1** by transient absorption spectroscopy.[23] For this reason, the stable form **1A** can be compared to *trans*-stilbene, while the higher energy isomer **1B** can be compared to *cis*-stilbene on the basis of relative stabilities.

In the case of the differences in the solvent-excluded volume between each isomer an opposite situation is observed; although the general shape of the excited state energy surface of stilbene is preserved in the molecular motors, the additional structural features of these molecular motors result in a reversal of the relative difference in the solvent-free volume of the stable and unstable forms.[19,24] As a consequence of the higher solvent-free volume of the stable isomers **A** relative to the unstable isomers **B**, photoisomerisation of **B** to **A** requires an expansion in volume and displacement of solvent molecules, which becomes increasingly difficult at increasing viscosities. By contrast, photochemical conversion of **A** to **B** can take place inside the same solvent cavity, and hence is much less affected by viscosity. Therefore the rate of the photochemical reaction of **1A** → **1B** is much less affected by the increase in viscosity than the conversion of **1B** → **1A**. Because the ratio of these two rates determines the position of the photostationary state, the equilibrium shifts towards higher conversion to the unstable form **1B** as solvent viscosity increases.

Analysis of thermal isomerization and changes in solvent-excluded volume

The retardation of thermal conversion of **2B** → **2A** is in agreement with the analysis in the previous section, as this step is associated with volume expansion. A linear relationship of the logarithm of the rate constant with the logarithm of viscosity is obtained when separating the dependence of viscosity from the dependence on temperature (Fig. 11). The retardation of thermal helix inversion in glassy environments can be accounted for in two ways:

■ The helix inversion, that involves an increase of the solvent-excluded volume, requires solvent molecules undergoing random thermal fluctuations to free space for molecular reorganisation; *i.e.* volume expansion can only take place after the cavity is enlarged by random thermal motion, and because reorganisation times are long in viscous environments helix inversion is retarded.

■ Helix inversion is associated with a release of free energy. As the solvent shell is part of the system undergoing this process, part of the free energy is used to reorganise the solvent shell, making room for volume expansion.

Although these two rationalisations are non-mutually exclusive, there is a fundamental distinction; in the first case, the molecule is passive, and can only wait for the solvent cavity to reorganise by random Brownian motion in such a way that volume expansion can occur. In this case, no work is performed on the environment, and Gibbs energy is converted completely into random thermal motion. In the second case, it is assumed that intermolecular interactions result in the excess Gibbs energy of the higher energy isomer influencing the solvent shell directly. It is not dissipated through solution as heat and translational motion because translation is slow in glassy environments. In this case, actual work is performed through volume expansion against an opposing force. These two models reflect two extremes in the possibilities that are conceivable for molecular solute–solvent interactions. Although it is likely that both models hold to a certain extent, from the empirical data available for compound **2** a fundamental limiting value to the amount of work performed by this molecular motor can be determined by considering the process as purely defined by the second option. Viscosity serves as the opposing force, and is thus quantified. The same holds for the rate of volume expansion, *i.e.* the rate of thermal helix inversion. Approximating **2A** and **2B** as spheres, the radius can be determined from the molecular volume: $r_{2A} \equiv r_2 = 4.019 \times 10^{-10}$ m, $r_{2B} \equiv r_1 = 3.985 \times 10^{-10}$ m. Approximating the expansion rate as linear, it can be written as

$$v_r = \frac{(r_2 - r_1)}{t_{1/2}} \quad (1)$$

An expression for the energy required to expand a sphere against an opposing viscosity can be derived from this (see Experimental section):

$$E_\eta = \int_{r_1}^{r_2} \tau_r \partial V = 8\pi\eta v_r \int_{r_1}^{r_2} r \partial r = 4\pi\eta v_r (r_2^2 - r_1^2) \quad (2)$$

Because the power output is the energy dissipated per unit time, the average power exerted on the viscous medium is given by:

$$P = 8\pi\eta r v_r^2 \quad (3)$$

Combining eqn (1–3), an expression for the maximum power output of a molecular motor is obtained:

$$P = \frac{4\pi\eta(r_2 - r_1)(r_2^2 - r_1^2)}{t_{1/2}^2} \quad (4)$$

The dependence of the rate on viscosity can be evaluated by determining the rate constants in glassy environments of known viscosities and subtracting these rates from the rates at low viscosity. The residual rate constants reflect the dependence of the rate on viscosity, with the effect of temperature excluded. Residual half-life times, $t_{1/2}^\eta$, can be derived from the residual rates, k^η (Table 3).

The data provide a constant value, $P = 3.8(1) \times 10^{-25}$ W, for all viscosity–rate combinations. This value is the maximum average power output per molecule, valid for the case where the medium is completely 'pushed away' by volume expansion of the motor, rather than for the case where volume expansion follows reorganisation of the solvent shell by random motion.

Table 3 Viscosities η and experimental residual half-lifes $t_{1/2}^\eta$ in relation to the expansion rate v_r against an opposing force, the energy required for volume expansion E_η and the power output P

η/kg m^{-1} s^{-1}	$t_{1/2}^\eta$/s	v_r/m s^{-1}	E_η/J	P/W
3.90×10^8	10.8	3.2×10^{-13}	4.2×10^{-24}	3.9×10^{-25}
2.25×10^8	8.3	4.1×10^{-13}	3.1×10^{-24}	3.8×10^{-25}
1.31×10^8	6.5	5.2×10^{-13}	2.3×10^{-24}	3.6×10^{-25}
8.63×10^7	5.0	6.8×10^{-13}	2.0×10^{-24}	4.0×10^{-25}
5.07×10^7	3.9	8.6×10^{-13}	1.5×10^{-24}	3.8×10^{-25}
3.15×10^7	3.1	1.1×10^{-12}	1.2×10^{-24}	3.8×10^{-25}
2.00×10^7	2.5	1.4×10^{-12}	9.4×10^{-25}	3.8×10^{-25}

The energy difference between the higher energy isomer and the most stable isomer represents the maximum possible amount of energy that can be dissipated by the volume expansion associated with conversion of the unstable form to the stable form. The energy difference determined by calculation is 14.9 kJ mol^{-1},[17] which corresponds to 2.47×10^{-20} J per molecule. It should be emphasized that only one specific step, *i.e.* the thermal helix inversion, is considered here and that the model focuses on solvent reorganization only. The photochemical isomerization step, and the power output associated with it, are not considered here but it is clear from the above analysis that this 'power stroke' step plays the major role in determining the overall efficiency in conversion of light to actual work done by the molecular motor.

Conclusion

The photochemistry of the molecular motors described here bear a not unexpected similarity with that of stilbene. A pronounced difference is observed between these systems in the solvent-excluded volume of the lower and higher energy isomers; for stilbenes the lowest-energy isomer has the smallest molar volume, whereas for molecular motors the higher energy isomer has the smallest volume. This allows for tuning the position of the photostationary state by adapting the viscosity of the medium at temperatures where thermal relaxation of the photochemical product does not occur. Since the efficiency of the photochemical reaction that causes an expansion of volume (*i.e.* from the higher energy to the lowest energy isomer) decreases much more than the efficiency of the pathway that causes a decrease in volume, the equilibrium shifts from approximately 20% unstable to 99% unstable in the glass-transition region of organic apolar binary solvent systems due to the increase in viscosity.

The process of thermal helix inversion that allows for recovery of the lowest energy conformation occurs with an increase in terms of molecular volume. Investigation of the thermal process in the glass-transition region of organic glasses reveals that the process is significantly retarded by increased viscosity. Comparison of the rates at normal viscosity in solution and at high viscosities allow for a separation of temperature- and viscosity contributions. The viscosity-dependence of the rate constant is assumed to reflect the mechanical stabilisation force of such media, and from that a fundamental limit on the power output associated with the thermal step of the molecular motor **2** is determined to be 3.8×10^{-25} J per molecule.

Experimental

All compounds were prepared following procedures reported previously.[12a] UV/Vis and CD spectroscopic experiments were performed using an Oxford Instruments OptistatDN variable temperature liquid nitrogen cryostat inserted in a Hewlett-Packard HP 8543 diode array spectrophotometer or a JASCO J-715 spectropolarimeter. Fluorescence-experiments were performed on a Horiba Fluorolog *R*3 modified to allow insertion of the cryostat, and equipped with a 400 nm cut-off filter immediately before the detector monochromator. Uvasol-grade solvents were used, except in the case of propane, which was received as a pressurized gas. Spectroscopically clear propane samples were prepared by liquefaction of propane at 200 K in a flame-dried dinitrogen atmosphere, and dried over $MgSO_4$. A 10^{-5} M isopentane solution containing the sample was evaporated to near-dryness and placed in the mouth of the nitrogen-flushed cryostat, in a continuous stream of nitrogen. The sample was diluted to the original volume by pouring the dried liquefied propane through a funnel-and- filter combination into the sample cell, which was then immediately inserted into the cryostat. Using this procedure, the strong nitrogen stream prevented condensation of water on the outside of the cuvette, and spectroscopically clear propane samples were obtained. Irradiations were performed using a 200W Oriel Xe(Hg) lamp or a 200 W Oriel Hg-lamp fitted with suitable bandpass filters (typical bandwidth 10 nm). Temperature equilibration was ensured by waiting periods of 15 min for the low viscosity (propane) experiments; periods of up to 90 min were required for equilibration of organic-glass-forming media before the start of the experiment. These periods were established separately by time-resolved UV/Vis spectroscopic experiments on the glass-forming media.

Derivation of eqn (4)

Models that take the discrete character of molecules into account are highly complex due to their statistical nature. Because in the literature as yet no agreement has been reached as to how to model viscosity as a molecular system, we do not consider application of any such model for use in calculations on intermolecular mechanical energy transfer. Instead, a model is adopted in which a discrete molecule is surrounded by a continuum matrix. This is justifiable because in a glassy environment, relaxation times are long compared to a low-viscosity solution. Volume expansion of the molecule during the thermal transition to the most stable isomer is assumed to be associated with a full active conversion of the energy into work, *i.e.* the pushing away of the solvent shell to make enough room for isomerisation. Thereby the result is an expression for the *maximum* power output, based on the arguments presented above.

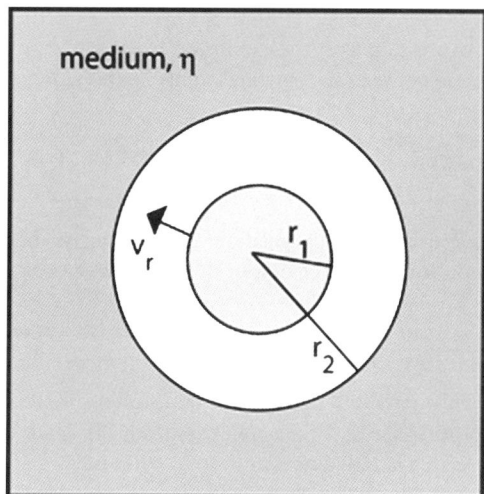

The energy required to 'push away' a medium of viscosity η can be obtained from fluid dynamics using the continuity equation, which is a means of saying that the mass that enters a system is equal to the mass that leaves a system. The continuity equation is written in a general form as

$$\frac{\partial \rho}{\partial t} = -\nabla \cdot (\rho v) \qquad (5)$$

in which ρ is density, v is speed and t is time. In an incompressible fluid ρ is considered a constant and the continuity equation reduces to

$$\nabla v = 0 \qquad (6)$$

Written in radial coordinates, and assuming only velocities in radial directions this is equal to stating:

$$\frac{1}{r^2}\frac{\partial}{\partial r}(r^2 v_r) = 0 \qquad (7)$$

differentiating by parts results in the relation

$$\frac{\partial v_r}{v_r} = -2\frac{\partial r}{r} \qquad (8)$$

Integration of 8 reveals that the rate of expansion v_r decreases exponentially with radius according to

$$v_r = \frac{c}{r^2} \qquad (9)$$

in which c is a constant. However, from 8 it is also clear that

$$\frac{\partial v_r}{\partial_r} = -2\frac{v_r}{r} \qquad (10)$$

According to the constitution equation the shear stress τ of a surface in a Newtonian medium of viscosity η is given by

$$\tau_r = -\eta \frac{\partial v_r}{\partial r} = \frac{2\eta v_r}{r} \qquad (11)$$

Integration of this expression over the expansion from r_1 to r gives the energy used for freeing the volume between the starting radius r_1 and r, E_η:

$$E_\eta = \int_{r_1}^{r} \tau_r \partial V = 8\pi \eta v_r \int_{r_1}^{r} r \partial r = 4\pi \eta v_r (r^2 - r_1^2) \qquad (2)$$

Because the power P exerted on the medium is equal to the time derivative of the energy, an expression for the power output during volume expansion is obtained:

$$P = \frac{\partial E_\eta}{\partial t} = 4\pi \eta \left((r^2 - r_1^2)\frac{\partial v_r}{\partial t} + 2v_r r \frac{\partial r}{\partial t} \right) = 4\pi \eta \left((r^2 - r_1^2) a_r + 2r v_r^2 \right) \qquad (12)$$

Although it is known from 9 that the rate of expansion decreases exponentially with radius, it is difficult to solve 12 for such an expansion. Instead, linear expansion is assumed, in which case the radial acceleration $a_r = 0$ and 12 reduces to

$$P = 8\pi \eta r v_r^2 \qquad (3)$$

Using this linear approximation, we can calculate the average power output during expansion by integration of 3 over the full expansion

$$<p> = \frac{1}{(r_2 - r_1)} \int_{r_1}^{r_2} P \partial r = \frac{8\pi \eta v_r^2 (r_2^2 - r_1^2)}{(r_2 - r_1)} \qquad (13)$$

Because a linear expansion rate is assumed, v_r can be written as the distance travelled by the average time it takes to cross this distance. This time is directly related to the rate constant k of the thermal process,[1] as the half-life time $t_{1/2}$ is defined as the time in which half of the particles have undergone transition. Therefore,

$$v_r = \frac{(r_2 - r_1)}{t_{1/2}} \quad (1)$$

Inserting eqn (1) in eqn (13) gives an expression for the average power exerted by a motor molecule in the thermal isomerization step in which both the initial and the final state are approximated by spheres, assuming a linear expansion rate in an incompressible Newtonian liquid that is modelled as a continuum:

$$P = \frac{4\pi\eta(r_2-r_1)(r_2^2-r_1^2)}{t_{1/2}^2} \quad (4)$$

Acknowledgements

NanoNed and the Nederlandse Organisatie voor Wetenschappelijk Onderzoek (Netherlands Organisation for Scientific Research) NWO-Vidi (WRB) are thanked for financial support.

References

1 (a) N. Koumura, R. W. J. Zijlstra, R. A. van Delden, N. Harada and B. L. Feringa, *Nature*, 1999, **401**, 152; (b) N. Koumura, E. M. Geertsema, M. B. van Gelder, A. Meetsma and B. L. J. Feringa, *J. Am. Chem. Soc.*, 2002, **124**, 5037; (c) W. R. Browne and B. L. Feringa, *Nat. Nanotechnol.*, 2006, **1**, 25; (d) B. L. Feringa, *J. Org. Chem.*, 2007, **72**, 6635; (e) M. M. Pollard, M. Klok, D. Pijper and B. L. Feringa, *Adv. Funct. Mater.*, 2007, **17**, 718.
2 M. Klok, N. Boyle, M. T. Pryce, A. Meetsma, W. R. Browne and B. L. Feringa, *J. Am. Chem. Soc.*, 2008, **130**(32), 10484.
3 D. H. Waldeck, *Chem. Rev.*, 1991, **91**, 415.
4 *Advances in Photochemistry*, Vol. 19, Neckers, D. C.; Volman, D. H.; von Bünau, G. (Eds), John Wiley & Sons, Inc (1995).
5 S. Malkin and E. Fischer, *J. Phys. Chem.*, 1964, **68**, 1153.
6 D. Gegiou, K. A. Muszkat and E. Fischer, *J. Am. Chem. Soc.*, 1968, **90**, 3907.
7 Y. Inoue, N. Yamasaki, A. Tai, Y. Daino, T. Yamada and T. Hakushi, *J. Chem. Soc., Perkin Trans. 2*, 1990, 1389.
8 D. Gegiou, K. A. Muszkat and E. Fischer, *J. Am. Chem. Soc.*, 1968, **90**, 12.
9 H. Greenspan and E. Fischer, *J. Phys. Chem.*, 1965, **69**, 2466.
10 G. A. von Salis and H. Labhart, *J. Phys. Chem.*, 1968, **72**, 752.
11 (a) A. C. Ling and J. E. Willard, *J. Phys. Chem.*, 1968, **72**(6), 1918; (b) A. C. Ling and J. E. Willard, *J. Phys. Chem.*, 1968, **72**(9), 3349.
12 (a) M. K. J. Ter Wiel, J. Vicario, S. G. Davey, A. Meetsma and B. L. Feringa, *Org. Biomol. Chem.*, 2005, **3**, 28; (b) Basic Molecular Devices, M. K. J. ter Wiel PhD thesis, University of Groningen, Groningen, the Netherlands, 2004.
13 The photostationary state was shifted slightly in favour of the unstable form **B** with irradiation at 365 nm in comparison with broadband light irradiation, due to the maximum in the absorption spectrum for the stable form **1A**, and the minimum for the unstable form **1B** at this wavelength.
14 Due to practical limitations on irradiation times in a cryostat, 365 nm irradiation could not be performed in high-viscosity media because of the lower intensity of light in combination with the decrease in photochemical quantum yield.
15 Differences in the absorption spectra between **1A** and **1B** were disregarded, due to the use of non-spectrally uniform broadband light.
16 M. J. Frisch, G. W. Trucks, H. B. Schlegel, G. E. Scuseria, M. A. Robb, J. R. Cheeseman, J. A. Montgomery, Jr., T. Vreven, K. N. Kudin, J. C. Burant, J. M. Millam, S. S. Iyengar, J. Tomasi, V. Barone, B. Mennucci, M. Cossi, G. Scalmani, N. Rega, G. A. Petersson, H. Nakatsuji, M. Hada, M. Ehara, K. Toyota, R. Fukuda, J. Hasegawa, M. Ishida, T. Nakajima, Y. Honda, O. Kitao, H. Nakai, M. Klene, X. Li, J. E. Knox, H. P. Hratchian, J. B. Cross, C. Adamo, J. Jaramillo, R. Gomperts, R. E. Stratmann, O. Yazyev, A. J. Austin, R. Cammi, C. Pomelli, J. W. Ochterski, P. Y. Ayala, K. Morokuma, G. A. Voth, P. Salvador, J. J. Dannenberg, V. G. Zakrzewski, S. Dapprich, A. D. Daniels, M. C. Strain, O. Farkas, D. K. Malick, A. D. Rabuck,

K. Raghavachari, J. B. Foresman, J. V. Ortiz, Q. Cui, A. G. Baboul, S. Clifford, J. Cioslowski, B. B. Stefanov, G. Liu, A. Liashenko, P. Piskorz, I. Komaromi, R. L. Martin, D. J. Fox, T. Keith, M. A. Al-Laham, C. Y. Peng, A. Nanayakkara, M. Challacombe, P. M. W. Gill, B. Johnson, W. Chen, M. W. Wong, C. Gonzalez, and J. A. Pople, *Gaussian 03, Revision C.02*, Gaussian, Inc., Wallingford CT, 2004.

17 Motors for use in Molecular Nanotechnology, M. Klok, PhD-thesis, University of Groningen, Groningen, the Netherlands, 2009.

18 E. Vogel, C. Küchenmeister, E. Bich and A. Laesecke, *J. Phys. Chem. Ref. Data*, 1998, **27**(5), 947.

19 G. Fischer, K. A. Muszkat and E. Fischer, *J. Chem. Soc. B*, 1968, 1156.

20 (a) T. W. J. Taylor and A. R. Murray, *J. Chem. Soc.*, 1938, 2078; (b) P. Bortolus and G. Cauzzo, *Trans. Faraday Soc.*, 1970, **66**, 1161; (c) G. B. Kistiakowsky and W. R. Smith, *J. Am. Chem. Soc.*, 1934, **56**, 638; (d) W. W. Schmiegel, F. A. Litt and D. O. Cowan, *J. Org. Chem.*, 1968, **33**, 3334; (e) A. V. Santoro, E. J. Barrett and H. W. Hoyer, *J. Am. Chem. Soc.*, 1967, **89**, 4545.

21 (a) J. Saltiel, *J. Am. Chem. Soc.*, 1967, **89**, 1036; J. Saltiel and E. D. Megarity, *J. Am. Chem. Soc.*, 1972, **94**, 2742; (b) J. Saltiel, J. T. D'Agostino, W. G. Herkstroeter, G. Saint-ruf and N. P. Buu-Hoi, *J. Am. Chem. Soc.*, 1973, **95**, 2543; (c) J. Saltiel and E. D. Megarity, *J. Am. Chem. Soc.*, 1969, **91**, 1265; (d) J. Saltiel, A. Marinari, D. W. -L. Chang, J. C. Mitchener and E. D. Megarity, *J. Am. Chem. Soc.*, 1979, **101**, 2982; (e) J. Saltiel, *J. Am. Chem. Soc.*, 1968, **90**, 6394; (f) J. Saltiel, E. D. Megarity and K. G. Kneipp, *J. Am. Chem. Soc.*, 1966, **88**, 2336; (g) J. Saltiel and B. Thomas, *Chem. Phys. Lett.*, 1976, **37**, 147.

22 (a) S. Malkin and E. Fischer, *J. Phys. Chem.*, 1962, **66**, 2482; (b) J. Saltiel and J. T. D'Agostino, *J. Am. Chem. Soc.*, 1972, **94**, 6445; (c) L. A. Brey, G. B. Schuster and H. G. Drickamer, *J. Am. Chem. Soc.*, 1979, **101**, 129; (d) J. Saltiel; J. D'Agostino; E. D. Megarity; L. Metts; K. R. Neuberger; M. Wrighton; O. C. Zafiriou; (e) *Organic Photochemistry*; Chapman, O. L., Ed.; Marcel Dekker: New York, 1973; (f) J. Saltiel, G. R. Marchand, E. Kirkor-Kaminska, W. K. Smothers, W. B. Mueller and W. L. Charlton, *J. Am. Chem. Soc.*, 1984, **106**, 3144; (g) J. Saltiel, S. Ganapathy and C. Werking, *J. Phys. Chem.*, 1987, **91**, 2755; (h) J. Saltiel, A. D. Rousseau and B. D. Thomas, *J. Am. Chem. Soc.*, 1983, **105**, 7631; (i) G. S. Hammond, J. Saltiel, A. A. Lamola, N. J. Turro, J. S. Bradshaw, D. O. Cowan, R. C. Counsell, V. Vogt and C. Dalton, *J. Am. Chem. Soc.*, 1964, **86**, 3197.

23 R. Augulis, M. Klok, B. L. Feringa and P. H. M. van Loosdrecht, *Phys. Status Solidi C*, 2009, **6**, 181.

24 K. Von Auwers, *Ber. Dtsch. Chem. Ges. B*, 1935, **68**, 1346.

Template sol–gel synthesis of mesostructured silica composites using metal complexes bearing amphiphilic side chains: immobilization of a polymeric Pt complex formed by a metallophilic interaction

Wataru Otani, Kazushi Kinbara and Takuzo Aida*

Received 23rd January 2009, Accepted 19th March 2009
First published as an Advance Article on the web 23rd July 2009
DOI: 10.1039/b904896k

A bipyridine-based ligand **1** bearing two amphiphilic side chains allowed for the formation of metal complexes Pt(**1**)Cl$_2$ (**2**), [Pt(**1**)$_2$](BF$_4$)$_2$ (**3**), and [Pt(**1**)$_2$][Pt(CN)$_4$] (**4**), which are capable of templating the sol–gel synthesis of mesostrucutred silica to give **2-MS**, **3-MS** and **4-MS**, respectively, whose silicate channels are filled with the corresponding platinum complexes. In particular, silica composite **4-MS**, upon spin-coating, gives a transparent film, which emits at 619 nm upon excitation at 495 nm, indicating that a weak Pt–Pt bond, formed between Pt(**1**)$_2^{2+}$ and Pt(CN)$_4^{2-}$ *via* a metallophilic interaction, is successfully incorporated into the silicate nanochannels.

Introduction

One-dimensional nanostructured organic materials[1] have attracted attention mostly in view of developing anisotropic optoelectronic properties. Immobilization of such aligned functional groups in optically transparent solid supports may guarantee certain functions arising from one-dimensionality. Mesoporous silicates with unidirectional nanoscopic channels are likely candidates for ensuring long-range ordering of columnar assemblies of functional groups.[2,3] In 2001, we[4] and Briker *et al.*[5] reported the first examples of silica–organic nanocomposite materials templated by rod-shaped micellar assemblies of 'functional' amphiphiles. In contrast with post-loading approaches, this method guarantees dense filling of silicate channels with functional organic molecules.[4–10] Not only ordinary amphiphiles but also disk-shaped amphiphilic molecules, such as triphenylene,[6] metallophthalocyanine,[7] and hexa-*peri*-benzocoronene derivatives[8] bearing oligo(ethylene glycol) side chains, are usable as templates for the sol–gel synthesis of mesoporous silica. Furthermore, we have also developed a "lizard" template containing a peptide unit with long alkyl tail, which can be cleaved off after formation of the mesostructured silica,[10] thereby providing fully organically functionalized silicate channels.

Here we report successful sol–gel synthesis of mesostructured silica templated by bipyridine-based amphiphilic Pt(II) complexes. Under appropriate conditions, square planar d^8 platinum complexes[11–14] are able to stack together *via* a Pt–Pt metal-lophilic interaction to form a luminescent one-dimensional supramolecular assembly. For example, [Pt(bpy)$_2$][Pt(CN)$_4$] (bpy = 2,2'-bipyridine) exhibits

Department of Chemistry and Biotechnology, School of Engineering, The University of Tokyo, 7-3-1 Hongo, Bunkyo-ku, Tokyo, 113-8656, Japan. E-mail: aida@macro.t.u-tokyo.ac.jp; Fax: +81 (0)3 5841 7310; Tel: +81 (0)3 5841 7251

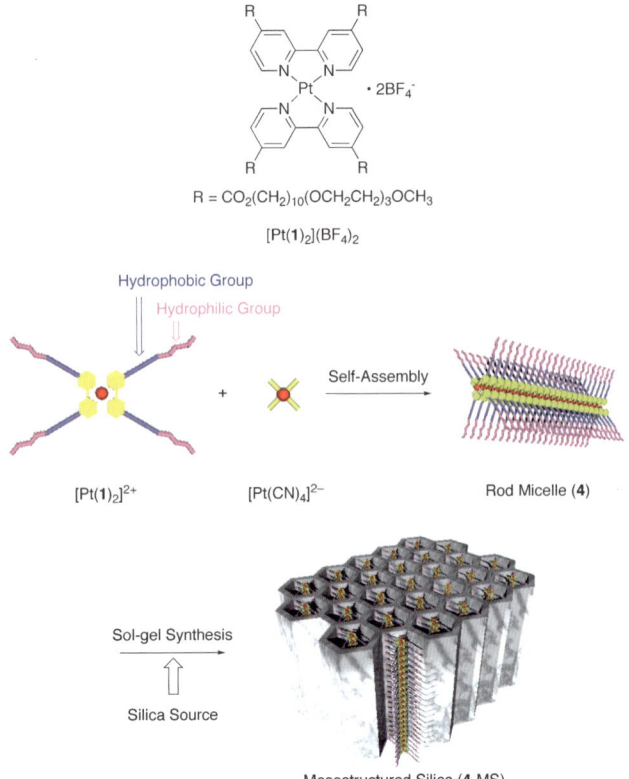

Scheme 1 Schematic representation of the sol–gel synthesis of mesostructured silica (**4-MS**) using linear-chain Pt(II) complex (**4**) as a template.

a characteristic phosphorescence emission (λ_{max} = 610 nm) upon exposure to visible light (λ_{ex} = 515 nm).[13] For the purpose of immobilizing such a luminescent polymeric assembly by a silicate nanochannel,[11,15] we designed bipyridine-based amphiphilic bidentate ligand **1**, which may allow for the formation of a linear-chain Pt(II) complex [Pt(**1**)$_2$][Pt(CN)$_4$] (**4**) (Scheme 1) surrounded by hydrophilic oxyethylene chains. Under appropriate conditions, complex **4** was able to template the sol–gel synthesis, affording luminescent hexagonal mesostructured silica, wherein linear-chain **4** was immobilized and confined in nanoscopic silicate channels.

Experimental

Measurements

Electronic absorption spectra were recorded on a JASCO U-best V-570 spectrophotometer. ^{1}H NMR spectra were recorded on a JEOL GSX-270 spectrometer, where chemical shifts were determined with respect to residual CHCl$_3$ (δ 7.24). Matrix-assisted laser desorption/ionization time-of-flight mass spectroscopy (MALDI-TOF-MS) was performed with dithranol as a matrix on Applied Biosystems BioSpectrometry Workstation STR Voyager-DE mass spectrometer. Electrospray ionization time-of-flight mass spectroscopy (ESI-TOF-MS) was performed on a JEOL AccuTOF JMS-T100L mass spectrometer. Infrared spectra of KBr pellet samples were recorded on a JASCO model FT/IR-600 Fourier transform infrared spectrophotometer. Luminescence spectra were recorded on a JASCO model FP-777W spectrofluorometer.

XRD analysis was carried out on a Rigaku model RINT 2400 X-ray diffractometer. Calcination of composite silica samples was carried out at 450 °C for 3 h with an ADVANTEC model KM-160 electric muffle furnace.

Materials

All reagents for synthesis were used as received from Tokyo Chemical Industry Co., Wako Pure Chemical Industries Ltd., Nacalai Tesque, Shin-etsu Chemical Industries Co., Kanto Chemical Co. and Aldrich. Surfactant $CH_3(CH_2)_{17}(OCH_2CH_2)_nOH$ (n = 10 (av.), $C_{18}EO_{10}$) was used as received from Aldrich.

Synthesis of 4,4′-bis[10-[2-[2-(2-methoxyethoxy)ethoxy]ethoxy]decyl]-2,2′-bipyridine (1)

Ligand **1** was synthesized according to Scheme 2. To $SOCl_2$ (30 mL) was added 2,2′-bipyridine-4,4′-dicarboxylic acid (2.50 g, 10 mmol) under dry Ar, and the resulting suspension was refluxed for 2 days. Then, the reaction mixture was evaporated to dryness, affording white powdery solid. To a $CHCl_3$ (50 mL) solution of the residue was dropwise added a $CHCl_3$ solution (80 mL) of 10-[2-[2-(2-methoxyethoxy)ethoxy]-ethoxy]decanol ($HOC_{10}EO_3$) (8.17 g, 25 mmol) followed by NEt_3 (6.0 mL, 43 mmol), and the mixture was refluxed for 4 h. After being stirred at room temperature for additional 12 h, the reaction mixture was evaporated to dryness, affording an oily residue, which was chromatographed on silica gel with CH_2Cl_2–MeOH (40 : 1 v/v) as an eluent. The second fraction containing mono- and di-substituted bipyridine derivatives was collected and subjected to preparative GPC with a JAI model LC908-C60 recycling preparative HPLC (JAIGEL-1H-40 and JAIGEL-2H-40). The first fraction was collected and evaporated to dryness under reduced pressure to allow isolation of **1** as colorless oil in 31% yield (2.74 g, 3.1 mmol). ^{1}H NMR (270 MHz, $CDCl_3$, 20 °C, ppm): δ 8.91 (s, 1H), 8.83 (d, 1H), 7.86 (d, 1H), 4.36 (t, 2H), 3.63–3.51 (m, 12H), 3.41 (t, 2H), 3.35 (s, 3H), 1.75 (m, 2H), 1.54 (m, 2H),

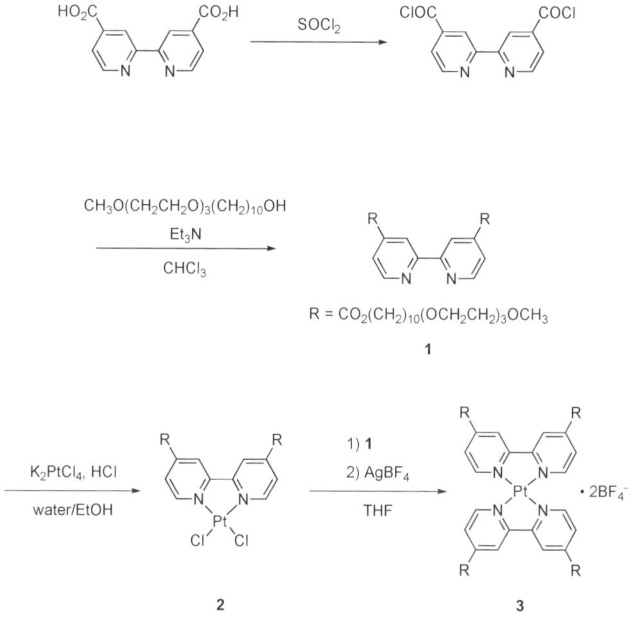

Scheme 2 Synthetic scheme of bipyridine-based bidentate amphiphilic ligand **1** and Pt(II) complexes **2** and **3**.

1.40–1.20 (m, 12H); IR (KBr, cm^{-1}): ν 2927 (s), 2856 (s), 1729 (s), 1459 (m), 1361 (m), 1286 (s), 1255 (s), 1126 (s) cm^{-1}; MALDI-TOF-MS (dithranol): m/z: calcd for $C_{46}H_{77}N_2O_{12}$: 849.54; found: 849.11 [M + H]$^+$.

Synthesis of Pt(1)Cl$_2$ (2).[16]

To an EtOH solution (4 mL) of **1** (200 mg, 240 μmol) was added an aqueous solution (3 mL) of K$_2$PtCl$_4$ (98 mg, 240 μmol), and the resulting mixture was heated for 2 h, whereupon the color of the reaction mixture turned from red to dark yellow. After the addition of water (30 mL), the reaction mixture was extracted three times with CHCl$_3$ (20 mL). The combined organic extract was evaporated to dryness, and the residue was subjected to preparative size-exclusion chromatography (Bio-Rad Laboratories Biobeads S-X1), where the second fraction was collected and evaporated to dryness. The residue was dissolved in a small amount of CH$_2$Cl$_2$ (2 mL) and subjected to reprecipitation with hexane (100 mL), where **2** was isolated as a yellow sticky solid in 83% yield (218 mg, 199 μmol). ^{1}H NMR (270 MHz, CDCl$_3$, 20 °C, ppm): δ 10.00 (d, 1H), 8.59 (s, 1H), 8.11 (d, 1H), 4.45 (t, 2H), 3.67–3.43 (m, 12H), 3.40 (t, 2H), 3.36 (s, 3H), 1.83 (m, 2H), 1.60–1.20 (m, 14H) ppm; IR (KBr, cm^{-1}) ν 2927 (s), 2856 (s), 1729 (s), 1461 (w), 1361 (w), 1288 (s), 1255 (s), 1109 (s); ESI-TOF-MS: m/z: calcd for $C_{46}H_{76}N_2O_{12}Cl_2PtNa$: 1136.43; found: 1136.51 [M + Na]$^+$.

Synthesis of [Pt(1)$_2$](BF$_4$)$_2$ (3).[17]

A THF (5 mL) solution of a mixture of **2** (35 mg, 32 μmol) and **1** (76 mg, 91 μmol) was refluxed for 5 h. To this solution was added AgBF$_4$ (12.1 mg, 62 μmol), and the mixture was refluxed for 24 h, allowed to cool to room temperature, and filtered off from insoluble substances. The filtrate was evaporated to dryness, and the residue was subjected to preparative size-exclusion chromatography (Bio-Rad Laboratories Biobeads S-X1), where the second fraction was collected and evaporated to dryness, affording **3** as a pale yellow sticky oil in 82% yield (53 mg, 26 μmol). ^{1}H NMR (270 MHz, CDCl$_3$, 20 °C, ppm): δ 9.15 (d, 1H), 8.79 (s, 1H), 8.39 (d, 1H), 4.45 (t, 2H), 3.66–3.51 (m, 12H), 3.41 (t, 2H), 3.34 (s, 3H), 1.82 (m, 2H), 1.60–1.20 (m, 14H); IR (KBr, cm^{-1}) ν 2935 s, 2866 s, 1731 s, 1598 s, 1456 m, 1296 w, 1254 m, 1124 s, 1084 s; ESI-TOF-MS: m/z: calcd for $C_{92}H_{152}N_4O_{24}B_2F_8PtNa$: 2089.04; found: 2089.34 [M + Na]$^+$; Anal. calcd for $C_{92}H_{152}N_4O_{24}B_2F_8Pt$: C 53.46, H 7.41, N 2.71; found: C 53.36, H 7.51, N 2.64.

Preparation of [Pt(1)$_2$][Pt(CN)$_4$] (4).[13]

Linear-chain Pt complex **4** was synthesized according to a literature method (Scheme 1).[13] To a MeOH (30 mL) solution of Pt complex **3** (62 mg, 30 μmol) was added an aqueous (30 mL) solution of K$_2$[Pt(CN)$_4$]·3H$_2$O (13 mg, 30 μmol). After the color of the solution turned from pale yellow to orange, a saturated aqueous solution of NaCl (30 mL) was added to the reaction mixture. Then, the resulting mixture was extracted three times with CH$_2$Cl$_2$ (30 mL, 3 times), and the combined organic extract was evaporated to dryness. The residue was suspended in acetone (3 mL) and filtered off from insoluble substances. Then, the filtrate was poured into hexane (100 mL), whereupon complex **4** precipitated out as orange sticky solid in 37% yield (25 mg, 11 μmol). IR (KBr, cm^{-1}) ν 2925 (s), 2854 (s), 2209 (w), 2151 (w), 1731 (s), 1634 (m), 1463 (w), 1259 (s), 1124 (s).

Results and discussion

A bipyridine derivative bearing two $C_{10}EO_3$ amphiphilic side chains at 4 and 4′ positions (**1**) was prepared and commonly used as a ligand for the preparation of Pt(II) complexes **2–4** (Scheme 2).

Sol–gel synthesis of mesostructured silica 2-MS templated by mono-bipyridyl Pt complex 2

Complex **2** was prepared by addition of an aqueous solution of K_2PtCl_4 to an EtOH solution of **1** (Scheme 2), according to a literature method[16] with a slight modification. Mesostructured silica films (**2-MS**) were synthesized in the presence of a mixture of amphiphilic Pt complex **2** and amphiphile $C_{18}EO_{10}$. Typically, tetramethylorthosilicate (1.0 g, 6.6 mmol) was added at 0 °C to a stirred mixture of hydrochloric acid (2.7 mM, 0.21 g; water 12 mmol and HCl 0.57 μmol) and MeOH (1.68 g, 52 mmol). The mixture was stirred for 10 min, Si–O bonds were hydrolyzed, and the resulting silanol functionalities were partially condensed to give a mixture of siloxane oligomers. Then, amphiphile **2** (6.7 mg, 6.0 μmol) and $C_{18}EO_{10}$ (10 mg, 12 μmol) were successively added, and the reaction mixture was subjected to sonication for 1 min at 20 °C. When the resulting mixture was spin-coated (3000 rpm) on a glass plate, a homogeneous and transparent yellow film (**2-MS**) resulted, which was then air-dried overnight at 25 °C to allow the silicate framework to develop.

In a powder X-ray diffraction (XRD) profile, **2-MS** exhibited peaks assignable to (100) and (200) diffractions of a hexagonal silicate structure with a d spacing of 50 Å (Fig. 1a). In a molecular modeling study, the distance between the Pt and carbon atoms at the end of the hydrophobic chain was estimated as 19 Å, which is roughly comparable to the pore radius of **2-MS**. Taking into account the lattice parameter (57 Å, $a = 2/\sqrt{3}d_{100}$) and pore diameter (19 × 2 = 38 Å), the silica wall was evaluated to be 19 Å thick, as often observable for conventional mesoporous silica (15 Å).[3] Upon calcination of **2-MS** at 450 °C for 3 h, the resultant substance still showed diffraction peaks, though with a reduction of the d spacing value from 50 to 41 Å, suggesting that **2-MS** adopts a hexagonal geometry rather than lamellar.

As a reference, when the sol–gel synthesis was carried out in the presence of $C_{18}EO_{10}$ (10 mg) alone as a template, the resulting silica showed a less intensive and broad XRD peak with the d spacing of 55 Å.

Infrared spectroscopy of **2-MS** (Fig. 2a) showed absorption bands due to C=O stretching and aromatic C=C vibrations at 1726 and 1459 cm^{-1}, respectively, indicating that **2-MS** indeed contains ligand **1**. Electronic absorption spectroscopy of **2-MS** showed an absorption band at 432 nm assignable to a metal-to-ligand charge transfer (MLCT) between Pt and bipyridine (Fig. 2b).[18] Together with the results of XRD analysis, these spectral profiles are considered to indicate that Pt complex **2** survives without decomposition under acidic sol–gel conditions, and is incorporated

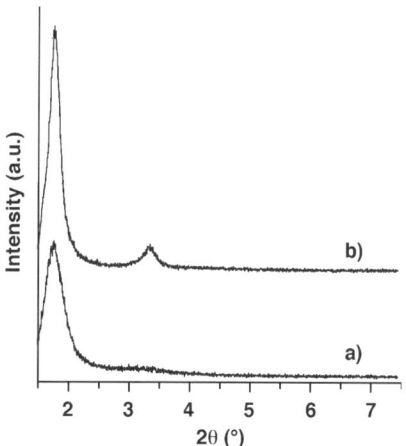

Fig. 1 Powder X-ray diffraction patterns of (a) **2-MS** and (b) **3-MS**.

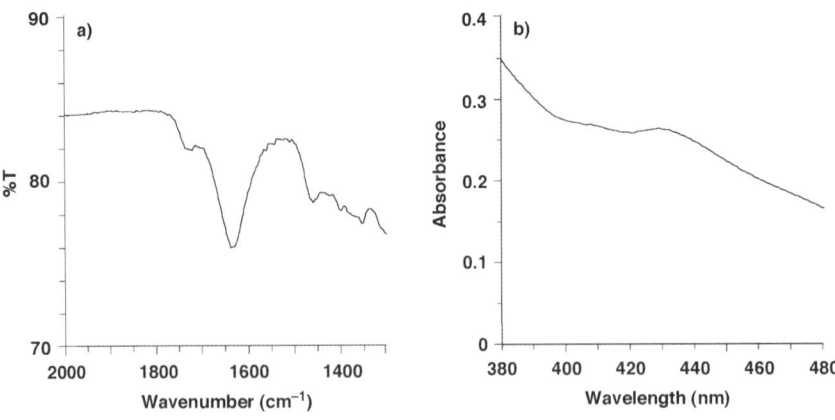

Fig. 2 (a) Infrared and (b) electronic absorption spectra of **2-MS**.

inside the hexagonally aligned silicate channels. Meanwhile, composite material **2-MS**, as expected, displayed no absorption band characteristic of a Pt–Pt metallophilic interaction.[16,19,20]

Sol–gel synthesis of mesostructured silica 3-Ms templated by di-bipyridyl Pt complex 3

For the synthesis of mesostructured silica **3-MS**, hydrolytic condensation of tetramethylorthosilicate was carried out using **3** (6.2 mg, 3.0 μmol) and $C_{18}EO_{10}$ (10 mg, 12 μmol) under conditions analogous to those for **2-MS**. The XRD pattern obtained for **3-MS** showed peaks corresponding to (100) and (200) with a d spacing of 50 Å, which are characteristic of mesoporous silica with a hexagonal geometry (Fig. 1b). After calcination at 450 °C for 3 h, **3-MS** still showed diffraction peaks, while the d spacing value (42 Å) observed was smaller than that before calcination. Infrared spectroscopy of **3-MS** showed bands at 1729 and 1458 cm^{-1} due to C=O stretching and aromatic C=C vibrations, respectively (Fig. 3a). Furthermore, electronic absorption spectroscopy displayed an absorption band at 340 nm, which is assignable to the π–π* electronic transition of a bipyridine unit coordinating to the Pt ion (Fig. 3b).[18,20] These spectral profiles indicate that the mesoporous silica contains amphiphilic Pt complex **3** inside their silicate channels. This also means

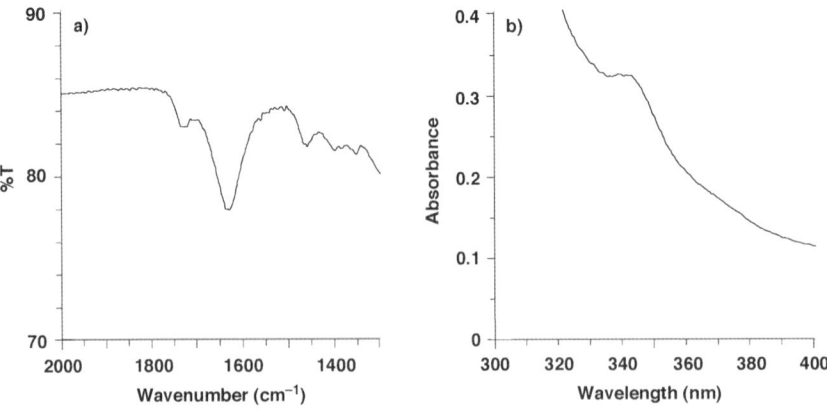

Fig. 3 (a) Infrared and (b) electronic absorption spectra of **3-MS**.

that complex **3** is tolerant under the acidic conditions for the sol–gel synthesis of mesostructured silica, although di-bipyridyl Pt complexes have been reported to dissociate into mono-bipyridyl Pt complexes in strongly acidic media.[16] Nevertheless, the absorption spectrum did not show any bands at the visible to near-infrared regions characteristic of Pt–Pt metallophilic interactions. Namely, it is unlikely that complex **3** in the silicate channel forms a polymeric structure.

Immobilization of linear-chain Pt double salt (4) within silicate nanochannels

By a procedure similar to the above, a transparent silica film exhibiting a characteristic orange color was successfully obtained by condensation of tetramethylorthosilicate in the presence of a mixture of **4** (6.8 mg, 3.0 μmol) and $C_{18}EO_{10}$ (10 mg, 10 μmol), followed by spin-coating (Scheme 2). XRD analysis of **4-MS** showed a set of diffraction peaks corresponding to the d spacing for (100) of 58 Å (Fig. 4). It is noteworthy that this value is obviously larger than those of **2-MS** (50 Å) and **3-MS** (50 Å). Compound **4-MS** most likely adopts a hexagonal structure, since the product obtained by calcination of **4-MS** at 450 °C for 3 h still displayed diffraction peaks. Again, this thermal treatment gave rise to a significant reduction of the d spacing value from 58 to 44 Å. Here it should be noted that compound **4-MS**, in contrast to **3-MS**, showed a visible absorption band centered at $\lambda = 467$ nm. Such a visible absorption is characteristic of a hydrated form of linear-chain **4** (485 nm).[13] As a reference, we spin-coated an aqueous solution of a mixture of **4** and $C_{18}EO_{10}$, where the resulting film (**4-film**) displayed an absorption band centered at 464 nm (Fig. 5a). Therefore, linear-chain double salt **4** survives under the applied sol–gel synthesis conditions for mesoporous silica and is successfully incorporated into the resulting silicate channels. In relation to this, of interest to recall is the fact that **4-MS** has a larger d-spacing value than the other two silica. This might be caused by a shorter Pt–Pt distance in **4-MS** than those in **2-MS** and **3-MS** that are devoid of metallophilic interactions. Such a close packing of the Pt centers could result in an extended conformation of their side chains. As described in the introductory part, Pt–Pt metallophilic interactions are known to give rise to phosphorescence. In fact, compound **4-MS** is luminescent. Upon excitation at 495 nm, this material showed an emission band centered at $\lambda = 619$ nm. This emission is red-shifted by 4 nm from that of **4-film** ($\lambda = 615$ nm) (Fig. 5b), suggesting a possible confinement effect of the silicate channel on the electronic properties of immobilized **4**.

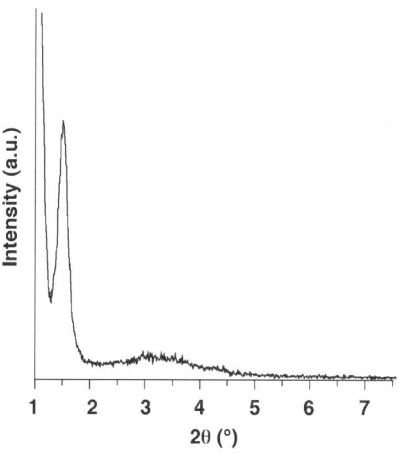

Fig. 4 Powder X-ray diffraction pattern of **4-MS**.

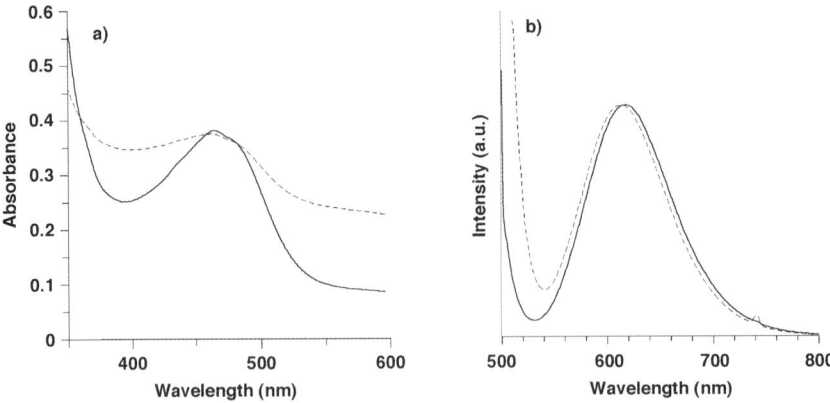

Fig. 5 (a) Electronic absorption and (b) emission spectra (λ_{ex} = 495 nm) of **4-MS** (solid curve) and **4-film** (broken curve).

Conclusions

We demonstrated that amphiphilic Pt complexes **2–4** in the presence of micelle-forming surfactant $C_{18}EO_{10}$ give rise to template-assisted sol–gel synthesis of mesoporous silica with a hexagonal geometry, whose silicate channels are densely filled with Pt complexes used as the templates. The most important achievement here is the successful immobilization of linear-chain **4**, formed by a Pt–Pt metallophilic interaction. While metallophilic interactions, in general, are just comparable in strength to hydrogen-bonding interactions, they are not broken by protonolysis and therefore survive under acidic conditions for the sol–gel synthesis of mesoporous silica. Luminescent mesoporous silica hybridized with Pt–Pt linear-chain metal complexes are not only interesting for sensing applications but also have the potential for catalysis.[21,22]

References

1 G. A. Ozin, *Adv. Mater.*, 1992, **4**, 612.
2 T. Yanagisawa, T. Shimizu, K. Kuroda and C. Kato, *Bull. Chem. Soc. Jpn.*, 1990, **63**, 988; C. T. Kresge, M. E. Leonowicz, W. J. Roth, J. C. Vartuli and J. S. Beck, *Nature*, 1992, **359**, 710; J. S. Beck, J. C. Vartuli, W. J. Roth, M. E. Leonowicz, C. T. Kresge, K. D. Schmitt, C. T.-W. Chu, D. H. Olson, E. W. Sheppard, S. B. McCullen, J. B. Higgins and J. L. Schlenker, *J. Am. Chem. Soc.*, 1992, **114**, 10834; E. Prouzet and T. J. Pinnavaia, *Angew. Chem., Int. Ed. Engl.*, 1997, **36**, 516; D. Zhao, P. Yang, N. Melosh, J. Feng, B. F. Chmelka and G. D. Stucky, *Adv. Mater.*, 1998, **10**, 1380.
3 M. Ogawa, *Chem. Commun.*, 1996, 1149.
4 T. Aida and K. Tajima, *Angew. Chem., Int. Ed.*, 2001, **40**, 3803.
5 Y. Lu, Y. Yang, A. Sellinger, M. Lu, J. Huang, H. Fan, R. Haddad, G. Lopez, A. R. Burns, D. Y. Sasaki, J. Shelnutt and C. J. Brinker, *Nature*, 2001, **410**, 913.
6 A. Okabe, T. Fukushima, K. Ariga and T. Aida, *Angew. Chem., Int. Ed.*, 2002, **41**, 3414.
7 M. Kimura, K. Wada, K. Ohta, K. Hanabusa, H. Shirai and N. Kobayashi, *J. Am. Chem. Soc.*, 2001, **123**, 2438.
8 A. Okabe, M. Niki, T. Fukushima and T. Aida, *Chem. Commun.*, 2004, 2572; J. Wu, J. Li, U. Kolb and K. Mullen, *Chem. Commun.*, 2006, 48.
9 M. Ikegame, K. Tajima and T. Aida, *Angew. Chem., Int. Ed.*, 2003, **42**, 2154.
10 Q. Zhang, K. Ariga, A. Okabe and T. Aida, *J. Am. Chem. Soc.*, 2004, **126**, 988; W. Otani, K. Kinbara, Q. Zhang, K. Ariga and T. Aida, *Chem.–Eur. J.*, 2007, **13**, 1731.
11 T. W. Thomas and A. E. Underhill, *Chem. Soc. Rev.*, 1972, **1**, 99; J. K. Bera and K. R. Dunbar, *Angew. Chem., Int. Ed.*, 2002, **41**, 4453.
12 J. S. Miller and A. J. Epstein, *Prog. Inorg. Chem.*, 1976, **20**, 1; K. Krogmann, *Angew. Chem., Int. Ed. Engl.*, 1969, **8**, 35.
13 V. H. Houlding and A. J. Frank, *Inorg. Chem.*, 1985, **24**, 3664.

14 R. Palmans, A. J. Frank, V. H. Houlding and V. M. Miskowski, *J. Mol. Catal.*, 1993, **80**, 327; R. Palmans and A. J. Frank, *J. Phys. Chem.*, 1991, **95**, 9438; C. A. Craig, F. O. Garces, R. J. Watts, R. Palmans and A. J. Frank, *Coord. Chem. Rev.*, 1990, **97**, 193; R. Palmans, D. B. MacQueen, C. G. Pierpont and A. J. Frank, *J. Am. Chem. Soc.*, 1996, **118**, 12647.
15 M. D. Ward, in *Electroanalytical Chemistry*, ed. A. J. Bard, Marcel Dekker, New York, 1989, p. 181.
16 G. T. Morgan and F. H. Burstall, *J. Chem. Soc.*, 1934, 965.
17 E. J. L. McInnes, R. D. Farley, C. C. Rowlands, A. J. Welch, L. Rovatti and L. J. Yellowlees, *J. Chem. Soc., Dalton Trans.*, 1999, 4203.
18 V. M. Miskowski and V. H. Houlding, *Inorg. Chem.*, 1989, **28**, 1529.
19 R. S. Osborn and D. Rogers, *J. Chem. Soc., Dalton Trans.*, 1974, 1002.
20 E. Bielli, P. M. Gidney, R. D. Gillard and B. T. Heaton, *J. Chem. Soc., Dalton Trans.*, 1974, 2133.
21 V. H. Houlding, A. J. Frank, in *Homogeneous and Heterogeneous Photocatalysis*, ed. E. Pelizetti and N. Serpone, NATO ASI Ser. C174, Reidel, Dordrecht, 1986, p 199.
22 K. Honda, K. Chiba, E. Tsuchida and A. J. Frank, *J. Mater. Sci. Lett.*, 1989, **24**, 4004.

PAPER

Self-assembled interpenetrating networks by orthogonal self assembly of surfactants and hydrogelators

Aurelie M. Brizard,[a] Marc C. A. Stuart[b] and Jan H. van Esch[*a]

Received 24th February 2009, Accepted 1st April 2009
First published as an Advance Article on the web 27th July 2009
DOI: 10.1039/b903806j

Interpenetrating networks (IPN) consist of two or more networks of different components which are entangled on a molecular scale and cannot be separated without breaking at least one of the networks. They are of great technological interest because they allow the blending of two or more otherwise incompatible properties or functions, and furthermore synergistic effects might arise from the simultaneous operation of the two networks. So far, the preparation of interpenetrating network gels by self-assembly approaches was doomed to fail because the conventional polymers and surfactant building blocks either phase separate or form mixed assemblies, respectively. Here we report on self-assembled interpenetrating networks obtained by the orthogonal self-assembly of small molecular hydrogelators and surfactants. Preliminary studies on the self-assembly behaviour and viscoelastic properties of these systems revealed that these self-assembled IPN have a number of intriguing properties. For instance, the presence of two coexisting networks offers new possibilities for compartmentalization, and will allow one to adjust the viscoelastic properties between 'soft' and 'hard' gels. The non-covalent character of such IPN makes their formation fully reversible, which can be exploited for dual responsive systems. Most interestingly, self-assembled IPN can also act as a very primitive, yet unique, model for biological interpenetrating networks like the extracellular matrix and the cytoskeleton, and thereby contribute to our understanding of these very complex systems.

1. Introduction

The self-assembly of small molecular building blocks is an attractive approach for the construction of nano-objects and nano-structured materials because their spontaneous and reversible formation under thermodynamic equilibrium conditions allows simple and large scale manufacturing methods compared to common top-down techniques.[1] Over the past decade, the self-assembly approach has been applied to many different building blocks, ranging from small molecules, proteins, nano-particles and colloids, up to even mesoscale building blocks, thereby enabling the fabrication of objects and materials with regular features from nano- to micro-metre dimensions.[2] Among the building blocks available, surfactants which share a common structure consisting of a hydrophilic segment and a hydrophobic segment, have been extensively used because they lead to well defined basic

[a]Self-Assembling Systems, Department of Chemical Engineering, Delft University of Technology, Julianalaan 136, 2628 BL, Delft, The Netherlands. E-mail: j.h.vanesch@tudelft.nl
[b]Groningen Biomolecular Sciences and Biotechnology Institute, University of Groningen, Nijenborg 4, 9747 AG, Groningen, The Netherlands

supramolecular architectures, like spherical or rod-like micelles, bilayers, vesicles or inverted micelles.[3] These aggregates encompass a large range of length scales (2 nm up to several μm), can easy integrate functional molecules due to the relatively low specificity of the hydrophobic interactions, and may encapsulate an aqueous compartment, which has led to widespread applications as detergents, catalysis, structuring agents, templates, and drug delivery.[4] Despite such diverse applications, surfactants however suffer from the limited scope of supramolecular architectures they can generate, justifying the need for the development of other building blocks. Small molecule gelators, peptides, block copolymers, and polymer–protein conjugates, have thus been used to construct nano-scale assemblies in aqueous environments, leading to novel architectures like ribbons, helices and tubes.[5,6] Considerable progress has even been made in controlling the diameter of 1-D aggregates, *e.g. via* the tuning of the chiral twist,[7] or adding functionality by incorporating biological entities or sensors for instance.[8]

This tremendous increase in the diversity of available molecular building blocks, however, does not allow the level of complexity or functionality found in nature to be met or the needs of future nanotechnology to be complied with. Interestingly, in nature, complex processes like replication, transcription, energy storage and several metabolic pathways rely more on the separation of incompatible structures in space and time, rather than on the diversity of the buildings blocks, which remain relatively limited. In cells for instance, these processes are achieved through compartmentalization by use of bilayer membranes, however always in conjunction with an internal cytoskeleton or an external cell wall. Remarkably, all these objects are self-assembled structures that can coexist because self-assembly of each individual constituent does not interfere with the others. It is this ability to self-sort and the tight integration of these self-assembled structures that gives cells their stability, shape and function.

The ability to integrate different structural elements is also essential for the fabrication of nano-structures with regular features in the 10–1000 nm regime.[2] These materials are expected to have superior mechanical properties like bone and wood, and are indispensable for photonic applications. The step-wise self-assembly of small components to such hierarchically structured objects and materials is also common in nature, like with viruses and ribosomes but also tissues such as bone and wood, and is receiving increasing attention as a valid approach to the fabrication of more complex nanostructures and materials.[9] Its successful application to generate well-defined objects or regular materials, however, requires tight control over the self-assembly process with regard to shape and size but also the strength and nature of the interactions between the different entities at each hierarchical level.

There is thus an urgent need to extend the current toolbox of available nanoarchitectures that can be obtained by self-assembly of structurally simple compounds. Just like in nature, a straightforward approach to increase the level of complexity in self-assembling systems consists of making use of multiple components that can display orthogonal self assembly, *i.e.* the independent formation of two supramolecular structures each with their own characteristic within a single system. Remarkably, the controlled phase separation between multiple different components has only recently been exploited to fabricate assemblies and materials with regular features at nanolength scale, *i.e.* nanostructure formation and interfacial patterning in phase separated polymer systems,[9,10] and low molecular weight organogels in liquid crystalline phases.[11]

Recently, we showed that the orthogonal self-assembly of surfactant molecules and small molecular hydrogelators is a versatile and powerful approach towards the formation of novel and more complex architectures.[12] For example, the orthogonal self-assembly of hydrogelators with micelle forming surfactants or phospholipids leads to the formation of micelles embedded in gel matrixes or vesicles encapsulated in a gel, respectively (Fig. 1). This last example has been exploited to successfully develop liposomes with an encapsulated self-assembling hydrogel

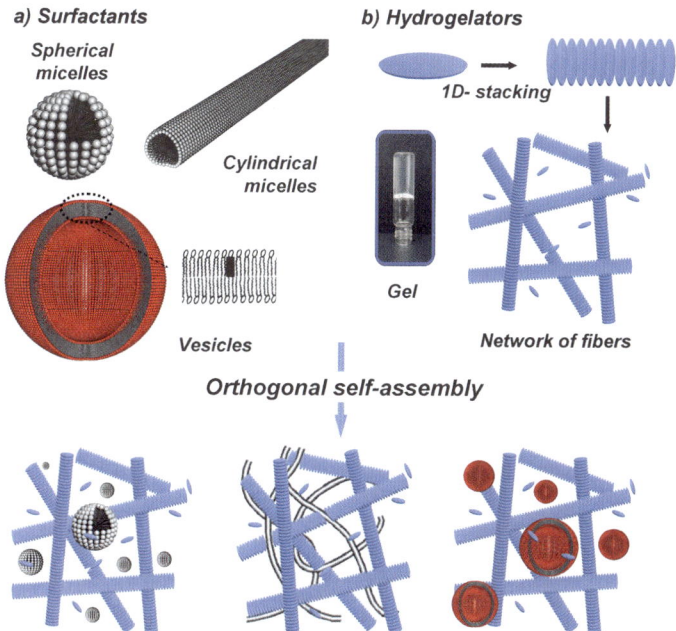

Fig. 1 Schematic illustration of orthogonal self-assembly between surfactants—which can form a variety of structures such as spherical micelles, wormlike micelles or vesicles—and hydrogelators, known to self-assemble in one dimension to give a network of fibers. Orthogonal self-assemblies between these two types of entities allows the independent formation of gel fibers in conjunction with micelles or vesicles.

("gellosomes"). The highly responsive character of the gelator makes it very attractive not only as a mimic of cytoskeleton[13] but also for drug delivery purposes.

Another system that is of great potential interest consists of self-assembled interpenetrating networks (IPN). Interpenetrating networks[14] consist of two or more networks of different components which are entangled on a molecular scale and cannot be separated without breaking at least one of the networks. They have been known for decades and are of great technological interest because they allow the blending of two or more otherwise incompatible properties or functions, and furthermore synergistic effects might arise from the simultaneous operation of the two networks.[15] So far, the preparation of interpenetrating network gels by self-assembly approaches was doomed to fail because the conventional polymers and surfactant building blocks either phase separate or form mixed assemblies, respectively. Recently, we found that the orthogonal self-assembly of low molecular weight (LMW) gelators together with surfactants allows the preparation of interpenetrating networks exclusively by self-assembly.[12] Here we report on our ongoing work on the self-assembly properties together with preliminary results on the viscoelastic properties of a prototype self-assembled IPN formed from a low molecular weight hydrogelator and a cylindrical micelle forming surfactant.

2. Experimental

Materials

Cetyltrimethylammonium tosylate (CTAT) was purchased from Sigma Aldrich and used without any further purification. The synthesis and full characterization of compound HG1 has been reported before.[16]

Gel preparation

CTAT is soluble in water above 23 °C ± 1 °C. At lower temperatures, CTAT exhibits a white appearance due to the presence of CTAT crystals dispersed in a viscous surfactant solution.

The gel samples were prepared by heating the gelator (typically 1 to 10 mM) in 1 mL of water or an aqueous solution containing the surfactants (binary systems) in a closed vial until a clear solution was obtained. Cooling to room temperature caused the formation of gels

Gel-to-sol transition temperature determination

All gel-to-sol transition temperatures (T_{GS}) were determined using the dropping ball method, which consists in carefully placing a stainless steel ball (65 mg, 2.5 mm in diameter) on top of a gel that had been prepared 8 h earlier in a 2 mL glass vial. These vials are subsequently placed in a heating block where the gel melting can be monitored by means of a CCD camera. The temperature of the heating block is increased by 5 °C h^{-1} and the T_{GS} is defined as the temperature at which the steel ball reaches the bottom of the vial.

Conductivity experiments

The micellization of CTAT in pure water or in a gel network of HG1 was studied by determining the concentration dependence of the specific conductivity at 25 °C. The conductivity measurements were carried out with a CONSORT C830 conductimeter, which was first calibrated with a 0.01 M KCl standard. For the conductivity experiments in an HG1 network (2 mM), each data point was obtained from a separate gel–CTAT mixture prepared by heating above the gel–sol transition of HG1 and subsequent cooling. To prevent any counter-ion effects, the HG1 powder was washed several times with double distilled water and freeze dried, prior to gel preparation. After this treatment, a value of 7.8 µS cm^{-1} was found in a 2 mM HG1 gel in pure water, instead of 1.4 µS cm^{-1} for water only. The difference (6.4 µS cm^{-1}) was therefore subtracted from each measurement for the binary systems.

CryoTEM

A few microlitres of suspension were deposited on a Quantifoil 3.5/1 holey carbon coated grid. After blotting away the excess of liquid the grids were plunged quickly in liquid ethane. Frozen-hydrated specimens were mounted in a cryo-holder (Gatan, model 626) and observed in a Philips CM 120 electron microscope, operating at 120 kV. Micrographs were recorded under low-dose conditions on a slow-scan CCD camera (Gatan, model 794).

Rheology

Oscillatory experiments were performed using a rheometer (AR G2, TA instruments) in a strain controlled mode, equipped with a steel cone-and-plate geometry of 2.0° and 40 mm in diameter. The temperature of the plates was controlled at 25 °C with an error of ±0.2 °C. In case of the pure CTAT samples, the linear regime existed up to 10% strain (strain sweep at 10 rad s^{-1}). For HG1 gels, the viscoelastic response was linear until 1% strain only. As a consequence, all the experiments were performed with a 0.5% strain, to ensure linearity of the viscoelastic response for the binary mixtures. The rheograms of CTAT solutions were highly reproducible, while the injection method influenced the results more with the HG1 gels. The best results were obtained by first ageing the gel for at least 24 h, pipette it slowly and inject it as such on the plate. All the experiments were performed between 0.005 Hz and 10 Hz, in a stiff bearing mode.

3. Results and discussion

In the present study, a low molecular weight gelator molecule HG1 based on 1,3,5-cyclohexyltricarboxamide was used because these molecules self-assemble into 1D arrays stabilized by both hydrogen bonding and hydrophobic interactions (Fig. 2).[16] A similar set of interactions serves to stabilize protein secondary structures, structures which are stable in the presence of weakly interacting surfactant. Previously it was shown that these compounds can be functionalized at the periphery with different solvophilic or pH sensitive groups, giving access to modular architectures and properties. In the case of HG1, hydrophilic oligoethylene oxide groups are attached leading to a nonionic hydrogelator.

Hydrogelator HG1 is insoluble in water at room temperature but gradually dissolves upon heating. Cooling of the hot solutions containing typically 0.1–1% w/v of gelator to room temperature leads to the formation of hydrogels, with a transparent to opalescent appearance depending on their concentration. This process can be repeated many times indicating that hydrogel formation by HG1 is thermoreversible. Cryo transmission electron microscopy (TEM) showed that HG1 self-assembles into thin fibers from about 500 nm up to several micrometres long, and with diameters between 5 nm to 25 nm (Fig. 3a). The long aspect ratio and polydispersity with regard to width and length indicate that fiber formation is the result of a kinetically controlled, highly anisotropic growth process.[5]

Remarkably, the fibers have different appearances depending on the processing conditions. Samples that are prepared simply by cooling from hot isotropic solutions consist of fibrils without fine structure (Fig. 3b,c). However, after mechanical agitation (vortexing) small bands of equally spaced stripes parallel to the fiber axis are clearly visible. Macroscopically, this mechanical agitation also leads to more fragile gels, which reform very slowly at high concentration (10 mM for instance) but are irreversibly disrupted close to the critical gelation concentration (0.4 mM). The width of a single stripe amounts to 4–5 nm which is in good agreement with the diameter of a single molecule of HG1. The striped bands appear at

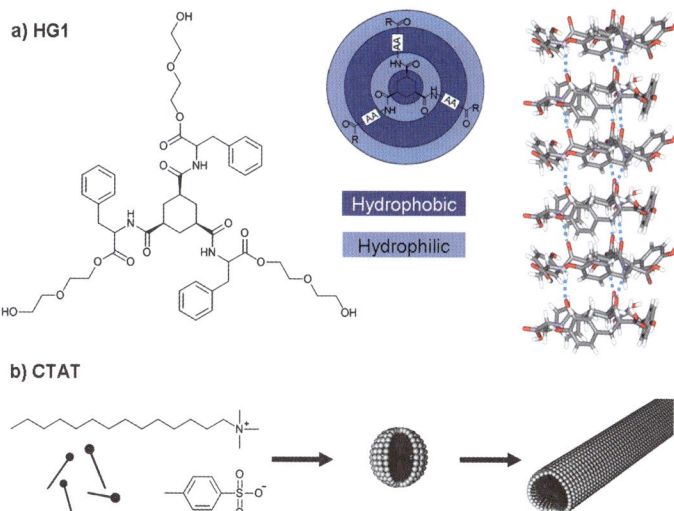

Fig. 2 Chemical structures of the molecules under study: (a) 1,3,5-triamide *cis*,*cis*-cyclohexane-based hydrogelator HG1 which self-assembles preferentially in one dimension. In the schematic representation, light blue regions correspond to hydrophilic groups and dark blue areas to hydrophobic entities (AA = amino acids). (b) Cetyltrimethylammonium tosylate surfactant (CTAT). This surfactant aggregates first in spherical micelles which transform into cylindrical micelles upon increasing concentration.

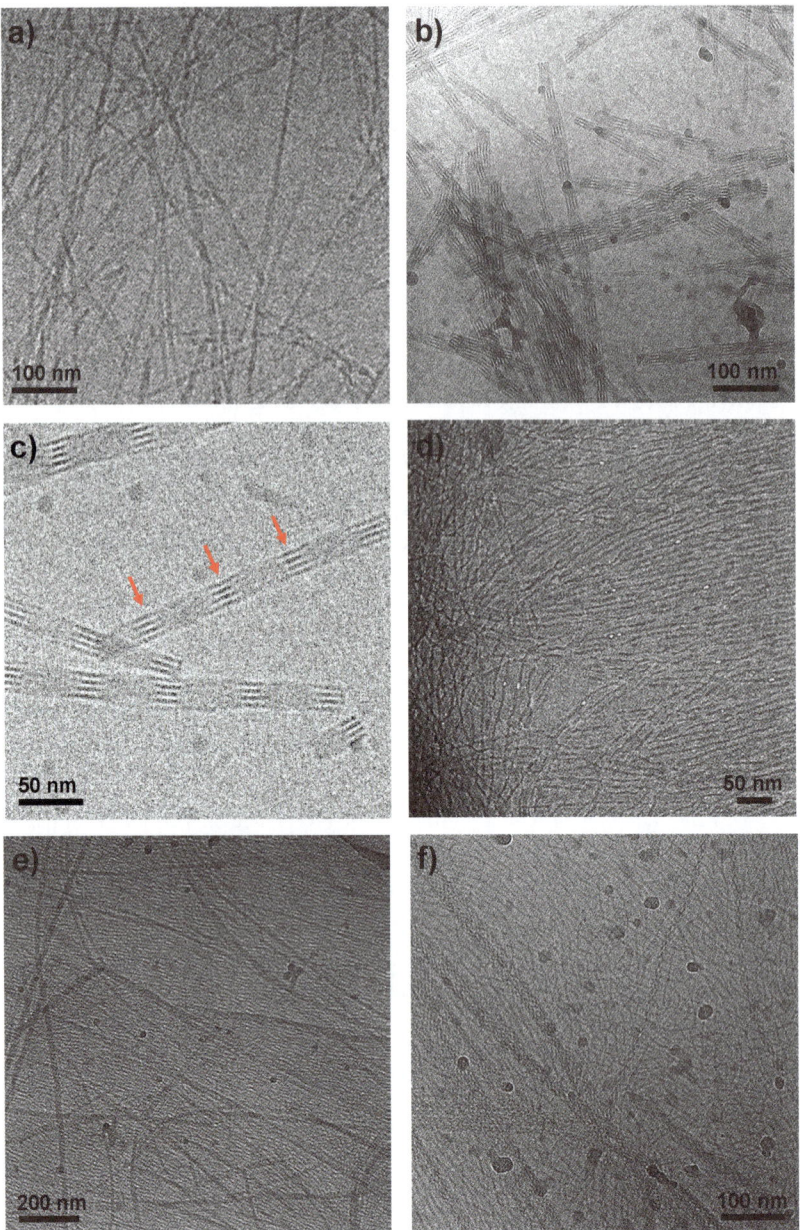

Fig. 3 CryoTEM pictures of fibrous networks of gelator HG1 (2mM) in water, (a) without vortexing or (b) and (c) after vortexing, which leads to the appearance of shorter fibers with striped bands. (d) Wormlike micelles of CTAT at 100 mM in water. (e) and (f) Binary system of CTAT at 100 mM and HG1 at 2 mM in water.

intervals of 60 nm which is about 300 times the hydrogen bonded stacking distance of 1,3,5-cyclohexyltricarboxamides.[16] Presumably, the fibers consist of twisted bundles of stacked molecules of HG1.

The persistence length of the fibers is much larger than their diameter and up to several hundreds of nanometres, and mechanical agitation of the samples results in the appearance of many broken fibers rather then bent ones, suggesting that they are rather stiff and brittle (Fig. 3b). Another striking observation is that these stripped bundles of fibers hardly split and fuse nor intertwine. Apparently, hydrogels formed from HG1 consist of an entangled network of very long, thin, and stiff fibers, and the network structure is stabilized by mechanical contacts between fibers rather then specific junction zones.

A very different type of network structure is formed by surfactants like cetyl-N,N,N-trimethylammonium tosylate (CTAT, Fig. 2). In water, CTAT first forms spherical micelles just above its critical micelle concentration (cmc) of 0.2 mM, due to hydrophobic interactions between the long alkyl chains.[17] The spherical micelles quickly transform into cylindrical micelles upon increasing the surfactant concentration. These elongated structures start to entangle around $c^* \approx 20$ mM, leading to highly viscous solutions which, however, still exhibit gravitational flow. The morphological transition from spherical to cylindrical micelles can be attributed to an increased binding of the tosylate counter-ions to the quarternary ammonium headgroups. This leads to a decrease of the headgroup repulsion and hence an increase of the packing parameter. Fig. 3d shows a cryoTEM micrograph of a 100 mM solution of CTAT, which consists of a dense network of highly entangled, thin cylindrical micelles. Their diameter is very uniform with a mean value of 5–6 nm, which nicely agrees with the literature.[18] Moreover, the persistence length is only 20–30 times the diameter of the micelles and many bifurcations can be observed.[19] These observations are in nice agreement with the general view of cylindrical micellar networks as highly dynamic equilibrium networks which are stabilized by specific interactions or junction zones between the cylindrical micelles.

Binary systems consisting of HG1 and CTAT in water could be prepared by just mixing the hydrogels and micellar solutions at room temperature, but macroscopically homogeneous samples could only be obtained by excessive vortexing. Moreover, the binary samples prepared in this way only converted slowly to a homogeneous gel state, typically within a few hours depending on the concentration. A much easier and reliable way to obtain macroscopically homogeneous aqueous samples is to exploit the thermoreversible nature of HG1 hydrogels. Thus, HG1 and CTAT could be dissolved together in water by heating at $T > T_{GS}$ (temperature of gel–sol transition) of HG1. After cooling to room temperature the samples were inspected for gel formation. Homogeneous, transparent gels were obtained if the HG1 concentration exceeded the minimum gelation concentration (cgc), in the presence of CTAT, at concentrations below and above the cmc and c^*. These observations suggest that at least gel formation by HG1 is compatible with the presence of spherical or cylindrical micelles. Fig. 3e and 3f show cryoTEM micrographs of gels of HG1 with CTAT above c^*. Elongated, almost straight fibers with a diameter of 5–25 nm are visible, reminiscent of the HG1 fibers morphology. In the areas in between these fibers, much thinner, curved fibers with a uniform diameter of 5 nm also can be distinguished clearly, similar to CTAT wormlike micelles. Moreover, both types of fibers appear to be homogeneously distributed over the samples and form a network structure of their own. Apparently, simple mixing of these two types of networks forming agents HG1 and CTAT leads to the formation of interpenetrating networks by self-assembly, a so-called SAIN. The similarity of both types of fibers to those observed in samples of the individual components strongly suggests that they originate from the self-assembly of HG1 and CTAT, respectively. At larger magnifications, however, the thicker fibers ascribed to HG1 seem to have a different fine structure then those in gels of HG1 alone. At this stage it is not clear whether this fine structure points to a different fiber structure, *e.g.* due to interactions with CTAT or to the overlay of the thicker fiber with much denser network of the thinner fibers.

These results clearly show that the morphological features of fibers of the separate components are largely preserved in the mixed systems, which suggests that self-assembly of HG1 and CTAT proceeds independently. However, the slight changes of fiber morphology of the thicker, presumably HG1, fibers might point to the

$$\text{HG1}_{sol} \xrightleftharpoons{\text{CTAT}_{sol,mic,cyl}} \text{HG1}_{gel}$$

$$\text{CTAT}_{sol} \xrightleftharpoons{\text{HG1}_{sol,gel}} \text{CTAT}_{mic} \xrightleftharpoons{\text{HG1}_{sol,gel}} \text{CTAT}_{-cyl}$$

$$\text{CTAT}_{sol} + \text{HG1}_{sol} \xrightleftharpoons{} \text{CTAT/HG1}_{agg}$$

Scheme 1 Existence of different aggregation states for (1) the hydrogelators that can exist in a monomeric or gel state in presence of surfactants, and (2) the surfactants that can exist in a monomeric state or in a micellar state (spherical or cylindrical). Situations 1 and 2 represent the case of truly orthogonal processes where the different aggregation states are not affected by the other component, while the last situation depicts the possible formation of mixed aggregates.

involvement of CTAT in HG1 fiber formation. Moreover, these conclusions are based on the assumption that the thicker and thinner fibers originate from HG1 and CTAT, respectively, and furthermore, the morphological investigations do not give insights to intermolecular interactions at the molecular or micellar level. In order to address the question of if self-assembly of the hydrogelator and surfactant are true orthogonal processes, one has to investigate the self-assembly phenomena in these mixed systems in more detail. Already a first glance at mixed systems of gelators and surfactants showed that this is a challenging task because the existence of different aggregation states for each component leads to many possible interactions (Scheme 1). A similar situation will exist in any other binary self-assembling system, and will become even more complicated with increasing number of components and/or aggregation states. Fortunately, in order to address the question of whether the self-assembly processes are orthogonal or not, it is sufficient to qualitatively assess whether the equilibria are affected by the presence of another component or not, rather then to investigate each equilibrium in detail.

Because gel fibers and micelles can be considered as separate phases, the self-assembly equilibria of gelators and surfactants can, in a first approximation, be described by the pseudo-phase model.[20] Hence, the free Gibbs energies for gel fiber or micelle formation from non-aggregated molecules can directly be determined from the critical aggregation concentrations (cac) or critical micelle concentrations (cmc), respectively. The cac and cmc of single component systems are easily measured by a wide variety of methods. For instance, a number of techniques have been implemented to obtain the cmc of surfactants,[20] among which are surface tension measurement, dynamic light scattering and NMR (line broadening or changes in the chemical shift can occur upon micellization). However, these latter methods obviously can not be applied to systems containing HG1 in the gel state, and therefore methods which are specific for the surfactant or micellar state have to be used. Fluorescence probes for hydrophobic microdomains[21] could be an option, but their fluorescence might still be affected by the gel fibers which could entrap some of them. In the specific case of HG1 and CTAT studied here, the non-ionic character of HG1 together with the ionic CTAT allows the application of conductimetry to measure the cmc (Fig. 4). Micellization is indeed usually accompanied by a sudden change in the slope of the specific conductivity as a function of the surfactant concentration, due to counter-ion condensation around the micelle, which decreases the free counter-ion concentration in the medium.[22] Using this property, a value of 0.2 ± 0.05 mM was found for CTAT in water, which is in agreement

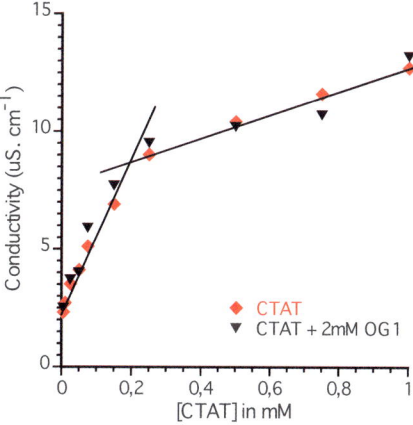

Fig. 4 Cmc determination by conductimetry of CTAT in pure water or in presence of 2 mM HG1. Extrapolation of the two linear domains gives a cmc of about 0.2 mM for CTAT alone, and 0.15 mM in presence of HG1.

with other methods as well as the literature. Remarkably, the conductivity of CTAT solutions was hardly influenced by the presence of HG1, and a comparable cmc of CTAT in presence of gel fibers was found (0.15 ± 0.05 mM). It should be noted that in this experiment HG1 was present at a concentration of 2 mM, *i.e.* well above its cac of 0.4 mM, implying that both monomeric HG1 and HG1 fibers were present. Therefore, it can be concluded that at least micelle formation of CTAT is hardly influenced by HG1 either as non-aggregated molecules or fibrous aggregates.

In order to establish whether binary systems of HG1 and CTAT display orthogonal self-assembly one also has to assess the effect of CTAT on HG1 self-assembly, *e.g.* on its cac or gel–sol phase transition temperatures (T_{GS}). Although it is straightforward to estimate the higher limit of the cac by simple, qualitative gelation experiments like the inverted tube test, precise determinations of the cac are much more difficult and time consuming because fiber and gel formation involve a nucleation step. On the other hand, the gel–sol phase transition temperatures can easily be determined by the dropping ball method.[23] This critical temperature increased steeply with the concentration of HG1 at lower concentrations, until a limiting value of approximately 130 °C was reached at concentrations of HG1 above 10 mM

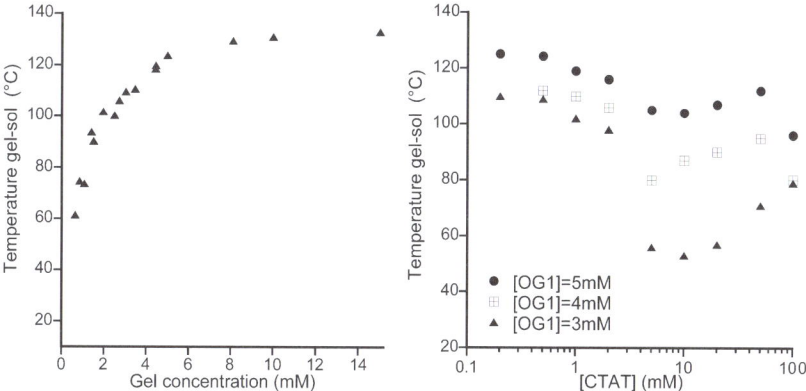

Fig. 5 (left) Temperature of gel–sol transition (T_{GS}) of HG1 as a function of its concentration. (right) Effect of CTAT concentration on the T_{GS} of HG1 at 3, 4 or 5 mM.

(Fig. 5). Extrapolation of the T_{GS} curve to room temperature indicated a minimum gelation concentration of 0.4 mM. Such phase behaviour is typical for LMW gelators and the T_{GS}–concentration curves have often been analyzed by the van't Hoff relationship for the melting temperature and enthalpy.[24] However, it should be noted that the dropping ball method applied here monitors the gel-to-sol phase transition. This transition can have different origins: (i) dissociation of the junction zones stabilizing the network and (ii) a decrease of the fiber concentration below a critical fiber concentration necessary for maintaining a self-supporting network, e.g. due to melting of individual fibers. These processes can coincide or take place independently, but in both cases fibers are still present at the observed T_{GS}, and therefore this temperature underestimates the actual melting temperature for a specific gelator concentration.[25]

While the presence of self-assembled gel fibers did not seem to significantly alter the cmc of the CTAT surfactants, reciprocally, the presence of surfactants in the hydrogels led to a significant destabilization of the gel network, manifested by a decrease of T_{GS} (Fig. 5). This effect was more pronounced at low concentrations of gel (3 mM HG1) than at higher concentration (HG1 = 5 mM) and occurred as soon as the first micelles of CTAT were formed. However, the T_{GS} typically increased back to its original value in water when the CTAT concentration was further increased, i.e. when the elongated micelles were progressively growing to finally entangle. The first decrease in the T_{GS} might be explained by a partial dissolution of monomers in the micelles, leading to a more limited number of gelator available for gel formation. In that case, a more significant shift in the cmc of the CTAT would be expected. The other possibility is that a small fraction of surfactants favours end-capping of the fibers, making them shorter and thus resulting in more fragile gels. While the outside of the fibers is mostly covered by ethylene glycol tails, the extremity of the fibers may indeed expose their more hydrophobic core possibly interacting with the hydrophobic tails of the surfactants. Such behaviour was in fact observed with a similar self-assembled gelator in the presence of spherical micelles. The further increase in T_{GS} with increasing CTAT concentrations is not clear yet. This effect might be due to the presence of more and more entangled micelles that could act as a mechanical support for the stiff gel fibers and thus prevent the gel to collapse. The other possibility is that the surfactant–surfactant interactions become stronger as the surfactant packing parameter increases, limiting as a consequence the surfactant–gel interactions.

In conclusion, the change in the critical parameters for gel formation suggests that a fraction of gelators and surfactants most probably interact at a molecular or supramolecular level. The fact that these binary systems are prepared from isotropic solutions and not simply mixed corresponds, of course, to the most drastic conditions to favour interactions between monomers of each entity. Different results may have been observed by only mixing the two species directly in an aggregated state. Still, even in conditions where the two components are given high opportunities to interact first from monomeric states, it appears that these molecules have a high tendency to self-sort in two different self-assembled networks, with characteristics very close to the original networks in water.

The development of such interpenetrating networks, with distinct rheological properties may also lead to the emergence of new viscoelastic properties, unattainable using the individual components separately. The rheological responses of the two distinct networks, as well as their behaviour when mixed together, were therefore investigated using oscillatory shear experiments. For each case, Fig. 6 shows the frequency dependence of the dynamic moduli, which provides insight into the relaxation and lifetime of the connections between the units forming a network. Frequency sweeps at 25 °C were performed within the common linear viscoelastic region of the two networks ($\gamma = 0.5\%$). For a 100 mM CTAT sample, in the low-frequency region, G' and G'' were found to scale with frequency with exponents of 2 and 1, respectively. At higher frequencies, G' and G'' crossed over at a characteristic

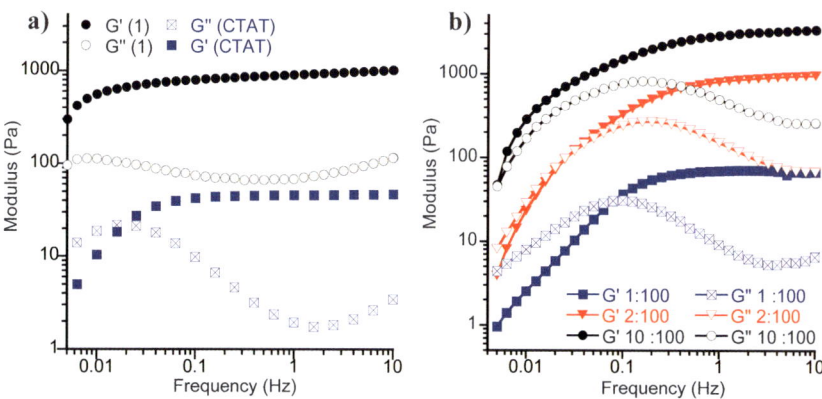

Fig. 6 (a) Moduli frequency dependence (G' and G'') of a 100 mM CTAT solution in water (wormlike micelles) or a hydrogel of HG1 at 2 mM in water. (b) Dynamic moduli G' (elastic) and G'' (viscous) as a function of frequency of a mixed system of cylindrical micelles (CTAT, 100 mM) and gel fibers of HG1 at 1 mM, 2 mM, and 10 mM respectively ([gelator]/[surfactant] = 1 : 100, 2 : 100 and 10 : 100).

frequency whose reciprocal corresponds to the main relaxation time (or disentanglement time) of the system, τ_R. A value of about 40 s was obtained at 25 °C, which is in agreement with literature.[26] Such behaviour is very characteristic of viscoelastic solutions of entangled cylindrical micelles.[27] By contrast, the gelators under study displayed a frequency response typical for gels with more permanent bonds,[28] as indicated by a small frequency dependence and G' being about 10 times higher than G''. Indeed, for a gel concentration of only 2 mM, G' and G'' displayed a plateau at around 1000 Pa and 100 Pa, respectively, over a wide range of frequencies (0.05–10 Hz), indicating a predominantly elastic response. The values of the plateaux reached by G' and G'' depended on the concentration of HG1: G' and G'' stabilized around 80 Pa and 10 Pa respectively for a 1 mM solution of HG1, while values of 4000 Pa (G') and 500 Pa (G'') were measured for 10 mM gels. Overall, the absence of a relaxation time, together with higher values of G' and G'' for hydrogel make this self-assembled network very different from solution of wormlike micelles.

Interestingly, when these two types of networks were combined, an intermediate and adjustable rheological response was observed depending on the concentration of HG1 in a CTAT solution (Fig. 6b). A 100 mM CTAT solution in presence of 1 mM of HG1 behaved mostly as a viscoelastic solution of entangled micelles. The values reached by G' and G'' were comparable to those observed with a pure solution of CTAT at the same concentration. The relaxation time measured was however smaller ($\tau_r \sim 10$ s), indicating a faster disentanglement of the micelles. When the concentration of HG1 was increased to 2 mM in a 100 mM CTAT solution, a viscous behaviour at low frequencies was still observed ($G' < G''$) followed by a more elastic response at high frequencies. Interestingly, the measured response was directly a combination of the two distinct behaviours of CTAT and HG1 alone. The response at low frequencies had a significant overlap with the typical response of CTAT micelles at the same concentration; while the response of the binary mixture at high frequencies appeared to be mostly governed by the quasi-elastic network of HG1. Increasing the concentration of HG1 further to 10 mM in CTAT solution resulted in a predominantly elastic behaviour. The rheological properties of the binary mixtures appeared thus to be easily tuneable from viscoelastic systems to more elastic with increasing concentrations of HG1, even at very low ratios and without any macroscopic phase separation. Apparently, the presence of stiff fibers of HG1 reinforces the otherwise soft gels of CTAT, whereas, reciprocally, rigid gels of HG1 become more dynamic due to the presence of cylindrical micelles of CTAT.

4. Summary

So far, the preparation of interpenetrating network gels by self-assembly approaches was doomed to fail because the conventional polymers and surfactant building blocks either phase separate or form mixed assemblies, respectively. In this work we showed that non-covalent interpenetrating networks can be easily prepared by the orthogonal self-assembly of small molecular hydrogelators and surfactants. The first studies showed that these self-assembled IPN have a number of intriguing properties. For instance, the presence of two coexisting networks offers new possibilities for compartmentalization, and will allow one to adjust the viscoelastic properties between 'soft' and 'hard' gels. The non-covalent character of such IPN makes their formation fully reversible, which can be exploited for dual responsive systems. Most interestingly, self-assembled IPN can also act as a very primitive, yet unique, model for biological interpenetrating networks like the extracellular matrix and the cytoskeleton, and thereby contribute to our understanding of these very complex systems.

Acknowledgements

This work was supported by Nanoned/STW and the Netherlands Organization for Scientific Research (NWO).

References

1 G. M. Whitesides, J. P. Mathias and C. T. Seto, *Science*, 1991, **254**, 1312.
2 G. M. Whitesides and B. Grybowski, *Science*, 2002, **295**, 2418; M. Antonietti and G. A. Ozin, *Chem.–Eur. J.*, 2004, **10**, 28.
3 J. N. Israelachvilli, *Intermolecular and surfaces forces*, 2nd edn, Academic Press, New York, 1992.
4 D. R. Karsa, *Industrial applications of surfactants IV*, Woodhead Publishing Limited, 1999, Cambridge; Y. Yamamoto, T. Fukushima, Y. Suna, N. Ishii, A. Saeki, S. Seki, S. Tagawa, M. Taniguchi, T. Kawai and T. Aida, *Science*, 2006, **314**, 1761.
5 R. G. Weiss and P. Terech, *Molecular gels – Materials with self-assembled Fibrillar Networks*, Springer, Dordrecht, The Netherlands, 2006; S. Jain and F. S. Bates, *Science*, 2003, **300**, 460; D. E. Discher and A. Eisenberg, *Science*, 2002, **297**, 967; S. Zhang, *Nat. Biotechnol.*, 2003, **21**, 1171; K. Velonia, A. E. Rowan and R. J. M. Nolte, *J. Am. Chem. Soc.*, 2002, **124**, 4224.
6 Z. Li, E. Kesselman, Y. Talmon, M. A. Hillmyer and T. P. Lodge, *Science*, 2004, **306**, 98–101; H. Cui, Z. Chen, S. Zhong, K. L. Wooley and D. J. Pochan, *Science*, 2007, **317**, 647–650; A. L. Parry, P. H. H. Bomans, S. J. Holder, N. A. J. M. Sommerdijk and S. C. G. Biagini, *Angew. Chem., Int. Ed.*, 2008, **47**, 8859–8862.
7 A. Aggeli, I. A. Nyrkova, M. Bell, R. Harding, L. Carrick, T. C. B. McLeish, A. N. Semenov and N. Boden, *Proc. Natl. Acad. Sci. U. S. A.*, 2001, **98**, 11857; R. Oda, I. Huc, M. Schmutz, S. J. Candau and F. C. MacKintosh, *Nature*, 1999, **399**, 566; H. Dong, S. E. Paramonov, L. Aulisa, E. L. Bakota and J. D. Hartgerink, *J. Am. Chem. Soc.*, 2007, **129**, 12468; D. Papapostolou, E. H. C. Brornley, C. Bano and D. N. Woolfson, *J. Am. Chem. Soc.*, 2008, **130**, 5124.
8 S. Bhuniya and B. H. Kim, *Chem. Commun.*, 2006, 1842; J. D. Hartgerink, E. Beniash and S. I. Stupp, *Science*, 2001, **294**, 1684.
9 O. Ikkala and G. ten Brinke, *Science*, 2002, **295**, 2407.
10 M. Boltau, S. Walheim, J. Mlynek, G. Krausch and U. Steiner, *Nature*, 1998, **391**, 877.
11 T. Kato, *Science*, 2002, **295**, 2414.
12 A. Heeres, C. van der Pol, M. Stuart, A. Friggeri, B. L. Feringa and J. H. van Esch, *J. Am. Chem. Soc.*, 2003, **125**, 14252; A. Brizard, M. Stuart, K. van Bommel, A. Friggeri, M. de Jong and J. van Esch, *Angew. Chem., Int. Ed.*, 2008, **47**, 2063.
13 A. M. Brizard and J. H. van Esch, *Soft Matter*, 2009, **5**, 1320.
14 L. H. Sperling, *Interpenetrating polymer networks and related materials*, Plenum, New York, 1981; Y. H. Lim, D. Kim and D. S. Lee, *J. Appl. Polym. Sci.*, 1997, **64**, 2647; A. Gutowska, Y. H. Bae, H. Jacobs, J. Feijen and S. W. Kim, *Macromolecules*, 1994, **27**, 4167; Y. Liua and M. B. Chan-Par, *Biomaterials*, 2009, **30**, 196; H. Katono, K. Sanui, N. Ogata, T. Okano and Y. Sakurai, *Polym. J.*, 1991, **23**, 1179.

15 *Advances in Interpenetrating Polymer Networks*, ed. D. Klempner and K. C. Frisch, Technomic: Lancaster, PA, 1994; J. J. M. Halls, C. A. Walsh, N. C. Greenham, E. A. Marseglia, R. H. Friend, S. C. Moratti and A. B. Holmes, *Nature*, 1995, **376**, 498.
16 K. J. C. van Bommel, C. van der Pol, I. Muizebelt, A. Friggeri, A. Herees, A. Meetsma, B. L. Feringa and J. H. van Esch, *Angew. Chem., Int. Ed.*, 2004, **43**, 1663–1667.
17 J. F. A. Soltero and J. E. Puig, *Langmuir*, 1995, **11**, 3337.
18 L. M. Walker, *Curr. Opin. Colloid Interface Sci.*, 2001, **6**, 451.
19 In solution, the persistence length is probably even shorter, since the preparation of the grids for cryoTEM most likely favours alignment of the elongated micelles, due to shear forces when blotting away the excess of liquid. For a critical analysis of this method applied to wormlike micelles see: Y. I. Gonzalez and E. W. Kaler, *Curr. Opin. Colloid Interface Sci.*, 2005, **10**, 256.
20 D. F. Evans and H. Wennerström, *The colloidal domain*, VCH, Weinheim, 1994.
21 M. C. A. Stuart, J. C. van de Pas and J. B. F. N. Engberts, *J. Phys. Org. Chem.*, 2005, **18**, 929; J. Aguiar, P. Carpena, J. A. Molina-Bolívar and C. Carnero Ruiz, *J. Colloid Interface Sci.*, 2003, **258**, 116; C. Carnero Ruiz, *Colloid Polym. Sci.*, 1995, **273**, 1033.
22 P. Mukerjee, *J. Phys. Chem.*, 1962, **66**, 1375; A. K. Jain and R. P. B. Singh, *J. Colloid Interface Sci.*, 1981, **81**, 536.
23 Takahashi, M. Sakai and T. Kato, *Polym. J.*, 1980, **12**, 335–341.
24 P. Terech and R. G. Weiss, *Chem. Rev.*, 1997, **97**, 3133.
25 P. Terech, C. Rossat and F. Volino, *J. Colloid Interface Sci.*, 2000, **227**, 363.
26 J. F. A. Soltero and J. E. Puig, *Langmuir*, 1996, **12**, 2654.
27 J. F. Berret, in *Rheology of worm-like micelles in Molecular gels – Materials with self-assembled Fibrillar Networks*, Springer, Dordrecht, The Netherlands, 2006, chapter 19.
28 P. Terech, A. Coutin and A. M. Giroud-Godquin, *J. Phys. Chem. B*, 1997, **101**, 6810; P. Terech, D. Pasquier, V. Bordas and C. Rossat, *Langmuir*, 2000, **16**, 4485; J. Brinksma, B. L. Feringa, R. M. Kellogg, R. Vreeker and J. H. van Esch, *Langmuir*, 2000, **16**, 9249.

General discussion

Professor Kornyshev opened the discussion of the paper by Professor Woolfson: What is the role of the surface charges? Charges must follow helical patterns on α-helices and must influence the shape of aggregates *e.g.* with regard to the skew angle between α-helices. Equally, counterions and buffer electrolytes affecting the electrostatic interactions must have a strong effect on the resulting structure and dynamics of their formation.

Professor Woolfson answered: This is a good point. The surfaces of the coiled coils play a major role in determining the thickness and morphology of the equilibrium fibres, and, indeed, electrostatics play a key role. We have used this to good effect to engineer thickened, thinner, and ordered fibres, and even to get fibres to interact to form gels. We have reported this in several papers.[1–4]

1 A. M. Smith, E. F. Banwell, W. R. Edwards, M. J. Pandya and D. N. Woolfson, *Adv. Mater.*, 2006, **16**, 1022–1030.
2 D. Papapostolou, A. M. Smith, E. D. T. Atkins, S. J. Oliver, M. G. Ryadnov, L. C. Serpell and D. N. Woolfson, *Proc. Natl. Acad. Sci. U. S. A.*, 2007, **104**, 10853–10858.
3 D. Papapostolou, E. H. C. Bromley, C. Bano and D. N. Woolfson, *J. Am. Chem. Soc.*, 2008, **130**, 5124–5130.
4 E. F. Banwell, E. S. Abelardo, D. J. Adams, M. A. Birchall, A. Corrigan, A. M. Donald, M. Kirkland, L. C. Serpell, M. F. Butler and D. N. Woolfson, *Nat. Mater.*, 2009, **8**, 596–600.

Dr Titmuss opened the discussion of the paper by Dr Vendruscolo: Your paper contains a very interesting statement in the section entitled Protein homeostasis in normal and aberrant biology: "It also indicates that protein molecules have co-evolved with their cellular environments to be sufficiently soluble for their biological roles, but no more so, and hence that aggregation can result from even minor changes in chemistry and in the regulation of otherwise harmless proteins." This is interesting to a non-biologist, as I understand that one of the interesting questions in cell biology is how proteins get switched on to be expressed in the region of the body that they are required. Is there a suggestion that aggregation might be a physical mechanism by which protein expression is regulated, such that if a cell is a part of the body where a protein is not required, the cell chemistry would be such as to cause aggregation should any such protein be expressed?

Dr Vendruscolo responded: The evidence that we have is that the physico-chemical properties of proteins have evolved so that their critical concentration under physiological conditions is very close to the maximal concentration that they can reach in any cell location where they carry out their functions. Therefore, rather than a physical mechanism of regulation of their expression, we believe that evolutionary processes are acting to maintain the proteins soluble even at the upper levels of concentrations required during a cell lifespan.

Mr Zayed continued the discussion of Professor Woolfson's paper: Regarding the trimeric DNA walker—what length of DNA will it tackle when packaged and filed as DNA usually is in the cell?

Professor Woolfson replied: The idea is to stretch the DNA out linearly and place it under tension in a microfluidics flow cell. This will also aid the addition of the fuel, *i.e.* the ligands for the repressor heads.

Professor Kornyshev asked: Protein mobility, like the motion of myosin, required ATP activity. No ATP, no mobility. What is the fuel in your "rotation wheel" driving on a DNA-track? And what is the chemical reaction behind the consumption of that fuel? For ATP, this is ATP hydrolysis. Which reaction works for you in your machine?

Professor Woolfson answered: In our case, the fuel will come from ligand binding to the three separate repressor heads in the tumbleweed motor. This ligand binding increases the affinity of binding to DNA and, so, allows the engagement of each head with DNA to be controlled. By having three different heads, A, B and C, the binding of each mediated by ligands, a, b and c, we envisage motion along DNA by adding paired combinations of the ligands *via* microfluidics: a+b, followed by b+c, then c+a, then a+b, and so on.

Professor Reinhoudt continued the discussion of the paper by Dr Vendruscolo: Firstly, can you predict from the primary sequence of a protein whether the folding is spontaneous into its native structure, or whether a chaperone is needed? To follow that, does the chaperone help the folding or prevent misfolding?

Dr Vendruscolo answered: Indeed, one of the major thrusts of our research is to develop methods to perform this type of prediction. We believe that such predictions are possible because in addition to the complex regulatory mechanisms that act at the cellular level, there is a second level of control of the behaviour of proteins that is based on the physico-chemical properties of their amino acid sequences. Evolutionary processes acting on these sequences can therefore contribute to the overall maintenance of protein homeostasis in the cell. We are thus working on the decoding of the information contained in the sequences that determine their ability to fold and to resist misfolding, as well as their levels of chaperone requirements.

One could say that many chaperones help the folding process by preventing misfolding, which is achieved by binding to misfolded conformations and facilitating their reconversion to the native form.

Professor Whitesides remarked: The cytosol (30% organic material, 1M ionic strength) is very different from dilute phosphate buffer. Does organic chemistry in simple buffers tell anything about chemistry in the cytosol?

Dr Vendruscolo answered: Despite recent advances, *in vivo* studies of protein behaviour remain very challenging. It is therefore very useful to consider models that can tell us how proteins can behave in the cell, although they might not necessarily do so, since the myriad interactions within the crowded cellular environment can actually profoundly influence their specific behaviours. There is evidence, however, that the activity of proteins is remarkably resilient to variations in the solvent composition of the type that is required to move from simple buffers to cytosol-mimetic solutions. Subject to careful design, proteins appear also to be capable of working even in inorganic solvents other than water, and in organic solvents other than the lipids making up biological membranes. Apart from supporting investigations in simple buffers, the versatility of proteins opens a range of opportunities for developing biocatalytic agents in "designer solvents" whose properties can be tuned according to economic and environmental requirements.

Dr Bittner opened the discussion of the paper by Professor Ulijn: First: can one control the diameter of the fibers? Can the diameter reach molecular dimensions, or are fiber bundles more common?

Second: apart from fluorescence spectroscopy, which methods can be used to analyse the peptide assembly, and what information do they provide?

Professor Ulijn replied: The answer to your first question is yes. Factors that control β-sheet curvature and therefore fibre/nanotube diameter are (i) the nature of the π-stacking ligand (fluorenyl, pyrenyl, naphthyl) and (ii) the peptide length and amino acid sequence. We have so far observed a range of dimensions, ranging from 3 nm for Fmoc-F2, approximately 20 nm for Fmoc-L3 to flat sheets (no curvature) for Fmoc-L5. Using molecular modelling we are making progress towards predicting β-sheet curvature which dictates these diameters. The tendency of individual tubes to aggregate depends again on the molecular composition, with Fmoc-F2 forming flat ribbons of side-by-side aligned tubes and Fmoc-L3 forming nanotubes that do not cluster.

Secondly, a range of complementary spectroscopic methods has been used. Our peptide nanostructures form *via* a combination of π-stacking interactions between aromatic stacking ligands (as analysed by fluorescence spectroscopy) and hydrogen bonding of peptide components (as analysed by FT-IR). Additional methods that have been used are circular dichroism (although interpretation of results can be complicated by linear dichroism contributions) which provides information about chiral orientation of fluorenyl groups as well as peptide backbone arrangements. A complementary technique is Raman optical activity (ROA); a first paper in this area will be published shortly.[1] X-Ray scattering techniques provide additional information on intermolecular distances (WAXS) and nanostructure dimensions.

1 A. M. Smith, R. J. Collins, R. V. Ulijn and E. Blanch, *J. Raman Spectrosc.*, 2009, in press.

Professor Dr van Esch asked: In the paper it is stated that the component exchange must be fully reversible in order to enable selection of the thermodynamically most stable component. I wonder whether this is still the case with the assemblies formed by the longer peptides? What is the typical time scale of exchange between solution and fibrous state for these longer peptides, and is there no risk with these longer peptides of running into a kinetically trapped state?

Professor Ulijn replied: In order for these system to reach thermodynamic equilibrium, fully reversible component exchange is naturally a requirement. In the Fmoc–dipeptide–ester systems described in this paper equilibrium distributions are reached within hours and formation of kinetically trapped aggregates is not observed (the same equilibrium distributions are achieved when using different starting points). However, for longer hydrophobic peptide sequences, reactions are much slower (over 3 months to reach an equilibrium distribution for Fmoc–leucine libraries) and formation of kinetically trapped aggregates is indeed a risk.

Professor Dr van Esch then continued the discussion of the paper by Professor Woolfson: To which extend are the peptide-based fibrous networks or gels dynamic materials, and do they still allow for remodeling during cell proliferation when used as an artificial extracellular matrix?

Professor Woolfson replied: We are currently exploring both the assembly kinetics and the dynamics of the standard (rigid fibrous) self-assembling fibres (SAFs) and the hydrogelating SAF networks (hSAFs). The SAFs are susceptible to proteolysis, and we are currently considering introducing matrix metalloprotease (MMP) cleavage sites to allow remodeling of the networks as the cells grow, as you suggest.

Professor Huskens returned to the discussion of the paper by Professor Ulijn: What is the role of the enzyme in the assembly process—does it only establish the equilibrium between assembling and non-assembling units or does it play a more active role in assembling the structure from the assembling units? The latter would constitute a powerful route for kinetically controlled assembly which is potentially more much diverse than thermodynamic control.

Professor Ulijn replied: The enzyme's role indeed goes beyond that of simply catalysing the amide synthesis/hydrolysis reaction—it provides a nanoscale enviroment where self-assembling building blocks are produced on-demand, allowing for localized structure nucleation and growth,[1] a property that relates to the self-assembly kinetics of the system. We are currently exploring the use of immobilised enzymes to exploit the features of kinetically controlled reactions further. One exciting recent observation was that kinetically controlled enzymatic reactions allow for controlled alignment of peptide nanofibres.

1 R. J. Williams, A. M. Smith, R. Collins, N. Hodson, A. K. Das and R. V. Ulijn, *Nat. Nanotechnol.*, 2009, **4**, 19–24.

Dr Titmuss continued the discussion of the paper by Dr Vendruscolo: Does a comparison of the relevant length-scales offer a way to reconcile Dr Vendruscolo's faith in the relevance of measurements made in phosphate buffer with the healthy scepticism of Prof. Whitesides, given that cellular medium is not phosphate buffer, but rather a high ionic strength medium comprising a high volume fraction of organic material? To be explicit, although the overall cellular medium may be 30% organic material, one could envisage that the local environment at the protein surface (the inner coordination sphere if you like) could be 100% water, in which case if the interactions that drive folding are short compared to the dimensions of this water layer, one might expect the result of the interactions (folding) to be similar to that measured in an aqueous phosphate buffer. Conversely, if the interactions are long-ranged on the dimensions of the cell, then it would be the overall dielectric composition that would be relevant.

Dr Vendruscolo responded: In addition to the ability to fold into their well defined native conformations, another remarkable property of proteins is that they remain soluble in the crowded cellular environment. Apart from the specific interactions with their functional partners, they essentially avoid the formation of stable contacts with other molecules. It is therefore likely that individual protein molecules are indeed surrounded by a water layer that may make their properties quite similar to those tested through *in vitro* experiments.

Professor Chau continued the discussion of the paper by Professor Woolfson: How valid is the assumption that the amino acid residues are independent in obtaining a consensus sequence for certain structural motif? How do you design the linker sequences between structural motifs? Are there general criteria?

Professor Woolfson responded: These are very good points. You are correct that linear consensus sequences are often not enough in themselves to deliver specific structures. So, we are also using pairwise information in the natural sequences to direct our designs. Regarding the second point on linker and turn design, this is a tough problem in protein design. We are using a combination of bioinformatics and empirical iterative design to do this.

Professor Chau returned to the discussion of the paper by Professor Ulijn: Are there potential applications in which specific enzymes can be used in screening self-assembling peptides?

Professor Ulijn answered: The use of non-specific enzymes with little kinetic bias in dynamic peptide libraries (such as thermolysin used in this paper) allows for the full range of possible peptide products to be accessed and is probably most amenable to identification of short peptide nanostructures and affinity ligands (5 amino acids or less). In some cases, it might be of interest to use a more 'modular' approach, where longer (structure-forming) peptide fragments are exchanged, in which case of more selective enzymes would be required. The modules could be, for example, α-helix-forming heptads for dynamic discovery of stable coiled-coil structures. In the context of identification of functional nanostructures one could think of exchanging sequences containing *e.g.* bioactive and structural regions.

Dr Steinke asked: Have there been any experiments to try and control the length of the formed fibres? If not, what possibilities are there to achieve control of fibre length?

Professor Ulijn responded: I am not aware of attempts to control the fibre length. I would imagine that control of average length of a population of fibres may be achieved by introducing certain ratios of 'capping molecules' (that are incorporated in the self-assembled structure, but prevent further assembly, *e.g.* by breaking up linearity of peptide chains).

Mr Aufderhorst-Roberts communicated: Firstly, your presentation mentioned the formation of nanospheres as a precursor to the formation of a hydrogel network. Such a two-stage gelation process has been reported previously for other Fmoc peptide derivatives. Do you think that this indicates a consistent behavior in the gelation process for this group of materials?

Secondly, given the proposed nanotubular structure of Fmoc peptide derivatives obtained by molecular modelling, how is the formation of spheres facilitated? Do you believe it indicates a new type of aggregate or some form of nanotube cluster?

Professor Ulijn communicated in reply: We and others have observed nanospheres as precursors for fibres in self-assembly, as well as the formation of nanospheres from fibres in dis-assembly so they indeed appear to be a consistent feature for at least some Fmoc peptides. For some peptide derivatives, spheres are stable for long periods of time (days) without evidence of fibre formation.

With regard to your second question, we have performed limited spectroscopic analysis on these nanospheres. So far, we have not found evidence for the π-β structure that underpins the fibrous/nanotubular architectures and the spheres therefore appear to represent a different type of aggregate.

Mr Ahangar opened the discussion of the paper by Dr Browne: What about the time scale limitations for switch between the glassy state and normal state? Also, can you alter the stereo conformation of the molecule to produce steric hindrance in it so that it spins only in one direction?

Dr Browne replied: For the first question—the time scale limitations for switching between the glassy and normal state—The switch between the glassy and normal state of glass itself takes, depending on temperature, anywhere between minutes and hours. The time required to reach a stable glass has been determined separately by UV-Vis spectrometry, and all experiments were performed in glasses that were fully relaxed. With regard to stereocontrol, these molecules are inherently chiral due to the methyl group at the stereogenic carbon adjacent to the central double bond. This together with the helix inversions which take place between the four separate states results in unidirectional rotation, *via* a series of four light–thermal–light–thermal steps. This has been described extensively in the literature already.[1]

1 B. L. Feringa, *J. Org. Chem.*, 2007, **72**, 6635–6652.

Dr Channon opened the discussion of the paper by Professor Dr van Esch: You have demonstrated the growth of the fibrous component of your system inside giant vesicles—could you comment on the efficiency of this phenomenon? For example, of all the lipid aggregates observed, roughly what fraction are individual, single-compartment giant vesicles like the examples in your presentation? How many of the vesicles have fibres inside them? And how often do you see fibres outside vesicles?

Professor Dr van Esch replied: So far we did not carry out quantitative studies on the efficiency of fiber formation inside vesicles, but cryo-transmission electron microscopy showed a significant fraction of vesicles without any fibers inside, and furthermore fibers outside of the vesicles are visible in all images. Our current hypothesis is that the process of fiber formation has caused disruption of a significant fraction of the vesicles leading to empty vesicles and fibers outside. The most likely reason for vesicle disruption are (i) an osmotic shock due to the addition of a reagent (base) to induce fiber formation or sudden decrease of gelator molecules, or (ii) mechanical stress exerted by the growing fiber on the vesicle membrane.

Mr Ahangar asked: Do you think that the encapsulation efficiencies can be related to microfluidics?

Professor Dr van Esch replied: I hope so because this remains a serious issue in our and other vesicle related work involving encapsulation. Recently, the group of Weitz has published very promising results on preparing large unilamellar vesicles by a double emulsion approach using microfluidics leading to high, in theory 100%, encapsulation efficiencies,[1] and I hope to implement this approach within the near future in our research as well.

1 H. C. Shum, D. Lee, I. Yoon, T. Kodger and D. A. Weitz, *Langmuir*, 2008, **24**, 7651–7653.

Dr Neto continued the discussion of the paper by Dr Browne: There is an experimental technique which allows the measurement viscosity down to the molecular level: the surface force apparatus (SFA). Using this technique, it has been shown that the viscosity of simple liquids remains unchanged down to a few molecular layers (see for example ref. 1 and 2).

1 J. N. Israelachvili, *J. Colloid Interface Sci.*, 1986, **110**, 263.
2 T. Becker and F. Mugele, *Phys. Rev. Lett.*, 2003, **91**, 166104.

Dr Browne responded: The surface force apparatus (SFA) is an excellent probe of the behaviour of solvent layers especially those in contact with atomically flat surfaces. However, it is most applicable to the study of the behaviour of fluids as discrete layers as I understand the technique. In the report of Becker and Mugele, it was noted that although the bulk behaviour of a liquid was retained right down to a few molecular layers, in very thin films layer slippage rather than solvent 'flow' becomes the dominant mode of removal of solvent between two approaching surfaces.

Extension of this technique to highly viscous media would be interesting however at the temperatures employed in the present study (90–140 K) this would present some technical challenges. In addition, although the technique suggests that the effect of surfaces on solvent viscous behaviour has a highly limited range, it is somewhat difficult to see how this can easily be related to viscosity as felt by molecular systems which is the subject here. In the present report we have taken data obtained by two other approaches, *i.e.* the dropping ball/rod and rotating disk techniques. We find that the best correlation with data obtained relating to molecular behaviour in the present study (*i.e.* photostationary states) is obtained with viscosity data determined using the dropping ball technique. Given the potential connection between the rotating disk and SFA techniques in terms of the data provided I guess that SFA-determined viscosity may not provide a good correlation either, were data to be available for the solvent systems employed here. It should be noted though that the aim of the present study was not to measure viscosity. As is very often the case in science, it was only during the course of our studies that we noted the limited amount of data available on viscosity under these conditions. The fact that we obtain a good correlation suggests that the PSS (photostationary state) of these systems could serve as a highly convenient tool to determine solvent viscosity in such generally experimentally difficult conditions. The questions that we now face, however, are (i) why do we get a correlation with the dropping ball technique, and (ii) in what sense does this correlation relate to viscosity at the molecular level. These remain open questions in our view but if we can establish that PSS does relate directly to solvent viscosity it does potentially offer a technique which is much simpler than, for example, fluorescence anisotropy.

Dr Neto then addressed Professor Dr van Esch: I agree with you that we are still a long way from achieving the complex self-assembly that is observed in nature. It seems to me that if we are to get any closer to that level of complexity we need to introduce "function" in our building blocks, rather than just several components. What is your view of how this can be achieved in the future?

Professor Dr van Esch answered: I do not yet see how complexity and function are directly related. I would be happy to discuss the issue of complexity related to orthogonal self-assembly if anyone is interested. With regard to function, in my opinion there are many avenues to reach this goal, and which route to take completely depends on the type of function that one aims to achieve. Our orthogonal self-assembling systems are clearly not aimed at catalytic functions or electronic functions, but they may lead to new functional systems for *e.g.* storage, transport, or actuation. In all cases, we will need to implement at least responsiveness and addressability, *e.g.* by triggers for the self-assembly processes or properties, to arrive at autonomous functional systems.

Professor Jones returned to the discussion of Dr Browne's paper: You posed the question "what is viscosity at the molecular level, and how do we measure it?" Another way of putting this would be to ask how small can one go before the continuum picture of a fluid implicit in the notion of a viscosity breaks down. An empirical answer to this comes from using the diffusion of probe particles or probe molecules of various sizes as a measure of viscosity *via* the Stokes–Einstein equation. For simple fluids, this shows that, for this purpose, the continuum pictures works surprisingly well right down to molecular dimensions. However, the fluids you are looking at aren't necessarily simple fluids—in glass forming liquids approaching a glass transition we would expect a growing degree of dynamic heterogeneity. Would you expect your kind of experiment to throw any light on this?

Dr Browne responded: I agree with you that measuring (translational) diffusion of fluorescent probes either particles or molecules of various sizes or of NMR active species by NMR spectroscopy are effective methods to determine solvent viscosity *via* the Stokes–Einstein equation. An alternative approach is to use fluorescence anisotropy to look at tumbling rates, *i.e.* rotational diffusion even at high viscosity as pointed out by Dr Howse. At very high viscosity, however, I am not confident that either method would be effective for several reasons (although *a priori* I would not claim that they would not work either, of course). However, although generally viewed as being a safe method, we then force ourselves to make an assumption that the Stokes Einstein equation holds at high viscosity for the system in question. A further complication is solvent inhomogeneity, which is very much an issue in glassy media, and this must be considered in the approach we have taken in the present study in the use of photoequilibria which is by default a statistical metric. Single molecule spectroscopy would be an interesting route to take but even there one could argue that the results obtained for a single molecule are only valid for the actual specific environment in which that molecule sits. I would note, though, that the assumption that low viscosity media are homogenous is also somewhat dangerous especially when one considers solvent mixtures.

The approach we have taken in this study was initially to use glassy media to achieve a particular end—shift the photoequilibrium in favour of the unstable form. Of course in any study one should be mindful of potential implications of results beyond our immediate goals—we should not be scientifically blinkered. Hence in answer to your last question, yes I think that our experiments could throw new light on the question of viscosity at the molecular level—in my view a highly pertinent question facing the field of molecular based nanomachines and devices. The reservations I have expressed in my talk and in the discussion are, I feel, justified—the next step that should be taken is to provide further demonstration of the validity of this approach and critically to understand why it correlates well with the dropping ball technique. As a final note I think your question reflects my question "what is viscosity at the molecular level, and how do we measure it?". The approach we take to measure viscosity itself has an inseparable link to the value for viscosity that we obtain—is viscosity the same when we consider rotation of an entire body, translation of a solvated or non-highly solvated body, or a conformational change of a small molecule? Unfortunately, I cannot give you an answer to the question I posed, but I still regard it as a question worth asking ourselves even if some may hold the view that viscosity is an essentially fully understood phenomenon.

Professor Huck continued the discussion of the paper by Professor Dr van Esch: On one of the slides, a TEM picture was shown that seemed to indicate a rod-like assembly distorting the spherical shape of a vesicle. Such a situation would suggest that the approach of orthogonal assembly leads to the generation of mechanical forces (after all, the spherical shape would be restored in the absence of other forces). These mechanical forces would then perhaps influence the nature of the self-assembly of the rods. My question to Professor Dr van Esch is to comment on the interplay between assembly and mechanics and his thoughts on the design of structures that would specifically address this intriguing phenomenon.

Professor Dr van Esch answered: I fully agree with Professor Huck that the mechanical interactions between self-assembled objects and their relation to the underlying self-assembly processes make for a very interesting issue. In nature such interactions occur frequently and are essential for processes like transport and organization within living cells, and in many cases interactions between the cytoskeleton biological membranes are involved. Certainly, the mechanical interaction between bilayer and fiber has an effect on the underlying self-assembly process because the fiber length is clearly limited by the confined space of the vesicles it has been growing in, probably involving a similar mechanism allowing the deformation of the biological membrane by microtubuli in natural systems. Here the gellosomes, *i.e.* fibers encapsulated in vesicles, can serve as a versatile and easy accessible model system in conjunction with more reliable approaches to produce them. Another possible subject is the natural cytoskeleton which is also a self-assembling interpenetrating network consisting of microtubuli, actin, and intermediate filaments. In this respect, the self-assembling interpenetrating networks developed by us would be a versatile and easily accessible model system to address for instance the interplay between network dynamics, the underlying self-assembly processes and the micro- and macroscopic viscoelastic properties. We are very keen to investigate these and related processes in more detail, ideally in collaboration with (bio)physical oriented groups.

Professor Kornyshev continued the discussion of the paper by Dr Browne: When you are going from macro to nano, there is a conceptual issue in distinguishing the object (moving) and the environment. In the case of a dropping ball, the ball is the ball, the water is water. Remember the Stokes formula for the motion of an ion in a liquid—it combines viscosity and the radius of an ion. If you want to reproduce experimental data using this formula with the macroscopic value of viscosity you have to fit the radius of the ion. It will be normally larger than the crystallographic radius, reflecting motion of the ion with its hydration shell or some part of it. It's the same with your motors: depending on the interaction with the solvent, part of the water may be associated with it, but then the rest will contribute something like the bulk viscosity.

Also, I would like to comment on the fluctuation–dissipation theorem (FDT). The relationship of viscosity to the correlation function is exact in the bulk. In a heterogeneous system, *i.e.* near an impurity, it can be distorted and, generally, has a much more complicated, non-universal form.

Dr Browne responded: With respect to the solvent shell and hydration, the present systems are studied in alkanes. However, the absence of a significant influence of solvent polarity on the photochemistry or thermal helix inversion rate in these systems suggests that solvation is not of major importance, *i.e.* the molecules do not impose significant order upon the first solvent sphere. Furthermore it is unlikely that there will be a significant difference in solvent–solute interaction between the two states of the molecule. This is the reason we have focused on solvent excluded volume as the origin of the effects we observe and feel we can justify defining the 'object' as only the molecule itself.

I agree with the viewpoint regarding FDT. The response of a system to an applied force is the same as the response to that same force when occurring randomly. In our view we need to increase the effective rotation rate of our motors, so that a net force will result from unidirectional rotation. In slow systems, the force resulting from rotation will be offset by random fluctuations; however, if we consider a system rotating at sufficiently high frequency we could obtain a force in a certain direction, as random encounters will not be able to offset the encounters enforced by rotation. Distortion of FDT by impurities, *i.e.* the motion of the different solvent layers with the rotation movement, could enhance this effect by increasing the effective size of the motor.

Mr Ahangar addressed Professor Kornyshev and Dr Browne: Do we need to redefine the equation of viscosity for microscopic systems as per the discussion, since it seems that the present viewpoint of viscosity is not complete?

Dr Browne answered: I think it would be somewhat premature to call for a redefinition of the equation for microscopic viscosity but it is certainly the case that the present viewpoint is incomplete. The difficulty, as we have noted, is that in measuring viscosity, be it using a bulk property such as with rotating disk methods or by a molecular property such as fluorescence anisotropy, we have to ask ourselves what exactly we are measuring. This is not an easy question to answer. Indeed, to quote the fairly recent report of Becker and Mugele:[1] "However, a consistent picture of dynamics in fluids at the boundary between molecular motion and continuum flow has yet to emerge."

In the present study we determine the effect of viscosity on a unimolecular process involving molecular motion and we see a correlation with macroscopic viscosity as determined by the dropping ball technique. It should be noted that in systems in which tanslation diffusion is a key component such as the triplet state decay of *e.g.* anthracene,[2] even though bulk viscosity in polymer–solvent mixtures can vary tremendously, molecular systems only 'see' the solvent viscosity. Hence it may be more to the point that when considering microscopic viscosity we must consider what the molecular systems see as the solvent environment.

I would like to note that the problem of viscosity is not trivial and presents a complete field in physical chemistry in itself; the reason for using a continuum model here is that it is not apparent which model for a molecular definition of viscosity is appropriate. Because we were interested in the mechanical effect of viscosity, the macroscopic continuum model seemed the best option in this case.

1 T. Becker and F. Mugele, *Phys. Rev. Lett.*, 2003, **91**, 166104.
2 G. E. Heppell, *Photochem. Photobiol.*, 2008, **4**, 7–12.

Professor Kornyshev replied: I do not think we need to redefine the notion of viscosity itself, but all forms of fluctuation–dissipation theorems in confined space and just near the surfaces may look substantially different than in the bulk. There is a vast amount in the literature about it; for a discussion of how it works in the dielectric response FDT, see *e.g.* ref. 1; the viscous response, compressibility and diffusion will, of course, require different treatment.

1 P. M. Platzman and P. A. Wolff, *Waves and interactions in solid state plasmas*, Solid State Phys., Suppl. 13, Academic Press, New York, 1973, section 14.

Professor Ikkala opened the discussion of the paper by Professor Aida: You showed some very interesting materials based on self-assembled supramolecules and thermally reversible changes of the structures upon heating and re-cooling that can be feasible in applications. My question is mostly semantic: you denoted the phenomenon as "self-healing after thermal stresses", but in many self-assembled systems there are different structures at different temperatures. On the other hand, self-healing often denotes a phenomenon where the material recovers after an imposed damage. Could you please comment on the selected wording?

Professor Aida answered: I appreciate this essential comment. I agree that there are different structures at different temperatures in self-assembled systems. However, we believe that our observation can be categorized as self-healing. In our system, the luminescence intensity immediately after the heated material was allowed to cool to 20 °C was much weaker than that before heating. However, when the material was held at 20 °C, the luminescence was gradually intensified and finally retrieved the original intensity after several hours. This means that heating indeed gave rise to some structural damage of the luminescence center which were mended, however, slowly but spontaneously.

Professor Colquhoun asked: Given the importance of aurophilic interactions in the gold-based materials discussed in your presentation, how do the structures of the analogous copper and silver compounds differ from those of the gold complexes?

Professor Aida replied: Copper and silver complexes behave in a manner similar to gold complexes; however, they are not very stable. Copper complexes are subject to oxidation, while silver complexes are photochemically labile. So, we need to consider some tricks for the stabilization of these complexes.

Professor Woolfson continued the discussion of the paper by Professor Dr van Esch: Related to the earlier question regarding engineering multi-component systems, this is being done in peptide-based fibres, but it is challenging. This is because many of these are 2D crystalline and are unforgiving of incorporating different building blocks, which means low incorporation or resorting to surface decoration. My question for Professor Dr van Esch is: do "softer" polymer systems suffer similar problems?

Professor Dr van Esch answered: I assume that by "softer" systems you refer to the fibrous networks formed by small molecule hydrogelators like our cyclohexane trisamide based gelators. We are frequently using modified gelators as probes *e.g.* to render a fibrous network fluorescent, and so far this has led to the desired result. That is, the fibers are stained with almost no fluorescent probe in solution. However, we did not investigate this issue in detail and cannot, for instance, exclude the clustering of probes or their preferential incorporation at early or late stages of the self-assembly process, both leading to an inhomogeneous distribution. In our research we considered these complications and tried to circumvent them by always using the same gelator scaffold for the probe and the parent fibers. However, it is expected that the distribution of probes within such fibers is only random when, for instance, the solubilities of probe and parent compound are comparable, and specific interactions between probe molecules are absent. This issue definitely requires more attention.

Professor Chan continued the discussion of Dr Browne's paper: Can you please clarify what it is you are trying to measure? For instance are you attempting to use the response of molecular motors to see if the idea of a (bulk) Newtonian viscosity of the solvent—as measured by a dropping ball experiment—remains valid down to the molecular level? Your results seem to suggest that this idea is not far wrong—in line with other experimental observations such as those relating the mobility of solutes in different solvents. You seem skeptical that your own experiments also seem to support this notion. What results are you anticipating?

Do you have an estimate of the time scale of the key mechanisms of your molecular motors? Is it well known that at short time scales (high frequencies) Newtonian liquids will exhibit viscoelastic behaviour? Is your system anywhere in this regime with respect to the solvent?

Dr Browne replied: Our primary goal is to use the effect of increased viscosity to move the position of photoequilibria of our molecular system towards the unstable state allowing us to study the thermal relaxation processes. That is, the influence of (mechanical) resistance, applied by viscosity, on the photochemical and the thermal step of rotation of a motor molecule. The fact that we use viscosity to apply resistance opens up several questions regarding viscosity at the molecular level: a key issue in the design of nanomachines. The data presented here indicates that Newtonian behaviour remains valid down to molecular dimensions as is the case where determined from solute diffusion experiments, *e.g.* fluorescence anisotropy. The primary issue we faced however was that for these high-viscosity glasses such molecularly determined data is not available to the best of our knowledge.

Newtonian behaviour has been directly established for these glasses before,[1] as is described in our ref. 10. Our hesitancy in making broader claims based on the present data set is that a relation between photoequilibria and macroscopically determined viscosity holds for one empirical data set and not for others. Hence, although the use of PSS as a metric for solvent viscosity in high viscosity regimes is highly promising, it is important to be critical and ask ourselves exactly what we are measuring and, more importantly, what we mean by viscosity.

The two steps (photochemical and thermally driven) are on a ps timescale. This has been verified experimentally for the photochemical step. For the thermal step, the common viewpoint is that this is represented by the rate constant for thermal helix inversion. If that were the case, the timescale would be in the microsecond regime. The rate constant, however, displays the average time for a motor to gain enough energy to pass the unstable-to-stable transition state barrier. In reality, the molecular motor moves around the potential energy surface, creating no net force during this time. This continues until it reaches the top of the barrier from where it will display a motion in a set direction, releasing energy and creating a net force. This intramolecular reorientation is limited again by the random motion of atoms, so it is reasonable to expect this motion to be at the same absolute rate as a photochemical step. It only seems slower due to the height of the barrier which enforces waiting times dependent on temperature (atomic motion rates).

Given the fact that the relaxation times of glasses at this viscosity are much longer than these timescales, we can assume the solvent cavity to be motionless during the half rotation of one half of the motor relative to the other. Viscoelastic properties will in that case be responsible for the resistance encountered by the upper and lower halves of the molecule moving in opposite rotational directions.

1 G. A. von Salis and H. Labhart, *J. Phys. Chem.*, 1968, **72**, 752.

Dr Titmuss said: In your presentation you repeatedly express concerns about the meaning of viscosity at the molecular level. You present an analysis that appears to be a hybrid of molecular thermodynamics/kinetics for the determination of the position of the photostationary state and continuum mechanics for the viscous dissipation of the molecular motor moving through a continuous fluid. I think that if you adopt a consistent viewpoint that is either thermodynamic/continuum or molecular and take care to properly consider what your 'system' is, then you will see that there is no mystery associated with viscosity. Viscous dissipation will break the time-reversal symmetry, yet you rely on microscopic reversibility to work out the position of the photostationary position. This is fine as long as you consider your microscopic system to be the molecular motor and all the degrees of freedom of the fluid. The fact that the microscopic Hamiltonian encompasses all the degrees of freedom of the fluid is why the macroscopic continuum description of viscosity is correct, as pointed out by Professor Jones and others in the discussion.

Dr Browne replied: My recollection of what I expressed repeatedly was concerns as to the level of experimental data available on viscosity in the low temperature glassy solvent regimes—as I said earlier to Dr Neto—the meaning of viscosity at a molecular level is highly dependent on the specific situation *i.e.* the thermodynamic system considered, as you pointed out. The two most complete reports available (as discussed in the paper) employ two distinct techniques to determine solvent viscosity: (i) the dropping rod and (ii) the rotating plates methods. Although the latter is generally perceived as being more reliable, we do not find a correlation between our experimental data and the data on viscosity determined by this method. By contrast an almost perfect relation is obtained with data obtained by the dropping rod method. The photoequilibrium under conditions of equal temperature but different solution viscosities is the empirical data employed in the study. The analytical determination of the photostationary state is by standard spectroscopic techniques. The overall aim of the study is to demonstrate the effect of viscosity on photoequilibria and to rationalise the origin of the effect in terms of differences in solvent excluded volume between the two molecular states. The system is a thermodynamic one and not kinetic although one would wonder where the boundary lies. The model, which was employed to attempt to understand the magnitude of the effect of differences in solvent excluded volume, assumes a continuum—it also assumes a spherical expansion. Whether these assumptions are justifiable is an open question; however, it does allow for a ball park approximation of the energies involved in expansion. With respect to microscopic reversibility, the transition state for the photoisomerisation (the so-called phantom state) is considered to be the same for the forward and reverse photoreactions. The "mystery"—this was inferred and not implied—or more appropriately the inconsistency with respect to viscosity is that two macroscopic methods of measurement give two different results, one of which correlates well with a molecular property and the other does not. As I stated repeatedly, such a correlation does not imply a true connection—as you noted correctly, a relation can also be a consequence of a circular argument. In this case, however, I do not see where such a circular argument would arise, although I suppose the fact that the position of the photoequilibrium is a statistical parameter could suggest that it reflects a bulk consideration of viscosity and not a molecular one. As noted by Professor Jones, glasses are inherently inhomogenous systems.

The goal of the paper was to understand why the photoequilibrium shifted in glassy media and to attempt to relate this to a physical phemonenon *i.e.* viscosity, solvent excluded volume. What we found is that the determination of viscosity is, in our view, not a done deal in highly viscous media. In my talk I highlighted that the synthetic variations which allow us to make unidirectional light driven molecular motors also allows us to take well understood paradigm systems such as stilbene and change their fundamental properties in sufficiently subtle ways that we can justifiably make comparisons. Your point regarding the breaking of time-reversal symmetry may be of considerable importance in understanding molecular motor function, especially where a 'ratcheted' Brownian motion is required to achieve net movment.

Professor Sen continued the discussion of the paper by Professor Aida: Can you carry out redox switching of the system? Can the system be used to sense small molecules that may bind to the metal?

Professor Aida replied: Although we have not tried yet, redox switching is possible in principle. As for sensory applications, we already have some preliminary results to suggest that our materials can sense small molecules. Nevertheless, for practical applications, one of the big challenges is how to realize a homeotropic alignment of the hexagonal columns upon spin coating to fabricate mesostructured composite films.

Dr Steinke asked: How could the columnar phase system using aurophilic interactions be integrated into a macroscopic self-healing system? As there are different levels of assembly and disassembly would the use of a columnar phase approach have benefits over other self-healing system such as that described by Hayes *et al.* in paper 16?

Professor Aida replied: Macroscopic self-healing, as reported by Hayes *et al.* is indeed important, particularly in a practical sense. However, the target of our research is not to realize self-healing at the macroscopic scale but at the molecular level, where a different type of challenge may exist. Silica-supported luminescent materials are potentially important for sensory and display applications. For use under harsh conditions, *e.g.* at high temperatures, such luminescent materials are required to be structurally robust or self-mendable from thermal damages. We highlight that the mesoporous silica framework allows the included organic material to possess both thermal robustness and self-healing capability.

Professor Chau continued the discussion of the paper by Professor Dr van Esch: Is it possible to control the number of fibers within each bundle and the diameter of the bundle?

Professor Dr van Esch responded: There are a number of examples known in the literature of molecular systems which self-assemble into fibers with defined diameter and discrete number of fibrils, but so far there is no recipe to design such systems. Most of the approaches relate back to the structure-shape concept of Israelachvili,[1] but some years ago, Ivan Huc and colleagues[2] presented some interesting ideas on the relation between chirality and fiber diameter in fibers formed by gemini surfactants. We also have our thoughts on this issue which I hopefully can share with you in the near future.

1 J. N. Israelachvili, S. Marcelja and R. G. Horn, *Q. Rev. Biophys.*, 1980, **13**, 121–200.
2 R. Oda, I. Huc, M. Schmutz, S. Candau and F. MacKintosh, *Nature*, 1999, **399**, 566–569.

Professor Huskens asked: Your orthogonal assemblies are still based on thermodynamic control, the absence of mutual influence on the molecular scale between the interaction motifs, and absence of energy dissipation. What do you see as the (first) options to make your systems truly more complex?

Professor Dr van Esch responded: Some of the systems discussed in our paper, for instance the vesicles with encapsulated fibers, are clearly metastable systems, but I agree with you that this condition does not yet make the systems complex, although they are the result of a kinetically controlled process. One possible first step towards complex self-assembling systems would be the development of a dissipative self-assembling system, *i.e.* a self-assembled state that is only stable by the continuous input of energy. The development of such a dissipative self-assembling system is already a great challenge and has not yet been accomplished, but it still would not yet lead to emergent properties, as complexity only arises in non-linear systems, which can be achieved by *e.g.* the introduction of autocatalytic formation and disassembly.

Dr Howse communicated to Dr Browne: On the topic of viscosity at the molecular level, have you looked at fluorescence depolarization? Here, a fluorescent probe molecule is excited in a single orientation using a linear polarized light source. As the excited probe molecule rotates it fluoresces. By monitoring the ratio of the emissions in the excitation orientation and at 90 degrees to this you are able to determine the tumbling speed of the probe molecule. This technique is also known as time-resolved fluorescence anisotropy. The viscosity is therefore linked into this tumbling speed through the Stokes–Einstein relationship. As an example of this technique see ref. 1.

1 M. K. Davies, A. L. Archer, A. O. S. Maczek, E. Manzanares-Papayanopoulos and I. A. McLure, *J. Chem. Eng. Data*, 2006, **51**(5), 1502–1508 (DOI: 10.1021/je0502593).

Dr Browne communicated in reply: Indeed fluorescence anisotropy is an exceptionally powerful tool which, as I understand it, can even be carried out at the single molecule level. We have considered using this technique to examine the present systems in the glassy state. A further advantage is that it would give us direct access to viscosity *via* the Stokes–Einstein equation. A major disadvantage is that it is not a simple technique and requires that the tumbling of the lumiphore occurs at a rate similar to the observed radiative rate constant. This precludes the application of a single chromophore to a wide range of solvent viscosities. Having said that, we are actually interested in looking at the effect of solvent viscosity on the position of the photoequilibrium. The observation of a correlation of PSS with bulk viscosity does suggest that we could use the PSS to determine solvent viscosity. The advantage is that the technique is steady state and not time resolved, thus yielding considerable practical benefits. Nevertheless, as with fluorescence anisotropy the range of viscosities that can be determined is limited by the ability to detect differences in PSS.

Soft nanotechnology: "structure" vs. "function"

George M. Whitesides* and Darren J. Lipomi

Received 25th August 2009, Accepted 25th August 2009
First published as an Advance Article on the web 1st September 2009
DOI: 10.1039/b917540g

This paper offers a perspective on "soft nanotechnology"; that is, the branch of nanotechnology concerned with the synthesis and properties of organic and organometallic nanostructures, and with nanofabrication using techniques in which soft components play key roles. It begins with a brief history of soft nanotechnology. This history has followed a path involving a gradual shift from the promise of revolutionary electronics, nanorobotics, and other futuristic concepts, to the realization of evolutionary improvements in the technology for current challenges in information technology, medicine, and sustainability. Soft nanoscience is an area that is occupied principally by chemists, and is in many ways indistinguishable from "nanochemistry". The paper identifies the natural tendency of its practitioners—exemplified by the speakers at this Faraday Discussion—to focus on synthesis and structure, rather than on function and application, of nanostructures. Soft nanotechnology has the potential to apply to a wide variety of large-scale applied (information technology, healthcare cost reduction, sustainability, energy) and fundamental (molecular biochemistry, cell biology, charge transport in organic matter) problems.

1. Introduction

1.1 What is "soft" nanoscience?

"Soft" is a word famously introduced into science by Pierre-Gilles de Gennes to refer to organic matter.[1] He was concerned primarily with the physics of polymers; nanotechnology can write its own definition. For the sake of this Discussion, we define it both in terms of what it is *not*—that is, *not* the nanoscience of "hard" materials such as metals, ceramics, or inorganic semiconductors; *e.g.*, the materials of primary concern in classical condensed matter physics—and in terms of what it *is*—the nanoscience of organic and organometallic matter, including molecules and structures in biology.

Within this definition, however, all nanostructures (hard and soft) share some characteristics: (i) some critical smallest dimension (*e.g.*, ≤100 nm); (ii) dimensions small enough to be "all—or mostly—surface"; (iii) the possible emergence of quantum properties at temperatures approaching room temperature; and (iv) small enough in number of particles to show non-Boltzmann statistics. The dimensions of nanostructures place soft nanoscience at the border between the science of large molecules and molecular aggregates (*e.g.*, polymers and vesicles) and small microfabricated structures such as the nanometer-scale wires in microprocessors, or biologically derived structures such as virus particles or the ribosome.

Harvard University, Department of Chemistry and Chemical Biology, 12 Oxford Street, Cambridge, MA, 02138, USA. E-mail: gwhitesides@gmwgroup.harvard.edu

1.2 History

The history of "soft" nanoscience in some ways recapitulates the history of nanoscience. In very general terms, all new technologies seem to follow a similar course. There is an initial period of exaggerated expectation (expectation usually unconstrained by experimental experience and reality), a following phase of disappointment as reality creeps in, and then a phase of growing focus on the areas in which the technology has immediate potential (Fig. 1). Both nanoscience and soft nanoscience have now passed through the first two periods, and have entered adolescence. Both are clearly interesting and important new fields of science and technology, and broad areas of application are emerging or have developed.[2] Still, much of the practical impact of the field of nanoscience has so far come in one area—information technology—and that impact is due to the extension of existing technologies by brilliant engineering, rather than from the introduction of radically new technologies.

The original expectations for nanoscience—including soft nanoscience—included revolutionary electronics (including the often controversial[3–9] topic of devices based on transport in single organic molecules,[10,11] ultradense microprocessors and memories, and quantum computers[12]); futuristic speculations concerning so-called "nanobots" and nanoscale machines; revolutionary materials with extraordinary applications (*e.g.*, buckytubes and the "space elevator",[13] or quantum dots and cancer-targeted drugs[14]); and applications in biomedicine relying on particles small on the scale of the cell.[15]

What has emerged is not a completely distinct list, but a list with substantial overlap and a completely distinct flavor.[16] At the top of the list is information technology and nanoelectronics. Developments in "conventional" microfabrication—shortwavelength light sources and immersion optics—have moved microfabrication to design rules in the deep nanoscale region (currently 45 nm);[17] extreme ultraviolet lithography,[18] double patterning lithography,[19] and multiple electron-beam (maskless) writing[20] have the potential to push the limit even further.[21] This technology is, of course, remaking the world through the internet and globalization. There is an enormous range of opportunities for nanoscience in technologies concerned with the production of energy and with global stewardship;[22] these range from heterogeneous catalysis to improved membranes for separation of water and gases.[23] The intuition that soft nanotechnology *must* be important in biology—based on the fact that the cell is micron-size in scale—seems inevitably to be correct,[24] although

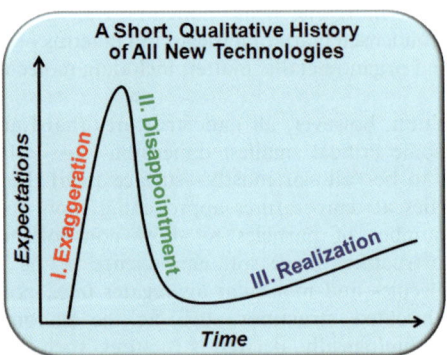

Fig. 1 The expectations of a new technology as a function of time. (I) In the beginning, there is a period of exaggerated expectations, during which exciting—but sometimes irreproducible—results and unrealistic claims are made. (II) When these high expectations go unmet, a period of disappointment sets in. (III) There is then a return to the fundamental aspects of the technology; science is linked with applications; new tools are developed; and real commercial investment begins.

the details of the applications in which nanoscience plays a role in fundamental biology remains to be established. The physical chemistry of systems containing small numbers of molecules or particles, and having most of these particles on or close to a surface, is just beginning to be explored.[25] And the quantum phenomena that emerge in small systems is very much *terra incognito*. So, the list of opportunities for the emerging and maturing field of soft nanoscience is long.

2. Focus of the Faraday Discussion

2.1 Nanochemistry and soft nanochemistry: *structure* and *function*

One of the characteristics of soft nanoscience is that it has developed in the hands of scientists with backgrounds in organic and organometallic chemistry; this group is particularly inclined philosophically to synthesis. As a consequence, the field has seen an explosion of work—often derived from prior themes in molecular synthesis—that has generated a broad range of new types of nanostructures: colloids, vesicles, polymers, molecular aggregates, self-assembled monolayers (SAMs),[26] and other small structures.[27] A characteristic of this work has been its focus on the *structure* rather than on *function*.

Historically, among the sciences, chemistry has been uniquely expert in synthesis of molecular and supramolecular structures, but less concerned with the functions of the structures that were synthesized. The enormous competence in organic synthesis that has emerged in the last century has made it possible to synthesize almost any small molecule. The justification for this work has often been its utility in pharmaceutical chemistry, although most of the work in that difficult and important field is done by biologists, pharmacologists, physiologists, and doctors, rather than by chemists. Similarly, the important field of polymer science has flourished because polymers are an important class of materials; chemists synthesized polymers, but others developed the applications for them. Soft nanochemistry is in a similar state. Chemistry is developing an exciting array of new synthetic techniques, but the field is still at the stage of "a solution chasing a problem". It is evident that one can make a wide variety of previously unmakable nanoscale structures.[27] The questions now are "Who cares?" and "What are they good for?".

2.2 Research interests of the attendees of this Faraday Discussion

This Faraday Discussion showed a broad spectrum of work, representative—in the usual non-representative way characteristic of a well-assembled conference—of the work going on in soft nanoscience. The range of topics offered a broad view of the topics being actively considered by the field. We classify these into eight groups; this classification certainly is not a uniquely correct sorting, but it gives a crude sense for what this group thought to be sufficiently important in soft nanoscience that it was worth pursuing.

(i) **Biology.** Jones[28] opened the conference around this theme, and a number of others reinforced his message concerning the almost unlimited opportunities offered by biology to nanoscience.

(ii) **Nanoactuation, nanomotion, and nanomechanics.** Sen[29] (redox motors), Sushko[30] (cantilevers and SAMs), and Schulten[31] (nanopores) focused on this subject (one that represents a *new* direction in chemistry).

(iii) **Vesicles.** Parnell[32] (encapsulation) and Matile[33] (polyarginine and anions) demonstrated some of the new possibilities that derive from one of the oldest fields in chemistry.

(iv) **Molecular recognition** was a theme that entered a number of discussions, with work by Matile,[33] Colquhoun[34] (non-biological sequences), and Mao[35] (DNA-based structural assembly) being prominent.

(v) **Polymers**—as expected for molecules at the border between "molecular" and "nano"—were components of many discussions, including those of Piñol[36] (structures formed from block copolymers), Hayes[37] (self-healing structures) and Ikkala[38] (fibrils).

(vi) **Synthesis** of molecules and aggregates showed the continuing skill of chemists in connecting atoms—both covalently and non-covalently—in new ways. Discussions including an important component of synthetic methods were those of Huskens[39] (templated molecular recognition), Mao[35] (DNA-based synthesis), Ikkala[38] (hierarchical self-assembly), Woolfson[40] (synthesis exploiting alpha-helices), Aida[41] (templated self-assembly), and van Esch[42] (self-assembly based on orthogonal interactions).

(vii) **Properties.** Chan[43] (hydrophobicity of surfaces), Hayes[37] (self-healing systems), Vendruscolo[44] (solubilities), Ulijn[45] (aggregation), and Aida[41] (self-healing) were all concerned with various aspects of the properties of soft-nanostructures.

(viii) **Interfaces.** Kornyshev[46] described his work on soft, charged interfaces.

Another conference, and another group of scientists, would certainly have provided a different distribution of interests among the attendees. Nonetheless, the focus in this group was clearly on large molecules, molecular aggregates and small particles, and on their synthesis and properties, rather than on the identification and exploitation of problems whose solutions might require such structures. It was, thus, a structure-focused discussion, rather than a function-focused discussion.

3. Expectations of soft nanotechnology

3.1 What does the rest of the world expect from nanotechnology? Does *soft* nanotechnology have a special role?

Technology proceeds both forward and backward: that is, from science and knowledge forward to applications, and from problems backward to technology for their solutions. There are many areas in which the capabilities offered by nanotechnology and the problems important to society and technology are sufficiently close that there is a clear opportunity for nanoscience to form the basis for nanotechnology, and nanotechnology to contribute to the solution of large-scale societal problems. Some of these will be particularly appropriate for soft, chemistry-based nanotechnology; others undoubtedly will be best treated by hard, physics-based nanotechnology.

Electronics. The extension of "conventional" microfabrication based on silicon and photolithography into the nanoscale region has been startlingly rapid, and has moved to sizes and complexities that were unimaginable twenty years ago.[21] When nanoscience emerged as a separate discipline, the perceived difficulties of making very small electronic systems were one of the sets of stimuli for its development, and the fabrication of microprocessors, computer memory, and other components of information systems was considered to be a possible area of application for *radical* nanotechnology, including soft forms of nanotechnology (*e.g.*, transistors—or even more exotic devices—based on single organic molecules).[5,47] It now seems unlikely that electronics and information technology will be an area in which radical soft nanotechnology will play an important role; technology based on extensions of the existing hard methods are so advanced, and the scale of investment required to develop new methods where they can be accepted in manufacturing so enormous, that it is improbable that it will make economic sense to develop an entire new technology for one or two generations of microprocessors. It is, however, always possible—in fact, probable—that soft nanotechnology and soft device physics will discover devices, processes, or functions that cannot be duplicated by hard methods. The use of step-and-flash imprint lithography, developed by Willson;[48] phase-shifting lithography using elastomeric masks;[49] and chemical-mechanical polishing[50] are three examples. There will almost certainly be others in the future.[51] Nonetheless, the theme of continuing development of classical methods of microfabrication—with chemical innovation, perhaps, in photoresists,[52] fluids with high indices of refraction,[17] and other chemical processes required by these methods, and especially now in technologies which lower costs—will continue, and, barring some presently unforeseen innovation, information technology will not represent the opportunity for radical nanotechnology that it once seemed.[21]

Biomedicine. If there *is* an important nanotechnological component to biomedicine, it seems almost inevitable that it will involve soft components.[53–56] The cell, and everything in it, are soft materials: that is, organic molecules in complex, functional configurations.[57] So far, however, the clear applications of nanostructures in biology and biomedicine have been surprisingly limited. One area that has demonstrated substantial promise is imaging. Quantum dots (for fluorescence microscopy at the cellular level),[58] and superparamagnetic colloids (for MRI contrast enhancement at the scale of localized tissues and organelles) have already demonstrated value;[59] other applications will certainly emerge. A second area where nanoscience and nanotechnology have already made important contributions is high-throughput analysis and array-based screening.[60,61] Although the components in these systems are now usually microscale—certainly larger than nanoscale as commonly defined—the future may well hold nanoscale components for biomedical assays,[62] and in any event the surface chemistry and surface functionalization that are required in the systems even now plausibly fall in the area of soft nanotechnology.[63] A third area with significant potential is in sensing based on either localized surface plasmonic effects in metallic nanostructures[64,65] or electrochemical gating of chemically prepared nanowires.[66]

New materials. Chemistry has always been uniquely skilled at synthesizing or generating new molecules, aggregates of molecules, and materials. There is every reason to think that the current high level of research activity in soft nanotechnology and soft nanomaterials science will produce new or improved systems. Among the successes of soft nanotechnology in materials science to date have been SAMs,[26] buckytubes,[67,68] quantum dots,[69] systems for chemical vapor deposition and atomic layer deposition,[70] methods for controlled synthesis of colloids,[71] and many others. This area is one in which chemistry will continue to excel, and in which the most important question is not "Can it be made?" but "What is the application?"

Energy. Nanotechnology is already ubiquitous in the production and use of energy.[72,73] Essentially all liquid fossil fuels are processed into useful forms by catalytic conversion over heterogeneous catalysts whose areas of catalytic activity are nanoscale.[74] Separations through nanoscale structures (usually fabricated in polymers) are crucial throughout the technology of energy production, and involve molecule-selective diffusion through polymer-based membranes, with or without discrete pores.[75] Thin-film solar photovoltaic cells—those based on amorphous silicon, CdTe, and other inorganic and organic semiconductors—rely on processes of charge separation and collection on the length scale of 5–100 nm.[76,77] All of these areas already constitute important fields of chemical technology and condensed matter science, but the rapid development of nanoscience offers the opportunity to re-examine them, and to develop new activities, new functions, and new levels of control.

4. Big problems

4.1 Inevitabilities

Beyond these well-defined technical problems, there is a bigger and more diffuse set—perhaps a dozen at any given time—of problems broadly important to society that require solutions that combine technology, economics, politics, sociology, law, and policy. Soft nanotechnology is now becoming sufficiently mature that it is appropriate to examine these very large-scale societal needs to see if some aspect of "nano" might offer options or partial solutions. We give four examples for illustration.

Sustainability: energy, water, and the environment. The problem of sustainability is often phrased and focused around energy. Sustainability is a large collection of different problems and is, of course, much more complicated than simply finding new methods of *producing* energy. There are many potential approaches to energy production, which include many untapped sources of fossil fuel (provided that problems of global climate change allow their use). Nuclear power generation can, in principle, also be expanded. Alternative sources of energy—especially renewable sources such as wind and solar—may fill important niches.[78]

Global warming and greenhouse gases (CO_2, CH_4, NO_x, and others) pose limitations to the growth of some of these technologies, and non-technical issues (*e.g.*, limiting proliferation of nuclear weapons) might make us wish to limit others. And if global warming is anthropogenic (and possibly even if it isn't), our ability to generate energy from fossil fuels will probably be severely limited.

Conservation has barely been exploited,[79] but it too will have technological solutions. Fig. 2 illustrates the flows of energy from sources to uses in the USA, and emphasizes the opportunities for conservation. While engineering development of this complex system continues, nanoscience has the potential to invent new materials (*e.g.* for coatings, catalysts, friction- and corrosion-resistant surfaces, and membranes). There could be as much potential for technological innovation in saving energy as there is in the production of more "new energy". Telecommuting and video conferencing are examples of technological alternatives to transportation, which is the largest consumer of distributed energy.

There presently are no solutions (other than the practically untested option of underground or underwater carbon sequestration) for disposal of carbon dioxide.[80] Alternative approaches may involve conversion of carbon dioxide into other materials, and large-scale processes based on chemical conversions will probably require new heterogeneous catalysts.[81]

Production of usable water (for agriculture and drinking) may, in fact, be a substantially more pressing short-term (<50 year) problem globally than energy production.[82] The separations involved in increasing the quality of water depend on membranes, and the design and fabrication of dramatically improved membranes (and perhaps of materials for other separations technologies) will probably have an important component of soft nanotechnology.

Sustainability may also require global-scale technologies. One of the most interesting and disquieting is "geoengineering"—the intentional modification of global climate.[83,84] Some of the technologies for geoengineering depend on creating small particles to scatter light—either in the atmosphere, in the ocean, or on land—and on the consequent modification of earth's albedo. The role of aerosols, particulates, and bubbles are ubiquitous in considerations of the reflectivity of the earth.[85]

The need for understanding nanoscale materials in thinking about the range of problems in sustainability is thus almost limitless, but in no case is nanotechnology by itself a complete solution to a problem. Understanding the constraints on the problem, and thinking it through from a systems perspective, would do a great deal to increase the impact of research in nanotechnology intended to apply to sustainability.

Healthcare and cost control. The distribution of benefits in healthcare (to the poor minority in the developed world, and to majority in the developing world) will require, and is causing, a radical rethinking of medicine and healthcare.[86] Access, cost, and effectiveness are all important issues, and to change all three, in a way that benefits consumers of healthcare, requires a movement from the high-technology, high-cost, end-of-life focus of the developed world to simple, low-cost, public health measures that make priorities of prevention and early detection of disease.[83] Again, throughout a shift of this sort, there will be opportunities for

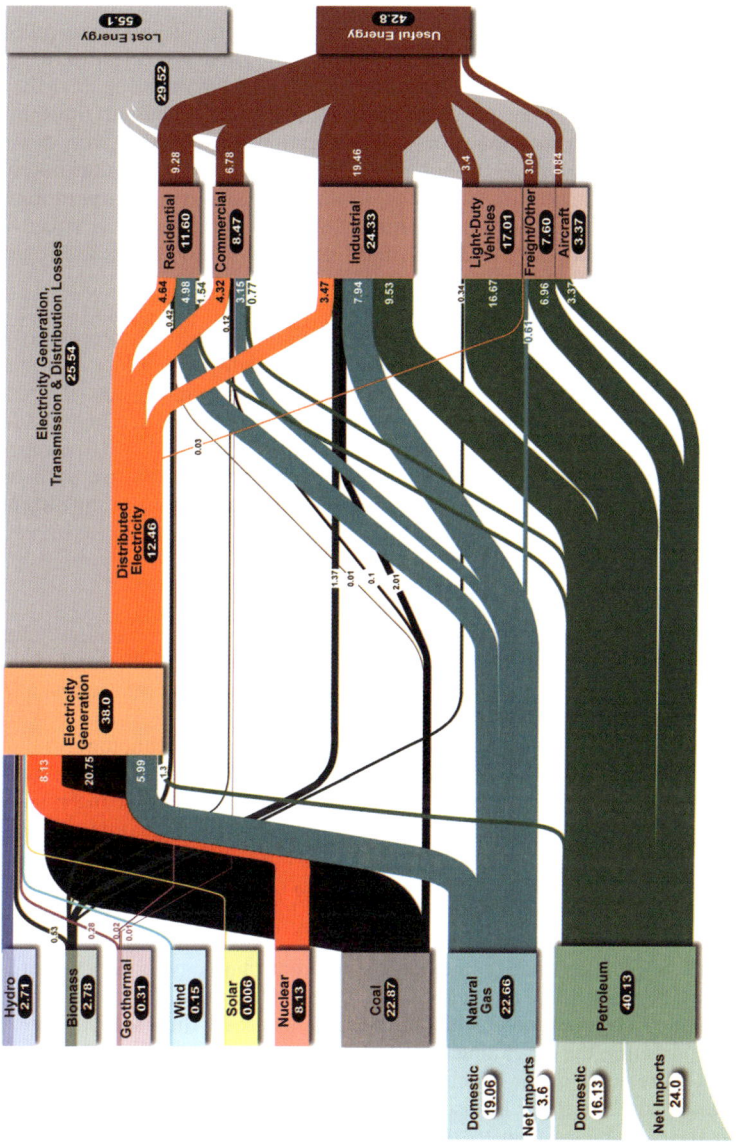

Fig. 2 Flowchart showing the origin and consumption of the energy produced in the US in 2005, including losses. The units are in quads; 1 quad = 10^{15} BTU = 1.055 EJ. Most of the energy is lost by transmission and conversion. The category "useful energy" conceals a large but incalculable fraction of energy that need not have been expended. (Source: Lawrence Livermore National Laboratory and US Department of Energy, 2005).

many new technologies, and no single one can make an enormous impact (although the availability of clean water and safe food, not smoking, exercise, weight control, and childhood vaccination provide a strong foundation).[87] For soft nanoscience, small systems and particles (for sensors, adjuvants, vaccines, systems for chronic healthcare monitoring, and others) have characteristics that suggest they could play an important role;[83] for nanotechnology in general, the marriage of information technology and medicine is probably the greatest opportunity.[88]

Megacities. In the future, much of the world's population will live in so-called "megacities": cities with populations of 20 to 40 million.[89] There is no clear technical strategy for managing populations of this magnitude, and the problems that they prose will be new. How does one provide clean water, clean air, safe food, transportation, personal security, education, work, and recreation for very large, very dense aggregates of people? We simply do not know, but it seems inevitable that micro- and nanoscience will be important parts of the services of such cities: for separations, inexpensive testing of water, food, and public health; as components of effective systems for public health; and as parts of many other technologies that are based in part upon small structures.

Information technology. We have offered the opinion that soft nanotechnology will probably not play a major part in radical change in high-performance information technology. There are, however, many other vitally important aspects of information technology where nanotechnology, and possibly soft nanotechnology, could plausibly play a role. These include consumer electronics for a wide variety of applications,[90] electronics for education, technologies related to healthcare such as implantable sensors,[91] very broadly distributed environmental sensors, intelligent buildings and cities for energy management,[92] and a range of others.

4.2 The next big thing?

Robotics. What will come after information technology and biotechnology? One possibility is robotics, which will, of course, have components from both information technology and biotechnology. The hardware of information technology (microprocessors, sensors, controllers) will unquestionably play a central role in the future development of robotic systems. Whether soft nanotechnology will be equally important will probably depend upon the emergence and ultimate importance of soft robotics.[93] Although most robotic systems are presently conceptually simple machines relying on sophisticated controllers, sensors, actuators, there is now great interest in developing robotic systems whose conceptual forbears are closer to squid or worms than to humans or donkeys.[92] These systems—with soft structures, and a requirement for flexibility—will benefit from soft electronics, and, we presume, from soft nanoscience.

The breadth of nanotechnology, including soft nanotechnology, is such that it is inevitable that these very large problems will involve solutions that incorporate some aspects of "nano". The question of whether chemistry can reach these solutions more efficiently by targeting the development of soft nanoscience to specific problems, or whether it is more efficient for chemists working in nanoscience to continue with relatively undirected programs in discovery, and for the engineers who work most directly on these problems then to reach back into nanoscience to find materials and components appropriate for their needs, remains to be seen.

5. Basic science: research problems for which soft nanoscience is uniquely suited

We have focused much of this summary and forecast on function and applications, in part because it seems that these areas are significantly less developed, and less of a focus, in the soft nanotechnology represented by this Faraday Discussion, than are exploratory academic studies focused on synthesis and properties. It is, however, important to understand that there are problems in fundamental science which will rely absolutely on understanding phenomenon that occur at the nanoscale for their solution. These problems constitute a long list, and include fundamental studies of heterogeneous catalysis, the nature of structures inside the cell, the role of nanoparticles in the environment, the relation between nanoscale structure and properties in materials, the nature and properties of nanoscale interfaces, options for actuation and sensing on the nanoscale, and synthetic techniques particularly appropriate to nanoscale science such as self-assembly and templated assembly. Simply for illustration, we briefly outline two problems in nanoscience that are especially important.

5.1 Molecular recognition and near-surface water

From a reductionist viewpoint, arguably the most important process in biology is molecular recognition. Everything—the formation of lipid membranes, the folding of proteins, the replication of DNA and reading of mRNA, the interaction of drugs with proteins, the activity of receptors and transporters, and so on—all rely on the recognition of one molecular surface by another. In the past, molecular recognition has been phrased as a scientific problem in terms of "lock and key". We now recognize that this formulation is fundamentally incorrect, in that it leaves out the medium, which is usually aqueous.[94,95] Understanding near-surface water (including, in particular, the water surrounding the functional surfaces of the drug, and the water filling the active site of its target protein) is perhaps the most important problem in biophysics.[96,97] The distances over which the phenomena involved extend, the interplay of enthalpy and entropy in these systems, the nature of interactions between hydrophobic surfaces and near-surface water, the interactions among electronically polarizable groups in an aqueous environment—all are nanoscale, soft, phenomenon. Soft nanoscience has an enormous opportunity to contribute to understanding the remarkable properties of liquid water within the first few nanometers of a molecular surface, and to do so—most importantly—using biological molecules and biological media.

5.2 Charge transport in organic matter

Photosynthesis supports almost all life on Earth. At the heart of photosynthesis is a series of energy and charge transport processes that convert the energy present in photons into reagents (ATP, NAD(P)(H)) and electrochemical gradients in photosynthetic plants, algae, and some bacteria.[98] Our understanding of charge transport in conductors and semiconductors is excellent; our corresponding understanding of charge transport in putative insulators (*e.g.*, within proteins and across membranes) is much less highly developed, both experimentally and theoretically.[99,100] Soft nanoscience offers the opportunity to study the fundamentals of charge transport across insulating systems relevant to photosynthesis, and to other areas of biology (*e.g.* oxidative metabolism)—and has begun to develop experimental systems relevant to these studies.[101,102]

6. Outlook: "nano" has moved; nano*chemistry* may need to adjust to keep up

This Faraday Discussion was successful in displaying a cross section of the interests and competencies of a central and distinguished group of soft nanoscientists, in laying out some of the opportunities in soft nanoscience, and in illustrating current activities of chemistry related to soft nanoscience and nanotechnology. It emphasized the skill of chemists in making things (whether on molecular or nanometer scale) and in studying chemical properties. It also betrayed a weakness of chemistry in thinking forward to function and applications in complex systems in which the chemistry of the starting materials was only a part of the problem. Synthesis, properties, function, applications—*all* are challenging to conceive and execute. Since chemistry is one of the central players in nanoscience, it could be very interesting, useful, important, and profitable for it to try to encompass the entire spectrum needed to go from fundamental understanding to societal benefit, rather than to leave the later links in this chain to other disciplines.

In the twenty-five years over which nanotechnology has grown as a recognizable subfield of science, its foci have shifted. Some of the original possibilities—radically new, small computers and machines—have been replaced by equally important but very different problems: focused extension or replacement of processes for fabricating silicon-based information processors, sustainability, energy production and conservation, imaging and genomics, and water. Ideally, soft (that is chemical) nanotechnology will become engaged in the most important of these problems, from their scientifically interesting beginning to their societally beneficial end.

References

1 P. G. de Gennes, *Soft Matter*, 2005, **1**, 16–16.
2 G. A. Ozin, A. C. Arsenault and L. Cademartiri, *Nanochemistry: A Chemical Approach to Nanomaterials*, Royal Society of Chemistry, London, 2009.
3 V. V. Zhirnov and R. K. Cavin, *Nat. Mater.*, 2006, **5**, 11–12.
4 H. Choi and C. C. M. Mody, *Soc. Stud. Sci.*, 2009, **39**, 11–50.
5 R. F. Service, *Science*, 2003, **302**, 556–558.
6 J. R. Heath, J. F. Stoddart and R. S. Williams, *Science*, 2004, **303**, 1136–1137.
7 E. A. Chandross, *Science*, 2004, **303**, 1137–1137.
8 P. S. Weiss, *Science*, 2004, **303**, 1137–1137.
9 R. F. Service, *Science*, 2004, **303**, 1137–1137.
10 C. A. Mirkin and M. A. Ratner, *Annu. Rev. Phys. Chem.*, 1992, **43**, 719–754.
11 *Molecular Electronics*, ed. J. Jortner and M. A. Ratner, Blackwell Science Ltd, Malden, MA, USA, 1997.
12 C. Day, *Phys. Today*, 2005, **58**, 21–23.
13 N. M. Pugno, *Nano Today*, 2007, **2**, 44–47.
14 K. K. Jain, *Drug Discovery Today*, 2005, **10**, 1435–1442.
15 S. E. A. Gratton, S. S. Williams, M. E. Napier, P. D. Pohlhaus, Z. L. Zhou, K. B. Wiles, B. W. Maynor, C. Shen, T. Olafsen, E. T. Samulski and J. M. Desimone, *Acc. Chem. Res.*, 2008, **41**, 1685–1695.
16 G. M. Whitesides, *Small*, 2005, **1**, 172–179.
17 J. Lopez-Gejo, J. T. Kunjappu, J. Zhou, B. W. Smith, P. Zimmerman, W. Conley and N. J. Turro, *Chem. Mater.*, 2007, **19**, 3641–3647.
18 B. Santo, *IEEE Spectrum*, 2007, **44**, 12–14.
19 C. A. Mack, *IEEE Spectrum*, 2008, **45**, 46–51.
20 R. F. Pease and S. Y. Chou, *Proc. IEEE*, 2008, **96**, 248–270.
21 C. G. Willson and B. J. Roman, *ACS Nano*, 2008, **2**, 1323–1328.
22 *Nanotechnology for the Energy Challenge*, ed. J. Garcia-Martinez, Wiley-VCH, in press.
23 W. Liu, D. King, J. Liu, B. Johnson, Y. Wang and Z. G. Yang, *JOM*, 2009, **61**, 36–44.
24 M. De, P. S. Ghosh and V. M. Rotello, *Adv. Mater.*, 2008, **20**, 4225–4241.
25 R. S. Berry, *Nature*, 1998, **393**, 212–213.
26 J. C. Love, L. A. Estroff, J. K. Kriebel, R. G. Nuzzo and G. M. Whitesides, *Chem. Rev.*, 2005, **105**, 1103–1169.
27 C. Burda, X. B. Chen, R. Narayanan and M. A. El-Sayed, *Chem. Rev.*, 2005, **105**, 1025–1102.
28 R. A. L. Jones, *Faraday Discuss.*, 2009, **143**, DOI: 10.1039/b916271m.
29 A. Sen, M. Ibele, Y. Hong and D. Velegol, *Faraday Discuss.*, 2009, **143**, DOI: 10.1039/b900971j.
30 M. L. Sushko, *Faraday Discuss.*, 2009, **143**, DOI: 10.1039/b900861f.
31 E. R. Cruz-Chu, T. Ritz, Z. S. Siwy and K. Schulten, *Faraday Discuss.*, 2009, **143**, DOI: 10.1039/b906279n.
32 A. J. Parnell, N. Tzokova, P. D. Topham, D. J. Adams, S. Adams, C. M. Fernyhough, A. J. Ryan and R. A. L. Jones, *Faraday Discuss.*, 2009, **143**, DOI: 10.1039/b902574j.
33 T. Takeuchi, N. Sakai and S. Matile, *Faraday Discuss.*, 2009, **143**, DOI: 10.1039/b900133f.

34 H. M. Colquhoun, Z. Zhu, C. J. Cardin, M. G. B. Drew and Y. Gan, *Faraday Discuss.*, 2009, **143**, DOI: 10.1039/b900684b.
35 C. Zhang, Y. He, M. Su, S. H. Ko, T. Ye, Y. Leng, X. Sun, A. E. Ribbe, W. Jiang and C. Mao, *Faraday Discuss.*, 2009, **143**, DOI: 10.1039/b905313c.
36 B. Xu, R. Piñol, M. Nono-Djamen, S. Pensec, P. Keller, P.-A. Albouy, D. Lévy and M.-H. Li, *Faraday Discuss.*, 2009, **143**, DOI: 10.1039/b902003a.
37 S. Burattini, H. M. Colquhoun, B. W. Greenland and W. Hayes, *Faraday Discuss.*, 2009, **143**, DOI: 10.1039/b900859d.
38 O. Ikkala, R. H. A. Ras, N. Houbenov, J. Ruokolainen, M. Pääkkö, J. Laine, M. Leskelä, L. Berglund, T. Lindström, G. ten Brinke, H. Iatrou, N. Hadjichristidis and C. F. J. Faul, *Faraday Discuss.*, 2009, **143**, DOI: 10.1039/b905204f.
39 X. Y. Ling, I. Y. Phang, D. N. Reinhoudt, G. J. Vancso and J. Huskens, *Faraday Discuss.*, 2009, **143**, DOI: 10.1039/b822156a.
40 C. T. Armstrong, A. L. Boyle, E. H. C. Bromley, Z. N. Mahmoud, L. Smith, A. R. Thomson and D. N. Woolfson, *Faraday Discuss.*, 2009, **143**, DOI: 10.1039/b901610d.
41 W. Otani, K. Kinbara and T. Aida, *Faraday Discuss.*, 2009, **143**, DOI: 10.1039/b904896k.
42 A. M. Brizard, M. C. A. Stuart and J. H. van Esch, *Faraday Discuss.*, 2009, **143**, DOI: 10.1039/b903806j.
43 D. Y. C. Chan, Md. H. Uddin, K. L. Cho, I. I. Liaw, R. N. Lamb, G. W. Stevens, F. Grieser and R. R. Dagastine, *Faraday Discuss.*, 2009, **143**, DOI: 10.1039/b901134j.
44 M. Vendruscolo and C. M. Dobson, *Faraday Discuss.*, 2009, **143**, DOI: 10.1039/b905825g.
45 A. K. Das, A. R. Hirst and R. V. Ulijn, *Faraday Discuss.*, 2009, **143**, DOI: 10.1039/b902065a.
46 M. E. Flatté, A. A. Kornyshev and M. Urbakh, *Faraday Discuss.*, 2009, **143**, DOI: 10.1039/b901253m.
47 A. K. Feldman, M. L. Steigerwald, X. F. Guo and C. Nuckolls, *Acc. Chem. Res.*, 2008, **41**, 1731–1741.
48 C. G. Willson, *J. Photopolym. Sci. Technol.*, 2009, **22**, 147–153.
49 D. J. Shir, S. Jeon, H. Liao, M. Highland, D. G. Cahill, M. F. Su, I. F. El-Kady, C. G. Christodoulou, G. R. Bogart, A. V. Hamza and J. A. Rogers, *J. Phys. Chem. B*, 2007, **111**, 12945–12958.
50 P. B. Zantye, A. Kumar and A. K. Sikder, *Mater. Sci. Eng., R*, 2004, **45**, 89–220.
51 B. D. Gates, Q. B. Xu, M. Stewart, D. Ryan, C. G. Willson and G. M. Whitesides, *Chem. Rev.*, 2005, **105**, 1171–1196.
52 H. Ito, *J. Photopolym. Sci. Technol.*, 2008, **21**, 475–491.
53 G. M. Whitesides and A. P. Wong, *MRS Bull.*, 2006, **31**, 19–27.
54 R. Langer and D. A. Tirrell, *Nature*, 2004, **428**, 487–492.
55 G. M. Whitesides, *Nat. Biotechnol.*, 2003, **21**, 1161–1165.
56 D. S. Goodsell, *Bionanotechnology: Lessons from Nature*, Wiley-Liss, Hoboken, NJ, 2004.
57 G. Colombo, P. Soto and E. Gazit, *Trends Biotechnol.*, 2007, **25**, 211–218.
58 I. L. Medintz and H. Mattoussi, *Phys. Chem. Chem. Phys.*, 2009, **11**, 17–45.
59 H. B. Na, I. C. Song and T. Hyeon, *Adv. Mater.*, 2009, **21**, 2133–2148.
60 J. Hong, J. B. Edel and A. J. deMello, *Drug Discovery Today*, 2009, **14**, 134–146.
61 R. J. Marinelli, K. Montgomery, C. L. Liu, N. H. Shah, W. Prapong, M. Nitzberg, Z. K. Zachariah, G. J. Sherlock, Y. Natkunam, R. B. West, M. van de Rijn, P. O. Brown and C. A. Ball, *Nucleic Acids Res.*, 2008, **36**, D871–D877.
62 K. B. Lee, E. Y. Kim, C. A. Mirkin and S. M. Wolinsky, *Nano Lett.*, 2004, **4**, 1869–1872.
63 N. L. Rosi and C. A. Mirkin, *Chem. Rev.*, 2005, **105**, 1547–1562.
64 M. E. Stewart, C. R. Anderton, L. B. Thompson, J. Maria, S. K. Gray, J. A. Rogers and R. G. Nuzzo, *Chem. Rev.*, 2008, **108**, 494–521.
65 K. Kneipp, H. Kneipp, I. Itzkan, R. R. Dasari and M. S. Feld, *J. Phys.: Condens. Matter*, 2002, **14**, R597–R624.
66 A. K. Wanekaya, W. Chen, N. V. Myung and A. Mulchandani, *Electroanalysis*, 2006, **18**, 533–550.
67 P. Avouris, *Phys. Today*, 2009, **62**, 34–40.
68 Q. Cao and J. A. Rogers, *Adv. Mater.*, 2009, **21**, 29–53.
69 C. B. Murray, C. R. Kagan and M. G. Bawendi, *Annu. Rev. Mater. Sci.*, 2000, **30**, 545–610.
70 H. Kim, H. B. R. Lee and W. J. Maeng, *Thin Solid Films*, 2009, **517**, 2563–2580.
71 Y. Xia, Y. J. Xiong, B. Lim and S. E. Skrabalak, *Angew. Chem., Int. Ed.*, 2009, **48**, 60–103.
72 G. M. Whitesides and G. W. Crabtree, *Science*, 2007, **315**, 796–798.
73 E. D. Williams, R. U. Ayres and M. Heller, *Environ. Sci. Technol.*, 2002, **36**, 5504–5510.
74 M. Zach, C. Hagglund, D. Chakarov and B. Kasemo, *Curr. Opin. Solid State Mater. Sci.*, 2006, **10**, 132–143.
75 N. B. McKeown and P. M. Budd, *Chem. Soc. Rev.*, 2006, **35**, 675–683.

76 G. W. Crabtree and N. S. Lewis, *Phys. Today*, 2007, **60**, 37–42.
77 B. C. Thompson and J. M. J. Frechet, *Angew. Chem., Int. Ed.*, 2008, **47**, 58–77.
78 N. S. Lewis and D. G. Nocera, *Proc. Natl. Acad. Sci. U. S. A.*, 2007, **104**, 20142–20142.
79 S. Kasahara, V. Paltsev, J. Reilly, H. Jacoby and A. D. Ellerman, *Environ. Resour. Econ.*, 2007, **37**, 377–410.
80 R. Lal, *Energy Environ. Sci.*, 2008, **1**, 86–100.
81 H. Arakawa, M. Aresta, J. N. Armor, M. A. Barteau, E. J. Beckman, A. T. Bell, J. E. Bercaw, C. Creutz, E. Dinjus, D. A. Dixon, K. Domen, D. L. DuBois, J. Eckert, E. Fujita, D. H. Gibson, W. A. Goddard, D. W. Goodman, J. Keller, G. J. Kubas, H. H. Kung, J. E. Lyons, L. E. Manzer, T. J. Marks, K. Morokuma, K. M. Nicholas, R. Periana, L. Que, J. Rostrup-Nielson, W. M. H. Sachtler, L. D. Schmidt, A. Sen, G. A. Somorjai, P. C. Stair, B. R. Stults and W. Tumas, *Chem. Rev.*, 2001, **101**, 953–996.
82 Royal Society of Chemistry, Sustainable Water: Chemical Science Priorities; Introduction to Report, http://www.rsc.org/water, accessed on 22 August 2009.
83 C. D. Chin, V. Linder and S. K. Sia, *Lab Chip*, 2007, **7**, 41–57.
84 D. W. Keith, *Annu. Rev. Energy Environ.*, 2000, **25**, 245–284.
85 U. Poschl, *Angew. Chem., Int. Ed.*, 2005, **44**, 7520–7540.
86 F. Gotch and J. Gilmour, *Nat. Immunol.*, 2007, **8**, 1273–1276.
87 M. G. Goldstein, E. P. Whitlock and J. DePue, *Am. J. Prev. Med.*, 2004, **27**, 61–79.
88 D. Blumenthal and J. P. Glaser, *N. Engl. J. Med.*, 2007, **356**, 2527–2534.
89 B. R. Gurjar and J. Lelieveld, *Atmos. Environ.*, 2005, **39**, 391–393.
90 B. D. Gates, *Science*, 2009, **323**, 1566–1567.
91 M. Frost and M. E. Meyerhoff, *Anal. Chem.*, 2006, **78**, 7370–7377.
92 J. E. Fernandez, *Science*, 2007, **315**, 1807–1810.
93 R. Pfeifer, M. Lungarella and F. Iida, *Science*, 2007, **318**, 1088–1093.
94 G. M. Whitesides, P. W. Snyder, D. T. Moustakas, K. A. Mirica, in *Physical Biology: From Atoms to Medicine*, ed. A. H. Zewail, Imperial College Press, London, 2008, pp. 189–215.
95 N. T. Southall, K. A. Dill and A. D. J. Haymet, *J. Phys. Chem. B*, 2002, **106**, 521–533.
96 D. Chandler, *Nature*, 2005, **437**, 640–647.
97 G. M. Whitesides and V. M. Krishnamurthy, *Q. Rev. Biophys.*, 2005, **38**, 385–395.
98 H. B. Gray and J. R. Winkler, *Proc. Natl. Acad. Sci. U. S. A.*, 2005, **102**, 3534–3539.
99 N. Tessler, Y. Preezant, N. Rappaport and Y. Roichman, *Adv. Mater.*, 2009, **21**, 2741–2761.
100 V. Coropceanu, J. Cornil, D. A. da Silva, Y. Olivier, R. Silbey and J. L. Bredas, *Chem. Rev.*, 2007, **107**, 926–952.
101 E. A. Weiss, J. K. Kriebel, M. A. Rampi and G. M. Whitesides, *Philos. Trans. R. Soc. London, Ser. A*, 2007, **365**, 1509–1537.
102 E. A. Weiss, R. C. Chiechi, G. K. Kaufman, J. K. Kriebel, Z. F. Li, M. Duati, M. A. Rampi and G. M. Whitesides, *J. Am. Chem. Soc.*, 2007, **129**, 4336–4349.

Poster titles

Design and synthesis of new nanocarriers for molecular imaging, **S. I. Pascu, R. M. J. Jacobs, Z. Hu, K. Jurkschat and A. Crossley**, *University of Bath, UK*

Controllable synthesis of conducting polymer nanostructures, **Y. Yan, Y. Huang, J. Huang and Z. Wie**, *National Centre for Nanoscience and Technology, Beijing*

HarmoniX™: AFM nanoscale material property mapping, **I. Armstrong**, *Veeco, UK*

Chemistry in nanoconfinement: characterization and application of block copolymer vescicles as nanocontainers, **Q. Chen, G. W. de Groot, H. Schönherr and G. J. Vancso**, *University of Twente, The Netherlands*

Local thermal activation of tert-butyl acrylate based polymer films: toward AFM based nanoscale thermal chemical lithography, **J. Duvigneau, S. Cornelissen, N. Bardaji Valls, H. Schönherr and G. J. Vancso**, *University of Twente, The Netherlands*

Nanoimprint lithography of functional polymers, **J. E. Slota and W. T. S. Huck**, *University of Cambridge, UK*

Highly tunable opto-electronic network structures from enzymatic directed hydrogelation, **A. R. Hirst, H. Xu, A. K. Das and R. V. Ulijn**, *University of Manchester, UK*

Aggregation and jamming of polymersomes induced by poly(ethyleneoxide) homopolymer, **T. P. Smart, O. O. Mykhaylyk, A. J. Ryan, J. R. Howse and G. Battaglia**, *University of Sheffield, UK*

Chiral quantum dots, **Y. K. Gun'ko, M. M. Maloney, J. Govan and S. Gallagher**, *University of Dublin, ROI*

Plastic viruses, **M. Massignani, C. LoPresti, I. Canton, J. Madsen, A. Blanazs, A. Lewis, S. Armes and G. Battaglia**, *University of Sheffield, UK*

Towards metallic nano-structures:patterned self-assembled monolayers as reactivity templates, **S. Stuart-Cole, A. K. Kumar, S. A. Ahmed, E. U. Haq, R. J. Bushby, G. J. Leggett and S. D. Evans**, *University of Leeds, UK*

Computer simulations of the interaction of fullerene with lipid membranes, **L. Monticelli, E. Salonen, P.-C. Ke and I. Vattulainen**, *INSERM, France*

Formation of nanopatterned heterojunction in organic photovoltaic devices by nanoimprinting, **X. He, F. Gao, G. Tu, D. Hasko, S. Hünter, U. Steiner, N. C. Greenman, R. Friend and W. Huck**, *University of Cambridge, UK*

Protein-conjugated, glucose-sensitive surface using fluorescent dendrimer porphyrin, **Y. Lee, J. Kim, S. Kim, W.-D. Jang and W.-G. Koh**, *Yonsei University, South Korea*

Synthesis and physical properties of high ionic conduction nanoparticles capped with an organic polymer, **R. Makiura**, *Kyushu University, Japan*

Synthesis and structural study of the novel Schiff-base macrocycle: swelling phenomena of the capsule compound, **A. Harano, M. Irie, T. Nakagaki, K. Ideta, M. Annaka, K. Goto and T. Shinmyozu**, *Kyushu University, Japan*

Mesophase modified electrodes:sensing films composed of self-assembled nanostructures materials, **E. H. Doeven, C. F. Hogan, B. W. Muir and A. Polyzos**, *La Trobe University, Australia*

2D luminescent interfaces for electrochemical sensing strategies, **G. J. Barbante, C. F. Hogan and A. B. Hughes**, *La Trobe University, Australia*

Cucurbit[8]uril-mediated supramolecular polymerisations in water, **J. M. Zayed and O. A. Scherman**, *University of Cambridge, UK*

Polymer brush driven aggregation phenomena, **M. Moglianetti, H. Jia, S. Edmondson, S. P. Armes and S. Titmuss**, *University of Oxford, UK*

Design of substrates with tunable compliance for epidermal stem cell patterning, **B. Trappmann, J. Gautrot, J. Connelly, M. Oyen, F. Watt and W. Huck**, *University of Cambridge, UK*

Biofunctional switchable azobenzene glycerol conjugates on surfaces, **C. Kördel and R. Haag**, *Freie Universität Berlin, Germany*

Polymer mediated dispersion of gold nanoparticles: using supramolecular moieties on the periphery, **A. D. Celiz, T.-C. Lee and O. A. Scherman**, *University of Cambridge, UK*

Surface modification study of plasma polymer thin films for DNA fixation by atmospheric pressure plasma system, **S.-J. Cho, I.-S. Bae, H. J. Kim, B. Hong and J.-H. Boo**, *Sungkyunkwan University, Korea*

Micro-patterning of TiO_2 thin films by MOCVD and study of its growth tendency, **B.-C. Kang, S.-H. Nam, S.-J. Cho, D.-Y. Jung Y.-T. Kim and J.-H. Boo**, *Sungkyunkwan University, Korea*

DNA-sensitive polymer hydrogel/gold nanoparticles composite for the detection of genetic cancer markers: an electrochemical study, **S. R. Flannery, O. Y. F. Henry, S. Kirwan and C. K. O'Sullivan**, *Universitat Rovira I Virgili, Spain*

Mechanism of the self-assembly of fractal structures from dipeptide derivatives, **W. Wang and Y. Chau**, *The Hong Kong University of Science and Technology, Hong Kong*

Hollow vesicles self-assembled by short peptide derivatives, **W. Wang and Y. Chau**, *The Hong Kong University of Science and Technology, Hong Kong*

Peptide-modified block copolymer for the assembly of stimuli-responsive vesicles, **W. K. Yeung and Y. Chau**, *The Hong Kong University of Science and Technology, Hong Kong*

Morphological control of self-assembling nanostructures by reverse microemulsion, **W. Wang and Y. Chau**, *The Hong Kong University of Science and Technology, Hong Kong*

Fabrication of micro and nano patterned surfaces using an automated microcontact printing tool, **C. Sporer, M. Huber, M. Machtler, K. Dietrich, C. A. Mills, R. Buser, A. Bernard and J. Samitier**, *Institute for Bioengineering of Catalonia (IBEC), Spain*

Photo-induced switchable gold nanoparticles based on imine derivatives, **Y. Luo, S. Korchak, H.-M. Vieth and R. Haag**, *Freie Universität Berlin, Germany*

Synthesis and characterization of surface-initiated binary mixed polymer brushes of poly(methacrylic acid) and poly(N-isopropylacrylamide), **X. Sui, S. Zapotoczny, E. Benetti, M. A. Hempenius and G. J. Vancso**, *University of Twente, The Netherlands*

Two self-assembly approaches towards compartmentalization, **J. Boekhoven, A. Brizard, P. van Rijn, M. Stuart and J. van Esch**, *Technical University Delft, The Netherlands*

Cucurbit-[n]-uril metal nanoparticle composites, **T.-C. Lee and O. A. Scherman**, *University of Cambridge, UK*

A membrane anchored DNA-based energy/electron transfer assembly, **K. Börjesson, J. Tumpane, T. Ljungdahl, L. M. Wilhelmsson, B. Nordén, T. Brown, J. Mårtensson and B. Albinsson**, *Chalmers University of Technology, Sweden*

Thermodynamics-based coarse-grained model of polystyrene, **G. Rossi, L. Monticelli, S. R. Puisto, I. Vattulainen and T. Ala-Nissila**, *Helsinki University of Technology, Finland*

The formation mechanism of bicontinuous cubic mesoporous materials using triblock-copolymer and butanol as structure directing agents, **L. Omer, D. Goldfarb and Y. Talmon**, *Technion—Israel Institute of Technology, Israel*

The Skinner Prize for the best poster was awarded to Miss Marzia Massignani of the University of Sheffield, UK, for her poster on plastic viruses.

List of Participants

Mr F. Ahangar, *University of Teesside, United Kingdom*
Professor T. Aida, *University of Tokyo, Japan*
I. Armstrong, *Veeco Instruments Ltd, United Kingdom*
Mr A. Aufderhorst-Roberts, *University of Cambridge, United Kingdom*
Mr G. Barbante, *La Trobe University, Australia*
Mr D. Barbero, *University of Cambridge, United Kingdom*
Mr M. Bird, *University of Reading, United Kingdom*
Dr A. Bittner, *CIC nanoGUNE, Spain*
Mr J. Boekhoven, *Technical University Delft, The Netherlands*
Mr K. Börjesson, *Chalmers University of Technology, Sweden*
Dr W. Browne, *University of Gröningen, The Netherlands*
Mr S. Burattini, *University of Reading, United Kingdom*
Dr H. Burch, *Royal Society of Chemistry, United Kingdom*
Mr D. Carew, *University of Bristol, United Kingdom*
Dr J. Carr, *Veeco Instruments Ltd, United Kingdom*
Professor A. Cass, *Imperial College London, United Kingdom*
Mr A. Celiz, *University of Cambridge, United Kingdom*
Professor D. Chan, *University of Melbourne, Australia*
Dr K. Channon, *University of Bristol, United Kingdom*
Professor Y. Chau, *The Hong Kong University of Science and Technology, Hong Kong*
Dr M. Chen, *CSIRO, Australia*
Mr Q. Chen, *University of Twente, The Netherlands*
Mr S. Cho, *Sungkyunkwan University, South Korea*
Mr S. Chuangchote, *Kyoto University, Japan*
Professor D. Clary, *University of Oxford, United Kingdom*
Professor H. Colquhoun, *University of Reading, United Kingdom*
Dr P. Cooper, *Royal Society of Chemistry, United Kingdom*
Professor L. Cronin, *University of Glasgow, United Kingdom*
Mr A. Das, *University of Strathclyde, United Kingdom*
Mr E. Doeven, *La Trobe University, Australia*
Mr J. Duvigneau, *University of Twente, The Netherlands*
Dr R. Eelkema, *Technical University Delft, The Netherlands*
Mrs S. Ewen, *Imperial College London, United Kingdom*
Dr J. Fahrenkamp-Uppenbrink, *Science, United Kingdom*
Dr C. Faul, *University of Bristol, United Kingdom*
Professor A. Fery, *University of Bayreuth, Germany*
Dr C. Finlayson, *Optoelectronics Group, United Kingdom*
Ms M. Gilbert, *Royal Society of Chemistry, United Kingdom*
Mrs S. Godfrey, *Royal Society of Chemistry, United Kingdom*
A. Graham, *Scientific and Medical Products Ltd, United Kingdom*
Dr B. Greenland, *University of Reading, United Kingdom*
Professor Y. Gunko, *Trinity College Dublin, Republic of Ireland*
Miss A. Harano, *Kyushu University, Japan*
Dr J. Harries, *Domino UK Ltd, United Kingdom*
Dr W. Hayes, *University of Reading, United Kingdom*
Miss X. He, *University of Cambridge, United Kingdom*
Dr O. Henry, *Universitat Rovira I Virgili, Spain*
Dr A. Hirst, *University of Manchester, United Kingdom*
Mrs C. Hodkinson, *Royal Society of Chemistry, United Kingdom*
Mr P. Hopkinson, *University of Cambridge, United Kingdom*
Dr J. Howse, *University of Sheffield, United Kingdom*
Professor W. Hu, *Nanjing University, China*

Professor W. Huck, *University of Cambridge, United Kingdom*
Professor J. Huskens, *University of Twente, The Netherlands*
Professor O. Ikkala, *Helsinki University of Technology, Finland*
Professor R. Jones, *University of Sheffield, United Kingdom*
Mr P. King, *University of Bristol, United Kingdom*
Mr C. Kördel, *Freie Universität Berlin, Germany*
Miss S. Latham, *Royal Society of Chemistry, United Kingdom*
Mr T. C. Lee, *University of Cambridge, United Kingdom*
Mr Y. Lee, *Yonsei University, South Korea*
Ms J. Li, *Nanjing University, China*
Ms Y. Luo, *Freie Universität Berlin, Germany*
Mr Y. Ma, *Nanjing University, China*
Miss R. Makiura, *Kyushu University, Japan*
Professor C. Mao, *Purdue University, U.S.A.*
Miss M. Massignani, *University of Sheffield, United Kingdom*
Mrs N. Masud, *Royal Society of Chemistry, United Kingdom*
Professor S. Matile, *University of Geneva, Switzerland*
Dr F. Meersman, *Katholieke Universiteit Leuven, Belgium*
Dr L. Monticelli, *Helsinki University of Technology, Finland*
Dr M. Munz, *National Physical Laboratory NPL, United Kingdom*
Dr C. Neto, *University of Sydney, Australia*
Mrs L. Omer, *Technion–Israel Institute of Technology, Israel*
Dr I. Osborne, *Science, United Kingdom*
Dr A. Parnell, *University of Sheffield, United Kingdom*
Dr S. Pascu, *University of Bath, United Kingdom*
Mr C. Patel, *Imperial College London, United Kingdom*
Dr R. Piñol, *Universidad de Zaragoza, Spain*
Miss T. Rattanavoravipa, *Kyoto University, Japan*
Professor D. Reinhoudt, *SMCT, Laboratories of Supramolecular Chemistry and Technology, The Netherlands*
Miss A. Roffey, *Royal Society of Chemistry, United Kingdom*
Dr G. Rossi, *Helsinki University of Technology, Finland*
Professor K. Schulten, *University Of Illinois At Urbana-Champaign, U.S.A.*
Professor A. Sen, *Pennsylvania State University, U.S.A.*
Mr J. Serginson, *Imperial College London, United Kingdom*
Dr M. Shaffer, *Imperial College London, United Kingdom*
Miss J. Slota, *University of Cambridge, United Kingdom*
Mr T. Smart, *University of Sheffield, United Kingdom*
Dr C. Sporer, *Institute for Bioengineering of Catalonia, Spain*
Professor U. Steiner, *University of Cambridge, United Kingdom*
Dr J. Steinke, *Imperial College London, United Kingdom*
Miss S. Stuart-Cole, *University of Leeds, United Kingdom*
Mr X. Sui, *University of Twente, The Netherlands*
Professor S. Sun, *National Nanoscience and Nanotechnology Center in China, China*
Dr M. Sushko, *University College London, United Kingdom*
Dr W. Teng, *SIRIM, Malaysia*
Dr S. Titmuss, *University of Oxford, United Kingdom*
Ms B. Trappman, *University of Cambridge, United Kingdom*
L. Trenberth, *IOP Publishing, United Kingdom*
Professor R. Ulijn, *University of Strathclyde, United Kingdom*
Professor Dr J. van Esch, *Technical University Delft, The Netherlands*
Dr M. Vendruscolo, *University of Cambridge, United Kingdom*
Professor Dr Z. Wei, *National Center for Nanoscience and Technoology, China*
Professor G. Whitesides, *Harvard University, U.S.A.*
Dr U. Wiesner, *Cornell University, U.S.A.*

Professor D. Woolfson, *University of Bristol, United Kingdom*
Dr A. Wotherspoon, *IOP Publishing, United Kingdom*
Mr W. Yuan, *Imperial College London, United Kingdom*
Mr J. Zayed, *University of Cambridge, United Kingdom*
Mr S. Zhou, *Imperial College London, United Kingdom*
Dr Z. Zhu, *University of Reading, United Kingdom*

Index of contributors*

Adams, D. J., 29
Adams, S., **29**
Ahangar, F., 81, 169, 265, 359
Aida, T., **21**, 359
Albouy, P.-A., **235**
Armstrong, C. T., **305**
Aufderhorst-Roberts, A., 169, 265, 359
Barbero, D., 81, 169
Berglund, L. A., **95**
Bittner, A., 81, 169, 359
Böker, A., **143**
Boyle, A. L., **305**
Brizard, A. M., **345**
Bromley, E. H. C., **305**
Browne, W. R., **20**, 169, 359
Burattini, S., **16**, 265
Carew, D., 81, 169, 265
Cai, T., **129**
Cardin, C. J., **205**
Chan, D. Y. C., 81, **151**, 169, 359
Channon, K., 81, 169, 265, 359
Chau, Y., A, 265, 359
Chiche, A., **143**
Cho, K. L., **151**
Colquhoun, H. M., 169, **205**, **251**, 265, 359
Cruz-Chu, E. R., **47**
Dagastine, R. R., **151**
Das, A. K., **293**
Dobson, C. M., **277**
Drew, M. G. B., **205**
Farrance, O., **129**
Faul, C. F. J., **95**, 169, 265
Feringa, B. L., **319**
Fernyhough, C. M., **29**
Fery, A., **143**, 169, 265
Flatté, M. E., **109**
Gan, Y., **205**
Greenland, B. W., **251**
Grieser, F., **151**
Hadjichristidis, N., **95**
Harries, J., 265
Hayes, W., **251**, 265
He, Y., **221**
Hiltl, S., **143**
Hirst, A. R., **293**
Hobbs, J. K., **129**
Hong, Y., **15**
Hopkinson, P., 81, 169
Horn, A., **143**
Houbenov, N., **95**
Howse, J. D., 81, 359
Hu, W., **129**, 265
Huck, W., 81, 169, 359
Huskens, J., 81, **118**, 169, 265, 359
Iatrou, H., **95**
Ibele, M., **15**
Ikkala, O., **95**, 169, 265, 359
Janssen, L. P. B. M., **319**
Jiang, W., **221**
Jones, R. A. L., **9**, **29**, 81, 359
Keller, P., **235**
Kinbara, K., **335**
Klok, M., **319**
Ko, S. H., **221**
Kornyshev, A. A., 81, **109**, 169, 359
Laine, J., **95**
Lamb, R. N., **151**
Leng, Y., **221**
Leskelä, M., **95**
Lévy, D., **235**
Li, M.-H., **235**
Liaw, I. I., **151**
Lindström, T., **95**
Ling, X. Y., **117**
Lipomi, D. J., **373**
Ma, Y., **129**
Mahmoud, Z. N., **305**
Mao, C., **221**, 265
Matile, S., 81, 169, **187**, 265
Munz, M., 169, 265
Neto, C., 169, 359
Nono-Djamen, M., **235**
Otani, W., **335**
Pääkkö, M., **95**
Parnell, A. J., **29**, 81
Pensec, S., **235**
Phang, I. Y., **117**
Piñol, R., **235**, 265
Ras, R. H. A., **95**
Reinhoudt, D. N., 81, **117**, 169, 265, 359
Reiter, G., **129**
Ribbe, A. E., **221**
Ritz, T., **47**
Ruokolainen, J., **95**
Ryan, A. J., **29**
Sakai, N., **187**
Schoberth, H. G., **143**
Schulten, K., **47**, 81
Schweikart, A., **143**

Sen, A., **15**, 81, 169, 265, 359
Shaffer, M., 169, 265
Siwy, Z. S., **47**
Smith, L., **305**
Sporer, C., 265
Steiner, U., 81, 169
Steinke, J., 81, 169, 265, 359
Stevens, G. W., **151**
Stuart, M. C. A., **345**
Su, M., **221**
Sun, X., **221**
Sushko, M. L., **63**, 81
Takeuchi, T., **187**
ten Brinke, G., **95**
Thomson, A. R., **305**
Titmuss, S., 81, 169, 265, 359
Topham, P. D., **29**

Tzokova, N., **29**
Uddin, Md. H., **151**
Ulijn, R. V., 265, **293**, 359
Urbakh, M., **109**
van Esch, J. H., 169, **345**, 359
Vancso, G. J., **117**
Velegol, D., **15**
Vendruscolo, M., **277**, 359
Wang, Q., **143**
Whitesides, G. M., 81, 169, 265, 359, **373**
Woolfson, D. N., 169, 265, **305**, 359
Xu, B., **235**
Ye, T., **221**
Zayed, J., 359
Zhang, C., **221**
Zhu, Z., **205**

* The page numbers in **bold** type indicate papers submitted for discussions.